Information Theory in Computer Vision and Pattern Recognition

Francisco Escolano • Pablo Suau
Boyán Bonev

Information Theory in Computer Vision and Pattern Recognition

Foreword by
Alan Yuille

Springer

Francisco Escolano
Universidad Alicante
Depto. Ciencia de la
Computación e
Inteligencia Artificial
Campus de San Vicente, s/n
03080 Alicante
Spain
sco@dccia.ua.es

Boyán Bonev
Universidad Alicante
Depto. Ciencia de la
Computación e
Inteligencia Artificial
Campus de San Vicente, s/n
03080 Alicante
Spain
boyan@dccia.ua.es

Pablo Suau
Universidad Alicante
Depto. Ciencia de la
Computación e
Inteligencia Artificial
Campus de San Vicente, s/n
03080 Alicante
Spain
pablo@dccia.ua.es

ISBN 978-1-4471-5693-2 ISBN 978-1-84882-297-9 (eBook)
DOI 10.1007/978-1-84882-297-9
Springer Dordrecht Heidelberg London New York

British Library Cataloguing in Publication Data
A catalogue record for this book is available from the British Library

Cover design: SPi Publisher Services

Printed on acid-free paper

Springer is part of Springer Science+Business Media (www.springer.com)

To my three joys: Irene, Ana, and Mamen.
Francisco

To my parents, grandparents, and brother. To Beatriz.
Pablo

To Niya, Elina, and Maria.
Boyan

Foreword

Computer vision and pattern recognition are extremely important research fields with an enormous range of applications. They are also extremely difficult. This may seem paradoxical since humans can easily interpret images and detect spatial patterns. But this apparent ease is misleading because neuroscience shows that humans devote a large part of their brain, possibly up to 50% of the cortex, to processing images and interpreting them. The difficulties of these problems have been appreciated over the last 30 years as researchers have struggled to develop computer algorithms for performing vision and pattern recognition tasks. Although these problems are not yet completely solved, it is becoming clear that the final theory will depend heavily on probabilistic techniques and the use of concepts from information theory.

The connections between information theory and computer vision have long been appreciated. Vision can be considered to be a decoding problem where the encoding of the information is performed by the physics of the world – by light rays striking objects and being reflected to cameras or eyes. Ideal observer theories were pioneered by scientists such as Horace Barlow to compute the amount of information available in the visual stimuli, and to see how efficient humans are at exploiting it. But despite the application of information theory to specific visual tasks, there has been no attempt to bring all this work together into a clear conceptual framework.

This book fills the gap by describing how probability and information theory can be used to address computer vision and pattern recognition problems. The authors have developed information theory tools side by side with vision and pattern recognition tasks. They have characterized these tools into four classes: (i) measures, (ii) principles, (iii) theories, and (iv) algorithms. The book is organized into chapters addressing computer vision and pattern recognition tasks at increasing levels of complexity. The authors have devoted chapters to feature detection and spatial grouping, image segmentation, matching, clustering, feature selection, and classifier design. As the authors address these topics, they gradually introduce techniques from information theory. These include (1) information theoretic measures, such as entropy and Chernoff information, to evaluate image features; (2) mutual information as a criteria for matching problems (Viola and Wells 1997); (3) minimal description length ideas (Risannen 1978) and their application to image segmentation (Zhu and Yuille 1996); (4) independent component analysis (Bell and Sejnowski 1995) and its use for feature extraction; (5) the use of rate distortion theory for clustering algorithms; (6) the method of types (Cover and Thomas 1991) and its application to analyze the convergence rates of vision algorithms (Coughlan and Yuille 2002); and (7) how entropy and infomax principles (Linsker 1988) can be used for classifier design. In addition; the book covers alternative information theory measures, such as Rényi alpha-entropy and Jensen–Shannon divergence, and advanced topics; such as data driven Markov Chain Monte Carlo (Tu and Zhu 2002) and information geometry (Amari 1985). The book describes these theories clearly, giving many illustrations and specifying the code by flow -charts.

Overall, the book is a very worthwhile addition to the computer vision and pattern recognition literature. The authors have given an advanced introduction to techniques from probability and information theory and their application to vision and pattern recognition tasks. More importantly, they have described a novel perspective that will be of growing importance over time. As computer vision and pattern recognition develop, the details of these theories will change, but the underlying concepts will remain the same.

UCLA, Department of Statistics and Psychology *Alan Yuille*
Los Angeles, CA
March 2009

Preface

Looking through the glasses of Information Theory (IT) has proved to be effective both for formulating and designing algorithmic solutions to many problems in computer vision and pattern recognition (CVPR): image matching, clustering and segmentation, salient point detection, feature selection and dimensionality reduction, projection pursuit, optimal classifier design, and many others. Nowadays, researchers are widely bringing IT elements to the CVPR arena. Among these elements, there are *measures* (entropy, mutual information, Kullback–Leibler divergence, Jensen–Shannon divergence...), *principles* (maximum entropy, minimax entropy, minimum description length...) and *theories* (rate distortion theory, coding, the method of types...).

This book introduces and explores the latter elements, together with the one of *entropy estimation*, through an incremental complexity approach. Simultaneously, the main CVPR problems are formulated and the most representative algorithms, considering authors' preferences for sketching the IT–CVPR field, are presented. Interesting connections between IT elements when applied to different problems are highlighted, seeking for a basic/skeletal research roadmap. This roadmap is far from being comprehensive at present due to time and space constraints, and also due to the current state of development of the approach. The result is a novel tool, unique in its conception, both for CVPR and IT researchers, which is intended to contribute, as much as possible, to a cross-fertilization of both areas.

The motivation and origin of this manuscript is our awareness of the existence of many sparse sources of IT-based solutions to CVPR problems, and the lack of a systematic text that focuses on the important question: *How useful is IT for CVPR?* At the same time, we needed a research language, common to all the members of the Robot Vision Group. Energy minimization, graph theory, and Bayesian inference, among others, were adequate methodological tools during our daily research. Consequently, these tools were key to design and build a solid background for our Ph.D. students. Soon we realized that IT was a unifying component that flowed naturally among our rationales for tackling CVPR problems. Thus, some of us enrolled in the task of writing a text in which we could advance as much as possible in the fundamental links between CVPR and IT. Readers (starters and senior researchers) will judge to what extent we have both answered the above fundamental question and reached our objectives.

Although the text is addressed to CVPR–IT researchers and students, it is also open to an interdisciplinary audience. One of the most interesting examples is the computational vision community, which includes people interested both in biological vision and psychophysics. Other examples are the roboticians and the people interested in developing wearable solutions for the visually impaired (which is the subject of our active work in the research group).

Under its basic conception, this text may be used for an IT-based one semester course of CVPR. Only some rudiments of algebra and probability are necessary. IT items will be introduced as the text flows from one computer vision or pattern recognition problem to another. We have deliberately avoided a succession of theorem–proof pairs for the sake of a smooth presentation. Proofs, when needed, are embedded in the text, and they are usually excellent pretexts for presenting or highlighting interesting properties of IT elements. Numerical examples with toy settings of the problems are often included for a better understanding of the IT-based solution. When formal elements of other branches of mathematics like field theory, optimization, and so on, are needed, we have briefly presented them and referred to excellent books fully dedicated to their description.

Problems, questions and exercises are also proposed at the end of each chapter. The purpose of the problems section is not only to consolidate what is learnt, but also to go one step forward by testing the ability of generalizing the concepts exposed in each chapter. Such section is preceded by a brief literature review that outlines the key papers for the CVPR topic, which is the subject of the chapter. These papers' references, together with sketched solutions to the problems, will be progressively accessible in the Web site *http://www.rvg.ua.es/ITinCVPR*.

We have started the book with a brief introduction (Chapter 1) regarding the four axes of IT–CVPR interaction (measures, principles, theories, and entropy estimators). We have also presented here the skeletal research roadmap (the ITinCVPR tube). Then we walk along six chapters, each one tackling a different problem under the IT perspective. Chapter 2 is devoted to *interest points, edge detection, and grouping*; interest points allow us to introduce the concept of entropy and its linking with Chernoff information, Sanov's theorem, phase transitions and the method of types. Chapter 3 covers *contour and region-based image segmentation* mainly from the perspective of model order selection through the minimum description length (MDL) principle, although the Jensen–Shannon measure and the Jaynes principle of maximum entropy are also introduced; the question of learning a segmentation model is tackled through links with maximum entropy and belief propagation; and the unification of generative and discriminative processes for segmentation and recognition is explored through information divergence measures. Chapter 4 reviews *registration, matching, and recognition* by considering the following image registration through minimization of mutual information and related measures; alternative derivations of Jensen–Shannon divergence yield deformable matching; shape comparison is encompassed through Fisher information; and structural matching and learning are driven by MDL. Chapter 5 is devoted to *image and pattern clustering* and is mainly rooted in three IT approaches to clustering: Gaussian mixtures (incremental method for adequate order selection), information bottleneck (agglomerative and robust with model order selection) and mean-shift; IT is also present in initial proposals for ensembles clustering (consensus finding). Chapter 6 reviews the main approaches to *feature selection and transformation*: simple wrappers and filters exploiting IT for bypassing the curse of dimensionality; minimax entropy principle for learning patterns using a generative approach; and ICA/gPCA methods based on IT (ICA and neg-entropy, info-max and minimax ICA, generalized PCA and effective dimension). Finally, Chapter 7, *Classifier Design*, analyzes the main IT strategies for building classifiers. This obviously includes decision trees, but also multiple trees and random forests, and how to improve boosting algorithms by means of IT-based criteria. This final chapter ends with an information projection analysis of maximum entropy classifiers and a careful exploration of the links between Bregman divergences and classifiers.

We acknowledge the contribution of many people to this book. In first place, we thank many scientists for their guide and support, and for their important contributions to the field. Researchers from different universities and institutions such as Alan Yuille, Hamid Krim, Chen Ping-Feng, Gozde Unal, Ajit Rajwadee, Anand Rangarajan, Edwin Hancock, Richard Nock, Shun-ichi Amari, and Mario Figueiredo, among many others, contributed with their advices, deep knowledge and highly qualified expertise. We also thank all the colleagues of the Robot Vision Group of the University of Alicante, especially Antonio Peñalver, Juan Manuel Sáez, and Miguel Cazorla, who contributed with figures, algorithms, and important results from their research. Finally, we thank the editorial board staff: Catherine Brett for his initial encouragement and support, and Simon Rees and Wayne Wheeler for their guidance and patience.

University of Alicante, Spain

Francisco Escolano
Pablo Suau
Boyan Bonev

Contents

1 Introduction 1
1.1 Measures, Principles, Theories, and More 1
1.2 Detailed Organization of the Book 3
1.3 The ITinCVPR Roadmap 10

2 Interest Points, Edges, and Contour Grouping 11
2.1 Introduction 11
2.2 Entropy and Interest Points 11
2.2.1 Kadir and Brady Scale Saliency Detector 12
2.2.2 Point Filtering by Entropy Analysis Through Scale Space 14
2.2.3 Chernoff Information and Optimal Filtering 16
2.2.4 Bayesian Filtering of the Scale Saliency Feature Extractor: The Algorithm 18
2.3 Information Theory as Evaluation Tool: The Statistical Edge Detection Case 20
2.3.1 Statistical Edge Detection 22
2.3.2 Edge Localization 23
2.4 Finding Contours Among Clutter 26
2.4.1 Problem Statement 27
2.4.2 A^* Road Tracking 29
2.4.3 A^* Convergence Proof 31
2.5 Junction Detection and Grouping 33
2.5.1 Junction Detection 33
2.5.2 Connecting and Filtering Junctions 35
Problems 38
2.6 Key References 41

3 Contour and Region-Based Image Segmentation 43
3.1 Introduction 43
3.2 Discriminative Segmentation with Jensen–Shannon Divergence 44

3.2.1 The Active Polygons Functional 44
3.2.2 Jensen–Shannon Divergence and the Speed Function ... 46
3.3 MDL in Contour-Based Segmentation 53
3.3.1 B-Spline Parameterization of Contours 53
3.3.2 MDL for B-Spline Parameterization 58
3.3.3 MDL Contour-based Segmentation 60
3.4 Model Order Selection in Region-Based Segmentation 63
3.4.1 Jump-Diffusion for Optimal Segmentation 63
3.4.2 Speeding-up the Jump-Diffusion Process 71
3.4.3 K-adventurers Algorithm 74
3.5 Model-Based Segmentation Exploiting The Maximum Entropy Principle 79
3.5.1 Maximum Entropy and Markov Random Fields 79
3.5.2 Efficient Learning with Belief Propagation 83
3.6 Integrating Segmentation, Detection and Recognition 86
3.6.1 Image Parsing 86
3.6.2 The Data-Driven Generative Model 91
3.6.3 The Power of Discriminative Processes 96
3.6.4 The Usefulness of Combining Generative and Discriminative 99
Problems 100
3.7 Key References 103

4 **Registration, Matching, and Recognition** 105
4.1 Introduction 105
4.2 Image Alignment and Mutual Information 106
4.2.1 Alignment and Image Statistics 106
4.2.2 Entropy Estimation with Parzen's Windows 108
4.2.3 The EMMA Algorithm 110
4.2.4 Solving the Histogram-Binning Problem 111
4.3 Alternative *Metrics* for Image Alignment 119
4.3.1 Normalizing Mutual Information 119
4.3.2 Conditional Entropies 120
4.3.3 Extension to the Multimodal Case 121
4.3.4 Affine Alignment of Multiple Images 122
4.3.5 The Rényi Entropy 124
4.3.6 Rényi's Entropy and Entropic Spanning Graphs 126
4.3.7 The Jensen–Rényi Divergence and Its Applications 128
4.3.8 Other Measures Related to Rényi Entropy 129
4.3.9 Experimental Results 132
4.4 Deformable Matching with Jensen Divergence and Fisher Information 132
4.4.1 The Distributional Shape Model 132
4.4.2 Multiple Registration and Jensen–Shannon Divergence 136

4.4.3 Information Geometry and Fisher–Rao Information 140
4.4.4 Dynamics of the Fisher Information Metric 143
4.5 Structural Learning with MDL 146
4.5.1 The Usefulness of Shock Trees 146
4.5.2 A Generative Tree Model Based on Mixtures.......... 147
4.5.3 Learning the Mixture 150
4.5.4 Tree Edit-Distance and MDL 151
Problems 153
4.6 Key References.. 156

5 Image and Pattern Clustering 157
5.1 Introduction 157
5.2 Gaussian Mixtures and Model Selection 157
5.2.1 Gaussian Mixtures Methods 157
5.2.2 Defining Gaussian Mixtures 158
5.2.3 EM Algorithm and Its Drawbacks 159
5.2.4 Model Order Selection 161
5.3 EBEM Algorithm: Exploiting Entropic Graphs 162
5.3.1 The Gaussianity Criterion and Entropy Estimation 162
5.3.2 Shannon Entropy from Rényi Entropy Estimation 163
5.3.3 Minimum Description Length for EBEM 166
5.3.4 Kernel-Splitting Equations 167
5.3.5 Experiments 168
5.4 Information Bottleneck and Rate Distortion Theory 170
5.4.1 Rate Distortion Theory Based Clustering............. 170
5.4.2 The Information Bottleneck Principle 173
5.5 Agglomerative IB Clustering 177
5.5.1 Jensen–Shannon Divergence and Bayesian Classification Error 177
5.5.2 The AIB Algorithm.................................. 178
5.5.3 Unsupervised Clustering of Images 181
5.6 Robust Information Clustering 184
5.7 IT-Based Mean Shift...................................... 189
5.7.1 The Mean Shift Algorithm............................ 189
5.7.2 Mean Shift Stop Criterion and Examples 191
5.7.3 Rényi Quadratic and Cross Entropy from Parzen Windows 193
5.7.4 Mean Shift from an IT Perspective 196
5.8 Unsupervised Classification and Clustering Ensembles........ 197
5.8.1 Representation of Multiple Partitions 198
5.8.2 Consensus Functions................................. 199
Problems 206
5.9 Key References.. 209

6 Feature Selection and Transformation 211
6.1 Introduction 211
6.2 Wrapper and the Cross Validation Criterion 212
6.2.1 Wrapper for Classifier Evaluation 212
6.2.2 Cross Validation 214
6.2.3 Image Classification Example 215
6.2.4 Experiments 219
6.3 Filters Based on Mutual Information 220
6.3.1 Criteria for Filter Feature Selection 220
6.3.2 Mutual Information for Feature Selection 222
6.3.3 Individual Features Evaluation, Dependence and Redundancy 223
6.3.4 The min-Redundancy Max-Relevance Criterion 225
6.3.5 The Max-Dependency Criterion 227
6.3.6 Limitations of the Greedy Search 228
6.3.7 Greedy Backward Search 231
6.3.8 Markov Blankets for Feature Selection 234
6.3.9 Applications and Experiments 236
6.4 Minimax Feature Selection for Generative Models 238
6.4.1 Filters and the Maximum Entropy Principle 238
6.4.2 Filter Pursuit through Minimax Entropy 242
6.5 From PCA to gPCA 244
6.5.1 PCA, FastICA, and Infomax 244
6.5.2 Minimax Mutual Information ICA 250
6.5.3 Generalized PCA (gPCA) and Effective Dimension 254
Problems 265
6.6 Key References 269

7 Classifier Design 271
7.1 Introduction 271
7.2 Model-Based Decision Trees 272
7.2.1 Reviewing Information Gain 272
7.2.2 The Global Criterion 273
7.2.3 Rare Classes with the Greedy Approach 275
7.2.4 Rare Classes with Global Optimization 280
7.3 Shape Quantization and Multiple Randomized Trees 284
7.3.1 Simple Tags and Their Arrangements 284
7.3.2 Algorithm for the Simple Tree 285
7.3.3 More Complex Tags and Arrangements 287
7.3.4 Randomizing and Multiple Trees 289
7.4 Random Forests 291
7.4.1 The Basic Concept 291
7.4.2 The Generalization Error of the RF Ensemble 291
7.4.3 Out-of-the-Bag Estimates of the Error Bound 294
7.4.4 Variable Selection: Forest RI vs. Forest-RC 295

7.5 Infomax and Jensen–Shannon Boosting 298
7.5.1 The Infomax Boosting Algorithm 299
7.5.2 Jensen–Shannon Boosting 305
7.6 Maximum Entropy Principle for Classification 308
7.6.1 Improved Iterative Scaling 308
7.6.2 Maximum Entropy and Information Projection 313
7.7 Bregman Divergences and Classification 324
7.7.1 Concept and Properties 324
7.7.2 Bregman Balls and Core Vector Machines 326
7.7.3 Unifying Classification: Bregman Divergences and Surrogates 331
Problems 339
7.8 Key References 341

References 343

Index 353

Color Plates 357

1

Introduction

Shannon's definition of channel capacity, information and redundancy was a landmark [...]. Since these measurable quantities were obviously important to anyone who wanted to understand sensory coding and perception, I eagerly stepped on the boat.

Horace Basil Barlow[1]

1.1 Measures, Principles, Theories, and More

The Information Theory (IT) approach has proved to be effective in solving many Computer Vision and Pattern Recognition (CVPR) problems (image matching, clustering and segmentation, extraction of invariant interest points, feature selection, optimal classifier design, model selection, PCA/ICA,[2] Projection Pursuit, and many others). The computational analysis of images and more abstract patterns is a complex and challenging task, which demands interdisciplinary efforts. Thus, the confluence of Bayesian/Statistical Learning, Optimization (Energy Minimization), Gibbs/Random Fields, and other methodologies yields a valuable cross-fertilization that is helpful both for formulating and solving problems.

Nowadays, researchers are widely exploiting IT elements to formulate and solve CVPR problems. Among these elements, we find *measures*, *principles*, and *theories*. Entropy, Mutual Information, and Kullback–Leibler divergence are well known *measures*, which are typically used as metrics or as optimization criteria. For instance, in ICA it is interesting to find the projection directions maximizing the independence of the outputs, and this is equivalent

[1] Redundancy reduction revisited, *Network: Comput. Neural Syst.* 12 (2001) 241–253.

[2] Principal Component Analysis, Independent Component Analysis.

to minimize their mutual information. On the other hand, some examples of IT *principles* are Minimum Description Length (MDL), and the Minimax Entropy principles. The first one, enunciated by Rissanen, deals with the selection of the simplest model in terms of choosing the shortest code (or number of parameters), explaining the data well enough. For example, when facing the region segmentation problem, it is convenient to partition the image in as few regions as possible, provided that the statistics of each of them may be explained by a reduced set of parameters.

Minimax Entropy, formulated by Christensen, has more to do with perceptual learning. Inferring the probability distribution characterizing a set of images may be posed in terms of selecting, among all distributions matching the statistics of the learning examples, the one with maximum entropy (ME) (Jaynes' maximum entropy principle). This ensures that the learned distribution contains no more information than the examples. However, the latter statistics depend on the features selected for computing them, and it is desirable to select the ones yielding the maximum information (minimal entropy).

Regarding IT *theories*, we refer to mathematical developments. Two examples are Rate Distortion Theory and the Method of Types. Rate Distortion Theory formalizes the question of what is the minimum expected distortion yielded by lowering the bit-rate, for instance, in lossy compression. From this point of view, clustering provides a compact representation (the prototypes) that must preserve the information about the example patterns or images. As classification implies some loss of individual identity in favor of the prototypes, a unique class for all patterns yields the maximal loss of information (minimizes mutual information between examples and prototypes). However, the compactness may be relaxed so that the average distortion is constrained to be under a given threshold. A more elaborated criterion is the Information-Bottleneck (IB) method in which the distorted upper bound constraint is replaced by a lower bound constraint over the relevant information. The key idea is to find a trade-off between maximizing information between relevant features and prototypes (relevance), and minimizing the mutual information between the examples and prototypes (compactness).

The Method of Types, developed by Csiszár and Körner, provides theoretical insights into calculating the probability of rare events. Types are associated to empirical histograms, and the Sanov's theorem yields bounds to the probability that a given type lies within a certain set of types (for instance those which indicate that the texture characterized by the histograms belongs to a given class instead of to another one). As a consequence of the Sanov's theorem, the expected number of the mis-classified textures depends on the Bhattacharyya distance, which, in turn, is bounded by Chernoff information. Chernoff information quantifies the overlapping between the distributions associated to different classes of patterns (textures, for instance). Bhattacharyya distance determines *order parameters* whose sign determines whether the discrimination problem may be solved or not (in this latter case, when the two pattern classes are too close). For the zero value, there is a *phase transition*. These results come from the analysis of other tasks like detecting a path (edge) among clutter, and the existence of order parameters, which allow to quantify the degree of success of related tasks like contour linking.

In addition to measures, principles and theories, there is a fourth dimension, orthogonal to the other three, to explore: the problem of *estimating entropy*, the fundamental quantity in IT. Many other measures (mutual information, Kullback–Leibler divergence, and so on) are derived from entropy. Consequently, in many cases, the practical application of the latter elements relies on a consistent estimation of entropy. The two extreme cases are the *plug-in* methods, in which the estimation of the probability density precedes the computing of entropy, and the *bypass* methods where entropy is estimated directly. For example, as the Gaussian distribution is the maximum entropy one among all the distributions with the same variance, the latter consideration is key, for instance, in Gaussian Mixture Models, typically used as classifiers, and also in ICA methods, which usually rely on the departure from Gaussianity. Thus, entropy estimation will be a recurrent topic along the book.

1.2 Detailed Organization of the Book

Given the above described introductory considerations, it proceeds now to review the organization of the book and the exposition of both the IT elements and entropy estimation approaches within the context of specific computer vision and pattern recognition problems. The first important consideration is that we will follow an increasing-complexity *problem driven* exposition: interest points and edges detection, contour grouping and segmentation, image and point matching, image and pattern clustering, feature selection and transformation, and finally, classifier design. The outcome is a total of six more chapters. The proposed increasing-complexity exposition allows us to preclude devoting this introductory chapter to mathematical formalisms. We sincerely believe in introducing the mathematical elements at the same time that each problem is formulated and tackled. IT beginners will understand the mathematical elements through the examples provided by CVPR problems, and researchers with an IT background will go ahead to the details of each problem. Anyway, basic notions of the Theory of Probability and Bayesian Analysis are needed, and additional formal details and demonstrations, lying beyond the understanding of each problem, will be embedded in text at the level necessary to follow the exposition. Let us give a panoramic view of each chapter.

The first topic dealt with in Chapter 2 is *interest points and edges detection*. A recent definition of saliency implies both the computation of the intensity information content in the neighborhood of a pixel, and how such information evolves along different scales. Information is quantified by entropy. However, the bottleneck of this method is the scale-space analysis, and also the huge amount of computation needed to extend it to the affine-invariant case or the multidimensional case. This is critical when these features are used for image matching. However, when information about the class of images is available, it is possible to reduce significantly the computational cost. In order to do so, we introduce concepts like Chernoff Information and Bhattacharyya coefficient, coming from the formulation of edge detection as a statistical inference problem, which is subsequently presented and discussed in detail. We emphasize the interest of this methodology for: (i) evaluating models for edges (and, in general, for other image features); (ii) evaluating and combining cues (filters); and (iii) adaptation between data sets (classes of images). Finally, we introduce the problem of edge linking when significant clutter is present, and the approach derived from applying the Method of Types.

Chapter 3 is devoted both to *contour-based and region-based segmentation.* We first present a modern approach to apply active contours (active polygons) to textured regions. This leads us to introduce the Jensen–Shannon (JS) divergence, as a cost function, because it can efficiently account for high-order statistics. However, as such divergence relies on entropy and this, in turn, relies on estimating the density function, the Jaynes' maximum entropy (ME) principle may be applied for estimating the shape of the density functions, though practical reasons (efficiency) impose the use of bypass approximations of entropy. Next key IT issue is how to measure the adequacy of the contour-based segmentation. The Minimum Description Length (MDL) principle is one of the answers. We firstly consider the problem of fitting a deformable contour to a shape in terms of finding the simplest contour that best explains its localization. Simplicity preclude a high number of descriptive parameters, and good localization implies enough contrast between the image statistics of inside and outside the shape (region model). From the algorithmic point of view, this can be solved by trying with all the number of parameters within a given range and finding, for each of them, the optimal placement of the contour and the statistics of the region model. Alternatively, one may use a jump-diffusion scheme to find the optimal number of regions in the image, as well as their descriptive parameters. In this case, MDL is implicit in the problem formulation and this optimal number of regions must be found. However, the algorithmic solution presented in this chapter (DDMCMC[3]) is a combination of high-level (top-down) hypothesis generator (Markov Chain Monte Carlo simulation), which suggests changing of the number and type of models (jumps), and a discriminative (bottom-up) computation that adds image-level knowledge to the jumps probabilities and accelerates convergence. The next interesting concept is the use of ME principle for learning a given intensity model. In this regard, our choice is a method with an interesting connection with belief propagation. Finally, in this chapter we look again at the relationship between discriminative and generative approaches, their possible

[3] Data driven Markov Chain Monte Carlo.

integration, and the connection between segmentation and object recognition. In this regard, an interesting fact is the quantification of convergence in terms of Kullback–Leibler divergences.

Chapter 4 addresses *registration, matching*, and *recognition* mainly rooted in the elements: (i) alignment using mutual information; (ii) point-sets registration with JS divergence; (iii) deformable shape matching and comparison with Fisher information; and (iv) structural learning with MDL. The first topic, alignment using mutual information, refers to a large body of work devoted to exploit statistics, through mutual information, for finding the optimal 2D transformation between two images, or for even aligning 3D data and model. Firstly, we present the classical approach consisting of a stochastic maximization algorithm. In this case, the proposed solution relies on density estimation in order to measure entropy, and thus, it is interesting to previously introduce the Parzen windows method. On the other hand, when histograms are used, it is important to analyze the impact of histogram binning in the final result. This is not a negligible problem, especially when noise and more general transformation arise. In this latter context, new ideas for representing joint distributions are very interesting. Another problem when using mutual information is the nature of the measure itself. It is far from being a *metric* in the formal sense. Thus, it is interesting to review both derived and alternative better conditioned measures, for instance, when affine deformations between model and target image arise. A good example is the sum of conditional entropies, which has been proved to work well in the multimodal case (matching between multiple images). It is also interesting to redefine the Jensen divergence in terms of the Rényi entropy. Furthermore, the estimation of such entropy is the target of the research in entropic graphs. Consequently, the redefinition of mutual information in the same terms yields a novel approach to image registration. Next, Jensen divergence is considered again, in this case as a metric for learning shape prototypes. The application of Fisher information to shape matching and comparison is the following topic. To conclude this chapter, we have focused on structural matching, mainly in trees (the simplest case) and, more specifically, in structural learning. This is one of the open problems in structural pattern recognition, and we present here a solution that considers the MDL principle as a driving force for shape learning and comparison.

Chapter 5 covers *pattern and image clustering* through many IT approaches/faces to/of the clustering problem, like: (i) mixture models; (ii) the Information Bottleneck method; (iii) the recent adaptation of Mean-Shift to IT; and (iv) consensus in clustering ensembles driven by IT. Mixtures are well-known models for estimation of probability density functions (pdf), as well as for clustering vectorized data. The main IT concern here is the *model selection problem*, that is, finding the minimal number of mixtures which adequately fit the data. Connection with segmentation (and with MDL) is immediate. The classic EM algorithm for mixtures is initialization dependent and prone to local maxima. We show that such problems can be avoided by starting with a unique kernel and decomposing it when needed. This is the entropy-based EM (EBEM) approach, and it consists of splitting a Gaussian kernel when bi-modality in the underlying data is suspected. As Gaussian distributions are the maximum-entropy ones among all distributions with equal variance, it seems reasonable to measure non-Gaussianity as the ratio between the entropy of the underlying distribution and the theoretical entropy of the kernel, when it is assumed to be Gaussian. As the latter criterion implies measuring entropy, we may use either a plug-in method or a bypass one (e.g. entropic graphs). Another trend in this chapter is the Information Bottleneck (IB) method, whose general idea has been already introduced. Here, we start by introducing the measure of distortion between examples and prototypes. Exploiting Rate Distortion Theory in order to constrain the distortion (otherwise information about examples is lost – maximum compression), a variational formulation yields an iterative algorithm (Blahut–Arimoto) for finding the optimal partition of the data given the prototypes, but not for obtaining the prototypes themselves. IB comes from another fundamental question: distortion measure to use. It turns out that trying to preserve the relevant information in the prototype about another variable, which is easier than finding a good distortion measure, leads to a new variational problem and the Kullback–Leibler divergence emerges as natural distortion measure. The new algorithm relies on deterministic annealing. It starts with one cluster and progresses by splitting. An interesting variation of the basic algorithm is its agglomerative version (start from as many clusters as patterns and build a tree through different levels of abstraction). There is also recent work (RIC[4]) that addresses the problem of model-order selection through learning theory (VC-dimension) and eliminates outliers by following a channel-capacity criterion. Next item is to analyze how the yet classic and efficient mean-shift clustering may be posed in IT terms. More precisely, the key is to minimize the Rényi's quadratic entropy. The last topic of the chapter, clustering ensembles, seeks for obtaining combined clusterings/partitions in an unsupervised manner, so that the resulting clustering yields better quality than individual ones. Here, combination means some kind of consensus. There are several definitions of consensus, and one of them, *median consensus*, can be found through maximizing the information sharing between several partitions.

After reviewing IT solutions to the clustering problem, Chapter 6 deals with a fundamental question, *feature selection*, which has deep impact both in clustering and in classifier design (which will be tackled in Chapter 7). We review filters and wrappers for feature selection under the IT appeal. Considering wrappers, where the selection of a group of features conforms to the performance induced in a supervised classifier (good generalization for unseen patterns), mutual information plays an interesting role. For instance, maximizing mutual information between the unknown true labels associated to a subset of features and the labels predicted by the classifier seems a good

[4] Robust Information Clustering.

criterion. However, the feature selection process (local search) is complex, and going far from a greedy solution, gets more and more impractical (exponential cost), though it can be tackled through genetic algorithms. On the other hand, filters rely on statistical tests for predicting the goodness of the future classifier for a given subset of features. Recently, mutual information has emerged as the source of more complex filters. However, the curse of dimensionality preclude an extended use of such criterion (maximal dependence between features and classes), unless a fast (bypass) method for entropy estimation is used. Alternatively, it is possible to formulate first-order approximations via the combination of simple criteria like maximal relevance and minimal redundancy. Maximal relevance consists of maximizing mutual information between isolated features and target classes. However, when this is done in a greedy manner it may yield redundant features that should be removed in the quasi-optimal subset (minimal redundancy). The combination is dubbed mRMR. A good theoretical issue is the connection between incremental feature selection and the maximum dependency criterion. It is also interesting to combine these criteria with wrappers, and also to explore their impact on classification errors. Next step in Chapter 5 is to tackle feature selection for generative models. More precisely, texture models presented in Chapter 3 (segmentation) may be learned through the application of the minimax principle. Maximum entropy has been introduced and discussed in Chapter 3 as a basis for model learning. Here we present how to specialize this principle to the case of learning textures from examples. This may be accomplished by associating features to filter responses histograms and exploiting Markov random fields (the FRAME approach: Filters, Random Fields and Maximum Entropy). Maximum entropy imposes matching between filter statistics of both the texture samples and the generated textures through Gibbs sampling. On the other hand, filter selection attending minimal entropy should minimize the Kullback–Leibler divergence between the obtained density and the unknown density. Such minimization may be implemented by a greedy process focused on selecting the next feature inducing maximum decrease of Kullback–Leibler divergence with respect to the existing feature set. However, as the latter divergence is complex to compute, the L_1 norm between the observed statistics and those of the synthesized texture, both for the new feature, is finally maximized. Finally in Chapter 6, we cover the essential elements necessary for finding an adequate projection basis for vectorial data and, specially, images. The main concern with respect to IT is the choice of the measures for quantifying the *interest* of a given projection direction. As we have referred to at the beginning of this introduction, projection bases whose components are as much independent as possible seem more interesting, in terms of pattern recognition, than those whose components are simply decorrelated (PCA). Independence may be maximized by maximizing departure from Gaussianity, and this is what many ICA algorithms do. Thus, the concept of *neg-entropy* (difference between entropy of a Gaussian and that of the current outputs/components distribution) arises. For instance, the well-known FastICA algorithm is a fixed-point method driven by neg-entropy maximization. The key point is how is neg-entropy approximated. Another classical approach, Infomax, relies on minimizing the mutual information between components, which can be done by maximizing the mutual information between inputs to neural network processors and their associated outputs. This is followed by the presentation of an improvement consisting of minimizing the sum of marginal entropies, provided that a density estimation is available. As we have seen in Chapter 3, the shape of the density function is determined by the maximum entropy principle. A simple gradient descent with respect to the parameterized version of the unknown matrix yields the best projection basis. Finally, we explore how generalized PCA (gPCA) leads to the simultaneous finding of the optimal subspaces fitting the data and their classification. In this regard we focus on the key concept of effective dimension which is the basic element of model-order selection in gPCA. This is a particularly interesting chapter end because of two reasons. First, it leaves some open problems, and second, it connects both with the topic of feature selection (Chapter 5) and the topic of classifier design (Chapter 7).

The last content-exposition chapter of the book is Chapter 7, and covers *classifier design*. We devote it to discuss the implications of IT in the design of classifiers. We will cover three main topics: (i) from decision trees to random forests; (ii) random forests in detail; and (iii) boosting and IT. Decision trees are well-known methods for learning classifiers, and their foundations emanate from IT. There has been a huge amount of research devoted to achieve remarkable improvements. In this book, we are not pursuing an exhaustive coverage of all of these improvements, but focusing on a couple of significant contributions: model-based trees and random forests. Model-based trees arise from changing the classical bottom-up process for building the trees and considering a top-down approach that relies on a statistical model. Such model is generated through global optimization, and the optimization criterion might yield a trade-off between accuracy and efficiency. The impact of the model-based approach is quite interesting. On the other hand, *random forests* are tree ensembles yielding a performance comparable to AdaBoost and also high robustness. We will review both Forest-RI (random input selection) and Forest-RC (linear combination of inputs). Despite random forests are good ensembles, they may be improved, for instance, by decreasing the correlation between individual trees in the forest. It is, thus, interesting to consider different improving strategies. Another class of classifier ensembles constitutes boosting methods. We will present recent developments including IT elements. The first one to be presented is infomax boosting, which relies on integrating good base classifiers, that is, very informative ones in the general AdaBoost process. Infomax features maximize mutual information with respect to class labels. For simplifying the process, the quadratic mutual information is chosen. On the other hand, in the JSBoost approach, the most discriminating features are selected on the basis of Jensen divergence. In the latter case, the classification function is modified in order to enforce discriminability.

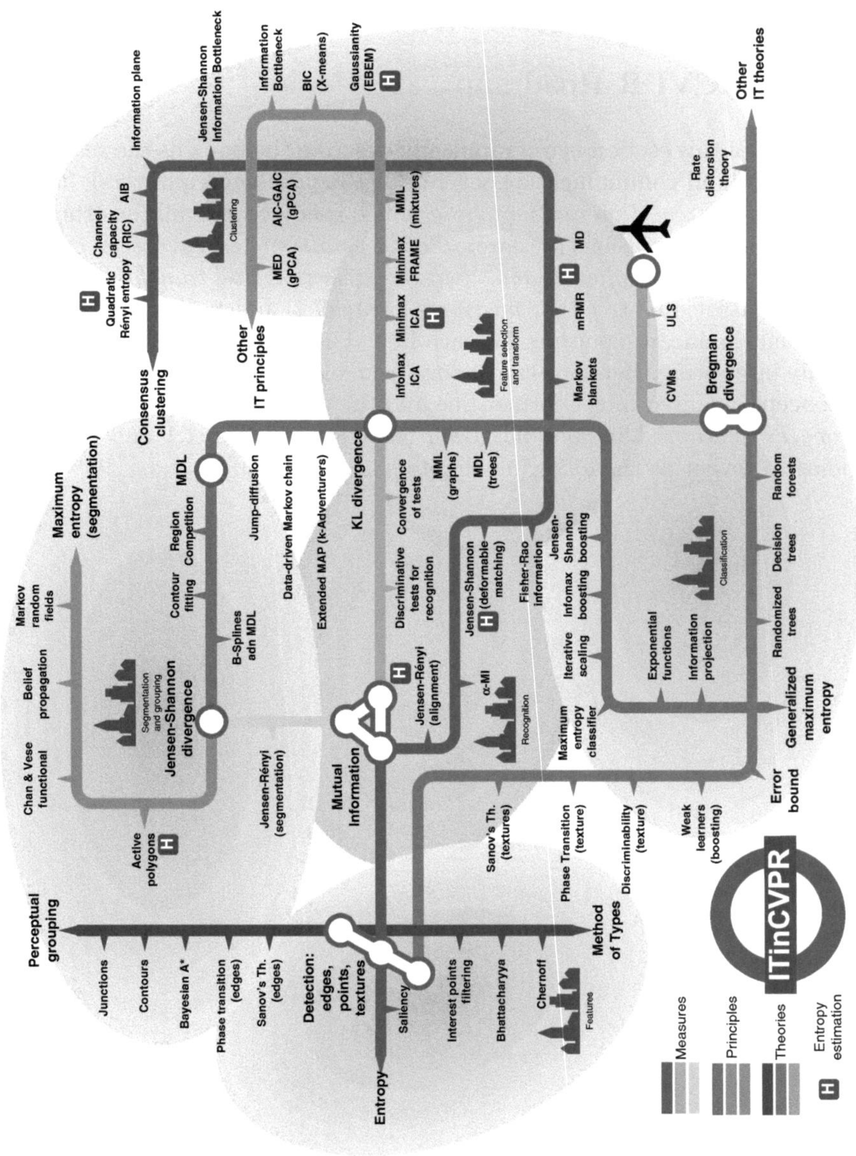

Fig. 1.1. The ITinCVPR tube/underground (lines) communicating several problems (quarters) and stopping at several stations. See Color Plates.

We finish this chapter, and the book, with an in depth review of maximum entropy classification, the exponential distributions family, and their links with information projection, and, finally, with the recent achievements about the implications of Bregman divergences in classification.

1.3 The ITinCVPR Roadmap

The main idea of this section is to graphically describe the book as the map of a tube/underground communicating several CVPR problems (quarters). The central quarter is *recognition and matching*, which is adjacent to all the others. The rest of adjacency relations ($\rightleftharpoons$) are: *interest points and edges* $\rightleftharpoons$ *segmentation* $\rightleftharpoons$ *clustering* $\rightleftharpoons$ *classifier design* $\rightleftharpoons$ *feature selection and transformation*. Tube lines are associated to either *measures*, *principles*, and *theories*. Stations are associated significant concepts for each line. The case of transfer stations is especially interesting. Here one can change from one line to another carrying on the concept acquired in the former line and the special stations associated to *entropy estimation*. This idea has been illustrated in Fig. 1.1 and it is a good point to revisit as the following chapters are understood.

2

Interest Points, Edges, and Contour Grouping

2.1 Introduction

This chapter introduces the application of information theory to the field of feature extraction in computer vision. Feature extraction is a low-level step in many computer vision applications. Its aim is to detect visual clues in images that help to improve the results or speed of this kind of algorithms. The first part of the chapter is devoted to the Kadir and Brady scale saliency algorithm. This algorithm searches the most informative regions of an image, that is, the regions that are salient with respect to their local neighborhood. The algorithm is based on the concept of Shannon's entropy. This section ends with a modification of the original algorithm, based on two information theoretical divergence measures: Chernoff information and mutual information. The next section is devoted to the statistical edge detection work by Konishi et al. In this work, Chernoff information and mutual information, two additional measures that will be applied several times through this book, are applied to evaluate classifiers performance. Alternative uses of information theory include the theoretical study of some properties of algorithms. Specifically, Sanov's theorem and the theory of types show the validity of the road tracking detection among clutter of Coughlan et al. Finally, the present chapter introduces an algorithm by Cazorla et al., which is aimed at detecting another type of image features: junctions. The algorithm is based on some principles explained here.

2.2 Entropy and Interest Points

Part of the current research on image analysis relies on a process called interest point extraction, also known as feature extraction. This is the first step of several vision applications, like object recognition or detection, robot localization, simultaneous localization and mapping, and so on. All these applications are based on the processing of a set of images (usually the query and database images) in order to produce a result. Repeating the same operations for all pixels in a whole set of images may produce an extremely high computational burden, and as a result, these applications may not be prepared to operate in real time. Feature extraction in image analysis may be understood as a preprocessing step, in which the objective is to provide a set of regions of the image that are informative enough to successfully complete the previously mentioned tasks. In order to be valid, the extracted regions should be invariant to common transformations, such as translation and scale, and also to more complex transformations. This is useful, for instance, when trying to recognize an object from different views of the scene, or when a robot must compare the image it has just took with the ones in its database in order to determine its localization on a map.

An extensive number of different feature extraction algorithms have been developed during the last few years, the most known being the multiscale generalization of the Harris corner detector and its multiscale modification [113], or the recent Maximally Stable Extremal Regions algorithm [110], a fast, elegant and accurate approach. However, if we must design an algorithm to search informative regions on an image, then it is clear that an information theory-based solution may be considered. In this section we explain how Gilles's first attempt based on entropy [68] was first generalized by Kadir and Brady [91] to be invariant to scale transformations, and then it was generalized again to be invariant to affine transformations. We will also discuss about the computational time of this algorithm, its main issue, and how it can be reduced by means of the analysis of the entropy measure.

2.2.1 Kadir and Brady Scale Saliency Detector

Feature extraction may be understood as the process of looking for visually salient regions of an image. Visual saliency can be defined as visual unpredictability; an image region is a salient region if it is locally uncommon. Think of a completely white image containing a black spot, or of an overview of highway containing a lonely car; both the black spot and the car are locally uncommon parts of their respective images. In information theory, unpredictability is measured by means of Shannon's entropy. Given a discrete random variable X that can take on possible values $\{x_1, \dots, x_N\}$, Shannon's entropy is calculated as

$$H(X) = -\sum_{i=1}^{N} p(x_i) \log_2 p(x_i) \tag{2.1}$$

and it is measured in bits. Low entropy values correspond to predictable or high informative random variables, that is, random variables in which the probability of a given random value is much higher than the probability of the rest of values. On the contrary, higher entropy values correspond to unpredictable random variables, in which the probability of all their possible random values is similar. If we translate feature extraction problem to the domain of information theory, it is obvious that a feature extraction algorithm must search highest entropy regions on the image.

This is the main idea of Gilles feature extraction algorithm, which formulates local saliency in terms of local intensity. Given a point x, a local neighborhood R_x, and a descriptor D that takes values on $\{d_1, \ldots, d_L\}$ (for instance, $D = \{0, \ldots, 255\}$ on a 8 bit gray level image), local entropy is defined as

$$H_{D,R_x} = -\sum_{i=1}^{L} P_{D,R_x}(d_i) \log_2 P_{D,R_x}(d_i) \tag{2.2}$$

where $P_{D,R_x}(d_i)$ is the probability of descriptor D taking the value d_i in the local region R_x. Highest entropy regions are considered salient regions and returned as the output of the algorithm. The most evident drawback of this approach is that the scale of the extracted regions, given by $|R_x|$, is a prefixed parameter, and as a consequence, only image features that lie in a small range of scales can be detected. Furthermore, Gilles approach is very sensitive to noise or small changes of the image, and extracted features rarely are stable over time.

The Gilles algorithm was extended by Kadir and Brady. The scale saliency algorithm considers salient regions not only in image space, but also in scale space, achieving scale invariance. A region is considered salient if it is salient in a narrow range of scales. A summary of this method is shown in Alg. 1. The result of this algorithm is a sparse three dimensional matrix containing weighted local entropies for all pixels at those scales where entropy is peaked. The highest values are selected as the most salient regions of the image, and then a clustering (non-maximum suppression) process is launched in order to join high overlapping regions. Figure 2.1 shows a comparison between the two methods.

This algorithm detects isotropic salient regions, meaning that it is not invariant to affine transformations like out of the plane rotations. However,

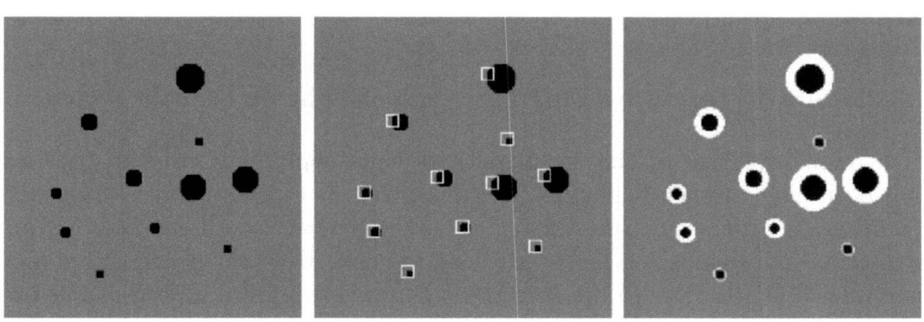

Fig. 2.1. Comparison of Gilles and scale saliency algorithms. *Left:* original synthetic image. *Center:* results for Gilles algorithm, using $|R_x| = 9$, and showing only one extracted region for each *black circle*. *Right:* scale saliency output, without clustering of overlapping results.

Algorithm 1: Kadir and Brady scale saliency algorithm

Input: Input image I, initial scale s_{min}, final scale s_{max}
for *each pixel x* **do**
  for *each scale s between* s_{min} *and* s_{max} **do**
    Calculate local entropy $H_D(s,x) = -\sum_{i=1}^{L} P_{s,x}(d_i)\log_2 P_{s,x}(d_i)$
  end
  Choose the set of scales at which entropy is a local maximum
  $S_p = \{s : H_D(s-1,x) < H_D(s,x) > H_D(s+1,x)\}$
  for *each scale s between* s_{min} *and* s_{max} **do**
    if $s \in S_p$ **then**
      Entropy weight calculation by means of a self-dissimilarity measure in scale space
      $W_D(s,x) = \frac{s^2}{2s-1}\sum_{i=1}^{L} \mid P_{s,x}(d_i) - P_{s-1,x}(d_i) \mid$
      Entropy weighting $Y_D(s,x) = H_D(s,x)W_D(s,x)$
    end
  end
end
Output: A sparse three dimensional matrix containing weighted local entropies for all pixels at those scales where entropy is peaked

this restriction is easily relaxed if we consider elliptical regions (anisotropic) rather than circular regions (isotropic). This change is achieved replacing the scale parameter s by a set of parameters (s, ϕ, θ), where θ is the orientation of the ellipse and ϕ the axis ratio. This way, major and minor axes are calculated as $s/\sqrt{(\phi)}$ and $s\sqrt{(\phi)}$, respectively. Indeed, this modification increases time complexity exponentially in the case of an exhaustive search, but an iterative approach may be applied. Starting from the regions detected by the original isotropic algorithm as seeds, their ellipse parameters are refined. First W_D is maximized modifying its ratio and orientation, and then H_D is also maximized modifying its scale. These two steps are repeated until there is no change. An example is shown in Fig. 2.2.

2.2.2 Point Filtering by Entropy Analysis Through Scale Space

The main drawback of this method is its time complexity, due to the fact that each pixel at each scale must be computed. As a consequence, Kadir and Brady scale saliency feature extractor is the slowest of all modern feature extraction algorithms, as recent surveys suggest. However, complexity may be remarkably decreased if we consider a simple idea: given a pixel x and two different neighborhoods R_x and R'_x, if R_x is homogeneous and $|R_x| > |R'_x|$,

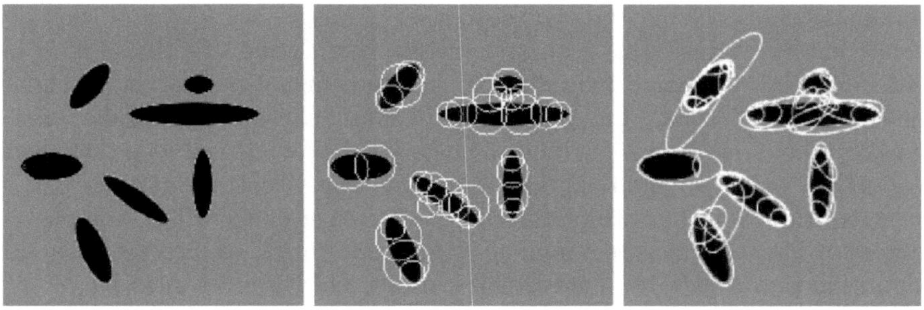

Fig. 2.2. *Left:* original synthetic image, formed by anisotropic regions. *Center:* output of the scale saliency algorithm. *Right:* output of the affine invariant scale saliency algorithm.

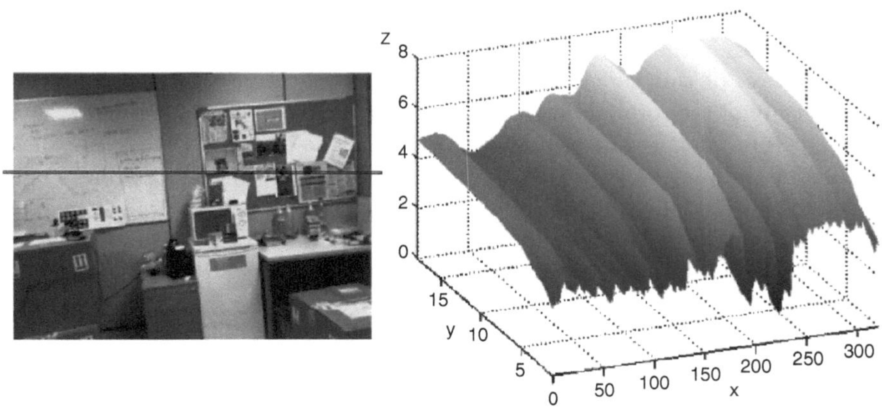

Fig. 2.3. Evolution of saliency through scale space. The graph on the right shows how the entropy value (z axis) of all pixels in the row highlighted on the left image (x axis) varies from $s_{\min} = 3$ to $s_{\max} = 20$ (y axis). As can be seen, there are not abrupt changes for any pixel in the scale range.

then the probability of R'_x of also being homogeneous is high. Figure 2.3 shows that this idea is supported by the evolution of entropy through scale space; the entropy value of a pixel smoothly varies through different scales.

Therefore, a preprocess step may be added to the original algorithm in order to discard several pixels from the image, based on the detection of homogeneous regions at $s_{\max}$. This previous stage is summarized as follows:

1. Calculate the local entropy H_D for each pixel at scale $s_{\max}$.
2. Select an entropy threshold $\sigma \in [0, 1]$.
3. $X = \{x \mid \frac{H_D(x, s_{\max})}{\max_x\{H_D(x, s_{\max})\}} > \sigma\}$.
4. Apply scale saliency algorithm only to those pixels $x \in X$.

It must be noted that the algorithm considers the relative entropy with respect to the maximum entropy value at $s_{\max}$ for the image in step 3. This way, an unique threshold may be applied to a set of images, regardless of their absolute entropy values. But a question arises: how should this threshold σ be chosen? A low threshold may discard small amount of points, and a higher one may discard points that are part of the most salient regions of the image. The threshold may be easily chosen a posteriori for an image if its most salient regions are detected, and then their minimum relative entropy at scale $s_{\max}$ is selected. However, the optimal threshold for another image will be completely different. The solution is to obtain an appropriate threshold for a set of several similar images by means of statistical learning and Bayesian analysis.

2.2.3 Chernoff Information and Optimal Filtering

Images belonging to the same image category or environment share similar intensity and texture distributions, so it seems reasonable to think that the entropy values of their most salient regions will lay in the same range. A proper identification of this range for a set of images may be performed by means of the study of two probability density functions known as $p_{\text{on}}(\theta)$ and $p_{\text{off}}(\theta)$. The $p_{\text{on}}(\theta)$ pdf defines the probability of a region to be part of the most salient regions of the image given that its relative entropy value is θ, while $p_{\text{off}}(\theta)$ defines the probability of a region to not to be part of the most salient regions of the image. This way, a previous learning stage should involve a training set of images from an image category; from these images, $p_{\text{on}}(\theta)$ and $p_{\text{off}}(\theta)$ are calculated. Then, the maximum relative entropy σ being $p_{\text{on}}(\sigma) > 0$ may be choosen as an entropy threshold for that image category. This way we avoid to select a threshold that would discard any of the most salient regions in training images. When a new image belonging to that category must be processed, all pixels whose relative entropy at $s_{\max}$ is lower than σ are discarded before applying scale saliency algorithm. However, Fig. 2.4 shows that this approach may be improved.

As stated in Fig. 2.4, more pixels may be discarded selecting a higher threshold if we assume the loss of some true salient regions, resulting in a decrease of computation time. It would be desirable to reach a trade-off between a high amount of discarded points and a low number of actual salient regions discarded. Seeing the examples in the previous figure, it is clear that the less overlapped the two distributions are, the easiest it is to find a threshold in order to discard larger nonsalient regions without removing too many salient ones; if the selected training images are homogeneous enough, the overlap of $p_{\text{on}}(\theta)$ and $p_{\text{off}}(\theta)$ is low. On the contrary, a heterogeneous set of images will result in similar $p_{\text{on}}(\theta)$ and $p_{\text{off}}(\theta)$ distributions, and it will be difficult to select an adequate threshold.

Thus, the splitting of training images in several homogeneous categories is a crucial step. Although addressing the point of selecting an optimal partition of the training images is out of the scope of this chapter, we should highlight a statistical measure that may help to decide if the partition is optimal: Chernoff

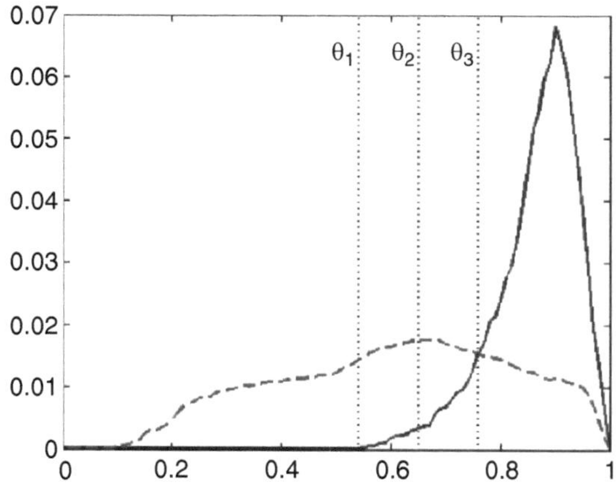

Fig. 2.4. $p_{\text{on}}(\theta)$ (*solid plot*) and $p_{\text{off}}(\theta)$ (*dashed plot*) distributions estimated from all pixels of a set of images belonging to a same image category. Threshold θ_1 ensures that the filter does not remove any pixel from training images that are not part of the most salient regions; however, in the case of new input images, we may find salient regions, the relative entropy of which is lower than this threshold. Choosing threshold θ_2 increases the amount of filtered points in exchange for increasing the probability of filtering salient regions of the image. Finally, threshold θ_3 assumes a higher risk, as far as more salient and not salient regions of the image will be filtered. However, the probability of a nonsalient pixel to be removed from the image is still higher than in the case of a salient one.

Information. The expected error rate of a likelihood test based on $p_{\text{on}}(\phi)$ and $p_{\text{off}}(\phi)$ decreases exponentially with respect to $C(p_{\text{on}}(\phi), p_{\text{off}}(\phi))$, where $C(p,q)$ is the Chernoff Information between two probability distributions p and q, and is defined by

$$C(p,q) = -\min_{0\leq\lambda\leq 1} \log\left(\sum_{j=1}^{J} p^{\lambda}(y_j)q^{1-\lambda}(y_j)\right) \tag{2.3}$$

where $\{y_j : j = 1, \ldots, J\}$ are the variables that the distributions are defined over (in this case, the probality of each relative entropy value in $[0,1]$). Chernoff Information quantifies the easiness of knowing from which of the two distributions came a set of values. This measure may be used as an homogeneity estimator for an image class during training. If the Chernoff Information of a training set is low, then images in that image class are not homogeneous enough and it must be splitted into two or more classes. A related measure is Bhattacharyya Distance. Bhattacharyya Distance is a particular case of Chernoff Information in which $\lambda = 1/2$:

$$BC(p,q) = -\log\left(\sum_{j=1}^{J} p^{\frac{1}{2}}(y_j)q^{\frac{1}{2}}(y_j)\right) \tag{2.4}$$

Figure 2.4 is again useful to understand how pixels on an image should be discarded. The log-likelihood ratio $\log(p_{\text{on}}(\theta)/p_{\text{off}}(\theta))$ is zero when $p_{\text{on}}(\theta) = p_{\text{off}}(\theta)$, that is, when the probability of a pixel with relative entropy θ at s_{max} to be part of the most salient regions is equal to the probability of not be part of them. In the other hand, positive values correspond to relative entropies that are more likely to be associated to the most salient regions. Thus, the log-likelihood ratio of these two densities may be used to discard points of an image. A threshold T must be chosen for an image class, so that any pixel from an image belonging to the same image class may be discarded if $\log(p_{\text{on}}(\theta)/p_{\text{off}}(\theta)) < T$.

Once again, information theory provides a tool capable of estimating a valid range in which this threshold should be chosen. The Kullback–Leibler divergence or relative entropy between two distributions p and q, given by

$$D(p||q) = \sum_{j=1}^{J} p(y_j) \log \frac{p(y_j)}{q(y_j)} \tag{2.5}$$

measures the dissimilarity between p and q. It estimates the coding efficiency loss of assuming that the distribution is q when the real distribution is p. The range of valid T values is given by (see Section 2.4.3)

$$-D(p_{\text{off}}(\theta)||p_{\text{on}}(\theta)) < T < D(p_{\text{on}}(\theta)||p_{\text{off}}(\theta)) \tag{2.6}$$

Selecting the minimum T value in this range ensures a good trade-off between a relatively high amount of discarded points and low error. More pixels can be filtered increasing T in this range and assuming that the error of the final results will increase depending on the Chernoff Information between $p_{\text{on}}(\theta)$ and $p_{\text{off}}(\theta)$. In fact, the Kullback–Leibler divergence and the Chernoff Information between these two distributions are related. If Chernoff value is low, distributions are similar and it is difficult to extract a good threshold to split points into homogeneous and non-homogeneous; as a consequence, the value of T must be selected from a narrower range.

2.2.4 Bayesian Filtering of the Scale Saliency Feature Extractor: The Algorithm

After introducing all the factors that take part in this approach to decrease the computation time of Kadir and Brady scale saliency algorithm, let us put them together to explain how it can be applied. First, we summarize how to extract a valid threshold from a set of training images belonging to the same image class or category:

1. Calculate $p_{\text{on}}(\theta)$ and $p_{\text{off}}(\theta)$ probability distributions from all points in a set of training images, considering if these points are part (on) or not (off) of the final displayed most salient features of its corresponding image, and being θ the relative entropy value of a pixel at s_{max} with respect to the maximum entropy value of any pixel of its image at the same scale.

2. Evaluate $C(p_{\text{off}}(\theta), p_{\text{off}}(\theta))$. A low value means that the image class is not homogeneous enough, and it is not possible to learn a good threshold. In this case, split the image class into new subclasses and repeat the process for each of them.
3. Estimate the range limits $-D(p_{\text{on}}(\theta)||p_{\text{off}}(\theta))$ and $D(p_{\text{off}}(\theta)||p_{\text{on}}(\theta))$.
4. Select a threshold in the range given by Kullback–Leibler divergence $-D(p_{\text{off}}(\theta)||p_{\text{on}}(\theta)) < T < D(p_{\text{on}}(\theta)||p_{\text{off}}(\theta))$. The minimum T value in this range is a conservative good trade-off between efficiency and low error rate. Higher T values will increase error rate accordingly to $C(p_{\text{off}}(\theta), p_{\text{off}}(\theta))$.

Then, new images belonging to the same image category can be filtered before applying the scale saliency algorithm, discarding points that probably are not part of the most salient features:

1. Calculate the local relative entropy $\theta_x = H_{D_x}/H_{\max}$ at $s_{\max}$ for each pixel x, where $H_{\max}$ is the maximum entropy value for any pixel at $s_{\max}$.
2. $X = \{x|\log\frac{p_{\text{on}}(\theta)}{p_{\text{off}}(\theta)} > T\}$, where T is the learned threshold for the image class that the input image belongs to.
3. Apply the scale saliency algorithm only to pixels $x \in X$.

In order to demonstrate the validity of this method, Table 2.1 shows some experimental results, based on the well-known *Object Categories* dataset from *Visual Geometry Group*, freely available on the web. This dataset is composed of several sets of images representing different image categories. These results were extracted following these steps: first, the training process was applied to each image category, selecting a random 10% of the images as training set. The result of the first step is a range of valid thresholds for each image category. Chernoff Information was also estimated. Then, using the rest of images from each category as test set, we applied the filtering algorithm, using two different thresholds in each case: the minimum valid value for each category and $T = 0$. Table 2.1 shows the results for each image category, including the mean amount of points (% points) filtered and the mean amount of time (% time) saved for each image category, depending on the used threshold. The last column shows the mean localization error of the extracted features (ϵ):

$$\epsilon = \frac{1}{N}\sum_{i=1}^{N}\frac{d(A_i, B_i) + d(B_i, A_i)}{2} \tag{2.7}$$

where N is the number of images in the test set, A_i represents the clustered most salient regions obtained after applying the original scale saliency algorithm to image i, B_i represents the clustered most salient regions obtained from the filtered scale saliency algorithm applied to image i, and

$$d(A, B) = \sum_{a\in A}\min_{b\in B}||a - b|| \tag{2.8}$$

Table 2.1. Application of the filtered scale saliency to the Visual Geometry Group image categories database.

Test set	Chernoff	T	% Points	% Time	ϵ
Airplanes_side	0.415	−4.98	30.79	42.12	0.0943
		0	60,11	72.61	2.9271
Background	0.208	−2.33	15.89	24.00	0.6438
		0	43.91	54.39	5.0290
Bottles	0.184	−2.80	9.50	20.50	0.4447
		0	23.56	35.47	1.9482
Camel	0.138	−2.06	10.06	20.94	0.2556
		0	40.10	52.43	4.2110
Cars_brad	0.236	−2.63	24.84	36.57	0.4293
		0	48.26	61.14	3.4547
Cars_brad_bg	0.327	−3.24	22.90	34.06	0.2091
		0	57.18	70.02	4.1999
Faces	0.278	−3.37	25.31	37.21	0.9057
		0	54.76	67.92	8.3791
Google_things	0.160	−2.15	14.58	25.48	0.7444
		0	40.49	52.81	5.7128
Guitars	0.252	−3.11	15.34	26.35	0.2339
		0	37.94	50.11	2.3745
Houses	0.218	−2.62	16.09	27.16	0.2511
		0	44.51	56.88	3.4209
Leaves	0.470	−6.08	29.43	41.44	0.8699
		0	46.60	59.28	3.0674
Motorbikes_side	0.181	−2.34	15.63	27.64	0.2947
		0	38.62	51.64	3.7305

is an Euclidean distance based measure between the most salient clustered regions. It must be noted that this distance is not only calculated in image space, but also in scale space; thus, errors in both localization and scale are considered. As can be seen, using the more conservative threshold yields a good trade-off between saved time and error, but using $T = 0$ produces noticeable improved results, keeping a low error rate. Generally, better results are obtained when Chernoff Information value is high, being more points discarded or decreasing error rate. Some examples of application are shown in Fig. 2.5.

2.3 Information Theory as Evaluation Tool: The Statistical Edge Detection Case

Edges are another kind of features extracted in early steps of several higher level computer vision algorithms. An edge is a part of the image where a sharp intensity discontinuity is present. These features are useful in the sense that they can give an idea of the shape or boundaries of the objects in the image,

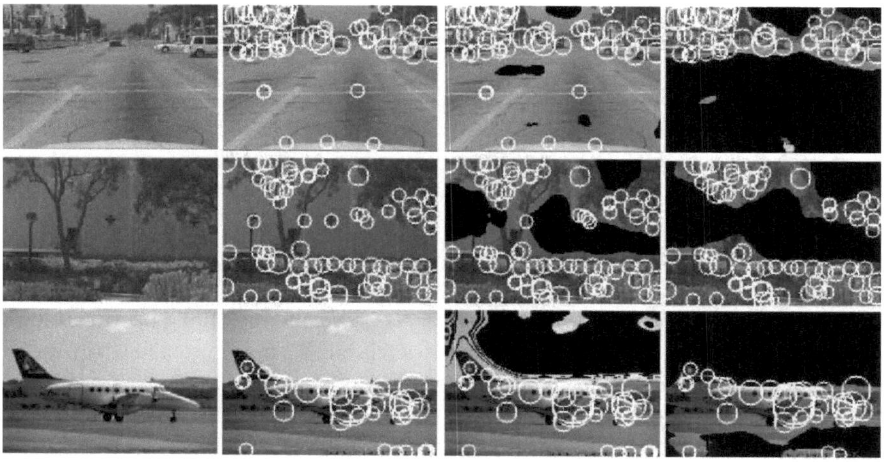

Fig. 2.5. Examples of filtering previous to scale saliency algorithm, applied to three images belonging to different image categories (*cars_brad_bg*, *background* and *airplanes_side*). From *left* to *right:* original image, results of the original scale saliency algorithm, results using the minimum valid threshold of the image category, results using $T = 0$. In all cases, *black* regions represent filtered points.

Fig. 2.6. *Left:* an example of natural scene containing textured regions. *Center:* output of the Canny algorithm applied to the left image. A high amount of edges appear as a consequence of texture and clutter; these edges do not correspond to object boundaries. *Right:* an example of ideal edge detection, in which most of the edges are part of actual object boundaries.

information that may be used in applications like object detection, medical imaging, or written characters recognition. The most common edge detection algorithms used in the literature are based on convolution masks, being Canny, Sobel and Prewitt some of these methods; however, their main drawback is that they do not provide an accurate response in the case of natural scenes with quite background clutter, as can be seen in Fig. 2.6.

We can deal with this problem by means of a statistical learning approach, quite similar to the method explained above to remove pixels from an image before applying the Kadir and Brady feature extractor. The work by Konishi et al. [99] is a good example. Their aim was to prove that it is possible to robustly extract edges from an image using simple filters at different scales.

But more interesting is that they use Chernoff Information and conditional entropy to evaluate the effect on edge extraction performance of several aspects of their approach: the use of a set of scales rather than only one scale, the quantization of histograms used to represent probability densities, and the effect of representing the scale space as an image pyramid, for instance. Thus, information theory is introduced as a valid tool for obtaining information about statistical learning processes.

Conditional entropy $H(Y|X)$ is a measure of the remaining entropy or uncertainty of a random variable Y, given another random variable X, the value of which is known. A low value of $H(Y|X)$ means that the variable X yields high amount of information about variable Y, making easier to predict its value. In the discrete case, it can be estimated as

$$H(Y|X) = \sum_{i=1}^{N} p(x_i)H(Y|X = x_i) = -\sum_{i=1}^{N} p(x_i) \sum_{j=1}^{M} p(y_j|x_i) \log p(y_j|x_i) \tag{2.9}$$

We will explain below how this measure may be applied to the context of classification evaluation.

2.3.1 Statistical Edge Detection

The Kadir and Brady feature extractor filter described above is inspired in statistical edge detection. It implies learning two probability distributions named $p_{\text{on}}(\phi)$ and $p_{\text{off}}(\phi)$, giving the probability of a pixel that is part of an edge or not, depending on the response of a certain filter ϕ. In the work by Konishi et al., the set of filters used to edge detection was a simple one (gradient magnitude, Nitzberg, Laplacian) applied to different data obtained from the images (gray-scale intensity, complete color intensities and chrominancy). Then, the log-likelihood ratio may be used to categorize any pixel $I(x)$ on an image I, classifying it as an edge pixel if this log-likelihood ratio is over a given threshold:

$$\log \frac{p_{\text{on}}(\phi(I(x)))}{p_{\text{off}}(\phi(I(x)))} > T \tag{2.10}$$

These distributions may be learnt from a image dataset that includes a groundtruth edge segmentation indicating the real boundaries of the objects in a sequence of images (like, for instance, the right image in Fig. 2.6). Two well-known examples are the *South Florida* and the *Sowerby* datasets, which, unfortunately, are not freely available. The presence of $p_{\text{off}}(\phi)$ is an improvement over traditional edge detection methods. The only information used by edge detection algorithms based on convolution masks is local information close to the edges; $p_{\text{off}}(\phi)$ may help to remove edges produced by background clutter and textures.

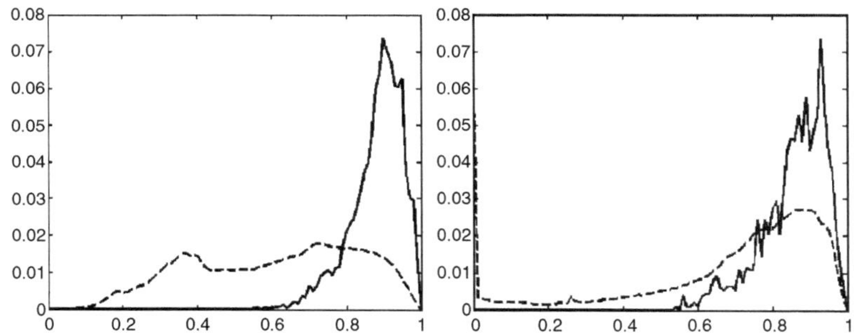

Fig. 2.7. $p_{\text{on}}(\phi)$ (solid line) and $p_{\text{off}}(\phi)$ (*dashed line*) for two example filter based classifiers. The overlap of the distributions on the left is lower than in the case on the right. In the first case, Chernoff Information value is 0.508, while in the second case, the value is 0.193. It can be clearly seen that the first classifier will distinguish better between on-edge and off-edge pixels.

As explained in previous sections, Chernoff Information is a feasible indicator of the performance of a classifier based on the log-likelihood ratio (see Eq. 2.10). When Chernoff Information values are high $p_{\text{on}}(\phi)$ and $p_{\text{off}}(\phi)$ are clearly separable, that is, it is easy to know if, given the response of a filter for a pixel x, this pixel is part or not of an edge. On other hand, when the Chernoff Information value is low, the overlapping of these two distributions is too high and it is difficult to characterize positive and negative samples. An example applied to edge detection problem is shown in Fig. 2.7. Several examples of classifier evaluation results by Konishi et al., based on Chernoff Information, can be seen in Fig. 2.8.

2.3.2 Edge Localization

The edge localization problem can be posed as the classification of all pixels on an image as being an edge or not depending on their distance to their nearest edge. This problem may be seen as a binary or a multiple class classification. Binary classification means that pixels are splitted into two categories: pixels whose nearest edge is below or over a certain distance. In the case of multiple class classification, each pixel is assigned a category depending on the distance to its nearest edge.

Let us first deal with *binary classification*. Given the function $w(x)$, that assigns to each pixel the distance to its nearest edge, and a threshold w, pixels can be splitted into two groups or classes:

$$\begin{cases} \alpha_1 = \{x : w(x) \leq w\} \\ \alpha_2 = \{x : w(x) > w\} \end{cases} \tag{2.11}$$

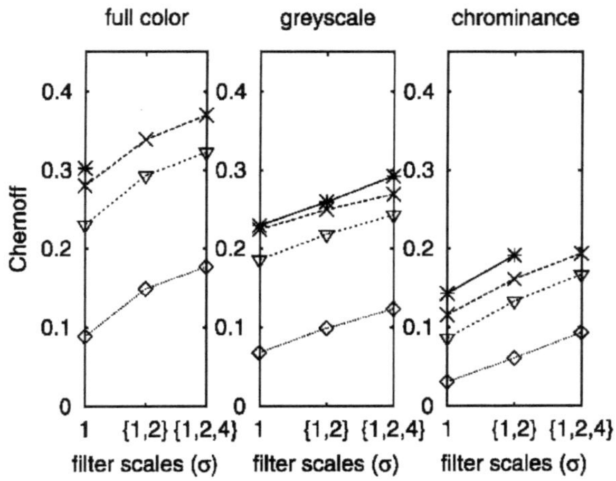

Fig. 2.8. Evolution of Chernoff Information for an edge classifier based on different filters when applied to full color, gray scale and chrominance information extracted from the images of the *South Florida* dataset, when using scale $\sigma = 1$, two scales $\sigma = \{1, 2\}$, and three scales $\sigma = \{1, 2, 4\}$. Chernoff Information increases as more scales are included; the conclusion is that a multiscale edge detection approach will perform better than a monoscale one. Furthermore, color information yields even higher Chernoff Information values, so this information is useful for this task. Further experiments of Chernoff Information evolution depending on scale in [98] show that if only one scale can be used, it would be better to choose an intermediate one. (Figure by Konishi et al. [99] ©2003 IEEE.)

Using the training groundtruth edge information, the $p(\phi|\alpha_1)$ and $p(\phi|\alpha_2)$ conditional distributions, as well as prior distributions $p(\alpha_1)$ and $p(\alpha_2)$ must be estimated. The classification task now is simple: given a pixel x and the response of the filter ϕ for that pixel $\phi(x) = y$, Bayes rules yield $p(\alpha_1|\phi(x) = y)$ and $p(\alpha_2|\phi(x) = y)$, allowing the classification algorithm to decide on the class of pixel x. Chernoff Information may also be used to evaluate the performance of the binary edge localization. A summary of several experiments of Konishi et al. [98] supports the coarse to fine edge localization idea: coarse step consists of looking for the approximate localization of the edges on the image using only an optimal scale σ^*. In this case, $w = \sigma^*$. Then, in the fine step, filters based on lower scales are applied in order to refine the search. However, the parameter σ^* depends on the dataset.

Multiclass classification differs from the previous task in the number of classes considered, which is now being greater than two. For instance, we could split the pixels into five different classes:

$$\begin{cases} \alpha_1 = \{x : w(x) = 0\} \\ \alpha_2 = \{x : w(x) = 1\} \\ \alpha_3 = \{x : w(x) = 2\} \\ \alpha_4 = \{x : 2 < w(x) \leq 4\} \\ \alpha_5 = \{x : w(x) > 4\} \end{cases} \tag{2.12}$$

Once again, and using the training groundtruth, we can estimate $p(\phi|\alpha_i)$ and $p(\alpha_i)$ for $i = 1 \ldots C$, C being the number of classes (5 in our example). Then, any pixel x is assigned class α^*, being

$$\alpha^* = arg \max_{i=1 \ldots C} p(\phi(x) = y|\alpha_i)p(\alpha_i) \tag{2.13}$$

In this case, conditional entropy is a useful tool to measure the performance of a filter ϕ:

$$H(\phi|y) = -\sum_y \sum_{i=1}^{C} p(\alpha_i|\phi = y)p(y) \log p(\alpha_i|\phi = y) \tag{2.14}$$

The conditional entropy quantifies the decrease of uncertainty about the class α_i a pixel belongs to *after* an observation $\phi = y$ was done. The performance of a filter ϕ_1 is better than the performance of a filter ϕ_2, if its conditional entropy is lower (in opposition to the Chernoff Information case, for which higher values mean higher performances); lower conditional entropy yields lower uncertainty, and as a consequence, it is easier to categorize a pixel. It must be noted that $H(\phi|y)$ is, in any case, lower or equal to $H(\alpha_i)$, that is, the response of a filter will decrease the a priori uncertainty that we have about the category of a pixel, or at least it will not increase it. A comparison of $H(\phi|y)$ with respect to $H(\alpha_i)$ may be a useful relative measure of the goodness of a filter. Several experiments driven by Konishi et al. are based on conditional entropy as an estimator of the performance of different filters in the multiclass edge localization problem [98].

Filter evaluation based on Chernoff Information and conditional entropy implies that the conditional distributions shown during this section must be represented as histograms. In the work by Konishi et al., decision trees are the base of probability distributions quantization. Since decision trees are introduced in Chapter 7, we do not give a detailed explanation on this part of the algorithm. However, we have mentioned it in order to remark an important aspect of Chernoff Information as classifier evaluation: its ability to detect overlearning.

A brief summary of how to perform quantization to build the histograms follows: starting from the whole filter space (the space representing all the possible filter responses; it could be a multidimensional space), a greedy algorithm looks for the cut in any axis that maximizes Chernoff Information. This cut produces two subspaces; for each of them, a new cut must be found that also maximizes Chernoff Information. This step is repeated with all the new subspaces until the desired number of cuts is reached. An advantage of this process is that histogram bins will be adapted to data: the parts of the filter space that need more discretization will be represented by a higher amount of bins. Another advantage is that a multidimensional filter can be represented using a notably lower number of bins than in the case of a regular quantization. As we can see in Fig. 2.9 (left), only a small number of cuts are needed to

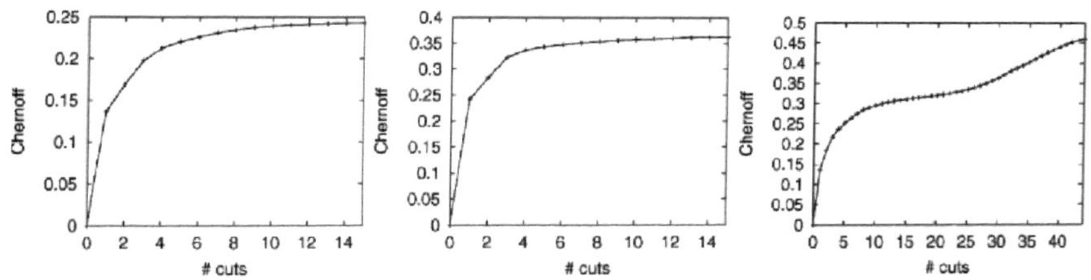

Fig. 2.9. *Left:* two examples of how Chernoff Information evolves depending on the number of cuts of the decision tree, that will define the probability distribution quantization. An asymptote is reached soon; most of the information can be obtained using a low amount of histogram bins. *Right:* example of overlearning. As long as the number of cuts increases an asymptote is reached, but, at certain point, it starts to increase again. This is an evident sign of overlearning. (Figure by Konishi et al. [99] ©2003 IEEE.)

reach an asymptote, that is, only a relative low amount of bins are needed to obtain a good quantization. On the other hand, in Fig. 2.9 (right), we can see overlearning effect. If the number of cuts is high enough, an abrupt increase of Chernoff Information value appears. This abrupt increase is produced by the fact that the chosen quantization is overfitted to the training data, for which the classification task will be remarkably easier; however, the obtained classifier will not be valid for examples different to those in the training dataset.

2.4 Finding Contours Among Clutter

In this section, we move to a related topic to introduce the Sanov's theorem and the method of types [43] in the context of contour finding. Sanov's theorem deals with rare events and their chance to occur. Given a set x^N of N i.i.d. samples obtained from an underlying distribution P_s, a histogram or type can be built from each sample. By the law of large numbers, as $N \to \infty$, this histogram will converge to the underlying distribution. The set of typical types $T^\epsilon_{P_s}$ for a distribution P_s is

$$T^\epsilon_{P_s} = \{x^N : D(P_{x^n}||P_s) \leq \epsilon\} \tag{2.15}$$

The probability that $x^N \in T^\epsilon_{P_s}$ tends to 1 as $N \to \infty$. On the other hand, the probability of a nontypical type $x^N \notin T^\epsilon_{P_s}$ tends to 0 as $N \to \infty$. Sanov's theorem puts a bound on the probability of these nontypical types (rare events).

Let $x_1, x_2, \ldots, x_N$ be a sample sequence ϕ i.i.d. from a distribution $P_s(x)$ with alphabet size J and let E be any closed set of probability distributions. Then

$$\frac{2^{-ND(\phi^*||P_s)}}{(N+1)^J} \leq p(\phi \in E) \leq (N+1)^J 2^{-ND(\phi^*||P_s)} \tag{2.16}$$

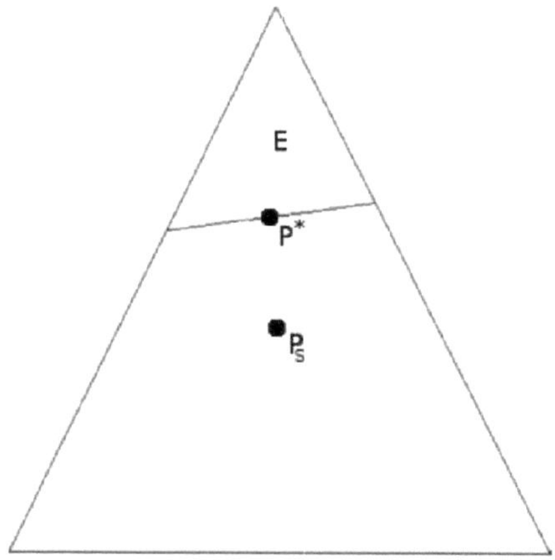

Fig. 2.10. The triangle represents the set of probability distributions, E being a subset within this set. P_s is the distribution which generates the samples. Sanov's theorem states that the probability that a type lies within E is determined by distribution P^* in E which is closer to P_s.

where $\phi^* = arg\min_{\phi \in E}(D\phi||P_s)$ is the distribution in E closest to P_s. Sanov's theorem yields two important implications. It states that when estimating the probability of a set of rare events, we only need to consider the most likely of these rare events (Fig. 2.10). It also states that the probability of rare events decreases exponentially with the divergence between the rare event (its type) and the true distribution.

2.4.1 Problem Statement

We base our discussion on the road tracking work of Coughlan et al. [42], which in turn, is based on a previous work by Geman and Jedymak. The underlying problem to solve is not the key question here, as long as the main objective of their work is to propose a Bayesian framework, and from this framework, analyze when this kind of problems may be solved. They also focus on studying the probability that the A^* algorithm yields an incorrect solution.

Geman and Jedymack's road tracking is a Bayesian Inference problem in which only a road must be detected in presence of clutter. Rather than detecting the whole road, it is splitted into a set of equally long segments. From an initial point and direction, and if road's length is N, all the possible Q^N routes are represented as a search tree (see Fig. 2.11), where Q is the tree's branching factor. There is only a target route, and thus in the worst case, the search complexity is exponential. The rest of paths in the tree are considered distractor paths.

We now briefly introduce problem's notation. Each route in the tree is represented by a set of movements $\{t_i\}$, where $t_i \in \{b_v\}$. The set $\{b_v\}$ forms an alphabet Q corresponding to the Q possible alternatives at each segment's

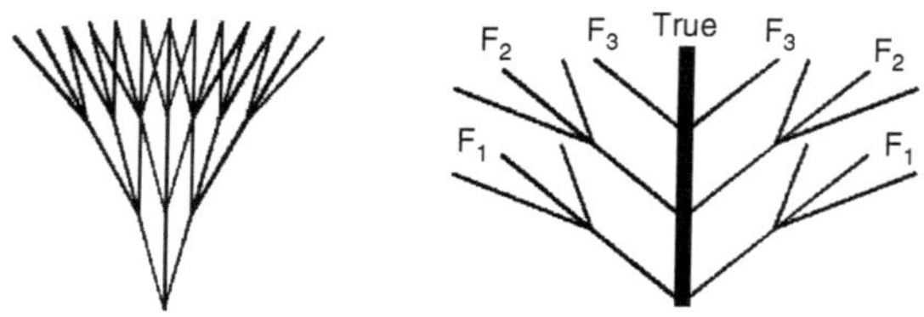

Fig. 2.11. *Left:* search tree with $Q = 3$ and $N = 3$. This search tree represents all the possible routes from initial point (at the bottom of the tree) when three types of movement can be chosen at the end of each segment: turning 15° to the left, turning 15° to the right and going straight. *Right:* Search tree divided into different sets: the target path (in *bold*) and N subsets $F_1, \ldots, F_N$. Paths in F_1 do not overlap with the target path, paths in F_2 overlap with one segment, and so on.

end (turn 15° left, turn 15° right, and so on). Each route has an associated prior probability given by

$$p(\{t_i\}) = \prod_{i=1}^{N} p_{\triangle G}(t_i) \tag{2.17}$$

where $p_{\triangle G}$ is the probability of each transition b_i. From now on, we assume that all transitions are equiprobable. A set of movements $\{t_i\}$ is represented by a set of tree segments $X = \{x_1, \ldots, x_N\}$. Considering $\mathcal{X}$ the set of all Q^N tree segments, it is clear that $X \in \mathcal{X}$. Moreover, an observation y_x is made for each $x \in \mathcal{X}$, being $Y = \{y_x : x \in \mathcal{X}\}$. In road tracking systems, observation values are obtained from a filter previously trained with road and non-road segments. Consequently, distributions $p_{\text{on}}(y_x)$ and $p_{\text{off}}(y_x)$ apply to the probability of y_x of being obtained from a road or not road segment. Each route $\{t_i\}$ with segments $\{x_i\}$ is associated with a set of observations $\{y_{x_i}\}$ that get values from an alphabet $\{a_\mu\}$ with size J.

Geman and Jedymak formulate road tracking in Bayesian Maximum a Posteriori (MAP) terms:

$$p(X|Y) = \frac{p(Y|X)p(X)}{p(Y)} \tag{2.18}$$

where prior is given by

$$p(X) = \prod_{i=1}^{N} p_{\triangle G}(t_i) \tag{2.19}$$

and

$$\begin{aligned} p(Y|X) &= \prod_{x \in X} p_{\text{on}}(y_x) \prod_{x \in \mathcal{X}/\mathcal{X}} p_{\text{off}}(y_x) \\ &= \prod_{i=1..N} \frac{p_{\text{on}}(y_{x_i})}{p_{\text{off}}(y_{x_i})} \prod_{x \in \mathcal{X}} p_{\text{off}}(y_x) \\ &= \prod_{i=1..N} \frac{p_{\text{on}}(y_{x_i})}{p_{\text{off}}(y_{x_i})} F(Y) \end{aligned}$$

In the latter equation, $F(Y) = \prod_{x \in \mathcal{X}} p_{\text{off}}(y_x)$ is independent of X, and can be ignored. In order to find the target route, $p(Y|X)p(X)$ must be maximized.

2.4.2 A^* Road Tracking

Coughlan et al. redesigned the algorithm to be solved by means of an A^* search. The aim of the algorithm is to search the route $\{t_i\}$ with associated observations $\{y_i\}$ that will be maximize a reward function. This route may be different to the one given by MAP estimate. We define reward function as

$$r(\{t_i\},\{y_i\}) = \sum_{i=1}^{N} \log\left(\frac{p_{\text{on}}(y_i)}{p_{\text{off}}(y_i)}\right) + \sum_{i=1}^{N} \log\left(\frac{p_{\triangle G}(t_i)}{U(t_i)}\right) \tag{2.20}$$

being $y_i = y_{x_i}$ and U the uniform distribution, so

$$\sum_{i=1}^{N} \log U(t_i) = -N \log Q \tag{2.21}$$

Reward function may be also expressed as

$$r(\{t_i\},\{y_i\}) = N\boldsymbol{\phi}\boldsymbol{\alpha} + N\boldsymbol{\psi}\boldsymbol{\beta} \tag{2.22}$$

where the components of vectors $\boldsymbol{\alpha}$ and $\boldsymbol{\beta}$ are

$$\alpha_\mu = \log \frac{p_{\text{on}}(\alpha_\mu)}{p_{\text{off}}(\alpha_\mu)}, \quad \mu = 1, \ldots, J \tag{2.23}$$

$$\beta_v = \log \frac{p_{\triangle G}(b_v)}{U(b_v)}, \quad v = 1 \ldots Q \tag{2.24}$$

and $\boldsymbol{\phi}$ and $\boldsymbol{\psi}$ are normalized histograms:

$$\phi_\mu = \frac{1}{M} \sum_{i=1}^{N} \delta_{y_i,\alpha_\mu}, \quad M = 1, \ldots, J \tag{2.25}$$

$$\psi_v = \frac{1}{N} \sum_{i=1}^{N} \delta_{t_i,b_v}, \quad v = 1, \ldots, Q \tag{2.26}$$

It must be noted that the Kronecker delta function $\delta_{i,j}$ is present in the latter equation.

Coughlan et al. address this question: given a specific road tracking problem, may the target road be found by means of MAP? Answering this question is similar to estimating the probability that the reward of any distractor path is higher than the one of the target path. In this case, the problem cannot be solved by any algorithm. For instance, let us consider the probability distribution $p_{1,\max}(r_{\max}/N)$ of the maximum normalized reward of all paths in F_1

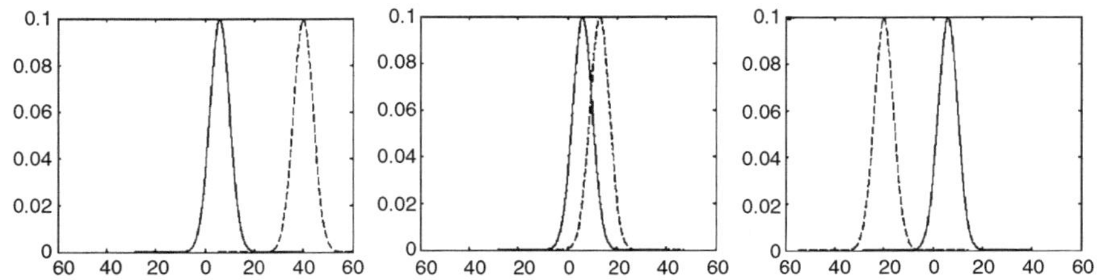

Fig. 2.12. Three cases of $p_{1,\max}(r_{\max}/N)$ (*solid line*) vs. $\hat{p}_T(r_T/N)$ (*dashed line*). In the first case, the reward of the target path is higher than the largest distractor reward, and as a consequence the task is straightforward. In the second case, the task is more difficult due to the overlapping of both distributions. In the last case, the reward of the target path is lower than the largest distractor rewards. Thus, it is impossible to find the target path.

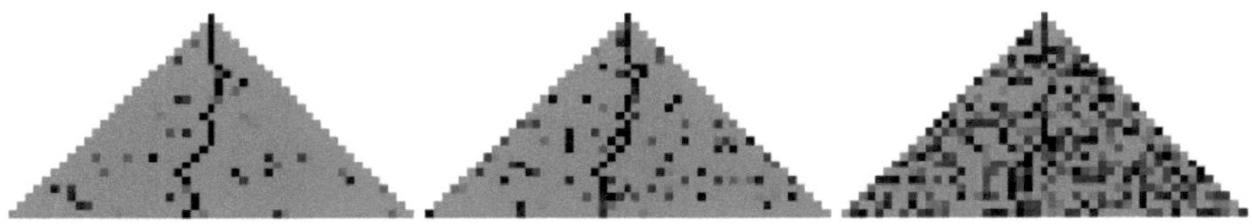

Fig. 2.13. Road tracking (*black pixels*) from initial point (at the top of each graph) in presence of increasing clutter. The value of the parameter K will increase from left to right, being the first example a simple problem and the last one an almost impossible one.

(see Fig. 2.11) and the probability distribution $\hat{p}_T(r_T/N)$ of the normalized reward of the target path. Figure 2.12 compares both. Using a similar method to Sanov's Theorem, it is possible to obtain a parameter K given by

$$K = D(p_{\text{on}}||p_{\text{off}}) + D(p_{\triangle G}||U) - logQ \tag{2.27}$$

that clarifies if the task can be solved. When $K > 0$, $\hat{p}_T(r_T/N)$ lies at the right of $p_{1,\max}(r_{\max}/N)$, thus finding the target path is straightforward. If $K \approx 0$, both distributions overlap and detecting the target path is not as simple. Finally, when $K < 0$, $\hat{p}_T(r_T/N)$ lies at the left of $p_{1,\max}(r_{\max}/N)$, and it is impossible to find the target path. An example is shown in Fig. 2.13.

The parameter K intuitively describes the difficulty of the task. For instance, the term $D(p_{\text{on}}||p_{\text{off}})$ is directly related to the filter quality. The higher this divergence is, the easier is to discriminate road and non-road segments. Clearly a better filter facilitates the road tracking task. The term $D(p_{\triangle G}||U)$, on the other hand, refers to the a priori information that is known about the shape of the road. If $p_{\triangle G} = U$, there is no prior information. Finally, Q measures the amount of present distractor paths. Therefore, K increases in the case of having a low error road detection filter, high a priori information and few distractor paths.

2.4.3 A^* Convergence Proof

Coughlan et al. adopt an inadmissible heuristic that converges faster than admissible heuristics, but may lead to incorrect results. This heuristic is $H_L + H_P$, L being the likelihood and P the geometric prior. Thus, a length M subpath has a reward of $(N - M)(H_L + H_P)$. Due to the fact that the $N(H_L + H_P)$ term does not vary for any subpath, it is dropped. This makes $-M(H_L + H_P)$ the heuristic in this problem.

In order to demonstrate the validity of the A^* algorithm results, two reward functions must be defined: a reward for length $M < N$ subpaths in the tree that completely overlap with the target path ($S_{\text{on}}(M)$), and a reward for length N paths in the tree that do not overlap at all with the target path ($S_{\text{off}}(N)$):

$$S_{\text{on}}(M) = M\{\boldsymbol{\phi}^{\text{on}}\boldsymbol{\alpha} - H_L\} + M\{\boldsymbol{\psi}^{\text{on}}\boldsymbol{\beta} - Hp\} \tag{2.28}$$

$$S_{\text{off}}N = N\{\boldsymbol{\phi}^{\text{off}}\boldsymbol{\alpha} - H_L\} + N\{\boldsymbol{\psi}^{\text{off}}\boldsymbol{\beta} - Hp\} \tag{2.29}$$

The following demonstration adopts the Bhattacharyya heuristic for the A^* algorithm. It simplifies the analysis and helps prove stronger convergence results. The heuristic should be lower than the normalized reward obtained from the target path and higher than the normalized reward obtained from any distractor path. If we consider H_L (the analysis is similar in the case of H_P), the expected reward for the target path is $D(p_{\text{on}}||p_{\text{off}})$, and it is $-D(p_{\text{off}}||p_{\text{on}})$ for distractor paths, and as a consequence:

$$-D(p_{\text{off}}||p_{\text{on}}) < H_L < D(p_{\text{on}}||p_{\text{off}}) \tag{2.30}$$

If we pretend to use Sanov's theorem, the heuristic must be expressed as the expected reward of data distributed according to the geometric mixture of $p_{\text{on}}, p_{\text{off}}$ given by $p_\lambda(y) = p_{\text{on}}^{1-\lambda}(y)p_{\text{off}}^{\lambda}(y)/Z[\lambda]$ ($Z[\lambda]$ being λ dependent normalization constant). Taking $\lambda = 1/2$ (that is, the Bhattacharyya heuristic), we are choosing a heuristic $H_L^* = \sum_y p_{\lambda=1/2}(y)\log\frac{p_{\text{on}}(y)}{p_{\text{off}}(y)}$ that is midway between target and distractors. Expressing the Bhattacharyya heuristic in reward notation (see Eq. 2.22) yields

$$H_L^* = \boldsymbol{\phi}_{Bh}\boldsymbol{\alpha}, \quad H_P^* = \boldsymbol{\psi}_{Bh}\boldsymbol{\beta} \tag{2.31}$$

where

$$\phi_{Bh}(y) = \frac{\{p_{\text{on}}(y)\}^{\frac{1}{2}}\{p_{\text{off}}(y)\}^{\frac{1}{2}}}{Z_\phi}, \quad \psi_{Bh}(t) = \frac{\{p_{\triangle G}(t)\}^{\frac{1}{2}}\{U(t)\}^{\frac{1}{2}}}{Z_\psi} \tag{2.32}$$

and Z_ϕ, Z_ψ are normalization constants.

We need to summarize two theorems, the proofs of which may be read elsewhere, in order to come to further conclusions in this section. The first one puts an upper bound on the probability that any false segment is searched, and is based on the fact that A^* algorithm searches the segment with the highest reward. This first theorem states that the probability that A^* searches the last segment of a particular subpath $A_{n,i}$ is less or equal to $P\{\exists m : S_{\text{off}}(n) \geq S_{\text{on}}(m)\}$. The second theorem bounds this probability by something that can be evaluated, and states that:

$$P\{\exists m : S_{\text{off}}(n) \geq S_{\text{on}}(m)\} \leq \sum_{m=0}^{\infty} P\{S_{\text{off}} \geq S_{\text{on}}(m)\} \tag{2.33}$$

The probability that A^* yields incorrect results is given by $P\{S_{\text{off}}(n) \geq S_{\text{on}}(m)\}$. The correctness of the algorithm can be analyzed searching a bound of this probability by means of Sanov's theorem. The bound is given by

$$P\{S_{\text{off}}(n) \geq S_{\text{on}}(m)\} \leq \{(n+1)(m+1)\}^{J^2 Q^2} 2^{-(n\Psi_1 + m\Psi_2)} \tag{2.34}$$

where $\Psi_1 = D(\boldsymbol{\phi}_{Bh}||p_{\text{off}}) + D(\boldsymbol{\psi}_{Bh}||U)$, and $\Psi_2 = D(\boldsymbol{\phi}_{Bh}||p_{\text{on}}) + D(\boldsymbol{\psi}_{Bh}||p_{\triangle G})$. The first step to prove this assertion is to define E, the set of histograms corresponding to partial off paths with higher reward than the partial on path:

$$E = \{(\boldsymbol{\phi}^{\text{off}}, \boldsymbol{\psi}^{\text{off}}, \boldsymbol{\phi}^{\text{on}}, \boldsymbol{\psi}^{\text{on}}) : n\{\boldsymbol{\phi}^{\text{off}}\boldsymbol{\alpha} - H_L^* + \boldsymbol{\psi}^{\text{off}}\boldsymbol{\beta} - H_P^*\} \tag{2.35}$$
$$\geq m\{\boldsymbol{\phi}^{\text{on}}\alpha - H_L^* + \boldsymbol{\psi}^{\text{on}}\boldsymbol{\beta} - H_P^*\}\} \tag{2.36}$$

By Sanov's theorem, we can give a bound in terms of ϕ^{off}, ψ^{off}, ϕ^{on} and ψ^{on} that minimizes

$$f(\phi^{\text{off}}, \psi^{\text{off}}, \phi^{\text{on}}, \psi^{\text{on}}) = nD(\boldsymbol{\phi}^{\text{off}}||p_{\text{off}}) + nD(\boldsymbol{\psi}^{\text{off}}||U) \tag{2.37}$$
$$+ mD(\boldsymbol{\phi}^{\text{on}}||p_{\text{on}}) + mD(\boldsymbol{\psi}^{\text{on}}||p_{\triangle G}) \tag{2.38}$$
$$+ \tau_1\{\sum_y \phi^{\text{off}}(y) - 1\} + \tau_2\{\sum_y \phi^{\text{on}}(y) - 1\} \tag{2.39}$$
$$+ \tau_3\{\sum_t \psi^{\text{off}}(t) - 1\} + \tau_4\{\sum_t \psi^{\text{on}}(t) - 1\} \tag{2.40}$$
$$+ \gamma\{m\{\boldsymbol{\phi}^{\text{on}}\boldsymbol{\alpha} - H_L^* + \boldsymbol{\psi}^{\text{on}}\boldsymbol{\beta} - H_P^*\} \tag{2.41}$$
$$- n\{\boldsymbol{\phi}^{\text{off}}\boldsymbol{\alpha} - H_L^* + \boldsymbol{\psi}^{\text{off}}\boldsymbol{\beta} - H_P^*\}\} \tag{2.42}$$

where the τ's and γ are Lagrange multipliers. The function f is convex, and so it has an unique minimum. It must be noted that f can be splitted into four terms of form $nD(\boldsymbol{\phi}^{\text{off}}||P_{\text{off}}) + \tau_1\{\sum_y \phi^{\text{off}}(y) - 1\} - n\gamma\boldsymbol{\phi}^{\text{off}}\boldsymbol{\alpha}$, coupled by shared constants. These terms can be minimized separately:

$$\boldsymbol{\phi}^{\text{off}*} = \frac{p_{\text{on}}^{\gamma} p_{\text{off}}^{1-\gamma}}{Z[1-\gamma]}, \boldsymbol{\phi}^{\text{on}*} = \frac{p_{\text{on}}^{1-\gamma} p_{\text{off}}^{\gamma}}{Z[\gamma]}, \boldsymbol{\psi}^{\text{off}*} = \frac{p_{\triangle G}^{\gamma} U^{1-\gamma}}{Z_2[1-\gamma]}, \boldsymbol{\psi}^{\text{on}*} = \frac{p_{\triangle G}^{1-\gamma} U^{\gamma}}{Z_2[\gamma]} \tag{2.43}$$

subject to the constraint given by Eq. 2.36. The value $\gamma = 1/2$ yields the unique solution. Moreover, $\gamma = 1/2$ yields

$$\boldsymbol{\phi}^{\text{off}*}\boldsymbol{\alpha} = H_L^* = \boldsymbol{\phi}^{\text{on}*}\boldsymbol{\alpha}, \boldsymbol{\psi}^{\text{off}*}\boldsymbol{\beta} = H_P^* = \boldsymbol{\psi}^{\text{on}*}\boldsymbol{\beta} \tag{2.44}$$

The solution occurs at $\boldsymbol{\phi}^{\text{on}*} = \boldsymbol{\phi}^{\text{off}*} = \boldsymbol{\phi}_{Bh}$, and $\boldsymbol{\psi}^{\text{on}*} = \boldsymbol{\psi}^{\text{off}*} = \boldsymbol{\psi}_{Bh}$. Thus, substituting into the Sanov bound gives Eq. 2.34. Another conclusion that can be extracted from Eq. 2.34 is that the probability of exploring a distractor path to depth n falls off exponentially with n:

$$P\{\exists m : S_{\text{off}}(n) \geq S_{\text{on}}(m)\} \leq \sum_{m=0}^{\infty} P\{S_{\text{off}}(n) \geq S_{\text{on}}(m)\} \tag{2.45}$$

$$\leq (n+1)^{J^2Q^2} C_2(\Psi_2) 2^{-n\Psi_1} \tag{2.46}$$

where

$$C_2(\Psi_2) = \sum_{m=0}^{\infty} (m+1)^{J^2Q^2} 2^{-m\Psi_2} \tag{2.47}$$

2.5 Junction Detection and Grouping

This section introduces a feature extraction algorithm by Cazorla et al. [36], which is based on the concepts and algorithms described in the two previous sections. The algorithm integrates junction detection and junction grouping. The aim of junction grouping is to remove false positives and detect false negatives. It may also be applied to obtain a mid-level representation of the scene.

2.5.1 Junction Detection

Based on the Kona junction representation, junctions are modeled as piecewise constant regions called wedges. A junction may be described by means of a parametric model (see Fig. 2.14) $\theta = (x_c, y_c, r, M, \{\theta_i\}, \{T_i\})$, where (x_c, y_c) and r are the center and the radius of the junction, respectively, M is the number of wedges incident with the center, $\{\theta_i\}$ is the wedge limits and $\{T_i\}$ is the intensity distribution for each wedge. SUSAN algorithm detects center candidates (x_c, y_c). It selects all candidates for which the number of close neighbors sharing its intensity is below a given threshold. The radius r is left as a free parameter that must be chosen by the user. In order to obtain the set of wedges $\{\theta_i\}$, the algorithm computes the averaged accumulated contrast for each $\theta \in [0, 2\pi]$ along the radius in these directions:

$$\tilde{I}_\theta = \frac{1}{r} \sum_{i=1}^{N} l_i I_i \tag{2.48}$$

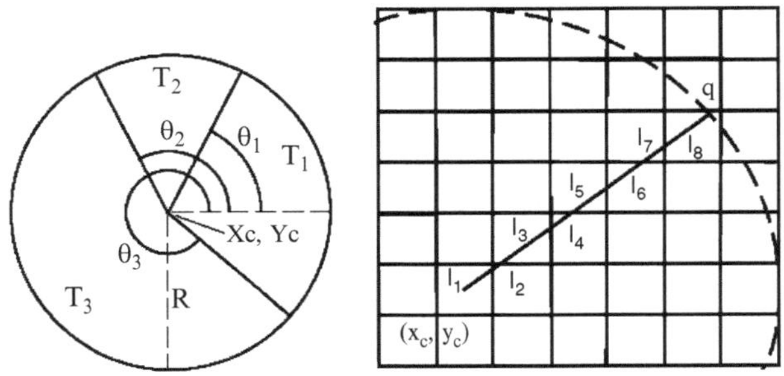

Fig. 2.14. *Left:* junction parametric model. *Right:* radius along direction θ_i discretized as a set of segments l_i. (Figure by Cazorla and Escolano ©2003 IEEE.)

Fig. 2.15. *Top left:* an example image. *Top right:* value of the log-likelihood ratio $\log(p_{\text{on}}/p_{\text{off}})$ for all pixels. *Bottom:* magnitude and orientation of the gradient. In the case of orientation, gray is value 0, white is π and black is $-\pi$. (Courtesy of Cazorla.)

where N is the number of segments l_i needed to discretize the radius in direction θ_i and I_i is the intensity of segment l_i (see Fig. 2.14).

Wedges may be found looking for peaks in Eq. 2.48. However, a better approach to wedge detection is given by the log-likelihood test seen in Section 2.3.1:

$$\tilde{I}_\theta = \frac{1}{r}\sum_{i=1}^{N} l_i \log\frac{p_{\text{on}}(I_i)}{p_{\text{off}}(I_i)} \tag{2.49}$$

Examples shown in Fig. 2.15 demonstrate that the filter can be improved adding gradient information. Thus, each pixel is represented as $\mathbf{I}_i = (I_i, \theta_i)$, where I_i is gradient magnitude and θ_i is a local estimation of the real orientation θ^* in which lies the pixel. Now, the orientations for which Eq. 2.50 is peaked and over a given threshold are selected as wedge limits:

$$\tilde{I}_\theta = \frac{1}{r}\sum_{i=1}^{N} l_i \log \frac{p_{\rm on}(\mathbf{I}_i|\theta^*)}{p_{\rm off}(\mathbf{I}_i)} \tag{2.50}$$

The $p_{\rm on}$ and $p_{\rm off}$ distributions in Eq. 2.50 are given by

$$p_{\rm on}(\mathbf{I}_i|\theta^*) = p_{\rm on}(I_i)p_{\rm ang}(\theta_i - \theta^*) \tag{2.51}$$

$$p_{\rm off}(\mathbf{I}_i) = p_{\rm off}(I_i)U(\theta_i) \tag{2.52}$$

where U is the uniform distribution and $p_{\rm ang}(\theta_i - \theta^*)$ is the probability that θ_i is the correct orientation. Although $p_{\rm ang}$ may be empirically estimated, in this case, its maximum when the difference is 0 or π. Finally, junctions given by Eq. 2.50 are pruned if $M < 2$ or $M = 2$, and the two wedges relative orientation is close to 0 or π. Figure 2.16 shows some examples of application.

2.5.2 Connecting and Filtering Junctions

The junction detection method introduced in the previous section is completed with a connecting paths search step. These paths connect junction pairs along edges between them. A connecting path P with length L, initial point at (x_c, y_c) and wedge limit θ is a set of segments $p_1, p_2, \ldots, p_L$ that may have variable or fixed length. Path curvature should be smooth, and thus, we also define orientations $\alpha_1, \alpha_2, \ldots, \alpha_{L-1}$ where $\alpha_j = \theta_{j+1} - \theta j$ is the angle between segments p_{j+1} and p_j.

The algorithm that searches paths is based on Bayesian A^* explained in Section 2.4. From junction center (x_c^0, y_c^0) and given a wedge orientation θ^0, the algorithm follows a search tree in which each segment p_j can expand Q branches, and thus, Q^N being possible paths. We recall the fact that the method described in Section 2.4 does not search the best path, but an unique path in clutter. The optimal path P^* is the one that maximizes

$$E(\{p_j, \alpha_j\}) = \sum_{j=1}^{L} \log \frac{p_{\rm on}(p_j)}{p_{\rm off}(p_j)} + \sum_{j=1}^{L-1} \log \frac{p_{\triangle G}(\alpha_{j+1} - \alpha j)}{U(\alpha_{j+1} - \alpha j)} \tag{2.53}$$

the first term being the intensity reward and the second term the geometric reward. The log-likelihood in the first term for a fixed length F segment is given by

$$\log \frac{p_{\rm on}(p_j)}{p_{\rm off}(p_j)} = \frac{1}{F}\sum_{i=1}^{N} l_i \log \frac{p_{\rm on}(\mathbf{I}_i|\theta^*)}{p_{\rm off}(\mathbf{I}_i)} \tag{2.54}$$

Fig. 2.16. Examples of junction detection. (Courtesy of Cazorla.)

Regarding the second term, $p_{\triangle G}(\alpha_{j+1} - \alpha j)$ models a first-order Markov chain of orientation variables α_j:

$$p_{\triangle G}(\alpha_{j+1} - \alpha j) \propto \exp\left\{-\frac{C}{2A}|\alpha_{j+1} - \alpha_j|\right\} \tag{2.55}$$

where A is the maximum angle between two consecutive segments and C models the path curvature. Finally, uniform distribution is added in geometric reward in order to express both terms of Eq. 2.53 in the same range of values. The algorithm evaluates the last L_0 segments of the path and prune them where both rewards are below a given threshold:

$$\frac{1}{L_0}\sum_{j=zL_0}^{(z+1)L_0-1}\log\frac{p_{\text{on}}(p_j)}{p_{\text{off}}(p_j)}<T \tag{2.56}$$

$$\frac{1}{L_0}\sum_{j=zL_0}^{(z+1)L_0-1}\log\frac{p_{\triangle G}(\alpha_{j+1}-\alpha j)}{U(\alpha_{j+1}-\alpha j)}<\hat{T} \tag{2.57}$$

given the threshold contraints (see Section 2.4):

$$-D(p_{off}||p_{\text{on}})<T<D(p_{\text{on}}||p_{\text{off}}) \tag{2.58}$$

$$-D(U||p_{\triangle G})<\hat{T}<D(p_{\triangle G}||U) \tag{2.59}$$

The thresholds may only be selected in a range of valid values. We recall here some of the properties given in Section 2.2.3. First of all, higher threshold values prune more segments than lower and conservative thresholds. On the other hand, the divergence between p_{on} and p_{off} distributions (and conversely, the divergence between U and $p_{\triangle G}$) also affects the amount of pruned segments. If these distributions are similar, the range of valid values is narrower, and as a consequence, this amount of pruned segments decreases.

In order to decrease computational burden, an inadmissible additional rule is introduced. This rule prioritizes the stability of longer paths that have survived more prunings, over shorter paths. Let L_{best} be the highest reward partial, path found. All length L_j paths for which $L_{\text{best}} - L_j > zL_0$ are pruned. Parameter z models the minimum allowed length difference between the best partial path and the rest. Lower z values remove more paths, increasing the risk to discard the true target path.

The junction connection algorithm searches a path from each junction center following each unvisited wedge limit, marking them as visited. The path exploration step finishes when the center (x_c^f, y_c^f) of another junction is in the neighborhood of the last segment of the best partial path. This last segment is linked to the wedge limit θ^f of the destination junction, and is marked as visited. Marking a wedge limit as visited prevents the algorithm to look for a path from it. If the A^* finishes and a junction is not found, two possible outcomes are defined, depending on the length of the last explored path. If it is below L_0, the algorithm discards it. Otherwise, the last position is stored and marked as terminal point, and this terminal point may link with another path in future steps of the algorithm. The algorithm finishes when there are no unvisited wedge limits. Unconnected junctions are considered false positives and removed. Similar images to the ones shown in Fig. 2.17 produce a 50% of false junctions; 55% of these false positives are removed due to junction grouping. As stated before, junction grouping also provides a mid-level visual representation of the scene that may be useful for high-level vision tasks.

Fig. 2.17. Examples of application of the connecting path search algorithm. (Courtesy of Cazorla.)

Problems

2.1 Understanding entropy

The algorithm by Kadir and Brady (Alg. 1) relies on a self-dissimilarity measure between scales. This measure is aimed at making possible the direct comparison of entropy values, in the case of different image region sizes. In general, and given a pixel on an image, how does its entropy vary with respect to the pixel neighborhood size? In spite of this general trend, think of an example in which the entropy remains almost constant in the range of scales between $s_{\min}$ and $s_{\max}$. In this last case, would the Kadir and Brady algorithm select this pixel to bė part of the most salient points on the image?

2.2 Symmetry property
Symmetry is one of the properties of entropy. This property states that entropy remains stable if data ordering is modified. This property strongly affects the Kadir and Brady feature extractor, due to the fact that it may assign the same saliency value to two visually different regions, if they share the same intensity distribution. Give an example of two visually different and equally sized isotropic regions, for which $p(0) = p(255) = 0.5$. For instance, a totally noisy region and a region splitted into two homogeneous regions. Although Eq. 2.2 assigns the same saliency to both regions, which one may be considered more visually informative? Think of a modification of Kadir and Brady algorithm that considers the symmetry property.

2.3 Entropy limits
Given Eq. 2.2, show why homogeneous image regions have minimum saliency. In which cases will the entropy value reach its maximum?

2.4 Color saliency
Modify Alg. 1 in order to adapt it to color images. How does this modification affect to the algorithm complexity? Several entropy estimation methods are presented through next chapters. In some of these methods, entropy may be estimated without any knowledge about the underlying data distribution (see Chapter 5). Is it possible to base Alg. 1 on any of these methods? And how this modification affects complexity?

2.5 Saliency numeric example
The following table shows the pixel intensities in a region extracted from an image. Apply Eq. 2.2 to estimate the entropy of the three square shape regions (diameters 3, 5 and 7) centered on the highlighted pixel. Did you find any entropy peak? In this case, apply self-dissimilarity to weight this value (see Alg. 1).

0	200	200	200	200	200	0
0	0	200	0	0	0	0
0	0	200	200	0	0	200
0	0	0	***0***	0	0	200
0	0	200	200	0	0	200
200	0	0	0	0	0	200
0	0	0	0	0	200	200

2.6 Conditional entropy as filter evaluator
Let ϕ_1 and ϕ_2 be two local filters applied to edge detection. The output of both filters for any pixel is a discrete value in the set $\{0, 1, 2\}$. Each pixel is labeled using an alphabet $\alpha = \{0, 1\}$; pixels belonging to an edge are labeled as 1, and pixels that are not part of any edge are labeled as 0. The following

table shows the real label of six pixels, and the output of both filters for each of them. Evaluate ϕ_1 and ϕ_2 by means of conditional entropy. Which filter discriminates better between on edge and off edge pixels?

α	ϕ_1	ϕ_2
0	0	0
0	1	2
0	1	2
1	1	1
1	2	0
1	2	1

2.7 Kullback–Leibler divergence as filter evaluator
Plot the p_{on} and p_{off} distributions corresponding to the two filters ϕ_1 and ϕ_2 from Prob. 2.6. Perform an evaluation of both filters, based on Kullback–Leibler divergence and Bhattacharyya distance. Is any of these two measures more informative than the other one with respect to the quality of the filters? Check that Kullback–Leibler divergence is not symmetric, that is, $D(p_{\text{on}}||p_{\text{off}}) \neq D(p_{\text{off}}||p_{\text{on}})$. Think of a simple divergence measure, based on Kullback–Leibler, which satisfies the symmetry property.

2.8 Conditional entropy and not informative classifiers
Conditional entropy may also be expressed as

$$H(Y|X) = H(Y,X) - H(X) \tag{2.60}$$

where the joint entropy is

$$H(X,Y) = -\sum_x \sum_y p(x,y) \log p(x,y) \tag{2.61}$$

Conditional entropy, $H(Y|X)$, is a measure of the remaining unpredictability of Y given X. Thus, $H(Y|X) = H(Y)$ is true only if X and Y are independent variables. In the case of edge detection, conditional entropy is a measure of the unpredictability of a pixel class (edge or non-edge), given a filter. Knowing all these facts, design the worst possible filter, a filter that does not give any information about the class of any pixel.

2.9 Sanov's theorem
Apply Sanov's theorem to estimate the possibility of obtaining more than 800 sixes when throwing 1,000 fair dices.

2.10 MAP and road tracking
Check, using Eq. 2.27, if the road tracking problem can be solved using any of the two filters shown in Prob. 2.6. Suppose a branching factor of 3 and $p_{\triangle G} = U$.

2.6 Key References

- T. Kadir and M. Brady, "Scale, Saliency and Image Description", *International Journal of Computer Vision*, 45(2): 83–105 (2001)
- S. Konishi, A. Yuille and J. Coughlan, "A Statistical Approach to Multi-scale Edge Detection", *Image and Vision Computing*, 21(1): 37–48 (2003)
- J. M. Coughlan and A. Yuille, "Bayesian A^* Tree Search with Expected O(N) Node Expansions for Road Tracking", *Neural Computation*, 14(8):1929–1958 (2002)
- M. Cazorla, F. Escolano, D. Gallardo and R. Rizo, "Junction Detection and Grouping with Probabilistic Edge Models and Bayesian A^*", *Pattern Recognition*, 35(9):1869–1881 (2002)
- M. Cazorla and F. Escolano, "Two Bayesian Methods for Junction Detection", *IEEE Transaction on Image Processing*, 12(3):317–327 (2003)

3

Contour and Region-Based Image Segmentation

3.1 Introduction

One of the most complex tasks in computer vision is segmentation. Segmentation can be roughly defined as optimally segregating the foreground from the background, or by finding the optimal partition of the image into its constituent parts. Here *optimal segregation* means that pixels (or blocks in the case of textures) in the foreground region share common statistics. These statistics should be significantly different from those corresponding to the background. In this context, active polygons models provide a discriminative mechanism for the segregation task. We will show that Jensen–Shannon (JS) divergence can efficiently drive such mechanism. Also, the maximum entropy (ME) principle is involved in the estimation of the intensity distribution of the foreground.

It is desirable that the segmentation process achieves good results (compared to the ones obtained by humans) without any supervision. However, such unsupervision only works in limited settings. For instance, in medical image segmentation, it is possible to find the contour that separates a given organ in which the physician is interested. This can be done with a low degree of supervision if one exploits the IT principle of minimum description length (MDL). It is then possible to find the best contour, both in terms of organ fitting and minimal contour complexity. IT inspires methods for finding the best contour both in terms of segregation and minimal complexity (the minimum description length principle).

There is a consensus in the computer vision community about the fact that the maximum degree of unsupervision of a segmentation algorithm is limited in the purely discriminative approach. To overcome these limitations, some researchers have adopted a mixed discriminative–generative approach to segmentation. The generative aspect of the approach makes hypotheses about intensity or texture models, but such hypotheses are contrasted with discriminative (bottom-up) processes. Such approaches are also extended to integrate

F. Escolano et al., *Information Theory in Computer Vision and Pattern Recognition*, segmentation with recognition. In both cases (bottom-up segmentation and/or recognition), information theory provides a formal framework to make these tasks feasible with real images.

3.2 Discriminative Segmentation with Jensen–Shannon Divergence

3.2.1 The Active Polygons Functional

Concerning the problem of capturing the contour of a region by means of an active contour [93], it is necessary to (i) devise a proper discrete representation of the contour, (ii) design differential equations that implement the motion flow of the contour points toward their optimal placement (the one minimizing a given energy function), and (iii) remove, as much as possible, the initialization-dependent behavior of the process. *Region competition* [184] unifies several region-based segmentation approaches by integrating statistics-based motion flows, smoothness-preserving flows, dynamics for region growing, and merging criteria.[1] Focusing here on the use of statistical models within active contours, a more evolved treatment of the question, besides the use of a simplistic polygonal discretization of the contour, consists in the *Active Polygons* [162, 163] computational model. This new model is driven by an information theory measure: the Jensen divergence.

In continuous terms, a contour is $\Gamma : [a,b] \subset \mathbb{R} \rightarrow \mathbb{R}^2$ defined around some region $\mathcal{R} \subset \mathbb{R}^2$. Considering there exists a function $f : \mathbb{R}^2 \rightarrow \mathbb{R}^2$, the divergence theorem for the plane states that *the integral of f inside $\mathcal{R}$ is equivalent to the integral of $\mathbf{F} \cdot \mathbf{n}$, being $f = \nabla \cdot \mathbf{F} = \partial F_1/\partial x + \partial F_2/\partial y$ the divergence, with $\mathbf{F} = (F_1, F_2)^T$, and $\mathbf{n}$ the unit normal to Γ*, that is

$$\int\int_{\mathcal{R}} f(x,y)dxdy = \oint_{\Gamma=\partial\mathcal{R}} \mathbf{F}\cdot\mathbf{n}ds \tag{3.1}$$

where ds is the Euclidean arc-length element, and the circle in the right-hand integral indicates that the curve is closed (the usual assumption), that is, if we define p as a parameterization of the curve, then $p \in [a,b]$ and $\Gamma(a) = \Gamma(b)$. Being $\mathbf{t} = (dx, dy)^T$ a vector pointing tangential along the curve, and assuming that the curve is positively-oriented (counterclockwise), we have $ds = \sqrt{dx^2 + dy^2}$. As the unit normal vector is orthogonal to the tangential one, we have $\mathbf{n} = (dy, -dx)^T$ and $||\mathbf{n}|| = 1 = ds$.

Alternatively, we may reformulate Eq. 3.1 in terms of the parameterization with p, and denote it $E(\Gamma)$:

$$E(\Gamma) = \int_a^b (\mathbf{F}\cdot\mathbf{n}) \underbrace{||\Gamma_p||dp}_{ds} \tag{3.2}$$

[1] Region competition contemplates also the MDL (Minimum Description Length) information-theoretic principle, which will be tackled later in this chapter.

where $||\Gamma_p|| = ||\partial\Gamma/\partial p|| = \sqrt{x_p^2 + y_p^2}$. As the unit tangent $\mathbf{t}$ may be defined in terms of $\Gamma_p/||\Gamma_p||$, the unit normal $\mathbf{n}$ may be similarly defined in the following terms:

$$\mathbf{n} = \mathbf{J}\mathbf{t} \;\text{ with }\; \mathbf{J} = \begin{pmatrix} 0 & 1 \\ -1 & 0 \end{pmatrix} \tag{3.3}$$

and consequently, we have

$$\mathbf{n}||\Gamma_p|| = \mathbf{J}\Gamma_p \tag{3.4}$$

and, thus, we may express $E(\Gamma)$ in the following terms:

$$E(\Gamma) = \int_a^b (\mathbf{F}\cdot\mathbf{J}\Gamma_p)dp \tag{3.5}$$

However, we need an expression for the *gradient-flow* of $E(.)$, that is, the partial derivatives indicating how the contour is changing. In order to do so, we need to compute the derivative of $E(.)$, denoted by $E_t(.)$, with respect to the contour. This can be usually done by exploiting the Green's theorem (see Prob. 3.2.). Setting $a = 0$, $b = 1$:

$$E_t(\Gamma) = \frac{1}{2}\int_0^1 (\mathbf{F}_t\cdot\mathbf{J}\Gamma_p + \mathbf{F}\cdot\mathbf{J}\Gamma_{pt})dp \tag{3.6}$$

where Γ_{pt} is the derivative of Γ_p. Integrating by parts the second term and applying the chain rule on the derivatives of $\mathbf{F}$, we remove Γ_{pt}:

$$E_t(\Gamma) = \frac{1}{2}\int_0^1 ((D\mathbf{F})\Gamma_t\cdot\mathbf{J}\Gamma_p - (D\mathbf{F})\Gamma_p\cdot\mathbf{J}\Gamma_p)dp \tag{3.7}$$

$D\mathbf{F}$ being the 2×2 Jacobian of $\mathbf{F}$. Then, isolating Γ_t in the left side of the dot product, we have:

$$E_t(\Gamma) = \frac{1}{2}\int_0^1 (\Gamma_t\cdot[\mathbf{J}^T(D\mathbf{F})^T - \mathbf{J}^T(D\mathbf{F})]\Gamma_p)dp \tag{3.8}$$

and exploiting the fact that the matrix [.] is antisymmetric ($\mathbf{A}$ is antisymmetric when $\mathbf{A} = -\mathbf{A}^T$), its form must be $\omega\mathbf{J}$ due to the antisymmetric operator $\mathbf{J}$:

$$E_t(\Gamma) = \frac{1}{2}\int_0^1 (\Gamma_t\cdot\omega\mathbf{J}\Gamma_p)dp = \frac{1}{2}\int_\Gamma (\Gamma_t\cdot\omega\mathbf{J}\Gamma_p)ds \tag{3.9}$$

Finally, setting $\omega = \nabla\cdot F = 2f$ we obtain

$$E_t(\Gamma) = \int_\Gamma (\Gamma_t\cdot\omega\mathbf{n})ds \tag{3.10}$$

which implies that the gradient flow (contour motion equation) is given by

$$\Gamma_t = f\mathbf{n} \tag{3.11}$$

Given the latter gradient flow, it is now necessary to translate it to a closed polygon defined by n vertices $\mathbf{V}_1, \ldots, \mathbf{V}_n$ belonging to $\mathbb{R}^2$. First of all, we must parameterize the contour Γ, which is associated to the polygon, by $p \in [0,1]$. Then, $\Gamma(p) = L(p - \lfloor p \rfloor, \mathbf{V}_{\lfloor p \rfloor}, \mathbf{V}_{\lfloor p \rfloor + 1})$, being $L(t, \mathbf{A}, \mathbf{B}) = (1-t)\mathbf{A} + t\mathbf{B}$ a parameterized line between vectors $\mathbf{A}$ and $\mathbf{B}$ (vertices in this case). The indices for the vertices are module n, which means that $\mathbf{V}_0 = \mathbf{V}_n$. Also, $\Gamma_p(p) = \mathbf{V}_{\lfloor p \rfloor + 1} - \mathbf{V}_{\lfloor p \rfloor}$. Now it is possible to have a *polygonal* version of Eq. 3.5:

$$E_t(\Gamma) = \int_0^1 (f(\Gamma(p))(\Gamma_t \cdot \mathbf{J}(\Gamma_p(p))dp \tag{3.12}$$

where the gradient flow *for each vertex* $\mathbf{V}_k$ is given by

$$\begin{aligned}\frac{\partial \mathbf{V}_k}{\partial t} &= \int_0^1 pf(L(p, \mathbf{V}_{k-1}, \mathbf{V}_k))dp\ \mathbf{n}_{k,k-1}\\ &+ \int_0^1 (1-p)f(L(p, \mathbf{V}_k, \mathbf{V}_{k+1}))dp\ \mathbf{n}_{k+1,k}\end{aligned} \tag{3.13}$$

$\mathbf{n}_{k,k-1}$ and $\mathbf{n}_{k+1,k}$ being the outward normals to edges $\mathbf{V}_{k-1}, \mathbf{V}_k$ and $\mathbf{V}_k, \mathbf{V}_{k+1}$, respectively. These equations allow to move the polygon as shown in Fig. 3.1 (top). What is quite interesting is that the choice of a *speed function* f adds flexibility to the model, as we will see in the following sections.

3.2.2 Jensen–Shannon Divergence and the Speed Function

The key element in the vertex dynamics described above is the definition of the speed function $f(.)$. Here, it is important to stress that such dynamics are not influenced by the degree of visibility of an attractive potential like the gradient in the classical snakes. On the contrary, to avoid such *myopic* behavior, region-based active contours are usually driven by *statistical forces*. More precisely, the contour Γ encloses a region $R_{\mathcal{I}}$ (inside) of image $\mathbf{I}$ whose complement is $R_{\mathcal{O}} = \mathbf{I} \setminus R_{\mathcal{I}}$ (outside), and the optimal placement Γ^* is the one yielding homogeneous/coherent intensity statistics for both regions. For instance, assume that we are able to measure m statistics $G_j(.)$ for each region, and let u_j and v_j, $j = 1, \ldots, m$ be respectively the expectations $E(G_j(.))$ of such statistics for the inside and outside regions. Then, the well-known functional of Chan and Vese [37] quantifies this rationale:

$$E(\Gamma) = \frac{1}{2|\mathbf{I}|}\sum_{j=1}^{m} \int_{R_{\mathcal{I}}} (G_j(\mathbf{I}(x,y) - u_j)^2 dxdy + \int_{R_{\mathcal{O}}} (G_j(\mathbf{I}(x,y) - v_j)^2 dxdy \tag{3.14}$$

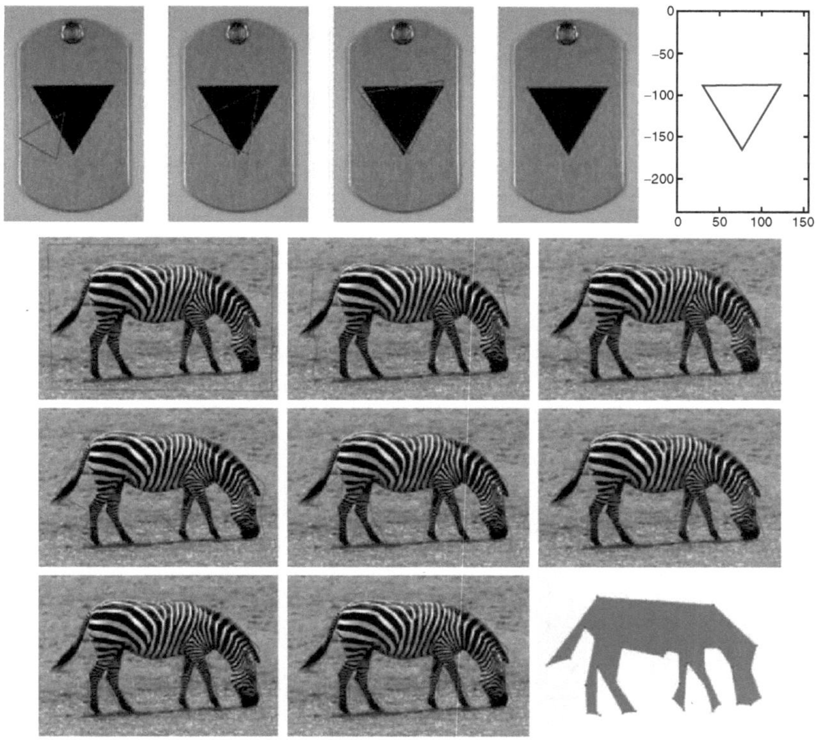

Fig. 3.1. Segmentation with active polygons. *Top*: simple, un-textured object. *Bottom*: complex textured object. Figure by G. Unal, A. Yezzi and H. Krim (©2005 Springer).

which is only zero when both terms are zero. The first term is nonzero when the background (outer) intensity/texture dominates the interior of the contour; the second term is nonzero when the foreground (inner) texture dominates the interior of the contour; and both terms are nonzero when there is an intermediate domination. Thus, the latter functional is zero only in the optimal placement of Γ (no-domination).

From the information-theory point of view, there is a similar way of formulating the problem. Given N data populations (in our case $N = 2$ intensity/texture populations: in and out), their *disparity* may be quantified by the *generalized Jensen–Shannon divergence* [104]:

$$JS = H\left(\sum_{i=1}^{N} a_i p_i(\xi)\right) - \sum_{i=1}^{N} a_i H(p_i(\xi)) \tag{3.15}$$

$H(.)$ being the entropy, a_i the prior probabilities of each class, $\sum_{i=1}^{N} a_i = 1$, $p_i(\xi)$ the ith pdf (corresponding to the ith region) of the random variable ξ (in this case pixel intensity $\mathbf{I}$). Considering the Jensen's inequality

$$\gamma\left(\frac{\sum a_i x_i}{\sum a_i}\right) \leq \frac{\sum a_i \gamma(x_i)}{\sum a_i} \tag{3.16}$$

$\gamma(.)$ being a convex function and the a_i positive weights, it turns out that, if X is a random variable, $\gamma(E(X)) \leq E(\gamma(X))$. In the two latter inequalities, changing to $\geq$ turns convexity into concavity. Let then P be the mixture of distributions $P = \sum_i a_i p_i$. Then, we have

$$\begin{aligned}
H(P) &= -\int P \log P d\xi \\
&= -\int \left(\sum_i a_i p_i\right) \log \left(\sum_i a_i p_i\right) d\xi \\
&= \int \left(\sum_i a_i p_i\right) \log \frac{1}{P} d\xi \\
&= \sum_i a_i \left(\int p_i \log \frac{1}{P} d\xi\right) \\
&= \sum_i a_i \left(\int p_i \log \frac{1}{p_i} d\xi + \int p_i \log \frac{1}{P} d\xi\right) \\
&= \sum_i a_i \left(H(p_i) + D(p_i||P)\right)
\end{aligned} \tag{3.17}$$

where $D(.||.)$ denotes, as usual, the Kullback–Leibler divergence. Therefore

$$JS = H\left(\sum_i a_i p_i\right) - \sum_i a_i H(p_i) = \sum_i a_i D\left(p_i || \sum_i a_i p_i\right) \tag{3.18}$$

As the KL divergence is nonnegative, this is also the case of JS. Therefore we have

$$H\left(\sum_i a_i p_i\right) \geq \sum_i a_i H(p_i) \tag{3.19}$$

and exploiting Jensen's inequality (Eq. 3.16), we obtain the concavity of the entropy. In addition, the KL-based definition of JS provides an interpretation of the Jensen–Shannon divergence as a weighted sum of KL divergences between individual distributions and the mixture of them.

For our particular case with $N = 2$ (foreground and background distributions), it is straightforward that JS has many similarities with Eq. 3.14 in terms of providing a useful divergence for contour-driving purposes. Furthermore, JS goes theoretically beyond Chan and Vese functional because the entropy quantifies higher order statistical interactions. The main problem here is that the densities $p_i(\xi)$ must be estimated or, at least, properly approximated. Thus, the theoretical bridge between JS and the Chen and Vese functional is provided by the following optimization problem:

$$\begin{aligned} p^*(\xi) &= \arg\max_{p(\xi)} - \int p(\xi)\log p(\xi)d\xi \\ \text{s.t.} \quad & \int p(\xi)G_j(\xi)d\xi = E(G_j(\xi)),\ j = 1,\dots,m \\ & \int p(\xi)d\xi = 1 \end{aligned} \tag{3.20}$$

where the density estimation $p^*(\xi)$ is the maximum entropy constrained to the verification of the expectation equations by each of the statistics. Thus, this is the first application in the book of the *principle of maximum entropy* [83]: given some information about the expectations of a set of statistics characterizing a distribution (this is ensured by the satisfaction of the first constraint), our choice of the pdf corresponding to such distribution (the second constraint ensures that the choice is a pdf) must be the *most neutral one* among all the $p(\xi)$ satisfying the constraints, that is, the more uninformed one that is equivalent to the maximum entropy distribution. Otherwise, our choice will be a pdf with more information than we have actually specified in the expectations constraints, which indicates all what is available for making the inference: our measurements $E(G_j(\xi))$.

The shape of the optimal pdf is obtained by constructing and deriving the Lagrangian (after assuming the natural logarithm):

$$\begin{aligned} \mathcal{L}(p,\Lambda) &= -\int p(\xi)\log p(\xi)d\xi + \lambda_0\left(\int p(\xi)d\xi - 1\right) \\ &\quad + \sum_{j=1}^{m}\lambda_j\left(\int p(\xi)G_j(\xi)d\xi - E(G_j(\xi))\right) \\ \frac{\partial\mathcal{L}}{\partial p} &= -\log p(\xi) - 1 + \lambda_0 + \sum_{j=1}^{m}\lambda_j G_j(\xi) \end{aligned} \tag{3.21}$$

$\Lambda = (\lambda_0,\dots,\lambda_m)$ being the $m+1$ Lagrange multipliers. Then, seeking for the maximum implies

$$\begin{aligned} 0 &= -\log p(\xi) - 1 + \lambda_0 + \sum_{j=1}^{m}\lambda_j G_j(\xi) \\ \log p(\xi) &= -1 + \lambda_0 + \sum_{j=1}^{m}\lambda_j G_j(\xi) \\ p^*(\xi) &= e^{-1+\lambda_0+\sum_{j=1}^{m}\lambda_j G_j(\xi)} \end{aligned} \tag{3.22}$$

where the multipliers are chosen in such a way that the constraints are satisfied. For instance, if we know $\lambda_1,\dots,\lambda_m$, it is straightforward to find λ_0 if we apply its constraint, and again considering the natural logarithm, we obtain

$$\begin{aligned}
1 &= \int p(\xi)^* d\xi \\
1 &= \int e^{-1+\lambda_0+\sum_{j=1}^m \lambda_j G_j(\xi)} dx \\
1 &= (e^{-1}e^{\lambda_0}) \underbrace{\int e^{\sum_{j=1}^m \lambda_j G_j(\xi)} dx}_{Z(\Lambda,\xi)} \\
\log 1 &= (\lambda_0 - 1) + \log Z(\Lambda, \xi) \\
1 - \lambda_0 &= \log Z(\Lambda, \xi)
\end{aligned} \tag{3.23}$$

$Z(\Lambda, \xi)$ being the so-called *partition function*, which is used to normalize the maximum entropy distribution. Therefore, we obtain the typical expression of such distribution:

$$p^*(\xi) = \frac{1}{Z(\Lambda, \xi)} e^{\sum_{j=1}^m \lambda_j G_j(\xi)} \tag{3.24}$$

The main problem is how to obtain the multipliers $\lambda_1, \ldots, \lambda_m$. This is usually done through iterative methods (there is as many nonlinear equations as constraints) as we will see along the book. Alternatively, some researchers have addressed the *approximation* of the maximum entropy (ME) distribution and the entropy itself. For instance, Hyvärinen [77] assumes that the ME distribution is not very far from a Gaussian distribution with the same mean and variance. Actually, the second Gibbs theorem states that the ME distribution among all the ones with the same mean and variance is the Gaussian one. Consequently, assuming that the random variable ξ has been standardized (zero mean and unit variance), we may assume that the ME distribution is close to $\varphi(\xi) = \exp(-\xi^2/2)/\sqrt{2\pi}$. This implicitly means that we must add two additional functions $G_{m+1}(\xi) = \xi$ and $G_{m+2}(\xi) = \xi^2$, and their respective constraints. Furthermore, in order to reduce the nonlinear equations derived from the constraints to linear ones, one may also choose $G_j(.)$ that are orthogonal among them and also to all polynomials of second degree. Thus, the new set of constraints, for $j, i = 1, \ldots, m$, are the following:

$$\begin{aligned}
\int \varphi(\xi) G_j(\xi) G_i(\xi) d\xi &= \begin{cases} 1 \text{ if } i = j \\ 0 \text{ otherwise} \end{cases} \\
\int \varphi(\xi) G_j(\xi) \xi^k d\xi &= 0 \text{ for } k = 0, 1, 2\ .
\end{aligned} \tag{3.25}$$

Thus, assuming near-Gaussianity leads to

$$p^*(\xi) = \tilde{Z}^{-1} \varphi(\xi) e^{\lambda_{m+1}\xi + (\lambda_{m+2}+\frac{1}{2})\xi^2 + \sum_{j=1}^m \lambda_j G_j(\xi)} \tag{3.26}$$

where $\tilde{Z}^{-1} = \sqrt{2\pi} Z^{-1}$. The latter equation may be simplified a little bit more if we take the first-order Taylor expansion $e^x \approx 1 + x$

$$p^*(\xi) \approx \tilde{Z}^{-1} \varphi(\xi) \left(1 + \lambda_{m+1}\xi + \left(\lambda_{m+2} + \frac{1}{2} \right) \xi^2 + \sum_{j=1}^m \lambda_j G_j(\xi) \right) \tag{3.27}$$

The orthogonalization assumption yields the linearization of the constraints. For instance, for $\int p^*(\xi)d\xi = 1$, we have

$$1 = \tilde{Z}^{-1}\left(\underbrace{\int \varphi(\xi)d\xi}_{1} + \lambda_{m+1}\underbrace{\int \varphi(\xi)\xi d\xi}_{0} + \right.$$
$$\left. + \left(\lambda_{m+2} + \frac{1}{2}\right)\underbrace{\int \varphi(\xi)\xi^2 d\xi}_{1} + \sum_{j=1}^{m}\lambda_j \underbrace{\int \varphi(\xi)G_j(\xi)d\xi}_{0}\right) \quad (3.28)$$

therefore

$$1 = \tilde{Z}^{-1}\left(1 + \left(\lambda_{m+2} + \frac{1}{2}\right)\right) \quad (3.29)$$

and, similarly, we also obtain

$$\begin{aligned}\int p^*(\xi)\xi d\xi &= \tilde{Z}^{-1}\lambda_{m+1} = 0\\ \int p^*(\xi)\xi^2 d\xi &= \tilde{Z}^{-1}(1 + 3(\lambda_{m+2} + 1/2)) = 1\\ \int p^*(\xi)G_j(\xi)d\xi &= \tilde{Z}^{-1}\lambda_j = E(G_j(\xi)), \quad j = 1, \ldots, m\end{aligned} \quad (3.30)$$

Then, we obtain: $\tilde{Z}^{-1} = 1$,$\lambda_{m+1} = 0$,$\lambda_{m+2} = -1/2$, and $\lambda_j = E(G_j(\xi))$ (that is, the λ_j are estimated from the samples). Consequently, the final expression for the pdf is

$$p^*(\xi) = \varphi(\xi)\left(1 + \sum_{j=1}^{m} u_j G_j(\xi)\right) \quad (3.31)$$

where $u_j = E(G_j(\xi))$.

In addition, the entropy $H(p^*(\xi))$ can be approximated by (i) exploiting Eq. 3.31, (ii) taking into account the Taylor expansion $(1 + x)\log(1 + x) = x + x^2/2$, (iii) grouping terms for expressing the entropy of $\varphi(\xi)$ in the first term, and (iv) applying the orthogonality constraints. Such approximation is

$$H(p^*(\xi)) = -\int p^*(\xi)\log p^*(\xi)d\xi \approx H(\nu) - \frac{1}{2}\sum_{j=1}^{m} u_j^2 \quad (3.32)$$

where $\nu = \varphi(\xi)$. At this point of the section, we have established the mathematical basis for understanding how to compute efficiently the Jensen–Shannon (JS) divergence between two regions ($N = 2$ in Eq. 3.15):

$$JS = H(a_1 p_1^*(\xi) + a_2 p_2^*(\xi)) - a_1 H(p_1^*(\xi)) - a_2 H(p_2^*(\xi)) \quad (3.33)$$

Here is the key to clarify the following approximation:

$$\begin{aligned}
P &= a_1 p_1^*(\xi) + a_2 p_2^*(\xi) \\
&\approx \left(a_1\varphi(\xi)\left(1+\sum_{j=1}^{m} u_j G_j(\xi)\right) + a_2\varphi(\xi)\left(1+\sum_{j=1}^{m} v_j G_j(\xi)\right)\right) \\
&= \varphi(\xi)\left(a_1\left(1+\sum_{j=1}^{m} u_j G_j(\xi)\right) + a_2\left(1+\sum_{j=1}^{m} v_j G_j(\xi)\right)\right) \\
&= \varphi(\xi)\left(\left(a_1 + \sum_{j=1}^{m} a_1 u_j G_j(\xi)\right) + \left(a_2 + \sum_{j=1}^{m} a_2 v_j G_j(\xi)\right)\right) \\
&= \varphi(\xi)\left(\underbrace{(a_1+a_2)}_{1} + \sum_{j=1}^{m}[G_j(\xi)(a_1u_j + a_2v_j)]\right)
\end{aligned} \tag{3.34}$$

where $u_j = \int_{R_\mathcal{I}} G_j(\mathbf{I}(x,y))/|R_\mathcal{I}|dxdy$ and $v_j = \int R_\mathcal{O} G_j(\mathbf{I}(x,y))/|R_\mathcal{O}|dxdy$. Then, exploiting the same rationale for deriving Eq. 3.32, we obtain

$$H(P) \approx H(\nu) - \frac{1}{2}\sum_{i=1}^{m}(a_1u_j + a_2v_j)^2 \tag{3.35}$$

Finally, we obtain the approximated JS:

$$\begin{aligned}
\hat{JS} &= H(\nu) - \frac{1}{2}\sum_{i=1}^{m}(a_1u_j + a_2v_j)^2 \\
&\quad - a_1\left(H(\nu) - \frac{1}{2}\sum_{j=1}^{m} u_j^2\right) - a_2\left(H(\nu) - \frac{1}{2}\sum_{j=1}^{m} v_j^2\right) \\
&= \frac{1}{2}a_1a_2\sum_{j=1}^{m}(u_j - v_j)^2
\end{aligned} \tag{3.36}$$

The negative of $\hat{JS}$ is denoted as an energy function E to minimize. Setting $a_1 = |R_\mathcal{I}|/|\mathbf{I}|$ and $a_2 = |R_\mathcal{O}|/|\mathbf{I}|$, we have that incorporating also the variations of $\hat{JS}$ with respect to the sizes of each area, we obtain

$$\begin{aligned}
\nabla\hat{JS} = \frac{\partial\Gamma}{\partial t} &= \frac{1}{2}\sum_{j=1}^{m}\frac{|R_\mathcal{I}||R_\mathcal{O}|}{|\mathbf{I}|^2}(2(u_j - v_j)(\nabla u_j - \nabla v_j))\mathbf{n} \\
&\quad \underbrace{+\frac{1}{2}\sum_{j=1}^{m}\frac{|R_\mathcal{I}|}{|\mathbf{I}|^2}(u_j - v_j)^2\mathbf{n} - \frac{1}{2}\sum_{j=1}^{m}\frac{|R_\mathcal{O}|}{|\mathbf{I}|^2}(u_j - v_j)^2\mathbf{n}}_{\frac{1}{2}\sum_{j=1}^{m}\frac{|R_\mathcal{I}|-|R_\mathcal{O}|}{|\mathbf{I}|^2}(u_j-v_j)^2\mathbf{n}},
\end{aligned} \tag{3.37}$$

being the partial derivatives of u_i and u_j with respect to the contour

$$\nabla u_j = \frac{G_j(I(x,y) - u_j)}{|R_{\mathcal{I}}|}\mathbf{n}$$
$$\nabla v_j = -\frac{G_j(I(x,y) - v_j)}{|R_{\mathcal{O}}|}\mathbf{n}\,, \tag{3.38}$$

and $\mathbf{n}$ the outward unit normal. Then, replacing the latter partial derivatives in Eq. 3.37, taking into account that $|\mathbf{I}| = |R_{\mathcal{I}}| + |R_{\mathcal{O}}|$, and rearranging terms, we finally obtain

$$\nabla \hat{JS} = \frac{\partial \Gamma}{\partial t} = f\mathbf{n} \tag{3.39}$$

where

$$f = \frac{1}{2|\mathbf{I}|}\sum_{j=1}^{m}(u_j - v_j)((G_j(\mathbf{I}(x,y)) - u_j) + (G_j(\mathbf{I}(x,y)) - v_j)) \tag{3.40}$$

f being the gradient flow of the Chan and Vese functional (Eq. 3.14). Thus f is defined in the terms described above, and the connection between JS and the contour dynamics is established. This f is the one used in Fig. 3.1 (bottom) when we use as generator functions such as $G_1(\xi) = \xi e^{-\xi^2/2}$, $G_2(\xi) = e^{-\xi^2/2}$ and $G_3(\xi) = |\xi|$.

3.3 MDL in Contour-Based Segmentation

3.3.1 B-Spline Parameterization of Contours

Considering the *snake-inspired* definition of a closed contour: $\Gamma(t) = (x(t), y(t))$, being, for instance, $t \in [0, 2\pi]$, in practice, the latter arc-length range is discretized and we understand $\Gamma = \{(x(t), y(t))^T : t = \frac{i2\pi}{(N-1)}\ , i = 0, \ldots, N-1\}$ as an *ordered sequence* of, say N, 2D points. What is more interesting, from the point of view of the complexity analysis of the contour, is that given the latter N points (samples), its continuous version $\Gamma(t)$ may be inferred from the samples by means of either Fourier analysis [149] or spline methods [6, 23]. These methods allow to parameterize the contours by implicitly introducing a smoothing constraint when a small number of parameters are selected (terms or control points, respectively). The important question here is to *infer the optimal number of parameters at the same time that the image is segmented, that is, the contour is optimally placed.*

For the sake of simplicity, let us discuss here one of the latter parameterizations, for instance, the B-spline one [58] (although both methods are explored in [57]). Let $\mathcal{B}^M = \{\mathbf{B}_k^M(t)\ : k = 0, \ldots, K - M - 1\}$ a set of N_B

B-splines, where $N_B = K - M$, and let $\{t_0 \le t_1 \le \ldots \le t_K\}$ be a set of *knots*. Each $\mathbf{B}_k^M(t)$ is a polynomial of typically low *order* $M \ge 2$ (degree $M-1$):

$$\mathbf{B}_k^M(t) = \mathbf{B}_k^{M-1}(t)\frac{t - t_k}{t_{k+M-1} - t_k} + \mathbf{B}_{k+1}^{M-1}(t)\frac{t_{k+M} - t}{t_{k+M} - t_{k+1}}$$
$$\mathbf{B}_k^1(t) = \begin{cases} 1 \text{ if } t_k \le t \le t_{k+1} \\ 0 \text{ otherwise} \end{cases} \quad (3.41)$$

whose *support* is $[t_k, t_{k+M}]$ and are *smoothly joined at knots*, that is, $M-2$ continuous derivatives exist at the joints, and the $(M-2)$th derivative must be equal in the joint (C^{M-2} continuity). B-splines satisfy $\mathbf{B}_k^M(t) \ge 0$ (non-negativity), $\sum_k \mathbf{B}_k^M(t) = 1$ for $t \in [t_M, t_{N_B}]$ (partition of unity), and periodicity. Finally, the N_B B-splines define a nonorthogonal basis for a linear space. With regard to the latter, one-dimensional functions are uniquely defined as

$$y(t) = \sum_{k=0}^{N_B-1} c_k \mathbf{B}_k^M(t) \quad t \in [t_{M-1}, t_{N_B}] \quad (3.42)$$

where $c_k \in \mathbb{R}$ are the so-called *control points*, where the kth control point influences the function only for $t_k < t < t_{k+M}$. If we have $N_B = K - M$ control points and K knots, then the order of the polynomials of the basis is exactly $K - N_B = M$, that is, we have $\mathcal{B}^M$ (for instance $M = 4$ yields *cubic B-splines*). For instance, in Fig. 3.2 (left), we have a B-spline composed from $N_B = 7$ cubic ($M = 4$) basis functions and $K = 11$ knots ($K - N_B = M$), where $t_0 = t_1 = t_2 = t_3 = 0$, $t_4 = 1.5$, $t_5 = 2.3$, $t_6 = 4$, and $t_7 = t_8 = t_9 = t_{10} = 5$. All the control points are 0 except $c_3 = 1$. Therefore, we have a plot of $\mathbf{B}_3^4(t)$ Furthermore, as the multiplicity of 0 and 5 is both 4, in practice, we have

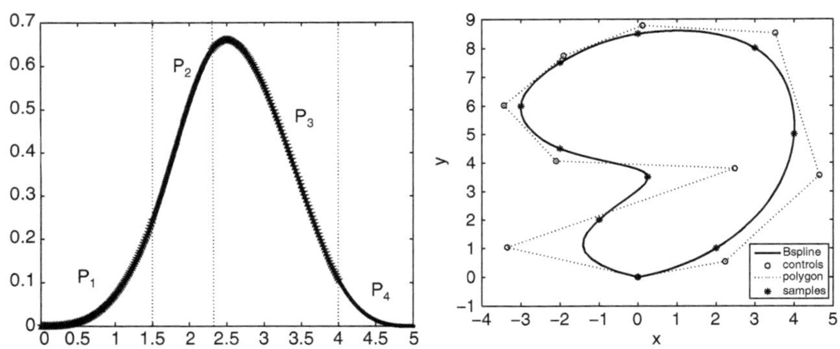

Fig. 3.2. *Left*: a degenerated one-dimensional example $y(t) = \mathbf{B}_3^4(t)$ with the four cubic polynomials $P_1 \ldots P_4$ describing it. *Right*: a 2D B-spline basis with $M = 4$ (*cubic*) for describing a contour (in *bold*). Such continuous contour is given by the interpolation of $N = 11$ samples (first and last coincide) that are indicated by $*$ *markers*. There are $N_B = 11$ control points, although two points of them are $(0,0)^T$, indicated by *circular markers* and we also draw the control polygon.

only $N_B = 4$ basis functions. If we do not use the B-form, we can also see the four polynomials $P_1 \ldots P_4$ used to build the function and what part of them is considered within each interval. As stated above, the value of $y(t)$ is 0 for $t < t_0 = 0$ and $t > t_{10} = 5$, and the function is defined in the interval $[t_{M-1} = t_3 = 0, t_{N_B} = t_7 = 5]$.

Following [58], we make the cubic assumption and drop M for the sake of clarity. Thus, given a sequence of N 2D points $\Gamma = \{(x_i, y_i)^T : i = 0, 1, \ldots, N-1\}$, and imposing several conditions, such as the periodicity cited above, the contour $\Gamma(t) = (x(t), y(t))^T$ is inferred through cubic-spline interpolation, which yields both the knots and the 2D control points $\mathbf{c}_k = (c_k^x, c_k^y)^T$. Consequently, the basis functions can be obtained by applying Eq. 3.41. Therefore, we have

$$\Gamma(t) = \begin{pmatrix} x(t) \\ y(t) \end{pmatrix} = \sum_{k=0}^{N_B-1} \mathbf{c}_k \mathbf{B}_k(t) = \sum_{k=0}^{N_B-1} \begin{pmatrix} c_k^x \\ c_k^y \end{pmatrix} \mathbf{B}_k(t) \quad t \in [t_{M-1}, t_{N_B}] \tag{3.43}$$

In Fig. 3.2 (right), we show the continuous 2D contour obtained by interpolating $N = 11$ samples with a cubic B-spline. The *control polygon* defined by the control points is also showed. The curve follows the control polygon closely. Actually, the convex hull of the control points contains the contour. In this case, we obtain 15 nonuniformly spaced knots, where $t_0 = \ldots = t_3 = 0$ and $t_{11} = \ldots = t_{15} = 25.4043$.

In the latter example of 2D contour, we have simulated the periodicity of the (closed) contour by adding a last point equal to the first. Periodicity in B-splines is ensured by defining bases satisfying $\mathbf{B}_k^M(t) = \sum_{j=-\infty}^{+\infty} \mathbf{B}_{k+j(t_K - t_0)}^M(t) : j \in \mathbb{Z}$, being $t_K - t_0$ the period. If we have $K + 1$ knots, we may build a B-splines basis of K functions $\mathbf{B}_0, \ldots, \mathbf{B}_{K-1}$ simply by constructing $\mathbf{B}_0$ and shifting this function assuming a periodic knot sequence, that is, t_j becomes $t_{j \bmod K}$. Therefore, $t_K = t_0$, so we have K *distinct* knots. Here, we follow the simplistic assumption that knots are uniformly spaced. The periodic basis functions for $M = 2$ and $M = 4$ are showed in Fig. 3.3. In the periodic case, if we set K, a closed contour can be expressed as

$$\Gamma(t) = \begin{pmatrix} x(t) \\ y(t) \end{pmatrix} = \sum_{k=0}^{K-1} \mathbf{c}_k \mathbf{B}_k(t) \quad t \in \mathbb{R} \tag{3.44}$$

As in practice what we have is a discrete description of the contour, our departure point is something like two columns with the x and y coordinates of the, say N points:

$$\Gamma = (\mathbf{x}\,\mathbf{y}) = \begin{pmatrix} x_0 & y_0 \\ \vdots & \vdots \\ x_{N-1} & y_{N-1} \end{pmatrix} = \begin{pmatrix} x(s_0) & y(s_0) \\ \vdots & \vdots \\ x(s_{N-1}) & y(s_{N-1}) \end{pmatrix} \tag{3.45}$$

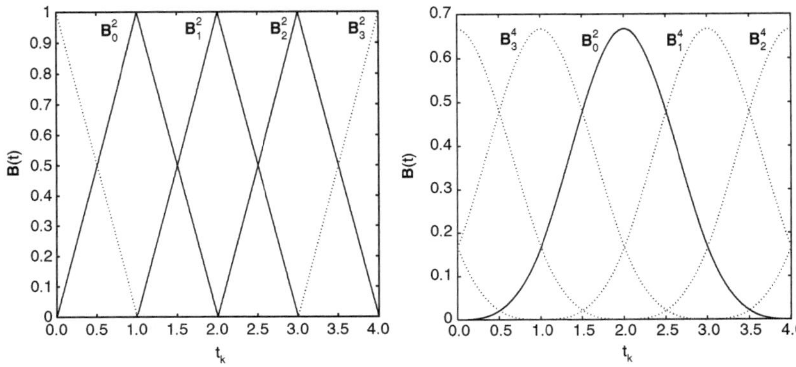

Fig. 3.3. Periodic basic functions for $M = 2$ (*left*) and for $M = 4$ (*right*). In both cases, the $K = 5$, distinct knots are $t_0 = 1 \ldots t_4 = 4$, and we show the $K - 1$ functions; those with a nonzero periodic fragment are represented in *dashed* lines.

where, under the assumption of periodical basis and uniformly spaced K distinct knots, we may set $t_j = j$, $j = 0, \ldots, K-1$, and thus then, $s_i = iK/N$, $i = 0, \ldots, N-1$ are the mapped positions of the discrete points in the contour parameterization. Therefore, we have the following two sets of association pairs: $\mathcal{X} = \{(s_i, \mathbf{x}_i)\}$ and $\mathcal{Y} = \{(s_i, \mathbf{y}_i)\}$, with $i = 0, \ldots, N-1$ for both cases. How to obtain the control points $\boldsymbol{\theta}_{(K)} = (\mathbf{c}_0^T \ldots \mathbf{c}_{K-1}^T)^T = (\boldsymbol{\theta}^x_{(K)}\ \boldsymbol{\theta}^y_{(K)})$? It is key to build the $N \times K$ matrix $\mathbf{B}_{(K)}$, which will satisfy

$$\Gamma = \mathbf{B}_{(K)}\boldsymbol{\theta}_{(K)} \equiv \{\mathbf{x} = \mathbf{B}_{(K)}\boldsymbol{\theta}^x_{(K)},\ \mathbf{y} = \mathbf{B}_{(K)}\boldsymbol{\theta}^y_{(K)}\} \tag{3.46}$$

The columns of $\mathbf{B}_{(K)}$ conceptually consist in the discretization, accordingly to the s_i, of each of the K (cubic) basic functions: $\mathbf{B}_{ij} = \mathbf{B}_j(s_i)$. This matrix is independent both on $\mathbf{x}$ and $\mathbf{y}$, and needs only to be calculated once. For instance, the least squares solution to $\mathbf{x} = \mathbf{B}_{(K)}\boldsymbol{\theta}^x_{(K)}$ consists of

$$\hat{\boldsymbol{\theta}}^x_{(K)} = \arg\min_{\boldsymbol{\theta}_{(K)}} ||\mathbf{x} - \mathbf{B}_{(K)}\boldsymbol{\theta}_{(K)}||^2 = \underbrace{(\mathbf{B}^T_{(K)}\mathbf{B}_{(K)})^{-1}\mathbf{B}^T_{(K)}}_{\mathbf{B}^\dagger_{(K)}}\mathbf{x} \tag{3.47}$$

$\mathbf{B}^\dagger_{(K)}$ being the pseudo-inverse, and the inverse $(\mathbf{B}^T_{(K)}\mathbf{B}_{(K)})^{-1}$ always exists. Furthermore, an approximation of the original $\mathbf{x}$ can be obtained by

$$\hat{x}_{(K)} = \mathbf{B}_{(K)}\hat{\boldsymbol{\theta}}^x_{(K)} = \underbrace{\mathbf{B}_{(K)}\mathbf{B}^\dagger_{(K)}}_{\mathbf{B}^\perp_{(K)}}\hat{\boldsymbol{\theta}}^x_{(K)} \tag{3.48}$$

where $\mathbf{B}^\perp_{(K)}$ is the so-called *projection matrix* because $\hat{x}_{(K)}$ is the projection of $\mathbf{x}$ onto the range space of K dimensions of $\mathbf{B}_{(K)}$: $\mathcal{R}(\mathbf{B}_{(K)})$. Such space is

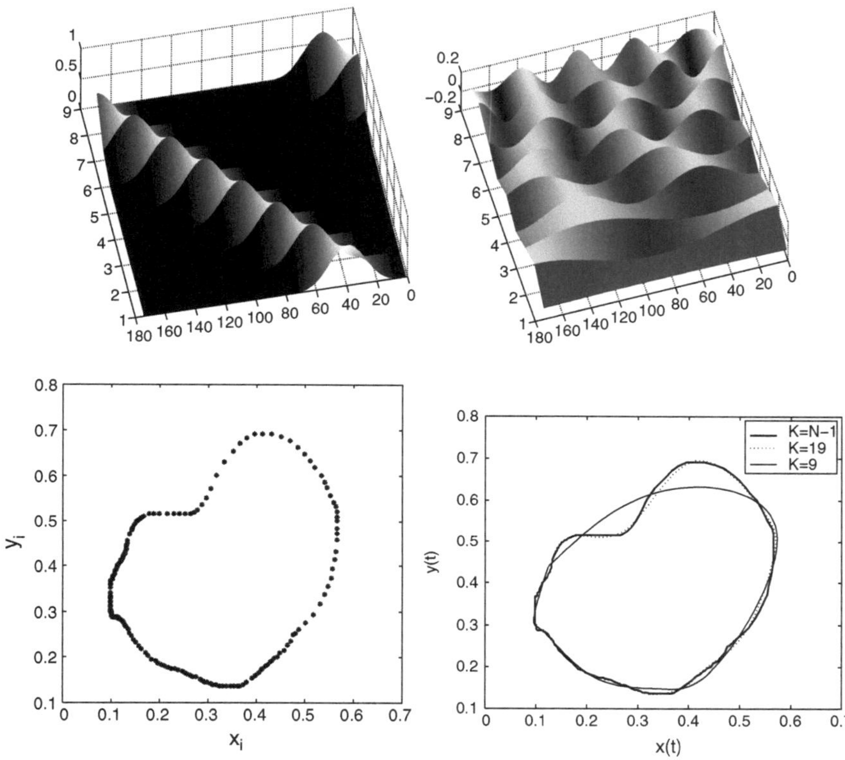

Fig. 3.4. *Top*: structure of $\mathbf{B}_{(K)}$ (*left*) for $N = 174$ samples and $K - 1 = 9$ basis, and structure of its orthogonal matrix spanning the range space. In both cases, each row is represented as a strip. *Bottom*: contours inferred for different values of K(see text).

the one expanded by an orthonormal basis with the same range of $\mathbf{B}_{(K)}$. The structure of both $\mathbf{B}_{(K)}$ and the corresponding orthonormal matrix is showed in Fig. 3.4 (top-left and top-right), respectively.

Applying to $\mathbf{y}$ and $\boldsymbol{\theta}^x_{(K)}$ the above rationale and fusing the resulting equations, we have that

$$\hat{\boldsymbol{\theta}}_{(K)} = (\hat{\boldsymbol{\theta}}^x_{(K)}\ \hat{\boldsymbol{\theta}}^y_{(K)}) = (\mathbf{B}^\dagger_{(K)}\mathbf{x}\ \ \mathbf{B}^\dagger_{(K)}\mathbf{y}) = \mathbf{B}^\dagger_{(K)}\Gamma$$
$$\hat{\Gamma} = (\hat{\mathbf{x}}\ \hat{\mathbf{y}}) = (\mathbf{B}^\perp_{(K)}\mathbf{x}\ \ \mathbf{B}^\perp_{(K)}\mathbf{y}) = \mathbf{B}^\perp_{(K)}\hat{\boldsymbol{\theta}}_{(K)} \tag{3.49}$$

Given, for example, $N = 174$ samples of a 2D contour (Fig. 3.4, bottom left), we may use the method described above for obtaining approximate contours for different values of K (Fig. 3.4, bottom right). For $K = N - 1$, we obtain a pretty good approximation. However, such approximation does not differ too much from the one obtained with $K = 19$ control points. Finally, the upper concavity of the contour is lost, and also the upper convexity is smoothed, with $K = 9$. Thus, two key questions remain: *What is the optimal value of* K*?*, and *What is the meaning of* optimal *in the latter question?* As we will see next, information theory will shed light on these questions.

3.3.2 MDL for B-Spline Parameterization

B-splines with K basis represent a *model* $\mathcal{M}_k$ (say of order K) for *encoding* a representative contour of N hand-drawn 2D points (the *crude data* $\mathcal{D}$). The minimum description length (MDL) principle introduced by Rissannen [138–140] states that *the optimal instance of a model of any order* $\mathcal{M}$ *is the one that minimizes*[2]*:*

$$\mathcal{L}(\mathcal{D}|\mathcal{M}) + \mathcal{L}(\mathcal{M}) \tag{3.50}$$

where $\mathcal{L}(.)$ denotes the length, in bits, of the *description* of the argument. Formally, a *description method* $\mathcal{C}$ is a way of mapping the source symbols or sequences $\mathbf{x}$ of one alphabet $\mathcal{A}$ with sequences of symbols $\mathcal{C}(\mathbf{x})$ (codewords) of another alphabet, typically $\mathcal{B} = \{0, 1\}$, in a way that not two different sources may be assigned to the same codeword. A *code* is a description method in which each source is associated to at most one codeword.*Prefix or instantaneous codes* are those wherein no codeword can be a prefix of another one. Let $\mathcal{L}_{\mathcal{C}}(a)$, $a \in \mathcal{A}$ denote the length in bits of $\mathcal{C}(a)$. Then, *given the alphabet* $\mathcal{A} = \{1, 2, \ldots, m\}$ *and a prefix code* $\mathcal{C}$ *whose codewords are defined over the alphabet* $\mathcal{B}$ *of length* D*, the following (Kraft) inequality is satisfied [43]:*

$$\sum_{a \in \mathcal{A}} D^{-\mathcal{L}_{\mathcal{C}}(a)} \leq 1 \tag{3.51}$$

and conversely, given a set of codewords satisfying this inequality, there exists a prefix code with these lengths. Moreover, when there is a probability distribution associated to the alphabet $P(a)$, the codewords' lengths satisfying the Kraft inequality and minimizing the expected length $\sum_{a \in \mathcal{A}} P(a)\mathcal{L}_{\mathcal{C}}(a)$ are $\mathcal{L}_{\mathcal{C}}(a) = \lceil -\log_D P(a) \rceil = \lceil \log_D \frac{1}{P(a)} \rceil$, and the expected length is exactly the entropy under $\log_D$ if we drop the rounding up to achieve integer lengths. Such code, the Shannon–Fano one (see the allocation of codewords in [43] – pp. 101–103), allows to relate lengths and probabilities, which, in turn, is key to understand MDL in probabilistic terms. Therefore, Eq. 3.50 can be rewritten in the following terms:

$$-\log P(\mathcal{D}|\mathcal{M}) + \mathcal{L}(\mathcal{M}) \tag{3.52}$$

so MDL relies on maximizing $P(\mathcal{D}|\mathcal{M})$, which is the probability density of $\mathcal{D}$ given M. In the case of contours, it is easy to see that, as $\mathcal{D} \equiv \Gamma$, the

[2] This is, in the Grünwald terminology [67], the *crude two-part* version of MDL.

probability of observing Γ, given an instance of a model $\mathcal{M} \equiv \boldsymbol{\theta}_{(K)}$, can be described by a Gaussian with white noise with given covariance Σ. Factorizing $P(\mathcal{D}|\mathcal{M})$, we have that

$$P(\Gamma|\boldsymbol{\theta}_{(K)}, \Sigma) = P(\mathbf{x}|\boldsymbol{\theta}^x_{(K)}, \sigma^2_x) \cdot P(\mathbf{y}|\boldsymbol{\theta}^y_{(K)}, \sigma^2_y) \; , \tag{3.53}$$

where, considering $z = \{x, y\}$ in the following, we have that

$$\begin{aligned} P(\Gamma|\boldsymbol{\theta}^z_{(K)}, \sigma^2_z) &= G(\mathbf{z} - \mathbf{B}_{(K)}\boldsymbol{\theta}^z_{(K)}, \sigma^2_z) \\ &\equiv (2\pi\sigma^2_z)^{-\frac{N}{2}} \exp\left\{ -\frac{||\mathbf{z} - \mathbf{B}_{(K)}\boldsymbol{\theta}^z_{(K)}||^2}{2\sigma^2_z} \right\} \end{aligned} \tag{3.54}$$

which is consistent with the maximum likelihood estimation through least squares: $\hat{\boldsymbol{\theta}}^z_{(K)} = \arg\max_{\boldsymbol{\theta}_{(K)}}\{\log P(\Gamma|\boldsymbol{\theta}^z_{(K)}, \sigma^2_z)\} = \mathbf{B}^{\dagger}_{(K)}\Gamma$. However, the MDL estimation is posed in the following terms:

$$(\hat{\boldsymbol{\theta}}_{(K)}, \hat{\sigma}^2_x, \hat{\sigma}^2_y) = \arg \min_{K, \boldsymbol{\theta}_{(K)}, \sigma^2_x, \sigma^2_y} \{-\log P(\Gamma|\boldsymbol{\theta}_{(K)}, \sigma^2_x, \sigma^2_y) + \mathcal{L}(\boldsymbol{\theta}_{(K)}, \sigma^2_x, \sigma^2_y)\} \tag{3.55}$$

although we may rewrite $\mathcal{L}(\boldsymbol{\theta}_{(K)}, \sigma^2_x, \sigma^2_y) = \mathcal{L}(\boldsymbol{\theta}_{(K)})$ if we assume that the variances have constant description lengths. The *MDL model order selection* problem consists in estimating

$$\begin{aligned} K^* = \arg\min_K \Bigg\{ & \mathcal{L}(\boldsymbol{\theta}_{(K)}) + \min_{\sigma^2_x} \left[\min_{\boldsymbol{\theta}^x_{(K)}} (-\log P(\mathbf{x}|\boldsymbol{\theta}^x_{(K)}, \sigma^2_x)) \right] \\ & + \min_{\sigma^2_y} \left[\min_{\boldsymbol{\theta}^y_{(K)}} (-\log P(\mathbf{y}|\boldsymbol{\theta}^y_{(K)}, \sigma^2_y)) \right] \Bigg\} \end{aligned} \tag{3.56}$$

Any of the two minimization problems between brackets may be solved as follows:

$$\begin{aligned} & \min_{\sigma^2_z} \left[\min_{\boldsymbol{\theta}^z_{(K)}} (-\log P(\mathbf{z}|\boldsymbol{\theta}^z_{(K)}, \sigma^2_z)) \right] \\ = \;& \min_{\sigma^2_z} \left[\frac{N}{2}\log(2\pi\sigma^2_z) + \frac{||\mathbf{z} - \mathbf{B}_{(K)}\hat{\boldsymbol{\theta}}^z_{(K)}||^2}{2\sigma^2_z} \right] \\ = \;& \min_{\sigma^2_z} \left[\frac{N}{2}\log(2\pi\sigma^2_z) + \frac{||\mathbf{z} - \mathbf{B}^{\perp}_{(K)}\mathbf{z}||^2}{2\sigma^2_z} \right] \\ = \;& \min_{\sigma^2_z} \underbrace{\left[\frac{N}{2}\log(2\pi\sigma^2_z) + \frac{||\mathbf{z} - \hat{\mathbf{z}}_{(K)}||^2}{2\sigma^2_z} \right]}_{f(\sigma^2_z)} = \frac{N}{2}\log(2\pi\hat{\sigma}^2_z(K)e) \end{aligned} \tag{3.57}$$

where $\hat{\sigma}_z^2(K) = \arg\min_{\sigma_z^2} f(\sigma_z^2) \equiv ||\mathbf{z} - \hat{\mathbf{z}}_{(K)}||^2/N$, which is consistent with being the optimal variance dependant on the approximation error. Therefore, defining $\hat{\sigma}_x^2(K)$ and $\hat{\sigma}_y^2(K)$ in the way described below, we have

$$
\begin{aligned}
K^* &= \arg\min_K \left\{ \mathcal{L}(\boldsymbol{\theta}_{(K)}) + \frac{N}{2}(\log(2\pi\hat{\sigma}_x^2(K)e) + \log(2\pi\hat{\sigma}_y^2(K)e)) \right\} \\
&= \arg\min_K \left\{ \mathcal{L}(\boldsymbol{\theta}_{(K)}) + \frac{N}{2}(\log(\hat{\sigma}_x^2(K)) + \log(\hat{\sigma}_y^2(K))) \right\} \\
&= \arg\min_K \left\{ \mathcal{L}(\boldsymbol{\theta}_{(K)}) + N\log\left(\sqrt{\hat{\sigma}_x^2(K)\hat{\sigma}_y^2(K)}\right) \right\} \qquad (3.58)
\end{aligned}
$$

As we have seen above, the Shannon–Fano coding is finally related to the goodness of fitting the data with the model of order K, but what about the definition of $\mathcal{L}(\boldsymbol{\theta}_{(K)})$? The simplistic assumption is considering that each parameter has a fixed description length, say λ, and thus, $\mathcal{L}(\boldsymbol{\theta}_{(K)}) = \lambda K$, but what is the more convenient value for λ? In this regard, another simplistic assumption is to adopt its asymptotical value for large N: $\frac{1}{2}\log N$. However, this asymptotical setting is not valid for control points because they do not depend too much on the number of hand-drawn points. However, considering that the parameters (control points) can be encoded with a given precision, let $\boldsymbol{\delta}^z$ be the difference between the parameters with two different precisions, being the more precise $\boldsymbol{\theta}_{(K)}$. Then, we define the error between the two versions of $\mathbf{z}$: $\boldsymbol{\epsilon}^z = \mathbf{B}_{(K)}\boldsymbol{\delta}^z$. Thus, $\max_i(\boldsymbol{\epsilon}_i) \leq \xi$, where ξ is the maximal absolute error for any of the coordinates. Considering finally that the control points will be inside the image plane, of dimensions $W_x \times W_y$ in pixels, the following criterion (simplified from [58]):

$$
\lambda = \log\left\{\frac{W_x}{\xi} + \frac{W_y}{\xi}\right\} = \log(W_x W_y) \text{ and, thus } \mathcal{L}(\boldsymbol{\theta}_{(K)}) = K\log(W_x W_y) \qquad (3.59)
$$

with $\xi = 1$ for discrete curves, reflects the fact that an increment of image dimensions is translated into a smaller fitting precision: the same data in a larger image need less precision. Anyway, the latter definition of λ imposes a logarithmically smoothed penalization to the increment of model order.

3.3.3 MDL Contour-based Segmentation

Given the latter B-spline adaptive (MDL-based) model for describing contours, the next step is how to find their ideal placement within an image, for instance, in the ill-defined border of a ultrasound image. Given an image $\mathbf{I}$ of $W_x \times W_y$ pixels containing an unknown contour $\Gamma = \mathbf{B}_{(K)}\boldsymbol{\theta}_{(K)}$, pixel intensities are the observed data, and, thus, their likelihood, given a contour (hypothesis or model), is defined as usual $P(\mathbf{I}|\boldsymbol{\theta}_{(K)}, \Phi)$, Φ being the (also unknown) parameters characterizing the intensity distribution of the image. For instance, assuming that the image consists of an object in the foreground, which is contained in the background, $\Phi = (\Phi_{in}, \Phi_{out})$ where Φ_{in} and Φ_{out} characterize, respectively, the homogeneous (in terms of intensity model) regions corresponding to the foreground and the background. This setting is adequate for medical image segmentations wherein we are looking for cavities defined by contours, and the intensity models (distributions) will depend on the specific application. Anyway, assuming that the pixel distribution is independent of the contour, the latter likelihood may be factorized as follows:

$$P(\mathbf{I}|\boldsymbol{\theta}_{(K)}, \Phi) = \left(\prod_{\mathbf{p} \in \mathcal{I}(\Gamma)} P(\mathbf{I_p}|\boldsymbol{\theta}_{(K)}, \Phi_{in}) \cdot \prod_{\mathbf{p} \in \mathcal{O}(\Gamma)} P(\mathbf{I_p}|\boldsymbol{\theta}_{(K)}, \Phi_{out}) \right) \quad (3.60)$$

where $\mathbf{I_p}$ is the intensity at pixel $\mathbf{p} = (i, j)$, and $\mathcal{I}(\Gamma)$ and $\mathcal{O}(\Gamma)$ denote, respectively, the regions inside and outside the closed contour Γ. Therefore, the segmentation problem can be posed in terms of finding

$$(\hat{\boldsymbol{\theta}}_{K^*}, \hat{\Phi}) = \arg \min_{K, \boldsymbol{\theta}_{(K)}, \Phi} \{ -\log P(\mathbf{I}|\boldsymbol{\theta}_{(K)}, \Phi) + K \log(W_x W y) \} \quad (3.61)$$

or equivalently in model-order complexity terms

$$K^* = \arg \min_K \left\{ K \log(W_x W_y) - \max_{\boldsymbol{\theta}_{(K)}, \Phi} \{ \log P(\mathbf{I}|\boldsymbol{\theta}_{(K)}, \Phi) \} \right\} \quad (3.62)$$

Apparently, an adequate computational strategy for solving the latter optimization problem consists of devising an algorithm for solving the inner maximization one for a fixed K and then running this algorithm within a range of K values. However, given that Φ is also unknown, the maximization algorithm must be partitioned into two intertwined algorithms: *Contour-fitting* and *Intensity-inference*. In the first one, K and Φ are fixed and we obtain $\hat{\boldsymbol{\theta}}_{(K)}$. In the second algorithm, K is fixed and both the refinement of $\hat{\boldsymbol{\theta}}_{(K)}$ and Φ are obtained.

Algorithm 2: GPContour-fitting

Input: $\mathbf{I}$, K, Φ, a valid contour $\hat{\Gamma}^{(0)} \in \mathcal{R}(\mathbf{B}_{(K)})$, and a stepsize ϵ
Initialization Build $\mathbf{B}_{(K)}$, compute $\mathbf{B}^{\perp}_{(K)}$, and set $t = 0$.
while $\neg$ *Convergence*($\hat{\Gamma}^{(t)}$) **do**
  Compute the gradient: $\delta\Gamma \leftarrow \nabla \log P(\mathbf{I}|\boldsymbol{\theta}_{(K)}, \Phi)|_{\Gamma = \hat{\Gamma}^{(t)}}$
  Project the gradient onto $\mathcal{R}(\mathbf{B}_{(K)})$: $(\delta\Gamma)^{\perp} \leftarrow \mathbf{B}^{\perp}_{(K)} \delta\Gamma$
  Update the contour (gradient ascent): $\hat{\Gamma}^{(t+1)} \leftarrow \hat{\Gamma}^{(t)} + \epsilon(\delta\Gamma)^{\perp}$
  $t \leftarrow t + 1$
end
Output: $\hat{\Gamma} \leftarrow \hat{\Gamma}^{(t)}$

Algorithm 3: MLIntensity-inference

Input: $\mathbf{I}$, K, and a valid contour $\hat{\Gamma}^{(0)} \in \mathcal{R}(\mathbf{B}_{(K)})$
Initialization Set $t = 0$.
while $\neg$ *Convergence*($\hat{\Phi}^{(t)}$, $\hat{\Gamma}^{(t)}$) **do**
  Compute the ML estimation $\hat{\Phi}^{(t)}$ given $\hat{\Gamma}^{(t)}$:
  $\hat{\Phi}_{in}^{(t)} = \arg\max_{\Phi_{in}} \left\{ \prod_{\mathbf{p}\in\mathcal{I}(\hat{\Gamma}^{(t)})} P(\mathbf{I_p}|\boldsymbol{\theta}_{(K)}, \Phi_{in}) \right\}$
  $\hat{\Phi}_{out}^{(t)} = \arg\max_{\Phi_{out}} \left\{ \prod_{\mathbf{p}\in\mathcal{O}(\hat{\Gamma}^{(t)})} P(\mathbf{I_p}|\boldsymbol{\theta}_{(K)}, \Phi_{out}) \right\}$
  Run the fitting algorithm: $\hat{\Gamma}(t+1)$ =GPContour-fitting($\mathbf{I}$, K, $\hat{\Phi}^{(t)}$, $\hat{\Gamma}^{(t)}$)
  $t \leftarrow t + 1$
end
Output: $\hat{\Phi} \leftarrow \hat{\Phi}^{(t)}$ and $\hat{\Gamma} \leftarrow \hat{\Gamma}^{(t)}$

Contour-fitting Algorithm

Given K and Φ, we have to solve the following problem:

$$\max_{\boldsymbol{\theta}_{(K)},\Phi} \left\{ \log P(\mathbf{I}|\boldsymbol{\theta}_{(K)}, \Phi) \right\} = \begin{pmatrix} \max_{\Gamma}\{P(\mathbf{I}|\Gamma, \Phi)\} \\ s.t.\ \Gamma \in \mathcal{R}(\mathbf{B}_{(K)}) \end{pmatrix} \tag{3.63}$$

that is, we must maximize the likelihood, but the solutions must be constrained to those contours of the form $\Gamma = \mathbf{B}_{(K)}\boldsymbol{\theta}_{(K)}$, and thus, belong to the range space of $\mathbf{B}_{(K)}$. Such constrained optimization method can be solved with a *gradient projection method* (GPM) [20]. GPMs consist basically in projecting successively the partial solutions, obtained in the direction of the gradient, onto the feasible region. In this case, we must compute in the $t-$th iteration the gradient of the likelihood $\delta\Gamma = \nabla \log P(\mathbf{I}|\boldsymbol{\theta}_{(K)}, \Phi)|_{\Gamma=\hat{\Gamma}(t)}$. Such gradient has a direction perpendicular to the contour at each point of it (this is basically the search direction of each contour point, and the size of this ℓ window one-pixel wide defines the short-sightness of the contour). Depending on the contour initialization, and also on ℓ, some points in the contour may return a zero gradient, whereas others, closer to the border between the foreground and the background, say p pixels, may return a gradient of magnitude p (remember that we are trying to maximize the likelihood along the contour). However, as this is a local computation for each contour point, the global result may not satisfy the constraints of the problem. This is why $\delta\Gamma$ is projected onto the range space through $(\delta\Gamma)^{\perp} = \mathbf{B}_{(K)}^{\perp}\delta\Gamma$, and then we apply the rules of the usual gradient ascent. The resulting procedure is in Alg. 2.

Intensity-inference Algorithm

This second algorithm must estimate a contour $\hat{\Gamma}$ in addition to the region parameters Φ, all for a fixed K. Therefore, it will be called Alg. 3. The estimation of Φ depends on the image model assumed. For instance, if it is Gaussian, then we should infer $\Phi_{in} = (\mu_{in}, \sigma_{in}^2)$ and $\Phi_{out} = (\mu_{out}, \sigma_{out}^2)$; this is easy to do if we compute these parameters from the samples. However, if the Rayleigh distribution is assumed, that is, $P(\mathbf{I_p}|\boldsymbol{\theta}_{(K)}, \Phi = \sigma^2) = (\mathbf{I_p}/\sigma^2) \cdot exp\{-\mathbf{I_p}/(2\sigma^2)\}$ (typically for modelling speckle noise in ultrasound images), then we should infer $\Phi_{in} = (\sigma_{in}^2)$ and $\Phi_{out} = (\sigma_{out}^2)$ (also from the samples). Given Alg. 3, we are able to obtain both $\hat{\boldsymbol{\theta}}_{(K)}$ and $\hat{\Phi}$ for a fixed K. Then, we may obtain the second term of Eq. 3.62:

$$\begin{aligned} \max_{\boldsymbol{\theta}_{(K)}, \Phi} \left\{ \log P(\mathbf{I}|\boldsymbol{\theta}_{(K)}, \Phi) \right\} &= \log P(\mathbf{I}|\hat{\boldsymbol{\theta}}_{(K)}, \hat{\Phi}) \\ \propto -\frac{|\mathcal{I}(\hat{\Gamma})|}{2} \log(\hat{\sigma}_{in}^2(K)) &- \frac{|\mathcal{O}(\hat{\Gamma})|}{2} \log(\hat{\sigma}_{out}^2(K)) \\ &= -\frac{1}{2} \left\{ log(\hat{\sigma}_{in}^2(K)\hat{\sigma}_{out}^2(K))^{|\mathbf{I}|} \right\} \\ &= \frac{1}{\sqrt{log(\hat{\sigma}_{in}^2(K)\hat{\sigma}_{out}^2(K))^{W_x W_y}}} \end{aligned} \tag{3.64}$$

independently of having a Gaussian or Rayleigh intensity model.

MDL Contour fitting and intensity inference

Once we have an algorithm for solving a fixed K, the MDL solution may be arranged as running this algorithm for a given range of K and then selecting the K^*. Thus, it is important to exploit the knowledge not only about the type of images to be processed (intensity), but also about the approximate complexity of their contours (in order to reduce the range of exploration for the optimal K). In Fig. 3.5, we show some summarizing results of the technique described above.

3.4 Model Order Selection in Region-Based Segmentation

3.4.1 Jump-Diffusion for Optimal Segmentation

Optimal segmentation is an open problem. A main issue is the model order selection. It consists of knowing the optimal number of classes. There is a trade-off between the simplicity of the model and its precision. Complex models fit better the data; however, we often prefer the simplest model that can describe data with an acceptable precision. George Box stated in Robustness in Statistics (1979), "All models are wrong, but some are useful." In computer vision and pattern recognition, unnecessarily complex models are not practical because they offer a poor generalization over new patterns.

The jump-diffusion strategy aims to solve both the problem of adjusting a model to the data, and selecting the model order. "Jump" refers to the latter,

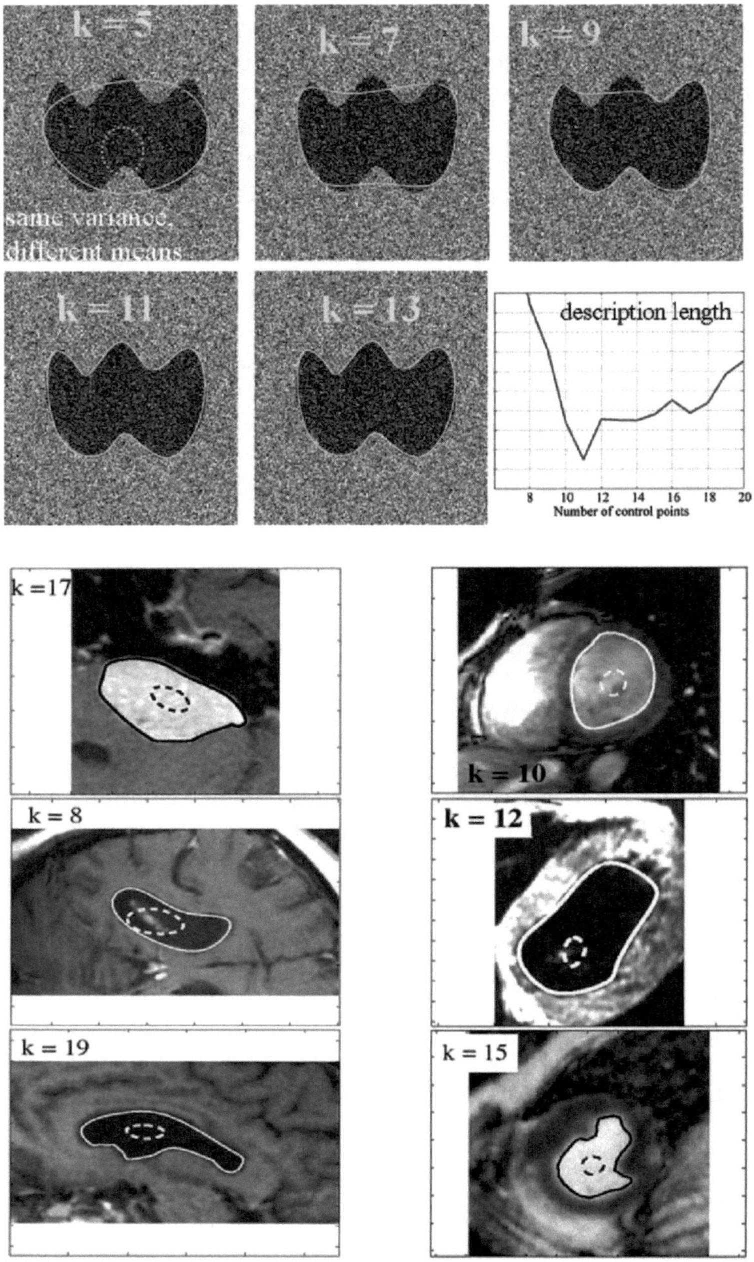

Fig. 3.5. Results of the MDL segmentation process. *Top*: synthetic image with same variances and different means between the foreground and background, and the optimal K (Courtesy of Figueiredo). *Bottom*: experiments with real medical images and their optimal K. In both cases, initial contours are showed in *dashed* lines. Figure by M.A.T. Figueiredo, J.M.N. Leitao and A.K. Jain (©2000 IEEE).

while "diffusion" refers to fitting the models. It is an optimization algorithm that iteratively simulates two types of moves [64, 65, 141]:

- Reversible jumps: move between subspaces of different dimensions, driven by a Metropolis–Hastings sampler. These jumps change both the type of region model for each segment, as well as the total number of models.
- Stochastic diffusions: Stochastic steepest descent described by the Langevin equations within each continuous subspace. This descent consists of adjusting the parameters of each region model (selected in the last jump) so that it fits the data better.

Bayesian formulation of the problem

In the following subsections, we explain the algorithm applied to a simple toy-example (Fig. 3.6). The data we want to segment are 1D synthetic signals with Gaussian noise with mean $\mu = 0$ and variance σ^2. Let us denote the data as $\mathbf{I}(x)$, $x \in [0, 1]$.

To formulate the problem, in the first place, we have to define how we want to model the image. The model has to be designed according to the nature of the data present in the image. For the 1D toy example, we could define some simple region models such as straight lines, arcs, and some more complex patterns like waves. Let us index the models with $l_i \in \{line, circle, wave, \ldots\}$, where i is the number of region given by the segmentation. Each one of the models is parameterized by some variables θ, which make possible its adjustment to the data. For example, the line is parameterized by $\theta = (a, b)$, where a is the slope and b is the intercept. A circular arc could be parameterized by $\theta = (c_x, c_y, r)$, where the center of the arc is (c_x, c_y) and r is the radius. Finally, the wave could be described by its slope, intercept, wave length and wave height. However, for simplicity, in the following examples we will use only the line model for fitting the data (Fig. 3.7). We will still use the index $l_i \in \{line\}$ to maintain generalization in the notation and formulation of the problem. For an example with two models (line and arc), see [159].

Each region model corresponds to some definite interval of the signal. Let us call $x_0 < x_1 < \cdots < x_k$ with $x_0 = 0$, $x_k = 1$ the change points, which limit each region (l_i, θ_i) to the description of the interval $[x_i - 1, x_i)$ of the signal.

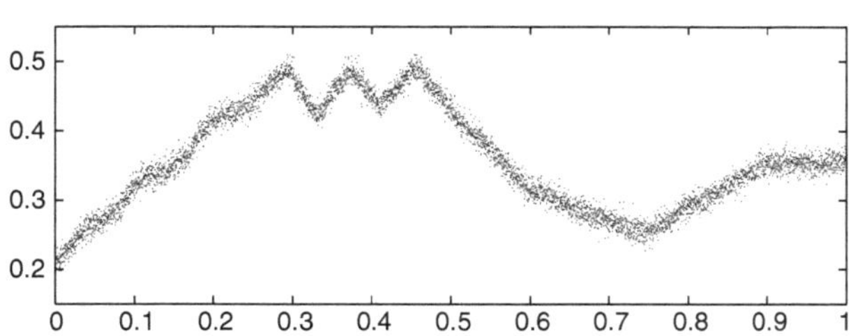

Fig. 3.6. A noisy signal composed of several segments.

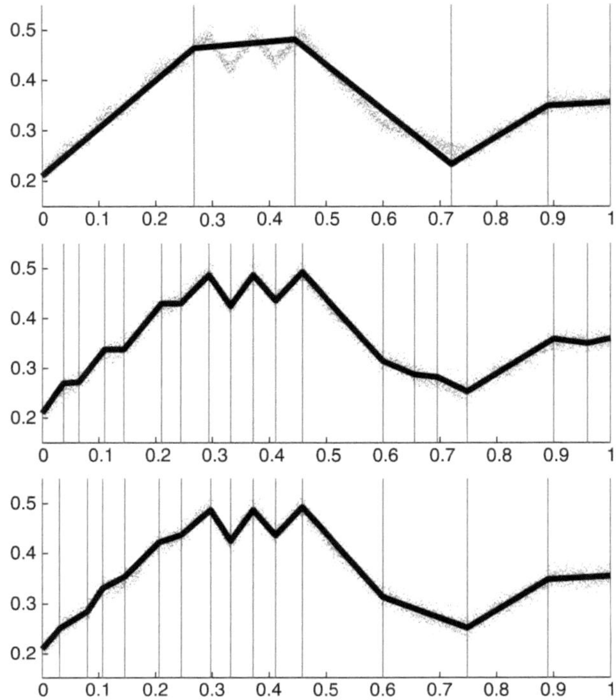

Fig. 3.7. Three different segmentations of the signal.

The global model $I_0(x; l_i, \theta_i)$ of the signal $I(x)$, or "world scene" is therefore described by the vector of random variables:

$$\mathbf{W} = (k, \{x_1, \ldots, x_{k-1}\}, \{(l_1, \theta_1), \ldots, (l_k, \theta_k)\}) \tag{3.65}$$

Assuming that

- In this 1D example the individual likelihoods for each region $I_0(x, l_i, \theta_i)$, $x_i - 1 \leq x < x_i$ decay exponentially with the squared error between the model and the actual signal:

$$P(I|l_i, \theta_i) = e^{\left(-\frac{1}{2\sigma^2}\int_{x_i-1}^{x_i}(I(x)-I_0(x;l_i,\theta_i))^2dx\right)} \tag{3.66}$$

- The prior $p(\mathbf{W})$ is given by penalizing
  - the number $k - 1$ of regions: $P(k) \propto e^{-\lambda_0 k}$
  - and the number $|\theta_i|$ of parameters: $P(\theta_i|l_i) \propto e^{-\lambda|\theta_i|}$
- All region models are equally likely a priori, that is, $p(l_i)$ is uniform.

Then, the standard Bayesian formulation of the posterior probability is

$$P(W|I) \propto e^{\left(-\frac{1}{2\sigma^2}\sum_{i=1}^{k}\int_{x_{i-1}}^{x_i}(I(x)-I_0(x;l_i,\theta_i))^2dx\right)} e^{-\lambda_0 k} e^{-\lambda\sum_{i=1}^{k}|\theta_i|} \tag{3.67}$$

The maximum a posteriori (MAP) solution comes from maximizing the posterior probability (Eq. 3.67). In energy minimization terms, the exponent of the posterior is used to define an energy function:

$$E(W) = \frac{1}{2\sigma^2}\sum_{i=1}^{k}\int_{x_{i-1}}^{x_i}(I(x) - I_0(x; l_i, \theta_i))^2 dx + \lambda_0 k + \lambda\sum_{i=1}^{k}|\theta_i| \qquad (3.68)$$

This energy function has to be minimized in a space with variable number of dimensions because the number of regions k is unknown. One way of solving the problem would be to use a greedy search, which starts with a high k and fuses regions, or to start with a low k and part regions, according to some metric, for example, the error. At each step, the models would have to be optimized for a fixed k in order to perform the evaluation. The jump-diffusion algorithm embodies the search of k in both directions and the optimization of the region models in a single optimization algorithm.

Reversible jumps

The posterior probability $P(W|I)$ (Eq. 3.67) is distributed over a countable number of solutions subspaces Ω_i of varying dimension. The union of these subspaces $\Omega = \cup_{n=1}^{\infty}\Omega_n$ forms the complete solution space.

To search over the solution space, there is the need to perform reversible jumps from one subspace to another. In the 1D example, we define two types of jumps: merge two adjacent regions, and split a region (see Fig. 3.8 for a graphical example). If there is more than one region model, then a third type of jump has to be defined, to change the model of a region, for example, from $l_i = line$ to $l_i = arc$. In the jump-diffusion formulation, the jumps are realized by a Metropolis move over a Markov chain of states. The world scene $W = (n, \psi)$ denotes the state of the stochastic process at time t, where $\psi \in \Omega_n$ are the state variables that define the intervals and parameters of a current

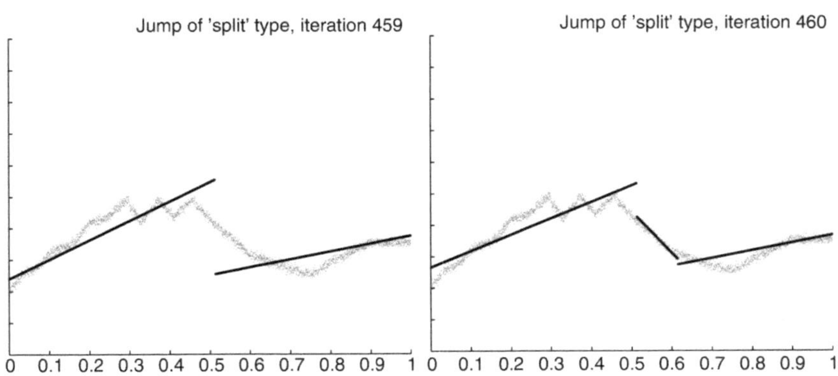

Fig. 3.8. A jump of "split" type. *Left*: plot of W and I. *Right*: plot of W' and I.

set of region models in a subspace Ω_n. Similarly, $W' = (m, \phi)$ with $\phi \in \Omega_m$ and $m \neq n$ is the destination state of the jump in the Markov chain. Then, the probability to accept the jump $W \to W'$ is given by:

$$\alpha(W'|W) = \min\left\{1, \frac{p(W'|I)d\phi}{p(W|I)d\psi} \times \frac{g(W' \to W)d\psi}{g(W \to W')d\phi}\right\} \tag{3.69}$$

where $g(\cdot)$ denotes the forward and backward proposal probabilities, which are given by the probability of choosing a destination space and the density of the parameters in that destination space:

$$\begin{aligned} g(W' \to W) &= q(W' \to W)q(\psi|n) \\ g(W \to W') &= q(W \to W')q(\phi|m) \end{aligned} \tag{3.70}$$

In the standard jumping scheme, the densities $q(\psi|n)$ and $q(\phi|m)$ are usually set to uniform distributions. This makes the process computationally very slow, which is an important drawback, and it is discussed in Section 3.4.2.

Let us see a numerical example with the jump $W \to W'$ plotted in Fig. 3.8. If uniform densities are supposed, then the probability of a jump will be given by the energies $E(W)$ and $E(W')$ of the states before and after the jump. The jump with the lowest energy will be the most probable one. The energy is calculated using Eq. 3.68, which consists, basically, of the error between the models and the actual signal and the penalizations to complex models. The trade-off between error and complexity is regularized by λ and λ_0. The variance σ^2 is approximately the variance of the signal, and $I(x)$ is the actual signal. The model is denoted as $I_0(x; l_i, \theta_i)$, and it depends on W. For the toy-example, we only have the linear region model $I_0(x, line, (a, b)) = ax + b$, and in the state W, there are two regions $k = 2$. The first region has parameters $\theta_1 = (0.6150, 0.2394)$, and the second region has parameters $\theta_2 = (0.2500, 0.1260)$. The change point between the regions is $x_2 = 0.5128$, so the state is denoted as

$$\begin{aligned} W = (2,\ &\{0, 0.5128, 1\}, \\ &\{(line, (0.6150, 0.2394)), (line, (0.2500, 0.1260))\}) \end{aligned} \tag{3.71}$$

and the error would be

$$\begin{aligned} &\sum_{i=1}^{k} \int_{x_i-1}^{x_i} (I(x) - I_0(x; l_i, \theta_i))^2 dx \\ &= \int_{0.5128}^{0} (I(x) - 0.6150x - 0.2394)^2 dx \\ &+ \int_{1}^{0.5128} (I(x) - 0.2500x - 0.1260)^2 dx \end{aligned} \tag{3.72}$$

The error is calculated in a similar way for W', which in the example is

$$\begin{aligned} W' = (3,\ &\{0, 0.5128, 0.6153, 1\}, \\ &\{(line, (0.6150, 0.2394)), \\ &(line, (-1.3270, 1.1078)), \\ &(line, (0.2493, 0.1204))\}) \end{aligned} \tag{3.73}$$

Table 3.1. Energies $E(W)$ and probabilities $P(W|I)$ of the destination states of the jumps from state W (Eq. 3.71 and Fig. 3.8, left). The jumps considered are splitting region 1, splitting region 2, merging regions 1 and 2, and remaining in the same state.

W	W_{actual}	W'_{split1}	W'_{split2}	$W'_{merge12}$	$\sum$
$E(W)$	0.1514	0.1987	0.0808	13.6823	14.1132
$P(W\|I)$	0.2741	0.2089	0.5139	0.0030	1.0000

For the signal $I(x)$ of the example, we obtain the energy values $E(W) = 0.0151$ and $E(W') = 0.0082$. The probabilities $p(W|I)$ and $p(W'|I)$ are given by the normalization of the inverse of the energy. There are several possible moves from the state W: splitting region 1, splitting region 2, merging regions 1 and 2, or remaining in the same state. In Table 3.1, the energies and probabilities of the considered destination states are shown. It can be seen that, considering uniform densities $q(\psi|n)$ and $q(\phi|m)$, the most probable jump is to the state in which region 2 is split.

Stochastic diffusions

In a continuous subspace with fixed region models, stochastic diffusions can be performed in order to adjust the parameters of the model so that it fits better the data. To do this, the energy function defined in Eq. 3.68 has to be minimized. The stochastic diffusion uses the continuous Langevin equations that simulate Markov chains with stationary density $p(\psi) \propto exp(-E(\psi)/T)$, where ψ is a variable of the model and T is a temperature that follows an annealing scheme. The equations have a term that introduces normal noise $\sqrt{2T(t)}N(0,(dt)^2)$ to the motion. This term, called Brownian motion, also depends on the temperature T and is useful for overcoming local minima. Then, the motion equation for a variable at time t is defined as

$$d\psi(t) = -\frac{dE(W)}{d\psi}dt + \sqrt{2T(t)}N(0,(dt)^2) \tag{3.74}$$

Given this definition, let us obtain the motion equations for the variables of the toy-example, which are the change points x_i and the parameters $\theta_i = (a, b)$ of the linear region models. We have to obtain the expression of the derivative of the energy $E(W)$ with respect to the time t for each one of these variables. Then, the motion equation for a change point x_i is calculated as

$$\frac{dx_i(t)}{dt} = \frac{dE(W)}{dx_i} + \sqrt{2T(t)}N(0,1) \tag{3.75}$$

Let us calculate the derivative $dE(W)/dx_i$. In the definition of $E(W)$ in Eq. 3.68, we can see that for a fixed k and fixed region models, the penalization terms are constant (c). Therefore, in the derivative of $E(W)$, these become null. On the other hand, the summation adds k terms and only two of them contain x_i, the rest are independent, so they are also null in the derivative. For compactness, let us denote the error between model and signal as f and its indefinite integral as F:

$$\begin{aligned} f_i(x) &= (I(x) - I_0(x; l_i, \theta_i))^2 \\ F_i(x) &= \int (I(x) - I_0(x; l_i, \theta_i))^2 dx \end{aligned} \tag{3.76}$$

Then the derivative of the energy is calculated as

$$\begin{aligned} \frac{dE(W)}{dx_i} &= \frac{d}{dx_i}\left(\frac{1}{2\sigma^2}\sum_{i=1}^{k}\int_{x_{i-1}}^{x_i} f_i(x)dx + c\right) \\ &= \frac{1}{2\sigma^2}\frac{d}{dx_i}\left(\cdots + \int_{x_{i-1}}^{x_i} f_i(x)dx + \int_{x_i}^{x_{i+1}} f_i(x)dx + \cdots\right) \\ &= \frac{1}{2\sigma^2}\frac{d}{dx_i}\left(F_i(x_i) - F_i(x_{i+1}) + F_{i+1}(x_{i+1}) - F_{i+1}(x_i)\right) \\ &= \frac{1}{2\sigma^2}\frac{d}{dx_i}\left(F_i(x_i) - F_{i+1}(x_i)\right) \\ &= \frac{1}{2\sigma^2}\left(I(x_i) - I_0(x_i; l_i, \theta_i))^2 - I(x_i) - I_0(x_i; l_{i-1,}\theta_{i-1}))^2\right) \end{aligned} \tag{3.77}$$

Finally, the expression obtained for $\frac{dE(W)}{dx_i}$ is substituted in the x_i motion equation, Eq. 3.75. The resulting equation is a 1D case of the region competition equation [184], which also moves the limits of a region according to the adjacent region models' fitness to the data.

The motion equations for the θ_i parameters have an easy derivation in the linear model case. The motion equation for the slope parameter a_i results in

$$\begin{aligned} \frac{da_i(t)}{dt} &= \frac{dE(W)}{da_i} + \sqrt{2T(t)}N(0,1) \\ &= \frac{1}{2\sigma^2}\frac{d}{da_i}\sum_{i=1}^{k}\int_{x_{i-1}}^{x_i}(I(x) - a_i x + b_i)^2 dx + \sqrt{2T(t)}N(0,1) \\ &= \frac{1}{2\sigma^2}\sum_{x=x_{i-1}}^{x_i} 2\left(ax^2 + bx - xI(x)\right) + \sqrt{2T(t)}N(0,1) \end{aligned} \tag{3.78}$$

Similarly, the motion equation for the intersect parameter b_i is

$$\begin{aligned} \frac{db_i(t)}{dt} &= \frac{dE(W)}{db_i} + \sqrt{2T(t)}N(0,1) \\ &= \frac{1}{2\sigma^2}\frac{d}{db_i}\sum_{i=1}^{k}\int_{x_{i-1}}^{x_i}(I(x) - a_i x + b_i)^2 dx + \sqrt{2T(t)}N(0,1) \\ &= \frac{1}{2\sigma^2}\sum_{x=x_{i-1}}^{x_i} 2\left(ax + b - I(x)\right) + \sqrt{2T(t)}N(0,1) \end{aligned} \tag{3.79}$$

In Fig. 3.9 the result of applying the motion equations over the time t, or number of iterations is represented. It can be seen that motion equations

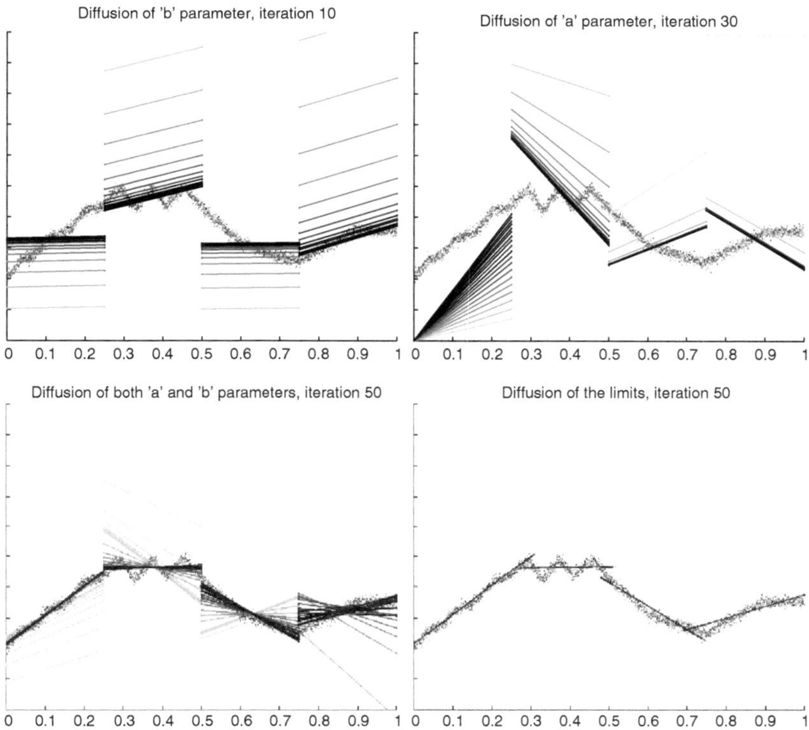

Fig. 3.9. Diffusion of 'b', 'a', both parameters together and the limits.

modify the parameters in the direction that approximates the model closer to the data. The parameters increments get larger when the model is far from the data and smaller when the model is close to the data.

As we already explained, diffusions only work in a definite subspace Ω_n and jumps have to be performed in some adequate moment, so that the process does not get stuck in that subspace. In the jump-diffusion algorithm, the jumps are performed periodically over time with some probability. This probability, referred to the waiting time λ between two consecutive jumps, follows a Poisson distribution, being κ the expected number of jumps during the given interval of time:

$$f_{Poisson}(\kappa; \lambda) = \frac{\lambda^{\kappa} e^{-\lambda}}{\kappa!} \tag{3.80}$$

A simulation of the random jumps with Poisson probability distribution in time is shown in Fig. 3.10.

3.4.2 Speeding-up the Jump-Diffusion Process

The jump-diffusion process simulates a Markov chain, which searches over the solution space Ω by sampling from the posterior probability $p(W|I)$.

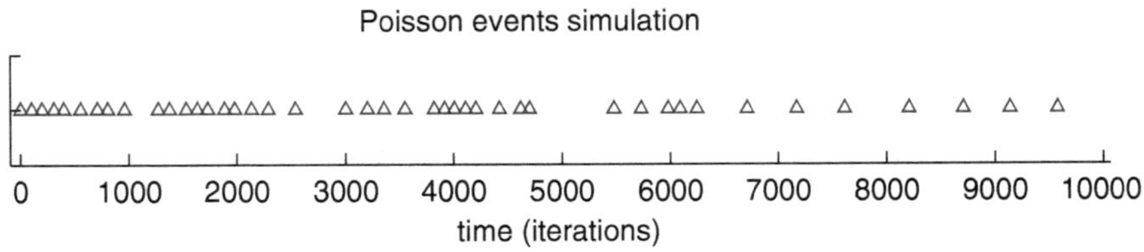

Fig. 3.10. Poisson events simulation.

Reversibility of the jumps is introduced as a tool for guaranteeing irreducibility of the Markov chain. A Markov chain is said to be irreducible if any state can be reached from any other state, after some sequence of states $\chi_0, \chi_1, \cdots, \chi_s$:

$$p(\chi_s = \Omega_m | \chi_0 = \Omega_n) > 0, \; \forall m, n/m \neq n \tag{3.81}$$

In theory, reversibility is not strictly necessary for the jump-diffusion process to achieve the optimal solution. Moreover, after a number of iterations, the jump-diffusion process can achieve the optimal solution with a probability close to one. However, the high number of iterations could make the process computationally unfeasible.

The speed bottlenecks

To make the jump-diffusion process viable, the search through the solution space has to be made more direct so that convergence to a good solution is achieved in a smaller number of iterations. The search can be tuned by using more suitable probability models for the proposal probabilities. In a general case, the forward proposal probability defined in Eq. 3.70 can be divided into three cases:

$$q(\phi|m) = \begin{cases} q(\theta_i|l_i, [x_{i-1}, x_i)) & \text{switch} \\ q(\theta|l, [x_{i-2}, x_i)) & \text{merge} \\ q(x|[x_{i-1}, x_i))q(\theta_a|l_a, [x_{i-1}, x))q(\theta_b|l_b, [x, x_i)) & \text{split} \end{cases} \tag{3.82}$$

where the first one is switching the interval $[x_{i-1}, x_i)$ from its associated region model to another (l_i, θ_i) model; the second case refers to merging two adjacent intervals into a single (l, θ) model; the third model is splitting the interval $[x_{i-1}, x_i)$ into two different region models, (l_a, θ_a) for the interval $[x_{i-1}, x)$, and (l_b, θ_b) for the interval $[x, x_i)$. In the toy-example, we only have one region model, so the proposal probabilities do not need to consider switching from one model to another, as the model parameters are already changed by the diffusions.

If the proposal probabilities are considered to follow a uniform distribution, then the jumps work as a random selection of new models, and as a consequence, the proposals are rejected most of the times. This increases very much the number of iterations, even though in theory, it is highly probable that the process finally achieves the optimal solution. In practice, a good design of the proposal probabilities is necessary.

Data-driven techniques

In [159], the data-driven Markov chain Monte Carlo scheme is explained and exploited. The idea is to estimate the parameters of the region models, as well as their limits or changepoints, according to the data that these models have to describe. These estimations consist of bottom-up strategies, for example, a simple strategy for estimating the changepoints x_i is to place them where edges are detected. Actually, taking edges as those points with a gradient above some threshold is not a good strategy. A better way is to take the edgeness measure as probabilities, and sample this distribution for obtaining the changepoints. For example, in Fig. 3.11, we show the edgeness of the toy-example data. If a split has to be performed of a region in the interval $[0.2, 0.4)$ to two new region models for the intervals $[0.2, x)$ and $[x, 0.4)$, the new changepoint x would be placed with a high probability on 0.3 because the edgeness corresponding to this interval defines such probability distribution.

The estimation of the parameters θ_i of the linear model of the toy-example can be performed by taking the slope and the intersect of the most voted line of the Hough transformation of the underlying data. However, a probabilistic approach is more suitable for the jump-diffusion process. It consists of computing the *importance proposal probability* by Parzen windows centered at the lines of the Hough transformation. When a new model is proposed in the interval $[x_{i-1}, x_i)$, its important proposal probability is

$$q(\theta_i | l_i, [x_{i-1}, x_i) = \sum_{j=1}^{N} \omega_j G(\theta_i - \theta_j) \tag{3.83}$$

where $G(x)$ is a Parzen window, N is the number of candidate lines of the Hough transformation, and ω_i are the accumulated weights of the Hough transformation votes.

The interesting concept that data-driven proposal probabilities design introduces is the fact that a top-down process is aided by bottom-up processes. This is a way of mixing generative and discriminative methods in a single algorithm. In Fig. 3.12, we compare the evolution of the toy-example energy $E(W)$ during the jump-diffusion process in four cases: the pure generative

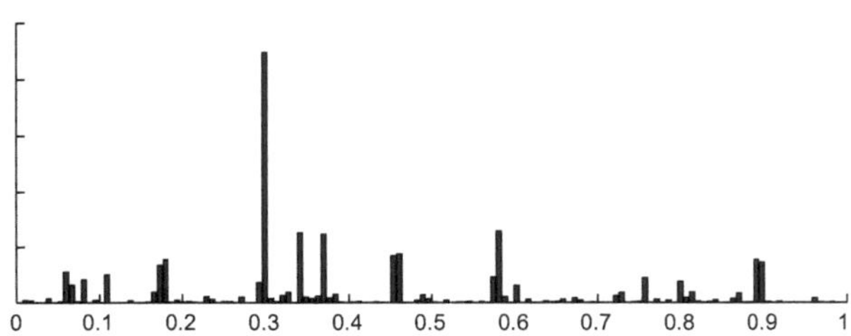

Fig. 3.11. Edgeness measure of the signal.

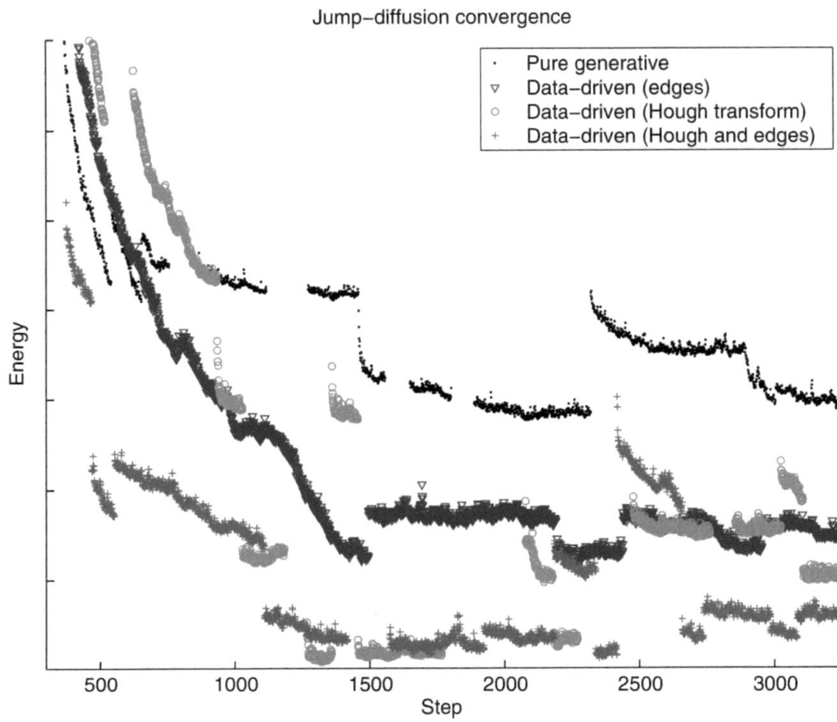

Fig. 3.12. Jump-diffusion energy evolution. See Color Plates.

scheme; the data-driven approach using edges; Hough transformation; and both of them. We can see that bottom-up methods help the jump-diffusion convergence and make it computationally more viable. In probabilistic terms, a good design of the proposal probabilities causes the increase of the ratio

$$\frac{p(m,\phi|I)}{p(n,\psi|I)} = e^{-\Delta E}, \tag{3.84}$$

and the jump proposals are more frequently successful.

3.4.3 K-adventurers Algorithm

The data-driven jump-diffusion process has the disadvantage of not defining a stop criterion. Although the energy of the model has a pronounced descent during the first iterations, a convergence to some low energy value cannot be detected. Low energies are obtained in several different subspaces Ω_n. This is due to the well-known fact that for most problems, segmentation has more than one good solution. Actually, as explained in [159], computing different solutions can be necessary for intrinsically ambiguous scenes, and for providing robustness to the segmentation process, given that the designed probability distributions are not perfect.

The solution space could contain a large amount of different solutions W_i, some of them very similar to each other. We are not interested in maintaining all of them, but in selecting the most important ones. The K-adventurers algorithm is designed for sampling from the probability distribution formed by the solutions, so that the most representative ones are selected. For example, if we have a trimodal distribution of solutions and we want a number of $K = 3$ solutions to represent the solution space, then there is a high probability that the K-adventurers algorithm yields the three solutions corresponding to the modes of the distribution. It is not necessary for K to be related to the modality of the distribution. We select the number of important solutions depending on how well we want to describe the solution space. Nevertheless, if the distribution of solutions is very complex, a larger number K of representative solutions is needed for obtaining a good sampling. This strategy is an implicit way of model order selection. It is probabilistic, and depending on the kinds of solutions present in their distribution, different model orders could be found.

The selection of important solutions is performed by defining the objective to minimize a Kullback–Leibler divergence $D(p||\hat{p})$ between the Bayesian posterior probability $p(W|I)$ (already defined in Eq. 3.67) and the non parametric probability $\hat{p}(W|I)$, which is represented by a fixed-size set S of K selected samples:

$$\hat{p}(W|I) = \sum_{i=1}^{K} \omega_i G(W - W_i, \sigma_i^2), \quad \sum_{i=1}^{K} \omega_i = 1 \tag{3.85}$$

where ω_i are weights proportional to $p(W|I)$ and they sum 1. G is a Gaussian window in the solution space Ω, with a σ_i^2 variance. The objective set S^* of selected solutions (ω_i, W_i), $i = 1, \ldots, K$ is therefore defined as

$$S^* = \arg \min_{S/|S|=K} D(p||\hat{p}) \tag{3.86}$$

where the Kullback–Leibler divergence (KL-divergence) consists of

$$D(p||\hat{p}) = \int p(W|I) \log \left(\frac{p(W|I)}{\hat{p}(W|I)} \right) dW \tag{3.87}$$

The K-adventurers algorithm iteratively calculates an approximation of S^*. Each time the jump-diffusion algorithm performs a successful jump and its energy is minimized by the diffusions process, the solution W is taken by the K-adventurers algorithm, and the KL-divergence is recomputed in order to update the $\hat{S^*}$ set of solutions if necessary. The algorithm is as follows:

The algorithm starts with fixed size set of K solutions, which initially are the same solution. This set is iteratively updated after each new solution yielded by the jump-diffusion process. The iterations are denoted with the *while* sentence in Alg. 4.

The stop criterion is not defined; however, the ergodicity of the Monte Carlo Markov chain process guarantees that significant modes will be visited

Algorithm 4: K-adventurers

Input: I, successive solutions (ω_{K+1}, W_{K+1}) generated by a jump-diffusion process
Initialize $\hat{S^*}$ with one initial (ω_1, W_1) solution K times
while $\exists\,(\omega_{K+1}, W_{K+1}) \leftarrow$ *jump-diffusion* **do**
  $S_+ \leftarrow \hat{S^*} \cup \{(\omega_{K+1}, W_{K+1})\}$
  for $i = 1, 2, \ldots, K+1$ **do**
    $S_{-i} \leftarrow S_+ / \{(\omega_{K+1}, W_{K+1})\}$
    $\hat{p} \leftarrow S_{-i}$
    $d_i = \hat{D}(p||\hat{p})$
  end
  $i^* = \arg\min_i d_i$
  $\hat{S^*} \leftarrow S_{-i^*}$
end
Output: $\hat{S^*}$

over time and there will be a convergence to the p distribution. The number of iterations necessary for a good approximation of S^* depends on the complexity of the search space. A good stop criterion is to observe whether S^* undergoes important changes, or on the contrary, remains similar after a significant number of iterations.

In each *while* iteration, a *for* sentence iterates the estimation of the KL-divergence $\hat{D}(p||\hat{p}_{-i})$ between p and each possible set of K solutions, considering the new one, and subtracting one of the former. The interesting point is the estimation of the divergence, provided that p consists of a solutions set whose size increases with each new iteration. The idea which Tu and Zhu propose in [159] is to represent $p(W|I)$ by a mixture of N Gaussians, where N is the number of solutions returned by the jump-diffusion process, which is the same as the number of iterations of the *while* sentence. These N solutions are partitioned into K disjoint groups. Each one of these groups is represented by a dominating solution, which is the closest one to some solution from the S^* set of selected solutions. The name of the algorithm is inspired by this basic idea: metaphorically, K adventurers want to occupy the K largest islands in an ocean, while keeping apart from each other's territories.

As shown in Eq. 3.85, the probability distributions are modelled with a sum of Gaussians centered at the collected solutions. These solutions are largely separated because of the high dimensionality of the solutions space. This is the reason for forming groups with dominating solutions and ignoring the rest of the solutions. More formally, for selecting $K << N$ solutions from the initial set of solutions S_0, a mapping function from the indexes of $\hat{S^*}$ to the indexes of S_0 is defined $\tau : \{1, 2, \ldots, K\} \rightarrow \{1, 2, \ldots, N\}$, so that

$$\hat{S^*} = \{(\omega_{\tau(i)}, W_{\tau(i)});\ i = 1, 2, \ldots, K\} \tag{3.88}$$

which, similarly to Eq. 3.85, encodes the nonparametric probability density model:

$$\hat{p}(W) = \frac{1}{\sum_{i=1}^{K} \omega_{\tau(i)}} \sum_{i=1}^{K} \omega_{\tau(i)} G(W - W_{\tau(i)}, \sigma^2_{\tau(i)}) \tag{3.89}$$

In the experiments performed in [159], the same variance is assumed for all Gaussians.

Given the former density model, the approximation of $D(p||\hat{p})$ can be defined. From the definition of the KL-divergence, we have

$$\begin{aligned} D(p||\hat{p}) &= \sum_{n=1}^{N} \int_{D_n} p(W) \log \frac{p(W)}{\hat{p}(W)} dW \\ &= \sum_{n=1}^{N} \int_{D_n} \sum_{i=1}^{N} \omega_i G(W - W_i; \sigma^2) \\ &\quad \times \log \frac{\sum_{i=1}^{N} \omega_i G(W - W_i; \sigma^2)}{\frac{1}{\sum_{j=1}^{N} \omega_{\tau(j)}} \sum_{j=1}^{N} \omega_{\tau(j)} G(W - W_{\tau(j)}; \sigma^2)} dW \end{aligned} \tag{3.90}$$

Where D_i, $i = 1, 2, \ldots, K$ are the disjoint groups, each one of them dominated by a single solution. A second mapping $c : \{1, 2, \ldots, N\} \rightarrow \{1, 2, \ldots, K\}$ is defined so that

$$\hat{p}(W) \approx \frac{\omega_{c(i)}}{\sum_{j=1}^{N} \omega_{\tau(j)}} G(W - W_{\tau(c(n))}; \sigma^2),\ W \in D_i,\ i = 1, 2, \ldots, N \tag{3.91}$$

Provided that the energy of each mode $p(W_i|I)$ is defined as $E(W_i) = -\log p(W_i)$, the approximation of $D(p||\hat{p})$ is formulated as

$$\begin{aligned} \hat{D}(p||\hat{p}) &= \sum_{n=1}^{N} \int_{D_n} \omega_n G(W - W_n; \sigma^2) \\ &\quad \cdot \left[\log \sum_{j=1}^{N} \omega_{\tau(j)} + \log \frac{\omega_n G(W - W_n; \sigma^2)}{\omega_{\tau(c(n))} G(W - W_{\tau(c(n))}; \sigma^2)} \right] dW \\ &= \sum_{n=1}^{N} \left[\log \sum_{j=1}^{N} \omega_{\tau(j)} + \log \frac{\omega_n}{\omega_{\tau(c(n))}} + \frac{(W_n - W_{\tau(c(n))})^2}{2\sigma^2} \right] \\ &= \log \sum_{j=1}^{N} \omega_{\tau(j)} + \sum_{n=1}^{N} \omega_n \left[E(W_{\tau(c(n))}) - E(W_n) + \frac{(W_n - W_{\tau(c(n))})^2}{2\sigma^2} \right] \end{aligned} \tag{3.92}$$

In [159], the goodness of the approximation $\hat{D}(p||\hat{p})$ is experimentally demonstrated and it is compared to the actual KL-divergence $D(p||\hat{p})$ and to $|p - \hat{p}|$. Finally, the definition of a distance measure between solutions W_1 and W_2 is a delicate question. For the 1D toy-example, a distance measure between solutions could be the error between the generated model and the actual data. When several region models are considered, the model type has to be considered in the distance measure too. Also, the number of regions is another important term to be considered in the measure.

In Fig. 3.13, we illustrate a solution space of the 1D toy-example. The $K = 6$ selected solutions are marked with red rectangles. These solutions are not necessarily the best ones, but they represent the probability distribution of the solutions yielded by the jump-diffusion process. In Fig. 3.14, the energies of these solutions are represented. Finally, in Fig. 3.15 the models of the

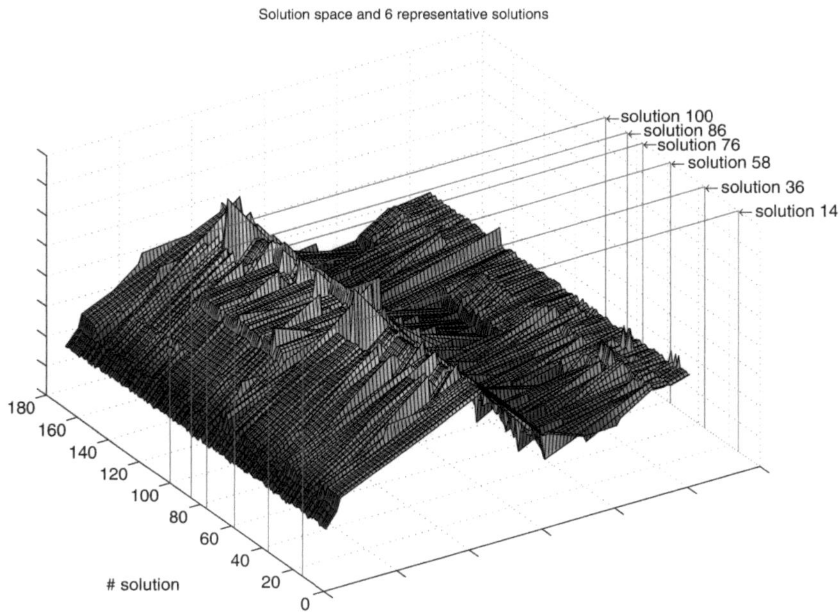

Fig. 3.13. The solution space in which the K-adventurers algorithm has selected $K = 6$ representative solutions, which are marked with *rectangles*. See Color Plates.

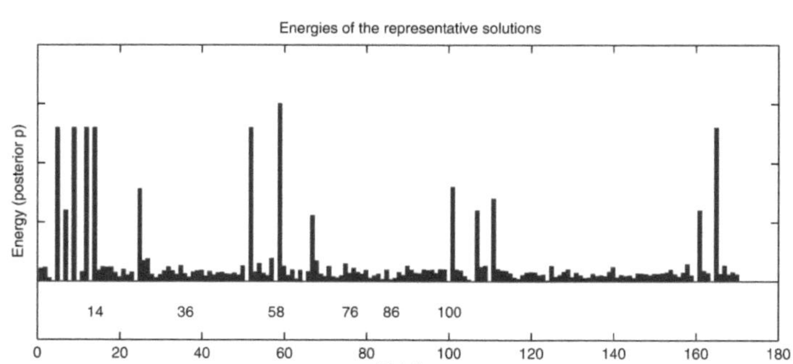

Fig. 3.14. The energies of the different solutions. The representative solutions selected by the K-adventurers algorithm are marked with *arrows*.

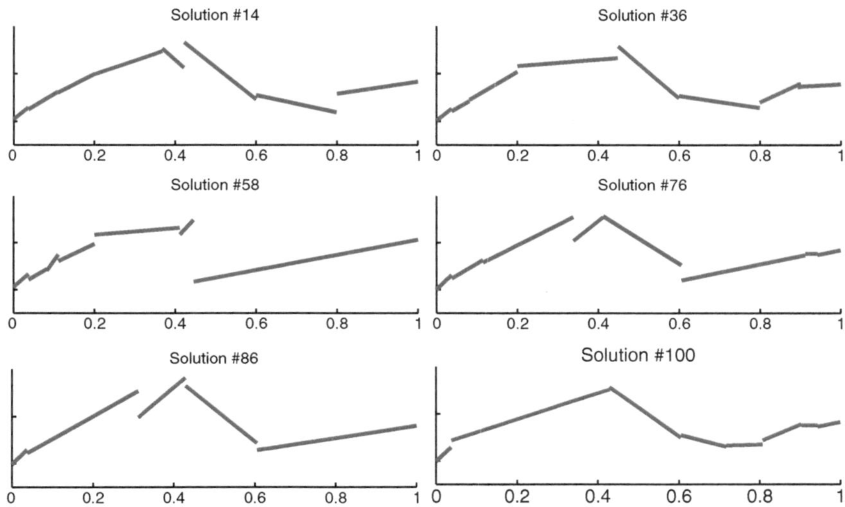

Fig. 3.15. Result of the K-adventurers algorithm for $K = 6$: the six most representative solutions of the solution space generated by jump-diffusion.

six solutions are shown. As already explained, an advantage of generative approaches is the possibility to generate data, once there is model with its parameters estimated. In conclusion, the jump-diffusion algorithm integrates both top-down and bottom-up processes. The K-adventurers algorithm, based on information theory, iteratively selects the most important solutions by estimating their probability distribution in the solution space.

3.5 Model-Based Segmentation Exploiting The Maximum Entropy Principle

3.5.1 Maximum Entropy and Markov Random Fields

In Section 3.2, we have introduced the maximum entropy (ME) principle and how it is used to find the approximated shape, through the less biased pdf, given the statistics of the sample. Here, we present how to use it for segmenting parts of a given image whose colors are compatible with a given ME model. Therefore, ME is the driving force of learning the model from the samples and their statistics. Such model is later used in classification tasks, such as the labeling of *skinness* (skin-color) of pixels/regions in the image [84]. This is a task of high practical importance like blocking adult images in webpages (see for instance [180]). A keypoint in such learning is the trade-off between the complexity of the model to learn and its effectiveness in ROC terms. Let $\mathcal{X} = \{(\mathbf{I}_i, y_i) : i = 1, \ldots, M\}$ be the training set, where $\mathbf{I}_i$ is a color image that is labeled as $y_i = 1$ if it contains skin, and $y_i = 0$ otherwise. In [84], the

Compaq Database [89] is used. In such database, skin is manually segmented and $M = 18{,}696$ RGB images[3] are used. Therefore, if $\mathbf{x}_s$ is the color of the sth pixel, then $\mathbf{x}_s \in S = \{0, \ldots, 255\}^3$.

The training set allows us to approximate $p(\mathbf{x}, y)$ with an independent model $\mathcal{C}_0$ called the *baseline model*:

$$\mathcal{C}_0 : \forall s \in S,\ \forall \mathbf{x}_s \in S, \forall y_s \in \{0,1\}\ : p(\mathbf{x}_s, y_s) = q(\mathbf{x}_s, y_s) \tag{3.93}$$

$q(\mathbf{x}_s, y_s)$ being the proportion of pixels in the training set with color $\mathbf{x}_s$ and skinness y_s (two tridimensional histograms, one for each skinness, typically quantized to 32 bins, or a four dimensional histogram in the strict sense). Using a modified version of the usual expectation constraints in ME, we have that

$$p(\mathbf{x}_s, y_s) = E_p[\delta_{\mathbf{x}_s}(\mathbf{x}_s)\delta_{y_s}(y_s)] \tag{3.94}$$

where $\delta_a(b) = 1$ if $a = b$ and 0 otherwise. Then, the shape of the ME distribution is

$$p(\mathbf{x}, y) = e^{\lambda_0 + \sum_{s \in S} \lambda(s, \mathbf{x}_s, y_s)}\ . \tag{3.95}$$

Thus, assuming $q(\mathbf{x}_s, y_s) > 0$, we find the following values for the multipliers: $\lambda_0 = 0$ and $\lambda(s, \mathbf{x}_s, y_s) = \log q(\mathbf{x}_s, y_s)$. Consequently, the ME distribution is

$$p(\mathbf{x}, y) = \prod_{s \in S} q(\mathbf{x}_s, y_s) \tag{3.96}$$

and the probability of belonging to a class is given by the Bayes theorem:

$$p(y|\mathbf{x}) = \prod_{s \in S} q(y_s|\mathbf{x}_s) = \prod_{s \in S} \frac{q(\mathbf{x}_s|y_s)q(y_s)}{q(\mathbf{x}_s)} = \prod_{s \in S} \frac{q(\mathbf{x}_s|y_s)q(y_s)}{\sum_{y_s=0}^{1} q(\mathbf{x}_s|y_s)q(y_s)} \tag{3.97}$$

Such probability is computed from two tridimensional histograms: $q(\mathbf{x}_s|y_s = 0)$ and $q(\mathbf{x}_s|y_s = 1)$. Such probability is expressed in terms of gray levels in the second column of Fig. 3.16.

The model in Eq. 3.96 is consistent with an unrealistic, but efficient *independence assumption* (the skinness of a pixel is independent of that of their neighbors). A more realistic assumption is the Markovian one: the skinness of a pixel s depends only on its neighbors $\mathcal{N}_s$ (for instance, it is usual to use a 4-neighborhood). This is coherent with the fact that skin pixels belong to larger regions. More formally, although there are excellent monographs on the subject (see [175]), the pixels in the image are the nodes (random variables) of an undirected graph and the edges are defined by the type of neighborhood chosen (for instance: north, south, east and west). Given two neighbors s and t, which are denoted by $< s, t >$, let $q(y_s, y_t)$ be the expected proportion of observations of $(y_s = a, y_t = b)$, that is, we have

[3] Although the skin hue is invariant to the ethnic group, skinness depends on illumination conditions, which are usually unknown.

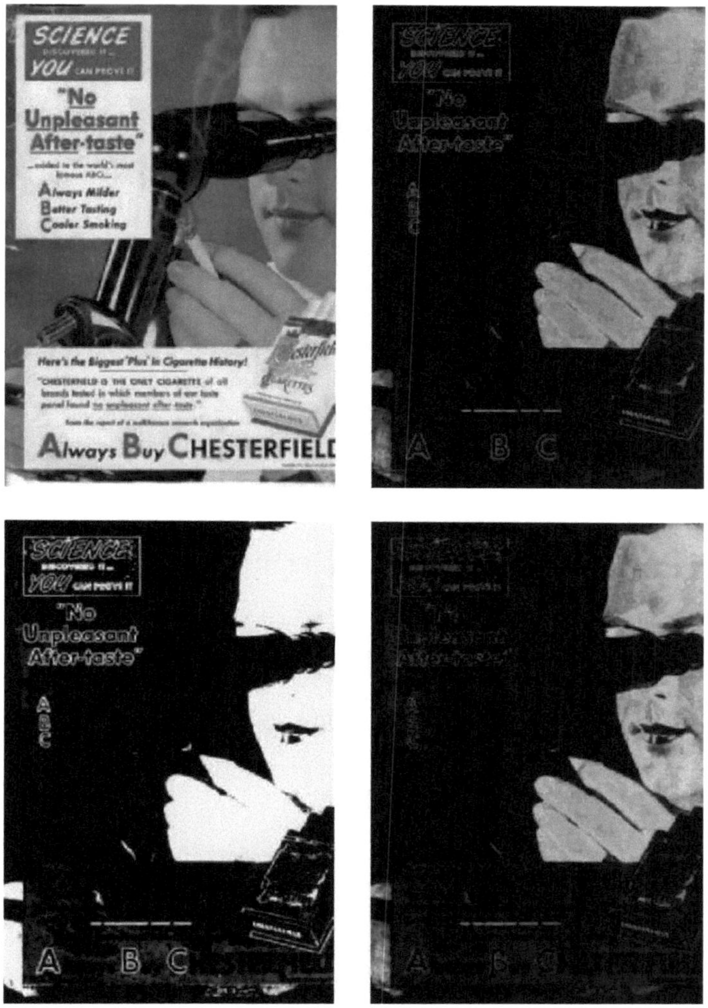

Fig. 3.16. Skin detection results. Comparison between the baseline model (*top-right*), the tree approximation of MRFs with BP (*bottom-left*), and the tree approximation of the first-order model with BP instead of Alg. 5. Figure by B. Jedynak, H. Zheng and M. Daoudi (©2003 IEEE). See Color Plates.

the four quantities: $q(y_s = 0, y_t = 0), q(y_s = 0, y_t = 1), q(y_s = 1, y_t = 0)$, and $q(y_s = 1, y_t = 1)$. Here, the aggregation of horizontal and vertical quantities yields an implicit assumption of isotropy, which is not an excessive unrealistic simplification. Let D be the following model:

$$\mathcal{D} : \forall < s, t > \in S \times S \ p(y_s = 0, y_t = 0) = q(0,0), \ p(y_s = 1, y_t = 1) = q(1,1) \tag{3.98}$$

The ME entropy model for $\mathcal{C}_0 \cap \mathcal{D}$ is obtained by starting the formulation of the expectation constraints:

$$\forall y_s \in \{0,1\}, \forall y_t \in \{0,1\}, p(y_s, y_t) = E_p[\delta_{y_s}(y_s)\delta_{y_t}(y_t)] \tag{3.99}$$

thus, the solution has the yet familiar exponential shape, but depending on more multipliers enforcing the constraints, labels of neighbors $< s,t >$ are equal:

$$\begin{aligned} p(x,y) &= e^{H(x,y,\Lambda)},\ \Lambda = (\lambda_0(.),\ldots,\lambda_3(.)) \\ H(x,y,\Lambda) &= \lambda_0 + \sum_s \lambda_1(s,\mathbf{x}_s,y_s) + \sum_{<s,t>} \lambda_2(s,t)(1-y_s)(1-y_t) \\ &\quad + \sum_{<s,t>} \lambda_3(s,t)y_s y_t \end{aligned} \tag{3.100}$$

Therefore, $p(\mathbf{x}_s, y_s)$, the marginal for $\mathbf{x}_s$, may be posed as

$$p(\mathbf{x}_s, y_s) = \sum_{\mathbf{x}_t, t\neq s} \sum_{y_t, t\neq s} p(x,y) = e^{\lambda_0 + \lambda_1(s,\mathbf{x}_s,y_s)} g(s,y_s) \tag{3.101}$$

where $g(s, y_s)$ is a function independent of $\mathbf{x}_s$. Thus, the marginal for y_s is

$$p(y_s) = \sum_{\mathbf{x}_s} p(\mathbf{x}_s, y_s) = e^{\lambda_0} g(s,y_s) \sum_{\mathbf{x}_s} e^{\lambda_1(s,\mathbf{x}_s,y_s)} \tag{3.102}$$

and, applying the Bayes theorem we have

$$\begin{aligned} p(\mathbf{x}_s|y_s) &= \frac{p(\mathbf{x}_s,y_s)}{p(y_s)} \\ &= \frac{e^{\lambda_0+\lambda_1(s,\mathbf{x}_s,y_s)} g(s,y_s)}{e^{\lambda_0} g(s,y_s) \sum_{\mathbf{x}_s} e^{\lambda_1(s,\mathbf{x}_s,y_s)}} \\ &= \frac{e^{\lambda_1(s,\mathbf{x}_s,y_s)}}{\sum_{\mathbf{x}_s} e^{\lambda_1(s,\mathbf{x}_s,y_s)}} = \frac{q(\mathbf{x}_s|y_s)}{\sum_{\mathbf{x}_s} q(\mathbf{x}_s|y_s)} \end{aligned} \tag{3.103}$$

because $p(\mathbf{x}_s|y_s) = q(\mathbf{x}_s|y_s)$, and $\lambda_1(s,\mathbf{x}_s,y_s) = \log q(\mathbf{x}_s|y_s)$ when positivity is assumed. Consequently, the resulting model is

$$p(x,y) \approx \prod_{s\in S} q(\mathbf{x}_s|y_s) e^{\sum_{<s,t>} a_0(1-y_s)(1-y_t)+a_1 y_s y_t} \tag{3.104}$$

where $a_0 = \lambda_2(s,t)$ and $a_1 = \lambda_3(s,t)$ are constants, which must be set to satisfy the constraints. Then,

$$p(y|x) \approx \prod_{s\in S} q(\mathbf{x}_s|y_s) p(y) \tag{3.105}$$

$$p(y) = \frac{1}{Z(a_0, a_1)} e^{\left[\sum_{<s,t>} a_0(1-y_s)(1-y_t)+a_1 y_s y_t\right]} \tag{3.106}$$

being $Z(a_0, a_1)$ the normalization (partition) function

$$Z(a_0, a_1) = \sum_y \left\{ e^{\left[\sum_{<s,t>} a_0(1-y_s)(1-y_t)+a_1 y_s y_t\right]} \right\} \tag{3.107}$$

Thus, the prior model $p(y)$ enforces that two neighboring pixels have the same skinness, which discards isolated points and, thus, smooths the result of the classification. Actually, such model is a version of the well-known Potts model.

An interesting property of the latter model is that for any $< s, t >$, we have $p(y_s = 1, y_t = 0) = p(y_s = 0, y_t = 1)$.

3.5.2 Efficient Learning with Belief Propagation

Following an increasing complexity (and more realism) in the proposed models for skin detection, the third one $\mathcal{C}_1 \subset (\mathcal{C}_0 \bigcap \mathcal{D}) \subset \mathcal{C}_0$ consists of imposing the following constraints:

$$\begin{aligned} &\mathcal{C}_1 : \forall < s, t > \in S \times S,\ \forall \mathbf{x}_t, \mathbf{x}_s \in C,\ \forall y_s, y_t \in \{0, 1\} \ : \\ &p(\mathbf{x}_s, \mathbf{x}_t, y_s, y_t) = q(\mathbf{x}_s, \mathbf{x}_t, y_s, y_t) \end{aligned} \tag{3.108}$$

$q(\mathbf{x}_s, \mathbf{x}_t, y_s, y_t)$ being the expected proportion of times in the training set that the two 4-neighboring pixels have the realization $(\mathbf{x}_s, \mathbf{x}_t, y_s, y_t)$ independently on their orientation. Therefore, the ME pdf must satisfy

$$p(\mathbf{x}_s, \mathbf{x}_t, y_s, y_t) = E_p[\delta_{\mathbf{x}_s}(\mathbf{x}_s)\delta_{\mathbf{x}_t}(\mathbf{x}_t)\delta_{y_s}(y_s)\delta_{y_t}(y_t)] \tag{3.109}$$

and, thus, the ME solution is

$$p(\mathbf{x}_s, \mathbf{x}_t, y_s, y_t) \approx e^{\left[\sum_{<s,t>} \lambda(\mathbf{x}_s, \mathbf{x}_t, y_s, y_t)\right]} \tag{3.110}$$

which implies estimating $256^3 \times 256^3 \times 2 \times 2$ Lagrange multipliers when assuming that a color pixel may take 256^3 values. Obviously, it is impossible to evaluate the partition function. The complexity of inference in MRFs is highly influenced by the fact that the neighborhoods $< s, t >$ define a undirected graph. However, if undirectedness is relaxed, a *tree approximation* is chosen. As a tree is a connected graph without loops, any pairwise MRF over a tree can be written in the following terms [123]:

$$p(z) \approx \prod_{<s,t>} \frac{p(z_s, z_t)}{p(z_s)p(z_t)} \prod_{s \in S} p(z_s) \tag{3.111}$$

Thus, setting $z = (x, y)$ we have

$$p(\mathbf{x}, y) \approx \prod_{<s,t>} \frac{q(\mathbf{x}_s, \mathbf{x}_t, y_s, y_t)}{q(\mathbf{x}_s, y_s)q(\mathbf{x}_t, y_t)} \prod_{s \in S} q(\mathbf{x}_s, y_s) \tag{3.112}$$

However, the latter model demands the computation of one 10-dimensional (sparse) histogram and, thus, it is quite prone to overfitting. This problem may be circumvented if we exploit the *color gradient* $\mathbf{x}_t - \mathbf{x}_s$ and apply it to the following approximation:

$$q(\mathbf{x}_s, \mathbf{x}_t | y_s, y_t) \approx q(\mathbf{x}_s | y_s) q(\mathbf{x}_t - \mathbf{x}_s | y_s, y_t) \tag{3.113}$$

whose evaluation requires six histograms of three dimensions. The above simplifications result in the following model:

$$\begin{aligned} &\mathcal{C}^* : \forall \mathbf{x}_t, \mathbf{x}_s \in C, \ \forall y_s, y_t \in \{0, 1\} \ : \\ &p(\mathbf{x}_s, y_s) = q(\mathbf{x}_s, y_s) \\ &p(\mathbf{x}_y, y_t) = q(\mathbf{x}_t, y_t) \\ &p(\mathbf{x}_s - \mathbf{x}_t, y_s, y_t) = q(\mathbf{x}_s - \mathbf{x}_t, y_s, y_t) \end{aligned} \tag{3.114}$$

As the entropy of $p(\mathbf{x}_s, \mathbf{x}_t, y_s, y_s)$ is given by

$$H(p) = - \sum_{\mathbf{x}_s, \mathbf{x}_t, y_s, y_s} p(\mathbf{x}_s, \mathbf{x}_t, y_s, y_s) \log p(\mathbf{x}_s, \mathbf{x}_t, y_s, y_s) \tag{3.115}$$

the ME solution is given by

$$p^*(\mathbf{x}_s, \mathbf{x}_t, y_s, y_t) = \mathcal{P}_\lambda \cap \mathcal{C}^* \tag{3.116}$$

that is, by the pdfs satisfying the above constrains in $\mathcal{C}^*$ and having the form

$$\mathcal{P}_\lambda = \frac{1}{Z_\lambda} e^{[\lambda(\mathbf{x}_s, y_s) + \lambda(\mathbf{x}_s, y_s) + \mathbf{x}_t, y_t) + \lambda(\mathbf{x}_s - \mathbf{x}_t, y_s, y_t)]} \tag{3.117}$$

being $Z_\lambda = \sum_{\mathbf{x}_s, \mathbf{x}_t, y_s, y_s} e^{[\lambda(\mathbf{x}_s, y_s) + \lambda(\mathbf{x}_s, y_s) + \mathbf{x}_t, y_t) + \lambda(\mathbf{x}_s - \mathbf{x}_t, y_s, y_t)]}$ the partition. In the latter formulation, we have reduced significantly the number of multipliers to estimate, but estimation cannot be analytic. The typical solution is to adapt to this context the *iterative scaling* method [47] (we will see an advanced version in the last section on the book, where we will explain how to build ME classifiers). The algorithm is summarized in Alg. 5.

Finally, there is an alternative mechanism based on *belief propagation* [177] using Bethe trees. Such a tree is rooted at each pixel whose color we want to infer, and we have trees $\mathcal{T}_k$ of different depths where k denotes the depth. Here, we assume a 4 neighborhood so that the root generates a child for each of its four neighbors. Then the children generate a node for each of its four

---

Algorithm 5: Iterative Scaling for Marginals

Input: Marginals: $q(\mathbf{x}_s, y_s), q(\mathbf{x}_t, y_t), q(\mathbf{x}_t - \mathbf{x}_s, y_s, y_t)$
Initialize
$\forall \mathbf{x}_t, \mathbf{x}_s \in C,\ \forall y_s, y_t \in \{0,1\}$:
Lambdas: $\lambda(\mathbf{x}_s, y_s) = \lambda(\mathbf{x}_t, y_t) = \lambda(\mathbf{x}_t - \mathbf{x}_s, y_s, y_t) = 0.0$,
repeat
  Marginal: $p(\mathbf{x}_s, y_s) = p(\mathbf{x}_t, y_t) = p(\mathbf{x}_s - \mathbf{x}_t, y_s, y_t) = 0.0$
  Partition function: $Z_\lambda = 0.0$
  foreach $(\mathbf{x}_s, \mathbf{x}_t, y_s, y_t)$ **do**
    Calculate: $g(\mathbf{x}_s, \mathbf{x}_t, y_s, y_t) = e^{[\lambda(\mathbf{x}_s, y_s) + \lambda(\mathbf{x}_s, y_s) + \lambda(\mathbf{x}_s - \mathbf{x}_t, y_s, y_t)]}$
    Update marginals and partition function:
    $p(\mathbf{x}_s, y_s) \leftarrow p(\mathbf{x}_s, y_s) + g(\mathbf{x}_s, \mathbf{x}_t, y_s, y_t)$
    $p(\mathbf{x}_t, y_t) \leftarrow p(\mathbf{x}_t, y_t) + g(\mathbf{x}_s, \mathbf{x}_t, y_s, y_t)$
    $p(\mathbf{x}_t - \mathbf{x}_t, y_s, y_t) \leftarrow p(\mathbf{x}_t - \mathbf{x}_t, y_s, y_t) + g(\mathbf{x}_s, \mathbf{x}_t, y_s, y_t)$
    $Z_\lambda \leftarrow Z_\lambda + g(\mathbf{x}_s, \mathbf{x}_t, y_s, y_t)$
  end
  foreach *Marginal* **do**
    $p(\mathbf{x}_s, y_s) \leftarrow \frac{p(\mathbf{x}_s, y_s)}{Z_\lambda}$
    $p(\mathbf{x}_t, y_t) \leftarrow \frac{p(\mathbf{x}_t, y_t)}{Z_\lambda}$
    $p(\mathbf{x}_t - \mathbf{x}_t, y_s, y_t) \leftarrow \frac{p(\mathbf{x}_t - \mathbf{x}_t, y_s, y_t)}{Z_\lambda}$
  end
  foreach λ **do**
    $\triangle\lambda(\mathbf{x}_s, y_s) = \ln \frac{q(\mathbf{x}_s, y_s)}{p(\mathbf{x}_s, y_s)}$
    $\triangle\lambda(\mathbf{x}_t, y_t) = \ln \frac{q(\mathbf{x}_t, y_t)}{p(\mathbf{x}_t, y_t)}$
    $\triangle\lambda(\mathbf{x}_t - \mathbf{x}_s, y_s, y_t) = \ln \frac{q(\mathbf{x}_t - \mathbf{x}_s, y_s, y_t)}{p(\mathbf{x}_t - \mathbf{x}_s, y_s, y_t)}$
    Update all $\lambda \leftarrow \lambda + \triangle\lambda$
  end
until *convergence of all* $\lambda()$;
Output: The λ parameters of the *marginal model solution* p^* (Eq. 3.116).

---

neighbors that is not yet assigned to a node, and so on. We may have the following general description for a pairwise model:

$$p(y|x) = \prod_{<s,t>} \psi(\mathbf{x}_s, \mathbf{x}_t, y_s, y_t) \prod_{s \in S} \phi(\mathbf{x}_s, y_s) \tag{3.118}$$

where $\psi(\mathbf{x}_s, \mathbf{x}_t, y_s, y_t)$ are pairwise compatibilities and $\phi(\mathbf{x}_s, y_s)$ are individual compatibilities between colors and their labels. The BP propagation algorithm for trees relies on computing k times (k is the tree depth) the following variables called messages

$$m_{ts}(y_s) \leftarrow \sum_{y_s} \phi(\mathbf{x}_s, y_s) \psi(\mathbf{x}_s, \mathbf{x}_t, y_s, y_t) \prod_{u \in \mathcal{N}(t), u \neq s} m_{ut}(y_t) \tag{3.119}$$

that is, messages m_{ab} is sent from a to b by informing about the consistency of a given label. These messages are initialized with value 1. There is one message per value of the label at a given node of the tree. Then, the probability of labeling a pixel as skin at the root of the tree is given by

$$p(y_s = 1|\mathbf{x}_s, s \in \mathcal{T}_k) \approx \phi(\mathbf{x}_s, y_s) \prod_{t \in \mathcal{N}(s)} m_{ts}(y_s) \tag{3.120}$$

This method is quite faster than the Gibss sampler. In Fig. 3.16 we show different skin detection results for the methods explored in this section.

3.6 Integrating Segmentation, Detection and Recognition

3.6.1 Image Parsing

What is *image parsing*? It is more than segmentation and more than recognition [154]. It deals with their unification in order to parse or decompose the input image $\mathbf{I}$ into its constituent patterns, say *texture*, *person* and *text*, and more (Fig. 3.17). Parsing is performed by constructing a *parsing graph* $\mathbf{W}$. The graph is hierarchical (tree-like) in the sense that the root node represents the complete scene and each sibling represents a pattern which that be, in turn, decomposed. There are also horizontal edges between nodes in the same level of hierarchy. Such edges define spatial relationships between patterns. Hierarchical edges represent *generative* (top-down) processes. More precisely, a graph W is composed of the root node representing the entire scene and a set of K siblings (one per pattern). Each of these siblings $i = 1, \ldots, K$ (intermediate nodes) is a triplet of attributes (L_i, ζ_i, Θ_i) consisting of the *shape descriptor* L_i determining the region $R(L_i) = R_i$ (all regions corresponding to the patterns must be disjoint and their union must be the scene); the *type (family) of visual pattern* ζ_i (faces, text characters, and so on); and the *model parameters* Θ_i (see Fig. 3.21). Therefore, the tree is given by $W = (K, \{(L_i, \zeta_i, \Theta_i)\ i = 1, \ldots, K\})$ where K is, of course, unknown (model order selection). Thus, the posterior of a candidate generative solution W is quantified by

$$\begin{aligned} p(W|\mathbf{I}) &= p(\mathbf{I}|W)p(W) \\ &= \prod_{i=1}^{K} p(\mathbf{I}_{R(L_i)}|L_i, \zeta_i, \Theta_i) \left\{ p(K) \prod_{i=1}^{K} p(L_i) p(\zeta_i|L_i) p(\Theta_i|\zeta_i) \right\}, \end{aligned} \tag{3.121}$$

where it is reasonable to assume that $p(K)$ and $p(\Theta_i|\zeta_i)$ are uniform and the term $p(\zeta_i|L_i)$ allows to penalize high model complexity and may be estimated from the training samples (learn the best parameters identifying samples of a given model type, as in the example of texture generation described in Chapter 5). In addition, the model $p(L_i)$, being $L_i = \partial R(L_i)$ the contour, is assumed to decay exponentially with its length and the enclosed area, when

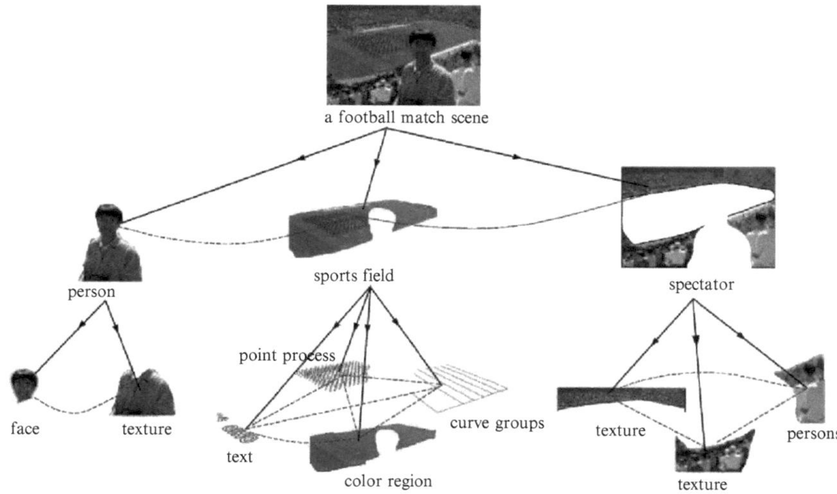

Fig. 3.17. A complex image parsing graph with many levels and types of patterns. (Figure by Tu et al. ©2005 Springer.) See Color Plates.

it is referred to generic visual patterns and faces. However, when considering the shape of an alphabetic character or digit, such model relies on the templates and the allowed deformations. In this latter case, severe rotations and distortions are penalized. In addition, assuming a B-spline representation of the contour, high elastic deformation with respect to the reference template is penalized also by an exponential decay.

On the other hand, several intensity models for computing the likelihood $p(\mathbf{I}_{R(L_i)}|L_i, \zeta_i, \Theta_i)$ are considered. The first one (p_1) is the *constant intensity* modeled by the two parameters of a Gaussian. The second one (p_2) is the nonparametric *clutter/texture model* given by the factorization of intensity frequencies inside a given region (n_i is the number of pixels with intensity value $j \in \{1, \ldots, G\}$, and h_j is the frequency of the jth histogram bin). The third is a *shading model* p_3 where each pixel intensity is characterized by a Gaussian defined over its difference between this intensity and a quadratic from. The same model is used for the 62 characters (including 10 digits and $26 \times 2 = 52$) in both lower and uppercase. Thus, $\mathbf{C} = (5, \ldots, 66)$ and the quadratic form is $J_{\mathbf{p}=(x,y)} = ax^2 + bxy + cy^2 + dx + ey + f$. Finally, the *face model* p_4 is given by a multidimensional Gaussian over the difference between the region and its de-projection over a face eigenspace (principal components $\{u_i\}$, and eigenvectors $(\phi_1, \ldots, \phi_n)$, that is $\mathbf{u} = \sum_{i=1}^{n} u_i \phi_i$) learnt from the samples.

$$
\begin{aligned}
p_1(\mathbf{I}_{R(L)}|L, \zeta = 1, \Theta) &= \prod_{\mathbf{p}\in R(L)} G(\mathbf{I}_\mathbf{p} - \mu; \sigma^2) \,, \quad \Theta = (\mu, \sigma) \\
p_2(\mathbf{I}_{R(L)}|L, \zeta = 2, \Theta) &= \prod_{j=0}^{G} h_j^{n_j} \,, \ \Theta = (h_1, \ldots, h_G) \\
p_3(\mathbf{I}_{R(L)}|L, \zeta \in \{3, \mathbf{C}\}, \Theta) &= \prod_{\mathbf{p}\in R(L)} G(\mathbf{I}_\mathbf{p} - J_\mathbf{p}; \sigma^2) \,, \Theta = (a, \ldots, f, \sigma) \\
p_4(\mathbf{I}_{R(L)}|L, \zeta = 4, \Theta) &= G(\mathbf{I}_{R(L)} - \mathbf{u}; \Sigma) \,, \Theta = (\phi_1, \ldots, \phi_n, \Sigma)
\end{aligned} \tag{3.122}
$$

Under a pure *generative strategy*, the inference of $\mathbf{W} \in \Omega$ may be posed in terms of sampling of the posterior $\mathbf{W} \sim p(\mathbf{W}|\mathbf{I}) \propto p(\mathbf{I}|\mathbf{W})p(\mathbf{W})$, for instance, as $\mathbf{W}^* = \arg\max_{\mathbf{W}\in\Omega} p(\mathbf{W}|\mathbf{I})$. However, the Ω is huge; consider the finite space of all parsing graphs, and imagine the possible types of transitions between each graph (create nodes, delete nodes, and change node attributes). Such hugeness recommends data-driven Markov chains (DDMCMC) to reduce the temporal complexity of the sampling, as we have seen in Section 3.4. Therefore, generative processes should be complemented by *discriminative* (bottom-up) ones for finding subgraphs and propose them to the top-down process in both an intertwined and competitive processing for composing the parts of the image. Discriminative processes are fast, but they are prone to errors and loose the global context. Let $\mathbf{W} = (w_1, \ldots, w_K)$, being w_i the nodes, denote a state of Ω. The *pure generative* approach let us synthesize the input image $\mathbf{I}$ through sampling the posterior $p(\mathbf{W}|I)$. However, *discriminative methods*, instead of providing global posteriors quantifying the probability that the image is generated, give conditional probabilities of how the parts are generated $q(w_j|T_j(\mathbf{I}))$. Part generation is driven by a set of bottom-up tests T_j, like the application of an boosted classifier (as we will see below and also, in more detail, in Chapter 7). Each test T_j relies on a set of local image features $F_{j,n}(\mathbf{I})$, that is, $T_j(\mathbf{I}) = (F_{j,1}(\mathbf{I}), \ldots, F_{j,n}(\mathbf{I}))$, $j = 1, \ldots, K$. Among these features, one may consider *edge cues*, *binarization cues*, *face region cues*, *text region cues*, *shape affinity cues*, *region affinity cues*, *model parameter cues*, and *pattern family cues* among others. Probabilities for edge cues may be learnt from data when using the statistical detectors presented in Chapter 2. Binarization cues are used to propose boundaries of text characters and rely on a binarization algorithm that run with different parameter settings; the discriminative probability is represented nonparametrically by a weighted set of particles. Regarding both face and text region cues, they are learnt with a probabilistic version of the Adaboost [61] algorithm. The output of such probabilistic version is the probability for the presence of a face/text in a region. In the case of text, edge and binarization cues are integrated in order to propose the boundary of characters. More precisely, given a set of training images, for instance with, say, text (the same reasoning follows for faces), one selects windows containing text as positive examples and nontext windows as negative ones; the purpose of Adaboost is to learn a binary-valued *strong classifier* $h(T(\mathbf{I}))$ for test $T(\mathbf{I}) = (h_1(\mathbf{I}), \ldots, h_n(\mathbf{I}))$ composed of n binary-valued *weak* classifiers (their performance is slightly better than chance). Regarding shape-affinity cues and region-affinity ones, they propose matches between shape boundaries and templates, and estimate the likelihood that two regions were generated by the same pattern family and model parameters. Model parameter and pattern family cues are based on clustering algorithms (mean-shift, for instance see Chapter 5), which depend on the model types. For the case of boosting, the hard classifier $T(\mathbf{I})$ can be learnt as a linear combination of the weak ones $h_i(\mathbf{I})$:

$$h_f(T(\mathbf{I})) = sign\left(\sum_{i=1}^{n} h_i(\mathbf{I})\right) = sign(\boldsymbol{\alpha} \cdot T(\mathbf{I})) \tag{3.123}$$

where $\boldsymbol{\alpha} = (\alpha_1, \ldots, \alpha_n)$ is the vector of weights, and n is the number of *selected features* (the elements of T) from a pool or dictionary $\mathcal{D}$ of say $m \geq n$ features (unselected features are assumed to have a zero weight). Then, given a *training set* of labeled images (positive and negative examples, for instance faces, nonfaces), $\mathcal{X} = \{(\mathbf{I}_i, \ell_i) : i = 1, \ldots, M \ \ell_i \in \{+1, -1\}\}$. The Adaboost algorithm (see Chapter 5) optimizes greedily the following cost function:

$$(\boldsymbol{\alpha}^*, T^*) = \arg \min_{\boldsymbol{\alpha}, T \subset \mathcal{D}} \sum_{i=1}^{M} e^{-\ell_i(\boldsymbol{\alpha} \cdot T(\mathbf{I}))} \tag{3.124}$$

that is, an exponential loss is assumed when the proposed hard classifier works incorrectly and the magnitude of the decay is the dot product. The connection between boosting and the learning of discriminative probabilities $q(w_j|T_j(\mathbf{I}))$ is the Friedman's theorem, which states that *with enough training samples* $\mathcal{X} = M$ *selected features* n, *Adaboost selects the weights and tests satisfying*

$$q(\ell = \zeta|\mathbf{I}) = \frac{e^{\zeta(\boldsymbol{\alpha} \cdot T(\mathbf{I}))}}{e^{(\boldsymbol{\alpha} \cdot T(\mathbf{I}))} + e^{-(\boldsymbol{\alpha} \cdot T(\mathbf{I}))}} \tag{3.125}$$

and the strong classifier converges asymptotically to the ratio test

$$h_f(T(\mathbf{I})) = sign(\boldsymbol{\alpha} \cdot T(\mathbf{I})) = sign\frac{q(\ell = +1|\mathbf{I})}{q(\ell = -1|\mathbf{I})} \tag{3.126}$$

Given this theorem, it is then reasonable to think that $q(\ell|T(\mathbf{I}))$ converge to approximations of the marginals $p(\ell|\mathbf{I})$ because a limited number of tests (features) are used. Regarding the training process for Adaboost, it is effective for faces and text. For learning texts, features of different computational complexities are used [38]. The simplest ones are means and variances of intensity and vertical or horizontal gradients, or of gradient magnitudes. More complex features are histograms of intensity, gradient direction, and intensity gradient. The latter two types of features may be assimilated to the statistical edge detection framework (Chapter 2) so that it is straightforward to design a

Fig. 3.18. Results of face and text detection. False positive and negative appear. (Figure by Tu et al. ©2005 Springer.)

weak classifier as whether the log-likelihood ratio between the *text* and *non-text* distributions is above a given threshold or not (here, it is important to consider Chernoff information and obtain peaked *on* empirical distributions as we remark in Prob. 3.12). More complex features correspond, for instance, to edge detection and linkage. When a high number of features are considered, weak classifiers rely either on individual log-likelihood ratios or on ratios over pairwise histograms. Anyway, it is impossible to set the thresholds used in the weak classifiers to eliminate all false positives and negatives at the same time (see Fig. 3.18). In a DDMCMC approach, such errors must be corrected by generative processes, as occurs in the detection of number 9, which will be detected as a shading region and latter recognized as a letter. Furthermore, in order to discard rapidly many parts of the image that do not contain a text/face, a *cascaded classifier* is built. A cascade is a degenerated tree with a classifier at each level; if a candidate region succeeds at a given level, it is passed to the following (deeper) one, and otherwise, is discarded. Considering the number of levels unknown beforehand, such number may be found if one sets the maximum acceptable false positive rate per layer, the minimum acceptance rate per layer, and the false overall positive rate [170]. Furthermore, as the classifiers in each layer may have different computational complexities, it is desirable to allocate the classifiers to the levels by following also this criterion. Although the problem of finding the optimal cascade, given the false positive rate, zero false negatives in the training set, and the average complexity of each classifier, is NP-complete, a greedy (incremental) solution has been proposed in [39]. The rationale of such approach is that, given a maximum time to classify, it is desirable to choose for a given layer the classifier maximizing the expected remaining time normalized by the expected number of regions remaining to be rejected. If the fixed time is not enough to succeed for a given maximum rate of false positives, additional time is used. Anyway, this strategy favors the positioning of simple classifiers at the first levels and speeds up 2.5 times the uniform-time cascade (see results in Fig. 3.19).

Once we have presented both the generative models and the discriminative methods, it is time to present the overall structure of the bidirectional

Fig. 3.19. Results of text detection in a supermarket. Good application for the visually impaired. (Figure by Chen et al. ©2005 IEEE.)

algorithm (see Fig. 3.20). Starting bottom-up, there are four types of computations of $q(w|T(\mathbf{I})$ (one for each type of discriminative task associated to a node w in the tree). The key insight of the DDMCMC is that these computations are exploited by the top-down processes, that is, these generative processes are not fully stochastic. More precisely, the top-down flow, that is the state transitions $\mathbf{W} \rightarrow \mathbf{W}'$, is controlled by a Markov chain $\mathcal{K}(\mathbf{W}, \mathbf{W}')$, which is the core of the Metropolis–Hastings dynamics. In this case, such kernel is decomposed into four subkernels $\mathcal{K}_a : a = 1, \ldots, 4$, each one activated with a given probability $\rho(a, \mathbf{I})$. In turn, each of the subkernels that alters the structure of the parsing tree (all except the model switching kernel and the region-competition kernel moving the borders, which is not included in the figure) is subdivided into two moves $\mathcal{K}_{ar}$ and $\mathcal{K}_{al}$ (the first one for node creation, and the second for node deletion). The corresponding probabilities ρ_{ar} and ρ_{al} are also defined.

3.6.2 The Data-Driven Generative Model

The computational purpose of the main kernel $\mathcal{K}$ with respect to the parse tree is to generate moves $\mathbf{W} \rightarrow \mathbf{W}'$ of three types (node creation, node deletion,

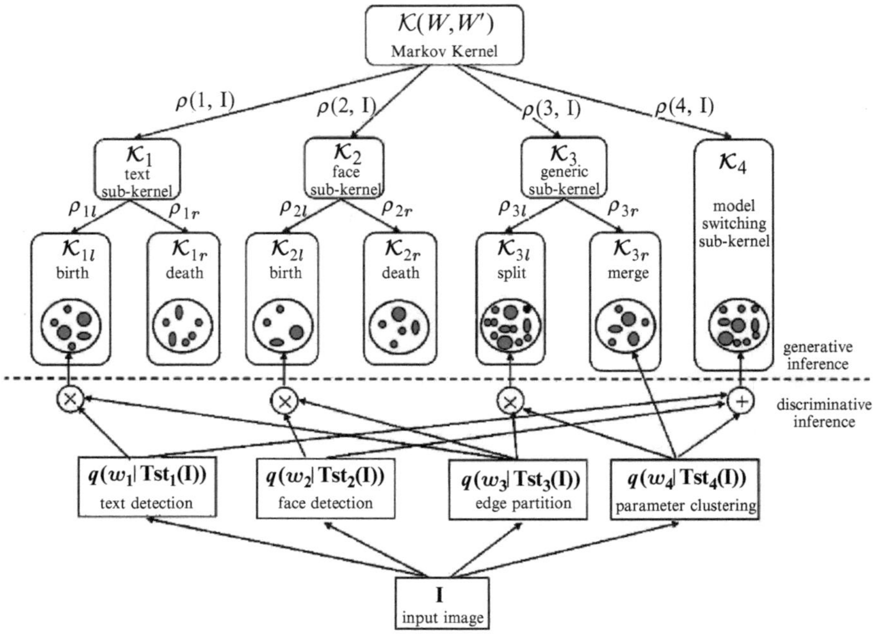

Fig. 3.20. Bidirectional algorithm for image parsing. (Figure by Tu et al. ©2005 Springer.)

and change of node attributes) and drive the search toward sampling the posterior $p(\mathbf{W}|\mathbf{I})$. The main kernel is defined in the following terms:

$$\mathcal{K}(\mathbf{W}'|\mathbf{W}:\mathbf{I}) = \sum_a \rho(a:\mathbf{I})\mathcal{K}_a(\mathbf{W}'|\mathbf{W}:\mathbf{I}) \text{ where } \sum_a \rho(a:\mathbf{I}) = 1 \quad (3.127)$$

and $\rho(a:\mathbf{I}) > 0$. In the latter definition, the key fact is that both the activation probabilities and the subkernels depend on the information in the image $\mathbf{I}$. Furthermore, the subkernels must be *reversible* (see Fig. 3.21) so that the main kernel satisfies such property. Reversibility is important to ensure that the posterior $p(\mathbf{W}|\mathbf{I})$ is the *stationary (equilibrium) distribution*. Thus, keeping in mind that $\mathcal{K}_a(\mathbf{W}'|\mathbf{W}:\mathbf{I})$ is a transition matrix representing the probability of the transition $\mathbf{W} \rightarrow \mathbf{W}'$ (obviously $\sum_{\mathbf{W}'} \mathcal{K}_a(\mathbf{W}'|\mathbf{W}:\mathbf{I}) = 1\ \forall\ \mathbf{W}$), kernels with creation/deletion moves must be grouped into reversible pairs (creation with deletion): $\mathcal{K}_a = \rho_{ar}\mathcal{K}_{ar}(\mathbf{W}'|\mathbf{W}:\mathbf{I}) + \rho_{al}\mathcal{K}_{al}(\mathbf{W}'|\mathbf{W}:\mathbf{I})$, being $\rho_{ar} + \rho_{al} = 1$. With this pairing, it is ensured that $\mathcal{K}_a(\mathbf{W}'|\mathbf{W}:\mathbf{I}) = 0 \Leftrightarrow \mathcal{K}_a(\mathbf{W}|\mathbf{W}':\mathbf{I}) = 1\ \forall\ \mathbf{W}, \mathbf{W} \in \Omega$, and after that pairing $\mathcal{K}_a$ is built in order to satisfy

$$p(\mathbf{W}|\mathbf{I})\mathcal{K}_a(\mathbf{W}'|\mathbf{W}:\mathbf{I}) = p(\mathbf{W}'|\mathbf{I})\mathcal{K}_a(\mathbf{W}|\mathbf{W}':\mathbf{I}) \quad (3.128)$$

which is the so-called *detailed balance equation* [175] whose fulfillment ensures reversibility. If all subkernels are reversible, the main one is reversible too.

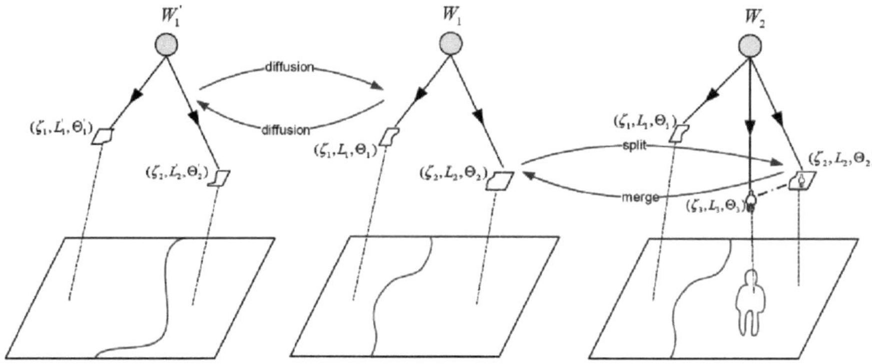

Fig. 3.21. Example of parsing transitions. (Figure by Tu et al. ©2005 Springer.)

Another key property to fulfill is *ergodicity* (it is possible to go from one state to every other state, that is, it is possible to escape from local optima). In this case, ergodicity is ensured provided that enough moves are performed. Reversibility and ergodicity ensure that the posterior $p(\mathbf{W}|\mathbf{I})$ is the invariant probability of the Markov chain. Being $\mu_t(\mathbf{W})$ the Markov chain probability of state $\mathbf{W}$, we have

$$\mu_{t+1}(\mathbf{W}) = \sum_{\mathbf{W}} \mathcal{K}_{a(t)}(\mathbf{W}'|\mathbf{W})\mu_t(\mathbf{W}) \tag{3.129}$$

Thus, given an initial state $\mathbf{W}_0$ with probability $\nu(\mathbf{W}_0)$, it happens that $\mu_t(\mathbf{W})$ approaches the posterior monotonically as time t increases:

$$\mathbf{W} \sim \mu_t(\mathbf{W}) = \nu(\mathbf{W}_0) \cdot [\mathcal{K}_{a(1)} \circ \cdots \circ \mathcal{K}_{a(t)}](\mathbf{W}_0, \mathbf{W}) \rightarrow p(\mathbf{W}|\mathbf{I}) \tag{3.130}$$

Monotonicity is important, but quantifying the *rate of convergence* for a given subkernel is key to measure its effectiveness or usefulness with respect to other subkernels. In this regard, *the Kullback–Leibler divergence between the posterior and the Markov chain state probability for a given kernel* $\mathcal{K}_a$ *decreases monotonically and such decreasing is*

$$\delta(\mathcal{K}_a) = D(p(\mathbf{W}|\mathbf{I})||\mu_t(\mathbf{W})) - D(p(\mathbf{W}|\mathbf{I})||\mu_{t+1}(\mathbf{W})) \geq 0 \tag{3.131}$$

and $\delta(\mathcal{K}_a) > 0$, *being only zero when the Markov chain becomes stationary, that is, when* $p = \mu$. Denoting by $\mu_t(\mathbf{W}_t)$ the state probability at time t, the one at time $t+1$ is given by

$$\mu_{t+1}(\mathbf{W}_{t+1}) = \sum_{\mathbf{W}_t} \mu_t(\mathbf{W}_t)\mathcal{K}_a(\mathbf{W}_{t+1}|\mathbf{W}_t) \tag{3.132}$$

and the joint probability $\mu(\mathbf{W}_t, \mathbf{W}_{t+1})$ may be defined as follows:

$$\mu(\mathbf{W}_t, \mathbf{W}_{t+1}) = \mu_t(\mathbf{W}_t)\mathcal{K}_a(\mathbf{W}_{t+1}|\mathbf{W}_t) = \mu_{t+1}(\mathbf{W}_{t+1})p_{MC}(\mathbf{W}_t|\mathbf{W}_{t+1}) \tag{3.133}$$

being $p_{MC}(\mathbf{W}_t|\mathbf{W}_{t+1})$ the posterior of state $\mathbf{W}_t$ at time t conditioned on state $\mathbf{W}_{t+1}$ at time $t+1$. On the other hand, the joint probability at equilibrium (when $p=\mu$) may be defined, after exploiting the detailed balance equation in the second equality, as

$$p(\mathbf{W}_t,\mathbf{W}_{t+1}) = p(\mathbf{W}_t)\mathcal{K}_a(\mathbf{W}_{t+1}|\mathbf{W}_t) = p(\mathbf{W}_{t+1})\mathcal{K}_a(\mathbf{W}_t|\mathbf{W}_{t+1}) \tag{3.134}$$

Then, the divergence between the latter two joint probabilities can be expressed in terms of $\mathbf{W}_t$:

$$\begin{aligned}
&D(p(\mathbf{W}_t,\mathbf{W}_{t+1})||\mu(\mathbf{W}_t,\mathbf{W}_{t+1}))\\
&= \sum_{\mathbf{W}_{t+1}}\sum_{\mathbf{W}_t} p(\mathbf{W}_t,\mathbf{W}_{t+1})\log\frac{p(\mathbf{W}_t,\mathbf{W}_{t+1})}{\mu(\mathbf{W}_t,\mathbf{W}_{t+1})}\\
&= \sum_{\mathbf{W}_{t+1}}\sum_{\mathbf{W}_t} p(\mathbf{W}_t)\mathcal{K}_a(\mathbf{W}_{t+1}|\mathbf{W}_t)\log\frac{p(\mathbf{W}_t)\mathcal{K}_a(\mathbf{W}_{t+1}|\mathbf{W}_t)}{\mu_t(\mathbf{W}_t)\mathcal{K}_a(\mathbf{W}_{t+1}|\mathbf{W}_t)}\\
&= \sum_{\mathbf{W}_t} p(\mathbf{W}_t)\log\frac{p(\mathbf{W}_t)}{\mu_t(\mathbf{W}_t)}\sum_{\mathbf{W}_{t+1}}\mathcal{K}_a(\mathbf{W}_{t+1}|\mathbf{W}_t)\\
&= D(p(\mathbf{W}_t)||\mu(\mathbf{W}_t))
\end{aligned} \tag{3.135}$$

But also, the divergence may be posed in terms of $\mathbf{W}_{t+1}$:

$$\begin{aligned}
&D(p(\mathbf{W}_t,\mathbf{W}_{t+1})||\mu(\mathbf{W}_t,\mathbf{W}_{t+1}))\\
&= \sum_{\mathbf{W}_{t+1}}\sum_{\mathbf{W}_t} p(\mathbf{W}_{t+1})\mathcal{K}_a(\mathbf{W}_t|\mathbf{W}_{t+1})\log\frac{p(\mathbf{W}_{t+1})\mathcal{K}_a(\mathbf{W}_t|\mathbf{W}_{t+1})}{\mu_{t+1}(\mathbf{W}_{t+1})p_{MC}(\mathbf{W}_t|\mathbf{W}_{t+1})}\\
&= \sum_{\mathbf{W}_{t+1}}\sum_{\mathbf{W}_t} p(\mathbf{W}_{t+1})\mathcal{K}_a(\mathbf{W}_t|\mathbf{W}_{t+1})\log\frac{p(\mathbf{W}_{t+1})}{\mu_{t+1}(\mathbf{W}_{t+1})}\\
&\quad+\sum_{\mathbf{W}_{t+1}}\sum_{\mathbf{W}_t} p(\mathbf{W}_{t+1})\mathcal{K}_a(\mathbf{W}_t|\mathbf{W}_{t+1})\log\frac{\mathcal{K}_a(\mathbf{W}_t|\mathbf{W}_{t+1})}{p_{MC}(\mathbf{W}_t|\mathbf{W}_{t+1})}\\
&= D(p(\mathbf{W}_{t+1})||\mu(\mathbf{W}_{t+1}))\\
&\quad+\sum_{\mathbf{W}_{t+1}} p(\mathbf{W}_{t+1})\sum_{\mathbf{W}_t}\mathcal{K}_a(\mathbf{W}_t|\mathbf{W}_{t+1})\log\frac{\mathcal{K}_a(\mathbf{W}_t|\mathbf{W}_{t+1})}{p_{MC}(\mathbf{W}_t|\mathbf{W}_{t+1})}\\
&= D(p(\mathbf{W}_{t+1})||\mu(\mathbf{W}_{t+1}))\\
&\quad+E_{p(\mathbf{W}_{t+1})}[D(\mathcal{K}_a(\mathbf{W}_t|\mathbf{W}_{t+1})||p_{MC}(\mathbf{W}_t|\mathbf{W}_{t+1}))]
\end{aligned} \tag{3.136}$$

Therefore, we have that

$$\begin{aligned}
\delta(\mathcal{K}_a) &\equiv D(p(\mathbf{W}_t)||\mu(\mathbf{W}_t)) - D(p(\mathbf{W}_{t+1})||\mu(\mathbf{W}_{t+1}))\\
&= E_{p(\mathbf{W}_{t+1})}[D(\mathcal{K}_a(\mathbf{W}_t|\mathbf{W}_{t+1})||p_{MC}(\mathbf{W}_t|\mathbf{W}_{t+1}))] \geq 0
\end{aligned} \tag{3.137}$$

that is, $\delta(\mathcal{K}_a)$ measures the amount of decrease of divergence for the kernel $\mathcal{K}_a$, that is, the *convergence power* of such kernel. It would be interesting to consider this information to speed up the algorithm sketched in Fig. 3.20, where each kernel is activated with probability $\rho(.)$. What is done for the moment is to make the activation probability dependent on the bottom-up processes. For texts and faces, $\rho(a \in 1,2 : \mathbf{I}) = \{\rho(a : \mathbf{I}) + kg(N(\mathbf{I}))/Z$, being $N(\mathbf{I})$ the number of text/faces proposals above a threshold t_a, $g(x) = x, x \leq T_b$, $g(x) = T_b, x \geq T_b$, and $Z = 1 + 2k$ (normalization). For the rest of kernels, there is a fixed value that is normalized accordingly with the evolution of activation probabilities of the two first kernels $\rho(a \in 3,4 : \mathbf{I}) = \rho(a \in 3,4 : \mathbf{I})/Z$.

Once we have specified the design requirements of the sub-kernels and characterized them in terms of *convergence power*, the next step is to design them according to the Metropolis–Hastings dynamics:

$$\mathcal{K}_a(\mathbf{W}'|\mathbf{W} : \mathbf{I}) = Q_a(\mathbf{W}'|\mathbf{W} : T_a(\mathbf{I})) \min\{1, \alpha(\mathbf{W}'|\mathbf{W} : \mathbf{I})\}\ , \mathbf{W}' \neq \mathbf{W} \tag{3.138}$$

being $Q_a(\mathbf{W}'|\mathbf{W} : T_a(\mathbf{I}))$ the *proposal probability* of the transition, and $\alpha(\mathbf{W}'|\mathbf{W} : \mathbf{I})$ the *acceptance probability* defined as

$$\alpha(\mathbf{W}'|\mathbf{W} : \mathbf{I}) = \min\left\{1, \frac{p(\mathbf{W}'|\mathbf{I})Q_a(\mathbf{W}|\mathbf{W}' : T_a(\mathbf{I}))}{p(\mathbf{W}|\mathbf{I})Q_a(\mathbf{W}'|\mathbf{W} : T_a(\mathbf{I}))}\right\} \tag{3.139}$$

The key elements in the latter definition are the proposal probabilities Q_a, which consist of a factorization of several discriminative probabilities $q(w_j|T_j(\mathbf{I}))$ for the elements w_j changed in the proposed transition $\mathbf{W} \rightarrow \mathbf{W}'$. Thus, we are assuming implicitly that Q_a are fast to compute because many of them rely on discriminative process. For the sake of additional global efficiency, it is desirable that Q_a proposes transitions where the posterior $p(\mathbf{W}'|\mathbf{I})$ is very likely to be high, and at the same time, that moves are as larger as possible. Therefore, let $\Omega_a(\mathbf{W}) = \{\mathbf{W}' \in \Omega : \mathcal{K}_a(\mathbf{W}'|\mathbf{W} : \mathbf{I}) > 0\}$ be the *scope*, that is, the set of reachable states from $\mathbf{W}$ in one step using $\mathcal{K}_a$. However, not only large scopes are desired, but also scopes containing states with high posteriors. Under this latter consideration, the proposals should be designed as follows:

$$Q_a(\mathbf{W}'|\mathbf{W} : T_a(\mathbf{I})) \sim \frac{p(\mathbf{W}'|\mathbf{I})}{\sum_{\mathbf{W}''} \in \Omega_a(\mathbf{W}) p(\mathbf{W}''|\mathbf{I})} \quad \text{if } \mathbf{W}' \in \Omega_a(\mathbf{W}) \tag{3.140}$$

and should be zero otherwise. For that reason, the proposals for creating/deleting texts/faces can consist of a *set of weighted particles* (Parzen windows were used in DDMCMC for segmentation in Section 3.4). More precisely, Adaboost, assisted by a binarization process for detecting character boundaries, yields a list of candidate text shapes. Each particle z (shape) is weighted by ω. We have two sets, one for creating and the other for deleting text characters:

$$\begin{aligned} S_{ar}(\mathbf{W}) &= \{(z_{ar}^{(\mu)}, \omega_{ar}^{(\mu)}) : \mu = 1, \ldots, N_{ar}\} \\ S_{al}(\mathbf{W}) &= \{(z_{al}^{(\nu)}, \omega_{al}^{(\nu)}) : \nu = 1, \ldots, N_{al}\} \end{aligned} \tag{3.141}$$

where $a = 1$ for text characters. Weights ω_{ar} for creating new characters are given by a similarity measure between the computed border and the deformable template. Weights ω_{al} for deleting characters are given by their posteriors. The idea behind creating and deleting weights is to approximate the ratios $p(\mathbf{W}'|\mathbf{I})/p(\mathbf{W}|\mathbf{I})$ and $p(\mathbf{W}|\mathbf{I})/p(\mathbf{W}'|\mathbf{I})$, respectively. Anyway, the proposal probabilities given by weighted particles are defined by

$$Q_{ar}(\mathbf{W}'|\mathbf{W}:\mathbf{I}) = \frac{\omega_{ar}(\mathbf{W}')}{\sum_{\mu=1}^{N_{ar}} \omega_{ar}^{(\mu)}}, \quad Q_{al}(\mathbf{W}|\mathbf{W}':\mathbf{I}) = \frac{\omega_{al}(\mathbf{W})}{\sum_{\nu=1}^{N_{al}} \omega_{al}^{(\nu)}} \qquad (3.142)$$

The latter definition is valid for $a = 1, 2, 3, 4$. What changes in each case is the way that particles and weights are built. For faces ($a = 2$), face boundaries are obtained through edge detection, and proposals are obtained in a similar way as in the case of text characters. For $a = 3$ (region split and region merge), the best region for splitting is the worse fitted to its model, and the best regions to merge are the ones with higher affinity in statistical terms (and also where a common border exists!). In terms of splitting, particles are obtained from Canny edge detectors at different scales (levels of details). On the other hand, merging relies on the proposal of removing boundary fragments. Finally, for $a = 4$ (model switching or changing the region type), the proposal probabilities rely on approximating the ratio $p(\mathbf{W}'|\mathbf{I})/p(\mathbf{W}|\mathbf{I})$ through particles. Roughly speaking, the weight is dominated by the ratio between the joint probability of the new label ζ' and new parameters Θ' and the joint probability of current model labels and parameters multiplied by the likelihood of each region given the current label, region label and region parameters. A similar ratio is key in the computation of the particles for split and merge regions, though in the region case, the ratio is dominated by an affinity measure between both regions (see Prob. 3.13).

3.6.3 The Power of Discriminative Processes

The key insight of the DDMCMC approach is exploiting bottom-up knowledge coming from discriminative computations. In the current version of the algorithm, the sequence of application of tests is not optimized at all. However, it is reasonable to think that $q(w_j|T(\mathbf{I})) \rightarrow p(w_j|\mathbf{I})$ (being $T(\mathbf{I})$ the test providing the optimal approximation of the true marginal of the posterior) as new tests are added. In terms of information theory, we should sequence, at least greedily, the tests, so that the maximum amount of information is obtained as soon as possible. Choosing such sequence, by selecting at each time the test with maximum information gain, is the driving mechanism for learning decision trees, as we will see in Chapter 7. Then, it proceeds to quantify the concept of *information gain* in the context of image parsing. Thus, the *information gained* for a variable w by a new test T_+ is *the decrease of*

Kullback–Leibler divergence between $p(w|\mathbf{I})$ and its best discriminative estimate, or the *increase of mutual information* between w and the tests:

$$\begin{aligned} &E_{\mathbf{I}}[D(p(w|\mathbf{I})||q(w|T(\mathbf{I})))] - E_{\mathbf{I}}[D(p(w|\mathbf{I})||q(w|T(\mathbf{I}),T_+(\mathbf{I})))] \\ &\quad = I(w;T(\mathbf{I}),T_+(\mathbf{I})) - I(w;T(\mathbf{I})) \\ &\quad = E_{T,T_+} D(q(w|T(\mathbf{I}),T_+(\mathbf{I}))||q(w|T(\mathbf{I}))) \geq 0 \end{aligned} \tag{3.143}$$

being $E_{\mathbf{I}}$ the expectation with respect to $p(\mathbf{I})$ and E_{T,T_+} the one with respect to the probability of test responses (T,T_+) induced by $p(\mathbf{I})$. The key insight behind the equalities above is twofold. Firstly, the fact that the divergence of $q(.)$ with respect to the marginal $p(.)$ decreases as new tests are added. Secondly, the degree of decreasing of the average divergence yields a useful measure for quantifying the effectiveness of a test with respect to other choices.

Regarding the proof of the equalities in Eq. 3.143, let us start by finding a more compact expression for the difference of expectations $E_{\mathbf{I}}$. For the sake of clarity, in the following we are going to drop the dependency of the tests on $\mathbf{I}$, while keeping in mind that such dependency actually exists, that is, $T = T(\mathbf{I})$ and $T_+ = T_+(\mathbf{I})$:

$$\begin{aligned} &E_{\mathbf{I}}[D(p(w|\mathbf{I})||q(w|T))] - E_{\mathbf{I}}[D(p(w|\mathbf{I})||q(w|T,T_+))] \\ &= \sum_{\mathbf{I}} p(\mathbf{I}) \left[\sum_w p(w|\mathbf{I}) \log \frac{p(w|\mathbf{I})}{q(w|T)}\right] - \sum_{\mathbf{I}} p(\mathbf{I}) \left[\sum_w p(w|\mathbf{I}) \log \frac{p(w|\mathbf{I})}{q(w|T,T_+)}\right] \\ &= \sum_{\mathbf{I}} p(\mathbf{I}) \left[\sum_w p(w|\mathbf{I}) \left\{\log \frac{p(w|\mathbf{I})}{q(w|T)} - \log \frac{p(w|\mathbf{I})}{q(w|T,T_+)}\right\}\right] \\ &= \sum_{\mathbf{I}} p(\mathbf{I}) \left[\sum_w p(w|\mathbf{I}) \log \frac{q(w|T,T_+)}{q(w|T)}\right] \\ &= \sum_{\mathbf{I}} \left[\sum_w p(w,\mathbf{I}) \log \frac{q(w|T,T_+)}{q(w|T)}\right] \end{aligned} \tag{3.144}$$

Next step consists of reducing the difference of mutual informations $I(w;T,T_+) - I(w;T)$ to the final expression in Eq. 3.144. To that end, we exploit the following properties:

$$\begin{aligned} p(\mathbf{I},T) &= p(T|\mathbf{I})p(\mathbf{I}) = p(\mathbf{I}) \\ p(w,\mathbf{I}) &= p(w,\mathbf{I},T,T_+) = q(w,T,T_+)p(x|T,T_+) \\ p(w,\mathbf{I}) &= p(w,\mathbf{I},T) = q(w,T)p(x|T) \end{aligned} \tag{3.145}$$

being x the dimensions of w that are independent of T and T_+. Therefore, the first property is derived by the fact that a test is a deterministic function of the image, and the other two factorize the joint distribution of w and the tests in terms of both the dependent and independent dimensions. Then, we have the following derivation:

$$\begin{aligned}
& I(w;T,T_+) - I(w;T) \\
&= \sum_{T,T_+}\left[\sum_w q(w,T,T_+)\log\frac{q(w,T,T_+)}{q(T,T_+)q(w)}\right] - \sum_T\left[\sum_w q(w,T)\log\frac{q(w,T)}{q(T)q(w)}\right] \\
&= \sum_{T,T_+}\left[\sum_w\left\{q(w,T,T_+)\log\frac{q(w,T,T_+)}{q(T,T_+)q(w)} - q(w,T)\log\frac{q(w,T)}{q(T)q(w)}\right\}\right] \\
&= \sum_{T,T_+}\left[\sum_w\{q(w,T,T_+)\log q(w|T,T_+) - q(w,T)\log q(w|T)\}\right] \\
&= \sum_{T,T_+}\left[\sum_w\left\{\frac{p(w,\mathbf{I})}{q(x|T,T_+)}\log q(w|T,T_+) - \frac{p(w,\mathbf{I})}{q(x|T)}\log q(w|T)\right\}\right] \\
&= \sum_{T,T_+}\left[\sum_w\left\{\frac{p(w,\mathbf{I})}{q(x)}\log q(w|T,T_+) - \frac{p(w,\mathbf{I})}{q(x)}\log q(w|T)\right\}\right] \\
&= \sum_{\mathbf{I}}\frac{1}{q(\mathbf{I}_x)}\left[\sum_w\{p(w,\mathbf{I})\log q(w|T,T_+) - p(w,\mathbf{I})\log q(w|T)\}\right] \\
&= \sum_{\mathbf{I}}\left[\sum_w p(w,\mathbf{I})\log\frac{q(w|T,T_+)}{q(w|T)}\right]
\end{aligned} \tag{3.146}$$

where the change of variables (T,T_+) by $\mathbf{I}$ is bidirectional and it is due to the fact that the tests distribution is induced by $p(\mathbf{I})$. Then, it is possible to sum out $q(x)$, being $\mathbf{I}_x$ the value of x in each image. Given the last expression in Eq. 3.146, and the latter considerations about the change of variables, we have

$$\begin{aligned}
& \sum_{\mathbf{I}}\left[\sum_w p(w,\mathbf{I})\log\frac{q(w|T,T_+)}{q(w|T)}\right] \\
&= \sum_{\mathbf{I}}\left[\sum_w q(w,T,T_+)p(\mathbf{I}_x|T,T_+)\log\frac{q(w|T,T_+)}{q(w|T)}\right] \\
&= \sum_{\mathbf{I}} p(\mathbf{I}_x|T,T_+)\left[\sum_w q(w,T,T_x)\log\frac{q(w|T,T_+)}{q(w|T)}\right] \\
&= \sum_{\mathbf{I}} p(\mathbf{I}_x|T,T_+)\left[\sum_w q(w|T,T_+)q(T,T_+)\log\frac{q(w|T,T_+)}{q(w|T)}\right] \\
&= \sum_{\mathbf{I}} \frac{p(\mathbf{I}_x,T,T_+)}{p(T,T_+)}\left[\sum_w q(w|T,T_+)q(T,T_+)\log\frac{q(w|T,T_+)}{q(w|T)}\right]
\end{aligned}$$

$$
\begin{aligned}
&= \sum_{\mathbf{I}} \frac{p(\mathbf{I}_x, T, T_+)}{p(T, T_+)} \left[\sum_w q(w|T, T_+) q(T, T_+) \log \frac{q(w|T, T_+)}{q(w|T)} \right] \\
&= \sum_{\mathbf{I}} \frac{p(\mathbf{I}_x) p(T, T_+)}{p(T, T_+)} q(T, T_+) \left[\sum_w q(w|T, T_+) \log \frac{q(w|T, T_+)}{q(w|T)} \right] \\
&= \sum_{\mathbf{I}} p(\mathbf{I}_x) q(T, T_+) \left[\sum_w q(w|T, T_+) \log \frac{q(w|T, T_+)}{q(w|T)} \right] \\
&= \sum_{\mathbf{I}} q(T, T_+) \left[\sum_w q(w|T, T_+) \log \frac{q(w|T, T_+)}{q(w|T)} \right] \\
&= \sum_{T, T_+} q(T, T_+) \left[\sum_w q(w|T, T_+) \log \frac{q(w|T, T_+)}{q(w|T)} \right] \\
&= E_{T, T_+} D(q(w|T, T_+) || q(w|T)) \qquad (3.147)
\end{aligned}
$$

As the Kullback–Leibler divergence is nonnegative, its expectation is also nonnegative.

3.6.4 The Usefulness of Combining Generative and Discriminative

As we have seen, discriminative processes are fast but prone to error, whereas generative ones are optimal but too slow. Image parsing enables competitive and cooperative processes for patterns in an efficient manner. However, the algorithm takes 10–20 min to process images with results similar to those presented in Fig. 3.22, where the advantage is having generative models for synthesizing possible solutions. However, additional improvements, like a better management of the segmentation graph, may reduce the overall computation time under a minute.

Bottom-up/top-down integration is not new either in computer vision or in biological vision. For instance, in the classical of Ullman [161], the *counter-stream* structure is a computational model applied to pictorial face recognition, where the role of integrating bottom-up/top-down processes is compensating for image-to-model differences in both directions: from pictures to models (deal with variations of position and scale) and from models to pictures (solve differences of viewing direction, expression and illumination). However, it is very difficult and computational intensive to build a generative model for faces, which takes into account all these sources of variability, unless several subcategories of the same face (with different expressions) are stored and queried as current hypothesis. Other key aspects related to the combination of the recognition paradigms (*pure pictorial recognition* vs *from parts to objects*) and the capability of generalization from a reduced number of views are classical topics both in computer and biological vision (see for instance [152]). Recently emerged schemas like *bag of words* (BoW) are in the beginning of incorporating information-theoretic elements (see for instance [181] where

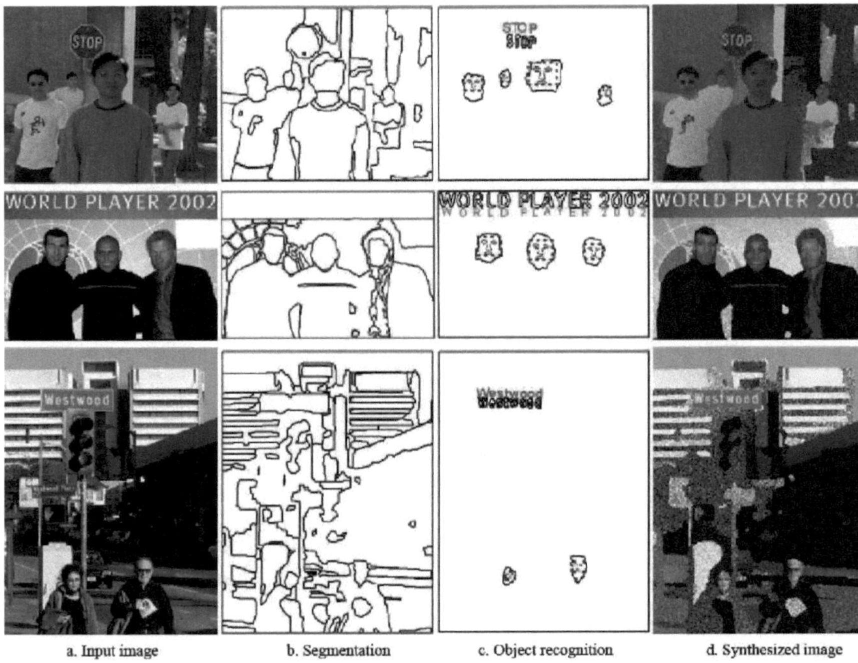

Fig. 3.22. Results of image parsing (*center column*) showing the synthesized images. (Figure by Tu et al. ©2005 Springer.)

information-bottleneck, a clustering criterion to be explained in Chapter 5, is exploited for categorization).

Problems

3.1 Implicit MDL and region competition

The region competition approach by Zhu and Yuille [184] is a good example of *implicit MDL*. The criterion to minimize, when independent probability models are assumed for each region, is the following:

$$E(\Gamma, \{\Theta_i\}) = \sum_{i=1}^{K} \left\{ \frac{\mu}{2} \int_{\partial R_i} ds - \int\int_{R_i} \log p(\mathbf{I}(x,y)|\Theta_i)dxdy + \lambda \right\}$$

where $\Gamma = \cup_{i=1}^{K} \partial R_i$, the first term is the length of curve defining the boundary ∂R_i, mu is the code length for a unit arc (is divided by 2 because each edge fragment is shared by two regions). The second term is the cost of coding each pixel inside R_i the distribution specified by Θ_i. The minimization attending MDL is done in two steps. In the first one, we estimate the optimal parameters by solving Θ_i^* as the parameters are maximizing $\prod_{(x,y)\in R_i} p(\mathbf{I}(x,y)|\Theta_i)$. In the second phase, we re-estimate each contour following the motion equation for the common contour Γ_{ij} between two adjacent regions R_i and R_j:

$$\frac{d\Gamma_{ij}}{dt} = -\mu\kappa_i \mathbf{n}_i + \log\left(\frac{p(\mathbf{I}(x,y)|\Theta_i)}{p(\mathbf{I}(x,y)|\Theta_j)}\right)\mathbf{n}_i$$

where κ_i is the curvature, and $\mathbf{n}_i = -\mathbf{n}_j$ is the normal ($\kappa_i\mathbf{n}_i = \kappa_j\mathbf{n}_j$). Find an analytical expression for the two steps assuming that the regions are characterized by Gaussian distributions. Then, think about the role of the log-likelihood ratio in the second term of the motion equation. For the Gaussian case, give examples of how decisive is the log-ratio when the distributions have the same variance, but closer and closer averages. Hint: evaluate the ratio by computing the Chernoff information.

3.2 Green's theorem and flow
The derivation of energy functions of the form $E(\Gamma) = \int\int_R f(x,y)dxdy$ is usually done by exploiting the Green's theorem, which states that for any vector $(P(x,y), Q(x,y))$ we have

$$\int\int_R \left(\frac{\partial Q}{\partial x} - \frac{\partial P}{\partial y}\right) dxdy = \int_{\partial R} (Pdx + Qdy) = \int_0^l (P\dot{x} + Q\dot{y})ds$$

We must choose P and Q so that $\frac{\partial Q}{\partial x} - \frac{\partial P}{\partial y} = f(x,y)$. For instance, setting

$$Q(x,y) = \frac{1}{2}\int_0^x f(t,y)dt \quad P(x,y) = -\frac{1}{2}\int_0^y f(x,t)dt$$

Using $L(x,\dot{x},y,\dot{y}) = Q(x,y)\dot{y} + P(x,y)\dot{x}$, show that we can write $E(\Gamma) = \int_0^l L(x,\dot{x},y,\dot{y})ds$. Prove also that using the Euler–Lagrange equations we finally find that

$$E_\Gamma(\Gamma) = f(x,y)\mathbf{n}$$

being $\mathbf{n} = (\dot{y}, -\dot{x})$ the normal, and $(P,Q) = (F_1, F_2)$

3.3 The active square
Given an image with a square foreground whose intensities follow a Gaussian distribution with (μ_1, σ_1) over a background also Gaussian (μ_2, σ_2). Initialize a square active polygon close to the convergence point and make some iterations to observe the behavior of the polygon. Test two different cases: (i) quite different Gaussians, and (ii) very similar Gaussians.

3.4 Active polygons and maximum entropy
What is the role of the maximum entropy principle in the active polygons configuration? Why maximum entropy estimation is key in this context? What is the main computational advantage of active polygons vs active contours?

3.5 Jensen–Shannon divergence
Why is the Jensen–Shannon divergence used for discriminating the foreground from the background?

3.6 Jensen–Shannon divergence and active polygons
What is the formal connection between the Jensen–Shannon divergence and the Chen and Vese functional? Think about connections with other kind of functionals.

3.7 Jensen–Shannon divergence and active contours
Establish a formal link between the Jensen–Shannon divergence as a discriminative tool and the observation model used in B-splines-based contour fitting. Think about the impact of replacing the observation model based on parametric distributions (used with B-splines) by a more general model-free criterion. What should be the impact of this change in the MDL formulation?

3.8 Alternative contour representations and MDL
The quantification of complexity used in the MDL-based algorithm for segmenting with contours results in a quite simple expression (firstly depending on the variances, and later assuming that the cost of a parameter is the logarithm of the image dimension). However, if the contour representation changes, it is necessary to reformulate the criterion. Consider, for instance, the alternative frequency-based representation of the contour in terms of a Fourier transform (low frequencies retain global aspects and high frequencies retain details). Think about reformulating the contour fitting problem using this kind of representation. Hint: use Fourier descriptors for describing the contour. Evaluate the relative flexibility of each representation (B-splines vs Fourier descriptors) to adapt to complex contours that are not uniformly smooth.

3.9 Jump-diffusion: energy function
The energy function guides the jump-diffusion process through Ω. Is there a parameter which controls the complexity of the generated solutions? Look for it in Eq. 3.68. What would happen if there was no parameter for limiting the complexity of the models?

3.10 Jump-diffusion: models and stochastic diffusions
Define the equations of an arc model. Which parameters ψ are needed? Derive the motion equations $d\psi(t)$ of the parameters (see Eq. 3.74).

3.11 Jump-diffusion: distance between solutions in K-adventurers
Propose a new distance between solutions $W_1 - W_2$. What properties are important in this distance measure? Is the distance measure dependent on the region models of the problem? Is it possible do define an independent one?

3.12 Using Chernoff information in discriminative learning
In the boosting learning process used for learning discriminative probabilities for text, consider the following types of features: horizontal, vertical, and diagonal edges, and their statistical edge models (including, for instance, gradient strength and angle) based on being *on* and *off* an edge of

the latter types (Chapter 1). Therefore, the weak classifiers are of the form $h_i(\mathbf{I}) = \log \frac{p_{\text{text}}}{p_{\text{non-text}}} > t_i$. As we will see in Chapter 7, the α_i are computed sequentially (greedily) within Adaboost (choosing a predefined order). What is more important is that such values depend on the effectiveness of the classifier h_i (the higher the number of misclassified examples, the lower becomes α_i). This virtually means that a classifier with $\alpha_i = 0$ is not selected (has no impact in the strong classifier). In this regard, what types of features will be probably excluded for text detection?

Considering now the thresholds t_i, incrementing them usually leads to a reduction of false positives, but also to an increment of false negatives. A more precise threshold setting would depend on the amount of overlap of the *on* and *off* distributions, that is, Chernoff information. How would you incorporate this insight into the current threshold setting?

3.13 Proposal probabilities for splitting and merging regions
Estimate the weights ω_{3r} and ω_{3l} for splitting and merging regions in the image parsing approach. Consider that these weights approximate (efficiently) the ratios $p(\mathbf{W}|\mathbf{I})/p(\mathbf{W}'|\mathbf{I})$ and $p(\mathbf{W}'|\mathbf{I})/p(\mathbf{W}|\mathbf{I})$, respectively. Consider that in a splitting move, region R_k existing in state $\mathbf{W}$ is decomposed into R_i and R_j in $\mathbf{W}$, whereas two adjacent regions R_i and R_j existing in $\mathbf{W}'$ are fused into R_k. Remember that in the approximation, a key ratio is the one between the compatibility measure of R_i and R_j and the likelihood of the R_k region for the split case, and the one of the two regions (assumed independent) for the fusion case.

3.14 Markov model and skin detection
What is the main motivation of using a Markov random field for modeling the skin color in images? How is the estimated extra computational cost of using this kind of models?

3.15 Maximum Entropy for detection
The maximum entropy principle is introduced in the context of skin color detection in order to introduce pairwise dependencies independently of the orientation of two 4-neighbors. Explain how the original ME formulation evolves to the consideration of color gradients and how the ME algorithm estimates the Lagrange multipliers.

3.7 Key References

- M. Figueiredo, J. Leitão, and A.K. Jain. "Unsupervised Contour Representation and Estimation Using B-splines and a Minimum Description Length Criterion". *IEEE Transactions on Image Processing* 9(6): 1075–1087 (2000)
- G. Unal, A. Yezzi, and H. Krim. "Information-Theoretic Active Polygons for Unsupervised Texture Segmentation". *International Journal of Computer Vision* 62(3):199–220 (2005)

- G. Unal, H. Krim, and A. Yezzi. “Fast Incorporation of Optical Flow into Active Polygons”. *IEEE Transactions on Image Processing* 14(6): 745–759 (2005)
- B. Jedynak, H. Zheng, and M. Daoudi. “Skin Detection Using Pairwise Models”. *Image and Vision Computing* 23(13): 1122–1130 (2005)
- Z.W. Tu and S.C. Zhu. “Image Segmentation by Data-Driven Markov Chain Monte Carlo”. *IEEE Transactions on Pattern Analysis and Machine Intelligence* 24(5): 657–673 (2002)
- Z.W. Tu, X.R. Chen, A.L. Yuille, and S.C. Zhu. “Image parsing: Unifying Segmentation, Detection and Recognition”. *International Journal of Computer Vision* 63(2):113–140 (2005)
- S.C. Zhu and A.L. Yuille. “Region Competition: Unifying Snakes, Region Growing, and Bayes/MDL for Multiband Image Segmentation”. *IEEE Transactions on Pattern Analysis and Machine Intelligence*, 18(9): 884–900, (1996)
- S. Geman and D. Geman. “Stochastic Relaxation, Gibbs Distributions, and the Bayesian Restoration of Images”. *IEEE Transactions on Pattern Analysis and Machine Intelligence*, 6 (pp. 721–741), (1984).
- X. Chen and A.L. Yuille. “Time-Efficient Cascade for Real Time Object Detection”. *First International Workshop on Computer Vision Applications for the Visually Impaired. Proceedings of IEEE Conference on Computer Vision and Pattern Recognition*, San Diego, CA (2004)
- G. Winkler. *Image Analysis, Random Fields and Markov Chain Monte Carlo Methods: A Mathematical Introduction.* Springer, New York (2003)

4

Registration, Matching, and Recognition

4.1 Introduction

This chapter mainly deals with the way in which images and patterns are compared. Such comparison can be posed in terms of registration (or alignment). Registration is defined as the task of finding the minimal amount of transformation needed to transform one pattern into another (as much as possible). The computational solution to this problem must be adapted to the kind of patterns and to the kind of transformations allowed, depending on the domain. This also happens with the metric or similarity measure used for quantifying the amount of transformation. In both cases, information theory plays a fundamental role, as we will describe along the present chapter.

In the case of images, it is reasonable to exploit the statistics of intensities. The concept of mutual information (MI), which quantifies statistical dependencies between variables, is a cornerstone here because such variables are instantiated to intensity distributions. Therefore, image registration can be posed as finding the (constrained) transformation that holds the maximal dependency between distributions. This rationale opens the door to the quest for new measures rooted in mutual information.

In the case of contours, registration stands for finding minimal deformations between the input and the target contours. Here, the space of deformations is less constrained, but these should be as smooth as possible. It is possible to represent a shape as a mixture or a distribution of keypoints. Such representation enables the use of other types of information theoretic measures, like the Jensen–Shannon (JS) divergence. Furthermore, it is possible to exploit interesting information geometry concepts, like the Fisher–Rao metric tensor.

Finally, in the case of structural patterns, like graphs, information theory is a key for driving the unsupervised learning of structural prototypes. For instance, binary shapes can be described by a tree-like variant of the skeleton, known as shock-tree. Then, shape information is collapsed into the tree, and shape registration can be formulated as tree registration using a proper tree distance. What is more important is that the MDL principle can be applied to find a prototypical tree for each class of shape.

4.2 Image Alignment and Mutual Information

4.2.1 Alignment and Image Statistics

In computer vision, alignment refers to finding the pose of an object in an image. The object can be represented as an image, given some imaging model and a pose of the object. The generated image has to be compared to the actual image in order to decide whether the pose is the correct one. One simple example is the alignment of two similar 2D images. Another example of an alignment problem is the estimation of the pose of a 3D model, given its 2D representation. Similarly, in medical imaging, multispectral images have to be aligned to 2D images, or to a 3D model.

The object's imaging model can be denoted as $u(x)$, and its image as $v(y)$. The model generates the image, given an imaging function and a noise.

$$v(T(x)) = F(u(x), q) + \eta \tag{4.1}$$

The image corresponds to a transformation T of the object, given the imaging function $F(u(x), q)$ and given a noise η. The imaging function not only depends on the model of the object, but also on some exogenous influences that are represented by the vector q.

Modeling the imaging function F is usually unfeasible and q is unknown. However, it is possible to find a consistent alignment of two images without knowing F and q. We can assume that the most consistent alignment is the one that produces less discontinuities in the relation between the two images. A way to evaluate the consistency $C(T)$ of some particular transformation T is to calculate the following sum, over points x_a and x_b, from the model:

$$C(T) = -\sum_{x_a \neq x_b} G_\psi \left(u(x_b) - u(x_a)\right) \left(v(T(x_b)) - v(T(x_a))\right)^2 \tag{4.2}$$

for a fixed standard deviation ψ of the Gaussian G_ψ. In this measure, T is considered consistent if points that have similar values in the model project to similar values in the image: $|u(x_a) - u(x_b)|$ and $|u(T(x_a)) - u(T(x_b))|$. The drawback of constancy maximization is that the most consistent transformation would be the one that matches the points of the model to a constant region of the image. Also, the assumption that the image is a function of the model may not always be useful.

If we do not assume the existence of any F, we can still assume that the best alignment is the one in which the model best predicts the image. Entropy is a concept that is related to predictability. The most predictable a random variable is, the lowest its entropy. In the alignment problem, the conditional entropy of the transformated image, given the model, has to be minimized by searching for a T along the space of transformations, Ω:

$$\arg\min_{T\in\Omega} H\big(v(T(x))|u(x)\big) \tag{4.3}$$

However, conditional entropy also has the constancy problem: it would be low for an image $v(T(x))$, which is predictable from the model $u(x)$, and it would also be low for an image that is predictable by itself, which is the case of constant images.

Adding a penalization to simple images solves the constancy problem. The first term of the following minimization objective is the conditional entropy. The second term awards images with a higher entropy than images with low entropy:

$$\arg\min_{T\in\Omega}\Big(H\big(v(T(x))|u(x)\big) - H\big(v(T(x))\big)\Big) \tag{4.4}$$

Given that $H(v(T(x))|u(x)) = H(v(T(x)), u(x)) - H(u(x))$, the latter formula (Eq. 4.4) corresponds with mutual information maximization:

$$\arg\min_{T\in\Omega}\Big(H\big(v(T(x)), u(x)\big) - H\big(u(x)\big) - H\big(v(T(x))\big)\Big) \tag{4.5}$$

$$= \arg\max_{T\in\Omega}\Big(H\big(u(x)\big) + H\big(v(T(x))\big) - H\big(v(T(x)), u(x)\big)\Big) \tag{4.6}$$

$$= \arg\max_{T\in\Omega} I\big(u(x), v(T(x))\big) \tag{4.7}$$

The third term in Eq. 4.6, $H(v(T(x)), u(x))$, is the one that awards transformations for which u better explains v. The term $H(v(T(x)))$ contributes to select transformations that make the model u correspond with more complex parts of the image v. The term $H(u(x))$ remains constant for any T, and so it does not condition the alignment.

An alignment example illustrating the meaning of conditional entropy can be seen in Fig. 4.1. In this simple example, the images were obtained from the same sensor, from two slightly different positions. Therefore, due to the differences of perspective, alignment cannot be perfect. Two cases are represented: a misalignment and a correct alignment. For the first case, the joint histogram is much more homogeneous than for the correct alignment, where the joint histogram is tightly concentrated along the diagonal. For more complex cases, involving different sensors, it is worth noting that even though the representation of the model and the actual image could be very different, they will have a higher mutual information when the alignment is the correct one.

A major problem of the latter approach is the estimation of entropy. Viola and Wells III [171] explain in detail the use of mutual information for alignment. They also present a method for evaluating entropy and mutual information called Empirical entropy manipulation and analysis (EMMA), which optimizes entropy, based on a stochastic approximation. EMMA uses the Parzen window method, which is presented in the next section.

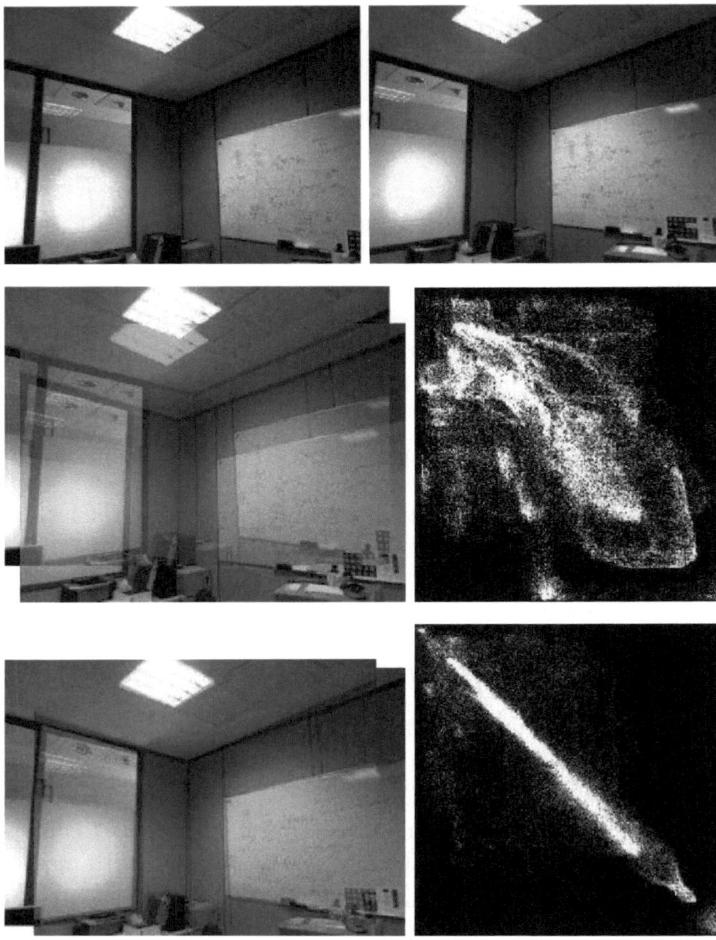

Fig. 4.1. Alignment problem example: *Top:* images obtained from the same sensor, from two slightly different positions. *Center-left:* a misalignment. *Center-right:* joint histogram of the misalignment. *Bottom-left:* a correct alignment. *Bottom-right:* joint histogram of the alignment.

4.2.2 Entropy Estimation with Parzen's Windows

The Parzen's windows approach [51, 122] is a nonparametric method for estimating probability distribution functions (pdfs) for a finite set of patterns. The general form of these pdfs is

$$P^*(Y, a) \equiv \frac{1}{N_a} \sum_{y_a \in a} K_\psi(y - y_a) \tag{4.8}$$

where a is a sample of the variable Y, N_a is the size of the sample, and $K(.)$ is a kernel of width ψ and centered in y_a. This kernel has to be a differentiable function, so that the entropy H can be derived for performing a gradient descent over it. A Gaussian kernel is appropriate for this purpose. Also, let us assume that the covariance matrix ψ is diagonal, that is, $\psi = Diag(\sigma_1^2, ..., \sigma_{N_a}^2)$. This matrix indicates the widths of the kernel, and by forcing ψ to be diagonal, we are assuming independence among the different dimensions of the kernel, which simplifies the estimation of these widths. Then, the kernel is expressed as the following product:

$$K_\psi(y - y_a) = \frac{1}{\prod_{i=1}^{d} \sigma_i (2\pi)^{d/2}} \prod_{j=1}^{d} \exp\left\{ -\frac{1}{2} \left(\frac{y^j - y_a^j}{\sigma_j} \right)^2 \right\} \tag{4.9}$$

where y^j represents the jth component of y, and y_a^j represents the jth component of kernel y_a. The kernel widths ψ are parameters that have to be estimated. In [171], a method is proposed for adjusting ψ using maximum likelihood.

The entropy of a random variable Y is the expectation of the negative logarithm of the pdf:

$$\begin{aligned} H(Y) &\equiv -E_Y[\log(p(Y))] \\ &\approx -\frac{1}{N} \sum_{y \in Y} \log(P(y_b)) = -\frac{1}{N_b} \log(\ell(b)) \end{aligned} \tag{4.10}$$

where b refers to the samples used for estimating entropy, and $\ell(b)$ is their likelihood.

In the alignment problem, the random variable Y has to be expressed as a function of a set of parameters T. The derivative of H with respect to T is

$$\frac{\partial}{\partial T} H^*(Y(T))$$

$$\approx \frac{1}{N_b} \sum_{y_b \in b} \sum_{y_a \in a} \frac{K_\psi(y_b - y_a)}{\sum_{y_a \in a} K_\psi(y_b - y_a)} (y_b - y_a)^T \psi^{-1} \frac{\partial}{\partial T} (y_b - y_a) \tag{4.11}$$

The first factor of the double summation is a weighting factor that takes a value close to one, if y_a is much closer to y_b than to the elements of a. However, if there is an element in which a is similar to y_a, the weighting factor approaches zero. Let us denote the weighting factor as W_Y, which assumes a diagonal matrix ψ_Y of kernel widths for the random variable Y:

$$W_Y(y_b, y_a) \equiv \frac{K_{\psi_Y}(y_b - y_a)}{\sum_{y_a \in a} K_{\psi_Y}(y_b - y_a)} \tag{4.12}$$

The weighting factor helps find a transformation T, which reduces the average squared distance between those elements that are close to each other, forming clusters.

4.2.3 The EMMA Algorithm

The abbreviation EMMA means "Empirical entropy Manipulation and Analysis." It refers to an aligment method proposed by Viola and Wells [171], and is based on mutual information maximization. In order to maximize the mutual information (Eq. 4.6), an estimation to its derivative with respect to T is needed.

$$\frac{\partial}{\partial T} I(u(x), v(T(x))) = \frac{\partial}{\partial T} H(v(T(x))) - \frac{\partial}{\partial T} H(u(x), v(T(x))) \tag{4.13}$$

$$\approx \frac{1}{N_b} \sum_{x_b \in b} \sum_{x_a \in a} (v_b - v_a)^T \left[W_v(v_i, v_j)\psi_v^{-1} - W_{uv}(w_i, w_j)\psi_{vv}^{-1} \right] \frac{\partial}{\partial T}(v_b - v_a) \tag{4.14}$$

Here, v_i denotes $v(T(x_i))$ and w_i denotes $[u(x_i), v(T(x_i))]^T$ for the joint density. Also, the widths of the kernels for the joint density are block diagonal: $\psi_{uv}^{-1} = Diag(\psi_{uu}^{-1}, \psi_{vv}^{-1})$. In the last factor, $(v(T(x_b)) - v(T(x_a)))$ has to be derived with respect to T, so the expression of the derivative depends on the kind of transformation involved.

Given the derivatives, a gradient descent can be performed. The EMMA algorithm performs a stochastic analog of the gradient descent. The stochastic approximation avoids falling into local minima. The following algorithm proved to find successful alignments for sample sizes N_a, N_b of about 50 samples.

$A \leftarrow N_a$ samples selected for Parzen
$B \leftarrow N_b$ samples selected for estimating entropy
Repeat:
  $T \leftarrow T + \lambda \frac{\partial}{\partial T} I^*$
  Estimate I_t for the current width ψ
Until: $|I_t - I_{t-1}| < \mu$

In the above algorithm, λ is the learning rate and μ is a threshold which indicates that convergence is achieved. $\frac{\partial}{\partial T} I^*$ is the approximation of the derivative of the mutual information (Eq. 4.14), and I_t and I_{t-1} are the mutual information estimations in the iteration t and in the previous one.

In Fig. 4.2, we represent two search surfaces generated by translating two images vertically and horizontally. The first surface is the resulting Normalized Cross Correlation of the alignments, while the second surface is the one resulting from mutual information estimation. The advantage of mutual information is that it produces smaller peaks. Therefore, the search algorithm will more easily avoid local maxima. Viola and Wells present some [171] illustrative alignment experiments. Apart from view-based recognition experiments with 2D images, they also present the alignment of a 3D object model to a

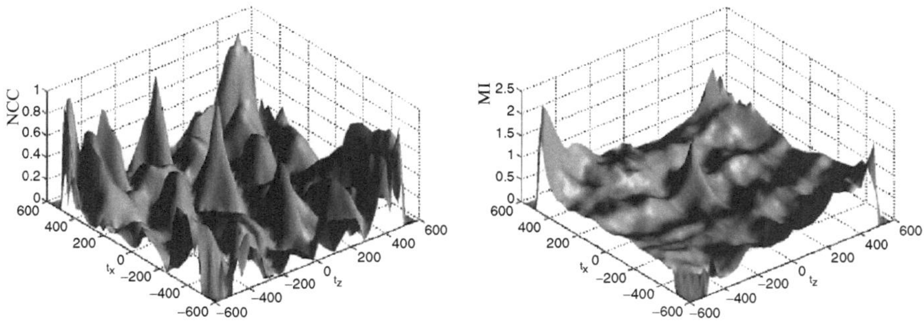

Fig. 4.2. Normalized Cross Correlation (*left*) and Mutual Information (*right*) of image alignments produced by translations along both axes. It can be observed that Mutual Information has smaller local maxima, then the maximum (in the *center* of the plot) is easier to find with a stochastic search. Figure by courtesy of J.M. Sáez.

2D image, as well as MRI (magnetic resonance imaging) alignment to data from another sensor, which is useful for medical applications.

4.2.4 Solving the Histogram-Binning Problem

Mutual information is very useful in computer vision as a similarity measure, given that it is insensitive to illumination changes. Mutual information (Eq. 4.15) can be calculated given a marginal entropy and the conditional entropy, or given the marginal entropies (Eq. 4.16) and the joint entropy (Eq. 4.17). The latter involves the estimation of joint probabilities.

$$
\begin{aligned}
I(X,Y) &= H(X) + H(Y) - H(X,Y) && (4.15)\\
&= -\sum_x p_x(x)\log p_x(x) - \sum_y p_y(y)\log p_y(y) && (4.16)\\
&\quad + \sum_x\sum_y p_{xy}(x,y)\log p_{xy}(x,y) && (4.17)
\end{aligned}
$$

The classical way to estimate the joint probability distributions is to calculate the joint histogram. A joint histogram of two images is a 2D matrix of $B \times B$ size, where each dimension refers to one of the images and B is the discretization used along the entire scale of intensities. For example, five bins along the scale, which goes from 0 to 255, would produce the intervals from 0 to 51, from 51 to 102, etc. Each histogram cell c_{ij} counts the times at which two pixels with the same coordinates on both images have intensity i on the first image and intensity j on the second image, as illustrated in Fig. 4.3.

The joint histogram represents an estimation of a joint distribution. For tuning the estimation, a crucial parameter is the number of bins B. Let us see an example. In Fig. 4.4, we can see three different joint histograms of

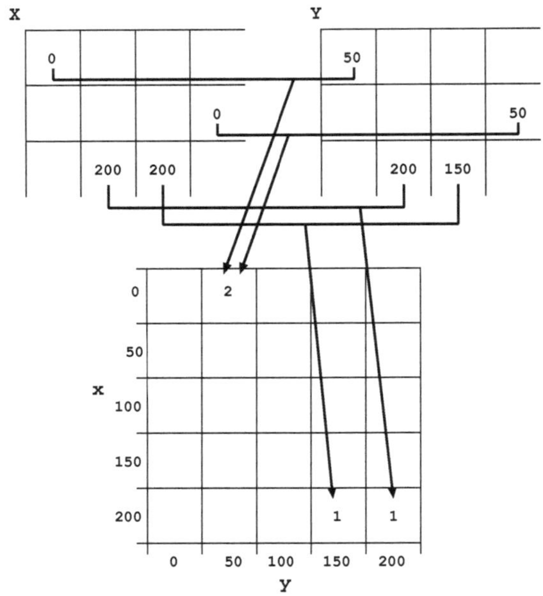

Fig. 4.3. How a classical joint histogram is built. Each cell of the histogram counts the times which some definite combination of intensities happens between two pixels in the two images, only for those pixels which have the same coordinates.

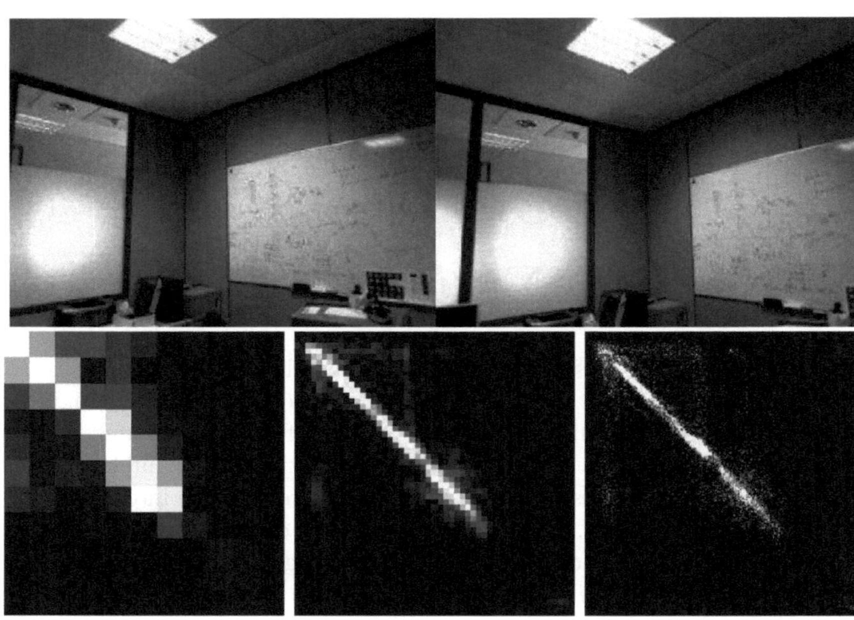

Fig. 4.4. Joint histograms of two different photographs which show the same view of a room. The numbers of bins of the histograms are 10 (*left*), 50 (*center*), and 255 (*right*).

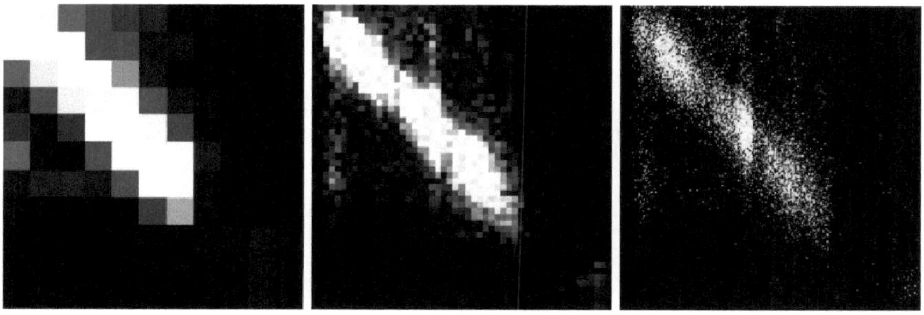

Fig. 4.5. The joint histograms, the photographs of Fig. 4.5, with the addition of Gaussian noise to one of the images. The numbers of bins of the histograms are 10 (*left*), 50 (*center*), and 255 (*right*).

two images, which show the same scene from a very similar point of view. The first histogram has $B = 10$, the second one $B = 50$, and the third one $B = 255$ bins. We can observe that 10 bins do not provide enough information to the histogram. However, 255 bins produce a very sparse histogram (the smaller the image is, the sparser the histogram would be). On the other hand, a large number of bins make the histogram very vulnerable to noise in the image. We can see the histograms of the same images with Gaussian noise in Fig. 4.5.

To deal with the histogram binning problem, Rajwade et al. have proposed a new method [133,134], which considers the images to be continuous surfaces. In order to estimate the density of the distribution, a number of samples have to be taken from the surfaces. The amount of samples determines the precision of the approximation. For estimating the joint distribution of the two images, each point of the image is formulated as the intersection of two-level curves, one-level curve per image. In a continuous surface (see Fig. 4.6), there are infinite level curves. In order to make the method computable in a finite time, a number Q of intensity levels has to be chosen, for example, $Q = 256$. A lower Q would proportionally decrease the precision of the estimation because the level curves of the surface would have a larger separation, and therefore, some small details of the image would not be taken into consideration.

In order to consider the image as a continuous surface, sub-pixel interpolation has to be performed. For the center of each pixel, four neighbor points at a distance of half pixel are considered, as shown in Fig. 4.7a. Their intensities can be calculated by vertical and horizontal interpolation. These four points form a square that is divided in two triangles, both of which have to be evaluated in the same way. For a definite image, for example, I_1, the triangle is formed by three points $p_i, i \in \{1, 2, 3\}$ with known positions (x_{p_i}, y_{p_i}) and known intensities z_{p_i}. Assuming that the intensities within a triangle are represented as a planar patch, it would be given by the equation

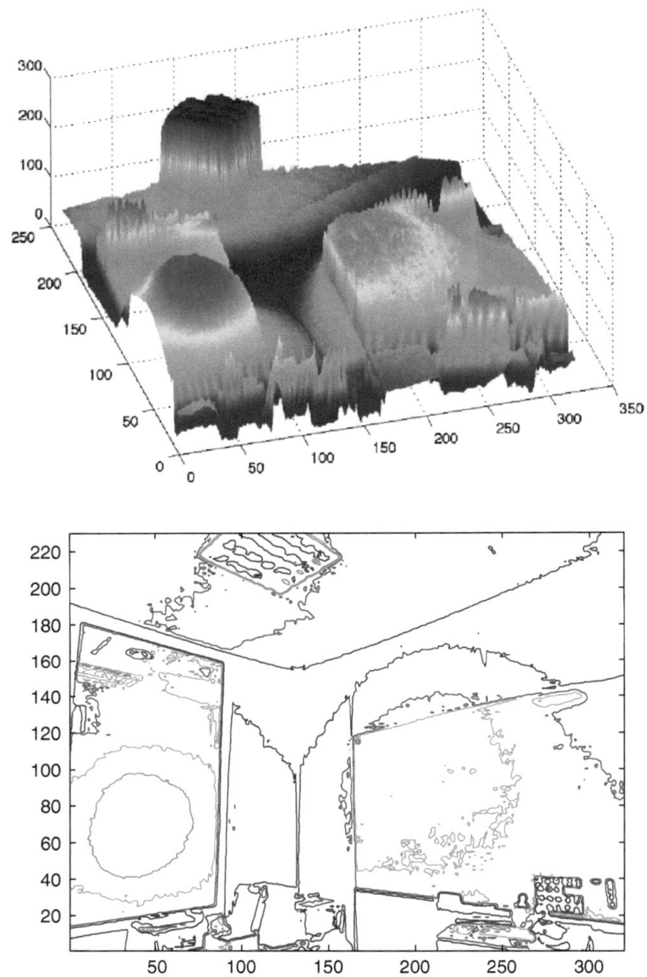

Fig. 4.6. *Top*: the continuous surface formed by the intensity values of the laboratory picture. *Bottom*: some of the level curves of the surface.

$z = A_1x + B_1y + C_1$ in I_1. The variables A_1, B_1, and C_1 are calculated using the system of equations given by the three points of the triangle:

$$\left.\begin{aligned} z_{p_1} &= A_1x_{p_1} + B_1y_{p_1} + C_1 \\ z_{p_1} &= A_1x_{p_1} + B_1y_{p_1} + C_1 \\ z_{p_1} &= A_1x_{p_1} + B_1y_{p_1} + C_1 \end{aligned}\right\} A_1 =?, B_1 =?, C_1 =? \tag{4.18}$$

For each triangle, once we have its values of A_1, B_1, and C_1 for the image I_1 and A_2, B_2, and C_2 for the image I_2, we can decide whether to add a vote for a pair of intensities (α_1, α_2). Each one of them is represented as a straight

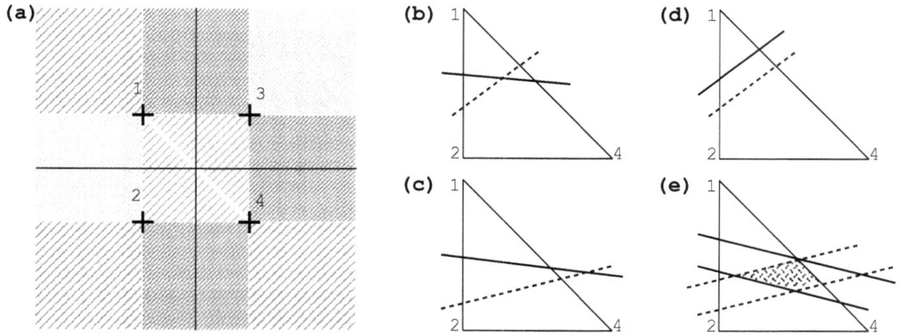

Fig. 4.7. (**a**) Subpixel interpolation: the intensities of four points around a pixel are calculated. The square formed is divided in two triangles. (**b**),(**c**) The iso-intensity lines of I_1 at α_1 (*dashed line*) and of I_2 at α_2 (*continuous line*) for a single triangle. In the first case (**b**), they intersect inside the triangle, so we vote for $p(\alpha_1, \alpha_2)$. In the cases (**c**) and (**d**), there is no vote because the lines do not intersect in the triangle; (**d**) is the case of parallel gradients. In (**e**), the iso-surfaces are represented, and their intersection area in the triangle represents the amount of vote for the pair of intensity ranges.

line in the triangle, as shown in Fig. 4.7b and c. The equations of these lines are given by A,B, and C:

$$\left.\begin{array}{l} A_1x + B_1y + C_1 = \alpha_1 \\ A_2x + B_2y + C_2 = \alpha_2 \end{array}\right\} x = ?, y = ? \tag{4.19}$$

The former equations form a system, which can be solved to obtain their intersection point (x, y). If it results to be inside the area of the triangle (p_1, p_2, p_3), a vote is added for the pair of intensities (α_1, α_2). All the triangles in the image have to be processed in the same way, for each pair of intensities. Some computational optimizations can be used in the implementation, in order to avoid repeating some calculations.

The results of this first approach can be observed in Figs. 4.8 and 4.9 (third row). This method, however, has the effect of voting more to zones with higher gradient. When the gradients are parallel, there is no vote, see Fig. 4.7d. In [134], Rajwade et al. have presented an improved method that, instead of counting intersections of iso-contours, sums intersection areas of iso-surfaces. A similar approach to density estimation had been previously taken by Kadir and Brady [90] in the field of image segmentation. An example of intersecting iso-surfaces is shown in Fig. 4.7e. When the gradients are parallel, their intersection with the triangle is still considered. In the case in which one image has zero gradient, the intersection of the iso-surface area of the other image is the considered area. This approach produces much more robust histograms. Examples can be observed in Figs. 4.8 and 4.9 (fourth row). The difference among the different methods is more emphasized in the images of the laboratory because the images are poorly textured, but have many edges.

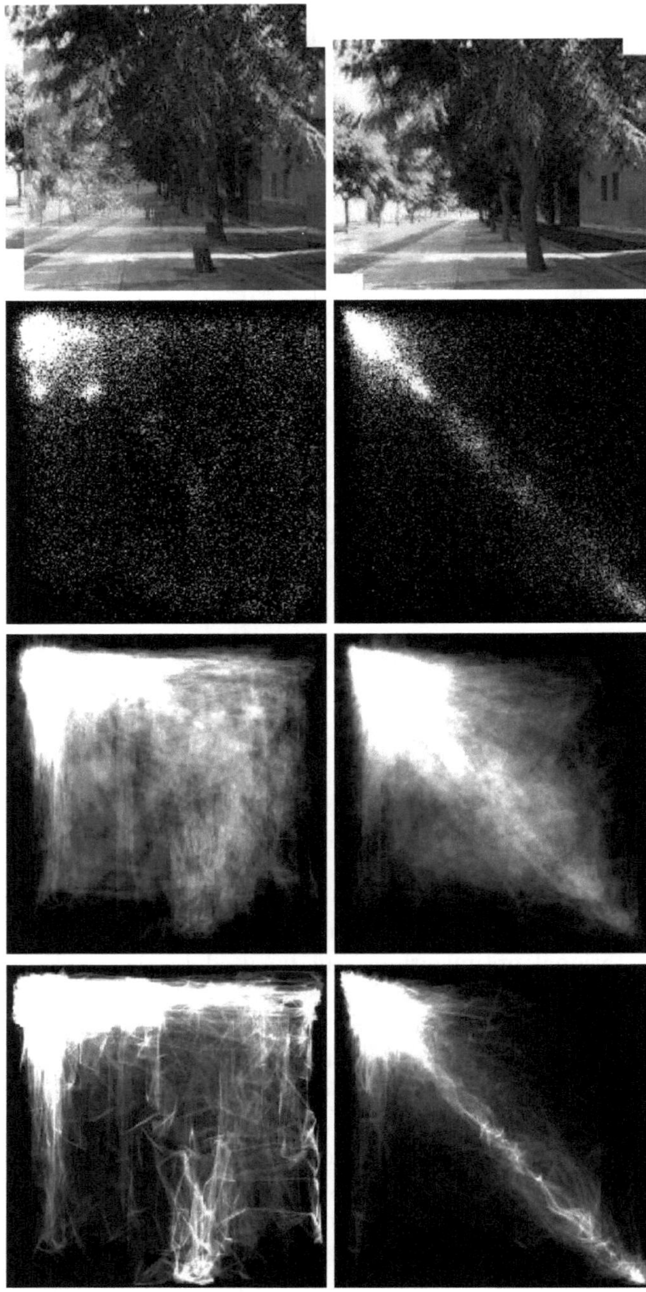

Fig. 4.8. *Left*: the disaligned case; *right*: the aligned case. *First row*: the alignment of the images; *second*, *third*, and *fourth row*: classical, point-counting, and area-based histograms, respectively.

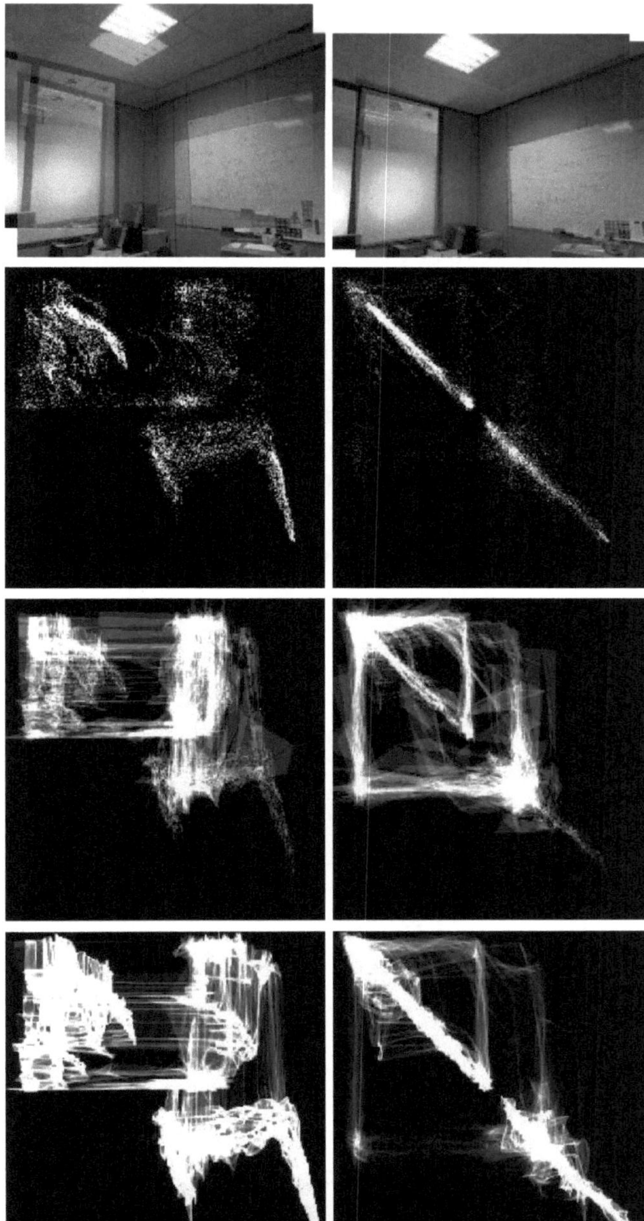

Fig. 4.9. *Left*: the disaligned case; *right*: the aligned case. *First row*: the alignment of the images; *second*, *third*, and *fourth row*: classical, point-counting, and area-based histograms, respectively.

This causes the method based on iso-contours to yield a histogram, which visibly differs from the classical histogram. The method based on iso-surfaces generates much more robust histograms due to its independence of the gradient. See [134] for more detailed explanations, for a generalization to multiple images, and for an extension to 3D images.

The joint histogram is necessary for not only calculating the joint entropy $H(X,Y)$ (Eq. 4.17), but also the marginal entropies of each image $H(X)$ and $H(Y)$ (Eq. 4.16) also have to be estimated in order to obtain the mutual information of two images. In the iso-contour method, the marginals are calculated by the addition of the lengths of the iso-contours inside the triangles, which is to say, by approximating the total length of each α-intensity level curve. In the iso-surfaces method, a similar procedure is implemented by the addition of iso-surface areas, instead of line lengths.

A comparison of both methods, applied to image alignment, can be seen in Fig. 4.10. In this example, the search spaces are represented as surfaces

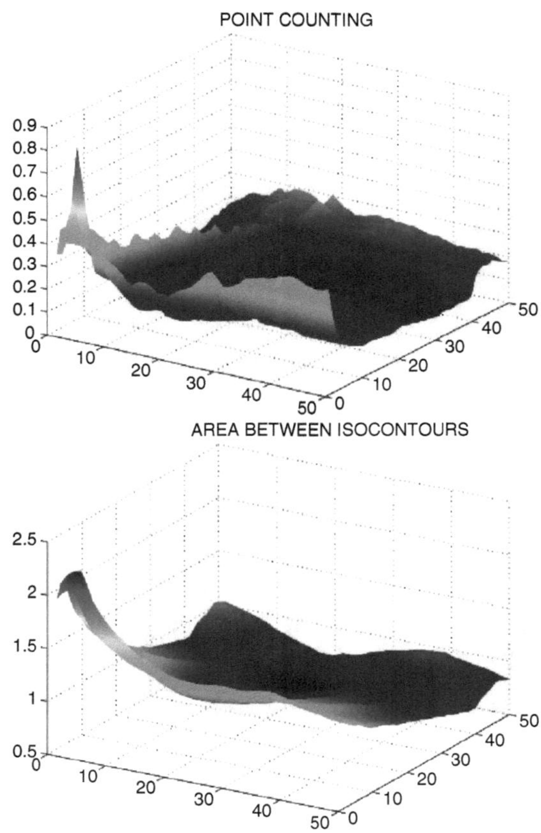

Fig. 4.10. Image alignment by means of Mutual Information – representation of the search spaces generated by each one of the following measures: iso-contours (*top*), iso-surfaces (*bottom*). Figure by courtesy of A. Rajwade and A. Rangarajan.

because only horizontal and vertical translations were used as parameters for the alignment. It can be seen that the method based on iso-contours, or point-counting yields a much more abrupt search surface than the iso-surfaces, or area-based method.

4.3 Alternative *Metrics* for Image Alignment

Mutual information is a widely used measure in image alignment (also referred to as image registration). However, it is not a metric as it does not satisfy the triangle inequality. A metric $d(x, y)$ should satisfy the following conditions:

1. Nonnegativity

$$d(x, y) \geq 0$$

2. Identity of indiscernibles

$$d(x, y) = 0 \Leftrightarrow x = y$$

3. Symmetry

$$d(x, y) = d(y, x)$$

4. Triangle inequality

$$d(x, z) \leq d(x, y) + d(y, z)$$

A pseudometric is a measure which relaxes the second axiom to

$$d(x, y) = 0 \Leftarrow x = y$$

This implies that two different points can have a zero distance.

4.3.1 Normalizing Mutual Information

Different mutual information normalizations are used in pattern recognition, and not all of them are metrics. Uncertainty coefficients are used to scale mutual information $I(X, Y)$, dividing it by the entropy of one of the terms, X or Y.

$$\frac{I(X;Y)}{H(Y)} \quad \text{or} \quad \frac{I(X;Y)}{H(X)}$$

However, the entropies $H(X)$ and $H(Y)$ are not necessarily the same. A solution to this problem would be to normalize by the sum of both entropies. The resulting measure is symmetric:

$$\begin{aligned} R(X,Y) &= \frac{I(X;Y)}{H(X)+H(Y)} \\ &= \frac{H(X)+H(Y)-H(X,Y)}{H(X)+H(Y)} \\ &= 1-\frac{H(X,Y)}{H(X)+H(Y)} \end{aligned}$$

This measure actually measures *redundancy* and is zero when X and Y are independent. It is proportional to the *entropy correlation coefficient* (ECC), also known as *symmetric uncertainty*:

$$\begin{aligned} ECC(X,Y) &= 2-\frac{2}{NI_1(X,Y)} \\ &= 2-\frac{2H(X,Y)}{H(X)+H(Y)} \\ &= 2R(X,Y) \end{aligned}$$

In the previous equation, $NI_1(X,Y)$ refers to another variant of mutual information normalization:

$$NI_1(X,Y) = \frac{H(X)+H(Y)}{H(X,Y)} \tag{4.20}$$

Another widely used mutual information normalization is defined as

$$NI_2(X,Y) = \frac{I(X;Y)}{\max(H(X),H(Y))} \tag{4.21}$$

It is not only symmetrical, but also satisfies positive definiteness (nonnegativity and identity of indiscernibles) as well as the triangle inequality, therefore it is a metric.

4.3.2 Conditional Entropies

In [179], Zhang and Rangarajan have proposed the use of a new information metric (actually a pseudometric) for image alignment, which is based on conditional entropies:

$$\rho(X,Y) = H(X|Y) + H(Y|X) \tag{4.22}$$

It satisfies nonnegativity, symmetry, and triangle inequality. However, $\rho(X,Y)$ can be zero not only when $X = Y$, but also when X is a function of Y, and therefore, it is not a metric, but a pseudometric.

Given the definition $H(X|Y) = H(X,Y) - H(Y)$, the relation of $\rho(X,Y)$ with mutual information is

$$\begin{aligned}\rho(X,Y) &= H(X|Y) + H(Y|X) \\ &= H(X,Y) - H(Y) + H(X,Y) - H(X) \\ &= H(X,Y) - (H(Y) - H(X,Y) + H(X)) \\ &= H(X,Y) - I(X;Y)\end{aligned} \tag{4.23}$$

In the alignment problem, $I(X,Y)$ has to be maximized. Contrarily, the pseudometric $\rho(X,Y)$ has to be minimized.

A useful normalization of $\rho(X,Y)$, which is also a pseudometric, is given by dividing it by the joint entropy of the variables:

$$\begin{aligned}\tau(X,Y) &= \frac{\rho(X,Y)}{H(X,Y)} \\ &= \frac{H(X|Y) + H(Y|X)}{H(X,Y)}\end{aligned}$$

Its relation with mutual information, according to Eq. 4.24, is

$$\begin{aligned}\tau(X,Y) &= \frac{H(X,Y) - I(X;Y)}{H(X,Y)} \\ &= 1 - \frac{I(X;Y)}{H(X,Y)}\end{aligned}$$

4.3.3 Extension to the Multimodal Case

Another important difference between the metric $\rho(X,Y)$ and the mutual information measure is that the new information metric is easily extensible to the multimodal case (registration of more than two images). The mutual information (MI) of three variables is be defined as

$$\begin{aligned}I(X;Y,Z) = &H(X) + H(Y) + H(Z) \\ &- H(X,Y) - H(X,Z) - H(Y,Z) \\ &+ H(X,Y,Z)\end{aligned} \tag{4.24}$$

There is another definition that uses the same notation, $I(X;Y;Z)$, called *interaction information*, also known as *co-information* [16]. It is commonly defined in terms of conditional mutual information, $I(X;Y|Z) - I(X;Y)$, which is equivalent to the negative MI as defined in Eq. 4.24. The conditional MI is defined as

$$\begin{aligned}I(X;Y|Z) &= H(X|Z) - H(X,Y,Z) \\ &= H(X,Z) + H(Y,Z) - H(Z) - H(X,Y,Z)\end{aligned}$$

and it is never negative, $I(X;Y|Z) \geq 0$. Provided that the MI between X and Y is

$$I(X;Y) = H(X) + H(Y) - H(X,Y)$$

then the co-information of the three variables X, Y, and Z is

$$\begin{aligned} I(X;Y|Z) - I(X;Y) &= H(X,Z) + H(Y,Z) - H(Z) - H(X,Y,Z) \\ &\quad -(H(X) + H(Y) - H(X,Y)) \\ &= -H(X) - H(Y) - H(Z) \\ &\quad +H(X,Z) + H(X,Y) + H(Y,Z) - H(X,Y,Z) \end{aligned}$$

which is equal to the negative MI of the three variables. The co-information and the MI between three variables can be positive, zero, or negative. The negativity of MI could be undesirable in some alignment approaches. To overcome this problem, some alternative MI definitions have been proposed in the literature. For example, even though it is not a natural extension of MI, the definition

$$I(X;Y;Z) = H(X) + H(Y) + H(Z) - H(X,Y,Z)$$

avoids the negativity problem. In information theory, it is known as total correlation.

The extension of $\rho(X,Y)$ to three or more random variables is straightforward:

$$\rho(X,Y,Z) = H(X|Y,Z) + H(Y|X,Z) + H(Z|X,Y)$$

$$\rho(X_1, X_2, \dots, X_n) = \sum_{i=1}^{n} H(X_i|X_1, \dots, X_{i-1}, X_{i+1}, \dots, X_n)$$

Similarly, the extension of $\tau(X,Y)$ to three or more random variables is

$$\tau(X,Y,Z) = \frac{\rho(X,Y,Z)}{H(X,Y,Z)}$$

$$\tau(X_1, X_2, \dots, X_n) = \frac{\rho(X_1, X_2, \dots, X_n)}{H(X_1, X_2, \dots, X_n)}$$

For a better understanding of $\rho(X,Y,Z)$, $\tau(X,Y,Z)$, and $I(X;Y;Z)$, see Fig. 4.11, where information is represented as areas. In [179], the computational complexity of estimating the joint probability of three variables is taken into consideration. The authors propose the use of an upper bound of the metric.

4.3.4 Affine Alignment of Multiple Images

In the case of alignment of two images, I_1 and I_2, a proper transformation $T^*(I_2)$ that aligns I_2 to I_1 has to be found. This proper transformation would

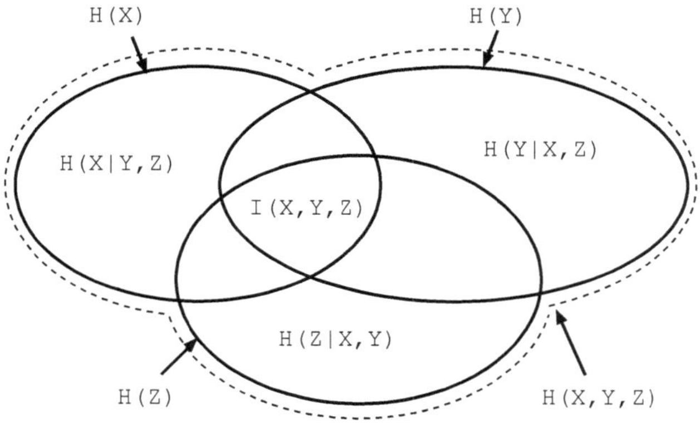

Fig. 4.11. Mutual Information $I(X,Y,Z)$, joint entropy $H(X,Y,Z)$, conditional entropies $H(\cdot\,|\cdot,\cdot)$, and marginal entropies $H(\cdot)$ for three random variables X, Y, and Z, represented in a Venn diagram.

be the one that maximizes or minimizes some criterion. In the case of the ρ metric, we search for a T^* that minimizes it:

$$T^* = \underset{T}{\operatorname{argmin}} \rho(I_1, T(I_2))$$

In the case of alignment of more than two or n images, $n-1$ transformations have to be found:

$$\{T_1^*, \ldots, T_{n-1}^*\} = \underset{\{T_1,\ldots,T_{n-1}\}}{\operatorname{argmin}} \rho(I_1, T_1(I_2), \ldots, T_{n-1}(I_n))$$

In the case of affine alignment, the transformations are limited to vertical and horizontal translations, rotations, scaling and shearing. Each transformation function T establishes a correspondence between the coordinates of an image I and the transformed image $I' = T(I)$. In the affine case, the transformation function can be expressed as a product of matrices between a transformation matrix M and the homogeneous coordinates in the original image, $(x_I, y_I, 1)$. The resulting homogeneous coordinates, $(x_{I'}, y_{I'}, 1)$, are the coordinates of the point in the transformed image I':

$$\begin{pmatrix} x_{I'} \\ y_{I'} \\ 1 \end{pmatrix} = \begin{pmatrix} a & b & e \\ c & d & f \\ 0 & 0 & 1 \end{pmatrix} \begin{pmatrix} x_I \\ y_I \\ 1 \end{pmatrix}$$

The transformation matrix M of size 3×3 defines the affine transformation. Concretely, e and f contain the translation:

$$\begin{pmatrix} x_{I'} \\ y_{I'} \\ 1 \end{pmatrix} = \begin{pmatrix} 1 & 0 & e \\ 0 & 1 & f \\ 0 & 0 & 1 \end{pmatrix} \begin{pmatrix} x_I \\ y_I \\ 1 \end{pmatrix} = \begin{pmatrix} x_I + e \\ y_I + f \\ 1 \end{pmatrix}$$

The scale is given by a and c:

$$\begin{pmatrix} x_{I'} \\ y_{I'} \\ 1 \end{pmatrix} = \begin{pmatrix} a\,0\,0 \\ 0\,c\,0 \\ 0\,0\,1 \end{pmatrix} \begin{pmatrix} x_I \\ y_I \\ 1 \end{pmatrix} = \begin{pmatrix} ax_I \\ cy_I \\ 1 \end{pmatrix}$$

Shearing is given by b:

$$\begin{pmatrix} x_{I'} \\ y_{I'} \\ 1 \end{pmatrix} = \begin{pmatrix} 1\,b\,0 \\ 0\,1\,0 \\ 0\,0\,1 \end{pmatrix} \begin{pmatrix} x_I \\ y_I \\ 1 \end{pmatrix} = \begin{pmatrix} x_I + by_I \\ y_I \\ 1 \end{pmatrix}$$

Finally, a rotation along the z axis of θ radians from the origin is given by a, b, c and d all together:

$$\begin{pmatrix} x_{I'} \\ y_{I'} \\ 1 \end{pmatrix} = \begin{pmatrix} \cos\theta & -\sin\theta & 0 \\ \sin\theta & \cos\theta & 0 \\ 0 & 0 & 1 \end{pmatrix} \begin{pmatrix} x_I \\ y_I \\ 1 \end{pmatrix} = \begin{pmatrix} x_I \cos\theta - Y_I \sin\theta \\ x_I \sin\theta + Y_I \cos\theta \\ 1 \end{pmatrix}$$

When simultaneously aligning n images, $n-1$ different matrices $M_1, \ldots, M_{n-1}$ have to be estimated.

4.3.5 The Rényi Entropy

Alfréd Rényi (20 March 1921–1 February 1970) was a Hungarian mathematician who mainly contributed in probability theory [137], combinatorics and graph theory. One of his contributions in probability theory is a generalization of the Shannon entropy, known as alpha-entropy (or α-entropy), as well as Rényi entropy.

A random variable $\mathbf{X}$ with possible values $x_1, \ldots, x_m$, has some distribution $\mathbf{P} = (p_1, p_2, \ldots, p_n)$ with $\sum_{k=1}^{n} p_k = 1$. The probability of $\mathbf{X}$ to be x_i is usually denoted as $P(\mathbf{X} = x_i)$ and sometimes it is abbreviatedly denoted as $P(x_i)$. The Shannon entropy definition can be formulated for the random variable, or for its probability distribution. In the following definition, E denotes the expected value function:

$$\begin{aligned} H(\mathbf{X}) &= -E_x(\log_b P(\mathbf{X})) && (4.25) \\ &= -\sum_{i=1}^{m} P(\mathbf{X} = x_i) \log_b P(\mathbf{X} = x_i) && (4.26) \\ &= -\sum_{i=1}^{m} P(x_i) \log_b P(x_i) && (4.27) \\ &= -\sum_{k=1}^{n} p_k \log_b p_k && (4.28) \end{aligned}$$

The base b of the logarithm $\log_b$ will be omitted in many definitions in this book because it is not significant for most pattern recognition problems. Actually, the base of the logarithm $\log_b$ determines the units of the entropy measure:

Logarithm	Units of H
log_2	bit
log_e	nat
log_{10}	dit/digit

Some pattern recognition approaches model the probability distributions as a function, called probability density function (pdf). In that case, the summation of elements of $\mathbf{P}$ becomes an integral. The definition of Shannon entropy for a pdf $f(z)$ is

$$H(f) = -\int_z f(z) \log f(z) dz \tag{4.29}$$

The Rényi entropy is a generalization of Shannon entropy. Its definition for a distribution $\mathbf{P}$ is

$$H_\alpha(\mathbf{P}) = \frac{1}{1-\alpha} \log \sum_{i=1}^{n} p_j^\alpha \tag{4.30}$$

and its definition for a pdf $f(z)$ is

$$H_\alpha(f) = \frac{1}{1-\alpha} \log \int_z f^\alpha(z) dz \tag{4.31}$$

The order α of the Rényi entropy is positive and it cannot be equal to 1, as there would be a division by zero in $\frac{1}{1-\alpha}$.

$$\alpha \in (0,1) \cup (1,\infty)$$

In Fig. 4.12, it can be seen that for the interval $\alpha \in (0,1)$, the entropy function H_α is concave, while for the interval $\alpha \in (1,\infty)$, it is neither concave, nor convex. Also, for the first interval, it is smaller or equal to the Shannon entropy, $H_\alpha(\mathbf{P}) \geq H(\mathbf{P})$, $\forall \alpha \in (0,1)$, given that H_α is a nonincreasing function of α.

The discontinuity at $\alpha = 1$ is very significant. It can be shown, analytically and experimentally, that as α approaches 1, H_α tends to the value of the Shannon entropy (see the proof in Chapter 5). This relation between both entropies has been exploited in some pattern recognition problems. As shown in the following subsection, some entropy estimators provide an approximation to the value of Rényi entropy. However, Shannon entropy could be necessary in some problems. In Chapter 5, we present an efficient method to approximate Shannon entropy from Rényi entropy by finding a suitable α value.

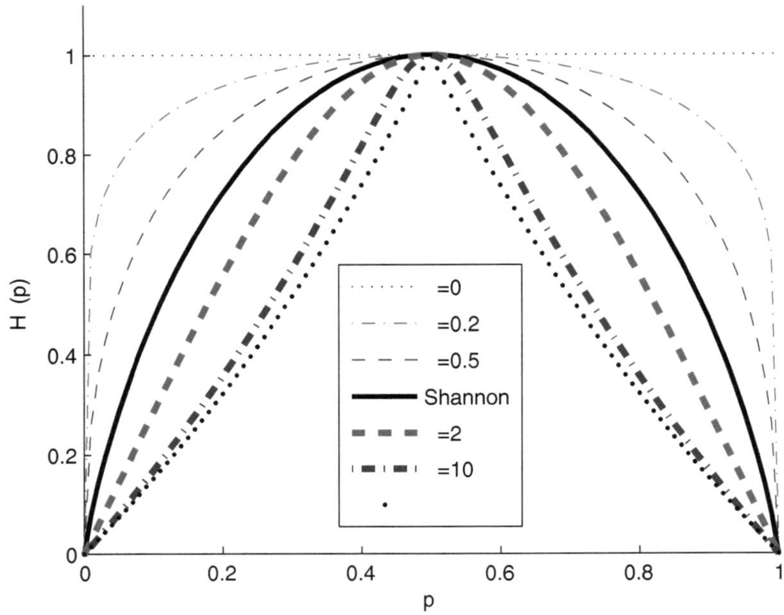

Fig. 4.12. Rényi and Shannon entropies of a Bernoulli distribution $\mathbf{P} = (p, 1-p)$.

4.3.6 Rényi's Entropy and Entropic Spanning Graphs

Entropy estimation is critical in IT-based pattern recognition problems. Entropy estimators can be divided into two categories: "plug-in" and "nonplug-in." Plug-in methods first estimate the density function, for example, the construction of a histogram, or the Parzen's windows method. The nonplug-in methods, on the contrary, estimate entropy directly from a set of samples.

The Rényi entropy of a set of samples can be estimated from the length of their *Minimal Spanning Tree* (MST) in a quite straightforward way. This method, based on Entropic Spanning Graphs [74], belongs to the "nonplug-in" methods of entropy estimation.

The MST graphs have been used for testing the randomness of a set of points (see Fig. 4.13). In [73], it was shown that in a d-dimensional feature space, with $d \geq 2$, the α-entropy estimator

$$H_\alpha(X_n) = \frac{d}{\gamma}\left[\ln \frac{L_\gamma(X_n)}{n^\alpha} - \ln \beta_{L_\gamma,d}\right] \tag{4.32}$$

is asymptotically unbiased and consistent with the PDF of the samples. Here, the function $L_\gamma(X_n)$ is the length of the MST, and γ depends on the order α and on the dimensionality: $\alpha = (d-\gamma)/d$. The bias correction $\beta_{L_\gamma,d}$ depends on the graph minimization criterion, but it is independent of the PDF. There are some approximations which bound the bias by (i) Monte Carlo simulation

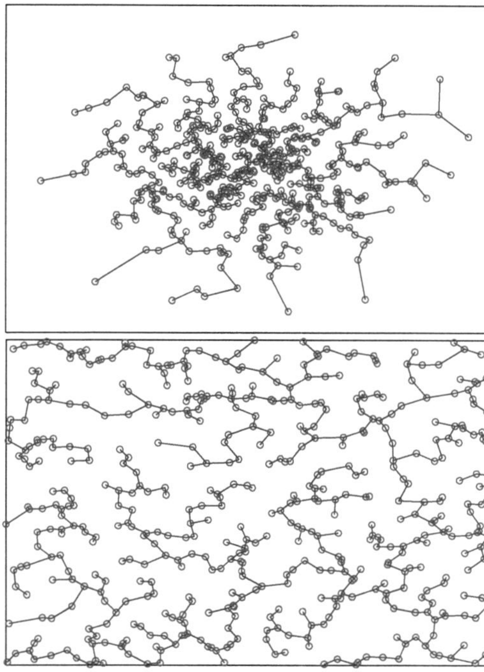

Fig. 4.13. Minimal spanning trees of samples with a Gaussian distribution (*top*) and samples with a random distribution (*bottom*). The length L_γ of the first MST is visibly shorter than the length of the second MST.

of uniform random samples on unit cube $[0,1]^d$ and (ii) approximation for large d: $(\gamma/2)\ln(d/(2\pi e))$ in [21].

The length $L_\gamma(X_n)$ of the MST is defined as the length of the acyclic spanning graph with minimal length of the sum of the weights (in this case the weights are defined as $\mid e \mid^\gamma$) of its edges $\{e\}$:

$$L_\gamma^{MST}(X_n) = \min_{M(X_n)} \sum_{e\in M(X_n)} \mid e \mid^\gamma \tag{4.33}$$

where $\gamma\in (0,d)$. Here, $M(X_n)$ denotes the possible sets of edges of a spanning tree graph, where $X_n = \{x_1,...,x_n\}$ is the set of vertices which are connected by the edges $\{e\}$. The weight of each edge $\{e\}$ is the distance between its vertices, powered the γ parameter: $\mid e \mid^\gamma$. There are several algorithms for building a MST, for example, the Prim's MST algorithm has a straightforward implementation. There also are estimators that use other kinds of entropic graphs, for example, the k-nearest neighbor graph [74].

Entropic spanning graphs are suitable for estimating α-entropy for $0 \leq \alpha < 1$, so Shannon entropy cannot be directly estimated with this method. In [185], relations between Shannon and Rényi entropies of integer order are discussed. In [115], Mokkadem constructed a nonparametric estimate of the Shannon entropy from a convergent sequence of α-entropy estimates (see Chapter 5).

4.3.7 The Jensen–Rényi Divergence and Its Applications

The Jensen–Rényi divergence is an information-theoretic divergence measure, which uses the generalized Rényi entropy for measuring the statistical dependence between an arbitrary number of probability distributions. Each distribution has a weight, and also the α order of the entropy can be adjusted for varying the measurement sensitivity of the joint histogram. The Jensen–Rényi divergence is symmetrical and convex. For n probability distributions $\mathbf{p}_1, \mathbf{p}_2, \ldots, \mathbf{p}_n$, it is defined as

$$JR_\alpha^{\boldsymbol{\omega}}(\mathbf{p}_1, \ldots, \mathbf{p}_n) = R_\alpha\left(\sum_{i=1}^{n} \omega_i \mathbf{p}_i\right) - \sum_{i=1}^{n} \omega_i R_\alpha(\mathbf{p}_i) \tag{4.34}$$

where R_α is the Rényi entropy and $\boldsymbol{\omega} = (\omega_1, \omega_2, \ldots, \omega_n)$ is the weight vector, and the sum of the weights (all of them positive) must be 1:

$$\sum_{i=1}^{n} \omega_i = 1, \quad \omega_i \geq 0$$

In [70], Hamza and Krim define the divergence and study some interesting properties. They also present its application to image registration and segmentation. In the case of image registration with the presence of noise, the use of the Jensen–Rényi divergence outperforms the use of mutual information, as shown in Fig. 4.14.

For segmentation, one approach is to use a sliding window. The window is divided in two parts W1 and W2, each one corresponding to a portion of data (pixels, in the case of an image). The divergence between the distributions of both subwindows is calculated. In the presence of an edge or two different regions A and B, the divergence would be maximum when the window is centered at the edge, as illustrated in Fig. 4.15. In the case of two distributions, the divergence can be expressed as a function of the fraction λ of subwindow W2, which is included in the region B.

$$JR_\alpha(\lambda) = R_\alpha(\mathbf{p}) - \frac{R_\alpha((1-\lambda)\mathbf{p_a} + \lambda\mathbf{p_b}) + R_\alpha(\mathbf{p_a})}{2}$$

where $\alpha \in (0,1)$ and

$$\mathbf{p} = \left(1 - \frac{\lambda}{2}\right)\mathbf{p}_a + \frac{\lambda}{2}\mathbf{p}_b$$

Results of segmentation are shown in Fig. 4.16. Different values of the α parameter have similar results in this case.

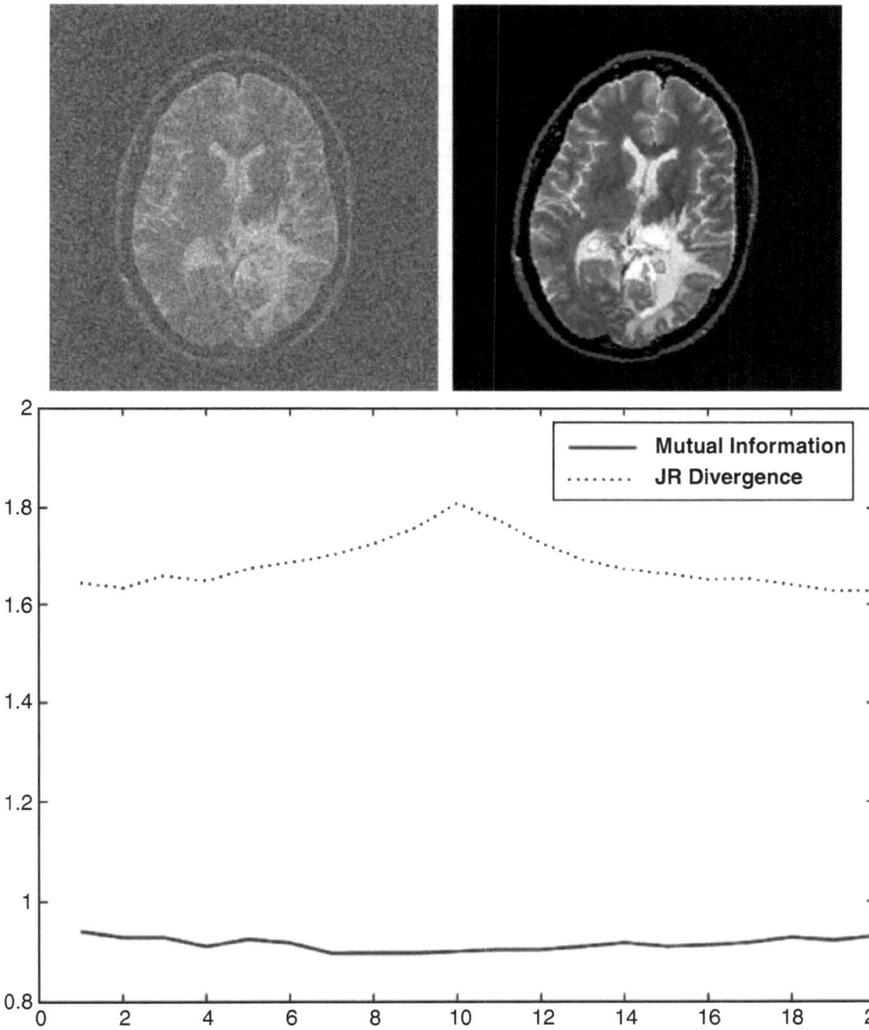

Fig. 4.14. Registration in the presence on noise. For the Jensen–Rényi divergence, $\alpha = 1$ and the weights are $\omega_i = 1/n$. Figure by A. Ben Hamza and H. Krim (©2003 Springer).

4.3.8 Other Measures Related to Rényi Entropy

In [137], Rényi introduces a measure of dissimilarity between densities, called Rényi α-divergence, or Rényi α-relative entropy. This divergence and some particular cases of it have been widely used for image registration (alignment), as well as in other pattern recognition problems. Given two densities $f(z)$ and

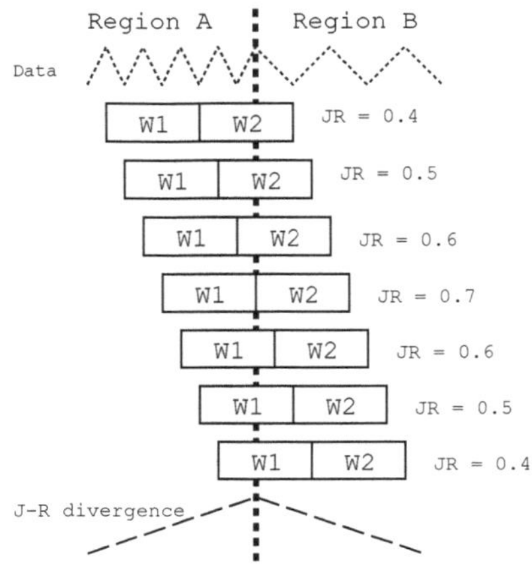

Fig. 4.15. Sliding window for edge detection. The divergence between the distributions of the subwindows W1 and W2 is higher when they correspond to two different regions of the image.

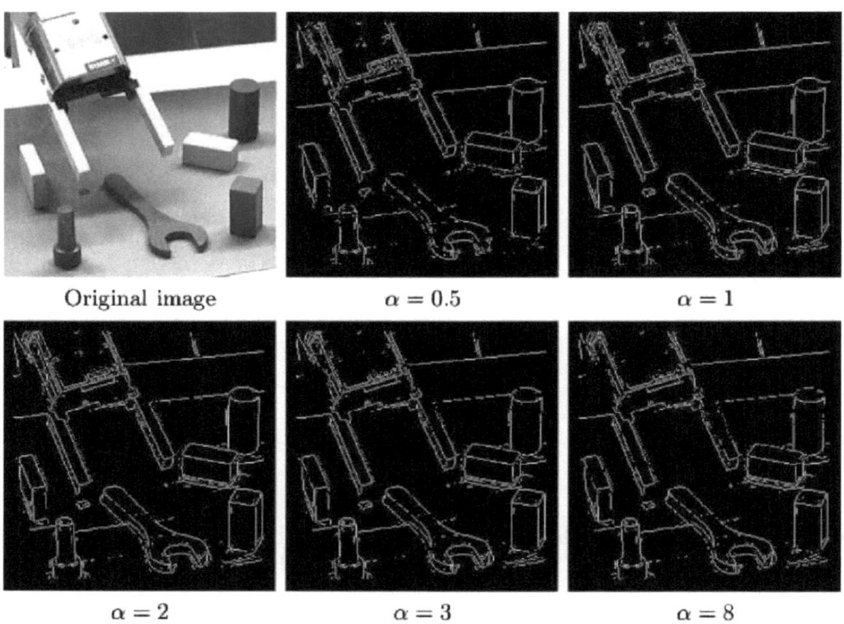

Fig. 4.16. Edge detection results using Jensen–Rényi divergence for various values of α. Figure by A. Ben Hamza and H. Krim (©2003 Springer).

$g(z)$, for a d-dimensional random variable z, the Rényi α-divergence of the density g from the density f is defined as

$$D_\alpha(f||g) = \frac{1}{\alpha - 1} \log \int g(z) \left(\frac{f(z)}{g(z)} \right)^\alpha dz \tag{4.35}$$

It is not symmetric, so it is not a distance, just a divergence. Depending on the order α, there are some special cases.

The most notable case is $\alpha = 1$, because there is a division by zero in $1/(\alpha - 1)$. However, in the limit of $\alpha \to 1$, the Kullback–Leibler (KL) divergence is obtained:

$$\lim_{\alpha \to 1} D_\alpha(f||g) = \int f(z) \log \frac{f(z)}{g(z)} dz = KL(f||g)$$

The KL-divergence is widely used in pattern recognition. In some problems it is referred to as "information gain," for example, in decision trees.

Another case is $\alpha = \frac{1}{2}$

$$D_{\frac{1}{2}}(f||g) = -2 \log \int \sqrt{f(z)g(z)} dz$$

which is related to the Bhattacharyya coefficient, defined as

$$BC(f,g) = \int \sqrt{f(z)g(z)} dz$$

and

$$D_B(f,g) = -\log BC(f,g)$$

is the Bhattacharyya distance, which is symmetric. The Hellinger dissimilarity is also related to the Bhattacharyya coefficient:

$$H(f,g) = \frac{1}{2}\sqrt{2 - 2BC(f,g)}$$

Then, its relation to the α-relative divergence of order $\alpha = \frac{1}{2}$ is

$$H(f,g) = \frac{1}{2}\sqrt{2 - 2\exp\left(-\frac{1}{2}D_{\frac{1}{2}}(f||g)\right)}$$

The divergences with $\alpha = -1$ and $\alpha = 2$ are also worth mentioning:

$$D_{-1}(f||g) = \frac{1}{2}\int \frac{(f(z) - g(z))^2}{f(z)} dz$$

and

$$D_2(f||g) = \frac{1}{2}\int \frac{(f(z) - g(z))^2}{g(z)} dz$$

From the definition of the α-divergence (Eq. 4.35), it is easy to define a mutual information that depends on α. The mutual information is a measure that depends on the marginals and the joint density $f(x,y)$ of the variables x and y. Defining g as the product of the marginals, $g(x,y) = f(x)f(y)$, the expression of the α-mutual information is

$$I_\alpha = D_\alpha(f||g) = \frac{1}{\alpha - 1} \log \int f(x)f(y) \left(\frac{f(x,y)}{f(x)f(y)}\right)^\alpha dxdy$$

which, in the limit $\alpha \to 1$, converges to the Shannon mutual information.

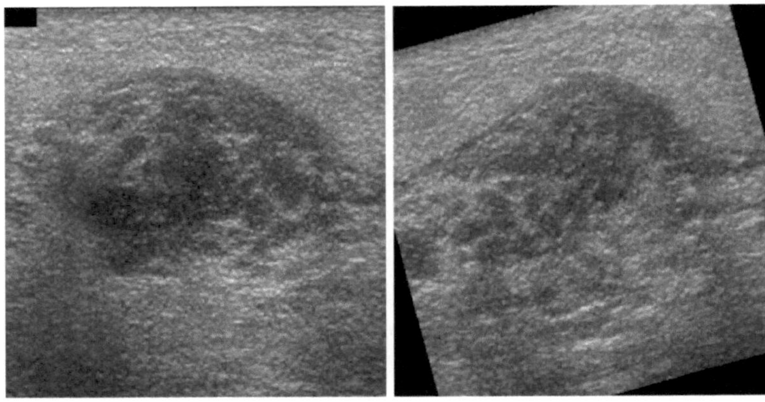

Fig. 4.17. Ultrasound images of breast tumor, separated and rotated. Figure by H. Neemuchwala, A. Hero, S. Zabuawala and P. Carson (©2007 Wiley).

4.3.9 Experimental Results

In [117], Neemuchwala et al. study the use of different measures for image registration. They also present experimental results, among which the one shown in Fig. 4.17 represented two images of a database of breast tumor ultrasound images. Two different images are shown, one of them is rotated. In Fig. 4.18, several alignment criteria are compared. For each one of them, different orders α are tested and represented.

4.4 Deformable Matching with Jensen Divergence and Fisher Information

4.4.1 The Distributional Shape Model

Let us focus on shape as a set (list) of points (like the rough discretization of a contour, or even a surface) (Fig. 4.19). We have a collection $\mathcal{S}$ of N shapes (data points sets): $\mathcal{C} = \{\mathbf{X}^c : \; c = 1, \ldots, N\}$ being $\mathbf{X}^c = \{\mathbf{x}_i^c \in \mathbb{R}^d : \; i = 1, \ldots, n_c\}$ a given set of n_c points in $\mathbb{R}^d$, with $d = 2$ for the case of 2D shapes. A *distributional model* [9,174] of a given shape can be obtained by modelling each point set as a *Gaussian mixture*. As we will see in more detail in the following chapter (clustering), Gaussian mixtures are fine models for encoding a multimodal distribution. Each mode (Gaussian cluster) may be represented as a multidimensional Gaussian, with a given average and covariance matrix, and the convex combination (coefficients add to one) of all the Gaussians. Gaussian centers may be placed at each point if the number of points is not too high. The problem of finding the optimal number of Gaussians and their placement and covariances will be covered in the first part of the next chapter. For the moment, let us assume that we have a collection of cluster-center point sets $\mathcal{V} = \{\mathbf{V}^c : \; c = 1, \ldots, N\}$, where each center point sets consists of

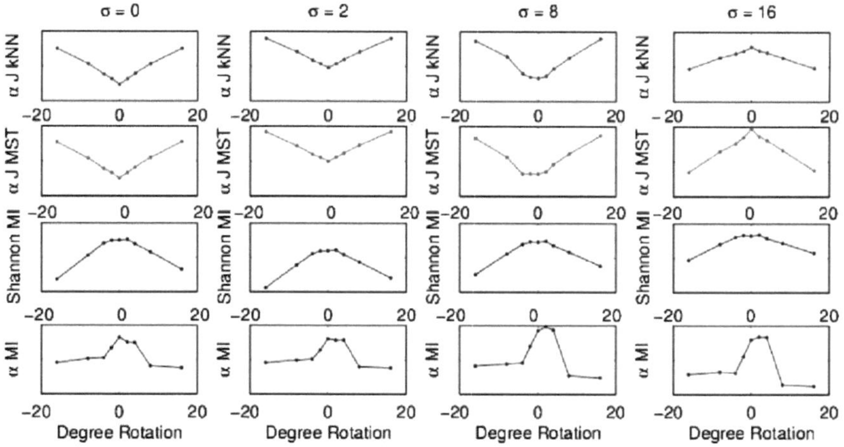

Fig. 4.18. Normalized average profiles of image matching criteria for registration of ultrasound breast tumor images. Plots are normalized with respect to the maximum variance in the sampled observations. Each column corresponds to a different α order, and each row to a different matching criterion. (Row 1) k-NN graph-based estimation of α-Jensen difference divergence, (row 2) MST graph-based estimation of α-Jensen difference divergence, (row 3) Shannon Mutual Information, and (row 4) α Mutual Information estimated using NN graphs. The features used in the experiments are ICA features, except for the row 3, where the histograms were directly calculated from the pixels. Figure by H. Neemuchwala, A. Hero, S. Zabuawala and P. Carson (©2007 Wiley).

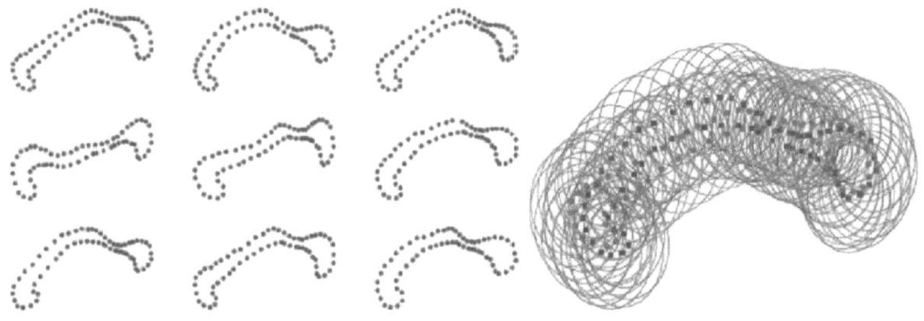

Fig. 4.19. Probabilistic description of shapes. *Left*: samples (denoted by numbers 1–3 (first row), 4–6 (second row) and 7–9 (third row). *Right*: alignment with the variance represented as a *circle* centered at each shape point. Figure by A. Peter and A. Rangarajan (©2006 IEEE).

$\mathbf{V}^c = \{\mathbf{v}_a^c \in \mathbb{R}^d : \ a = 1, \ldots, K^c\}$, K^c being the number of clusters (Gaussians) for the cth point set. For simplicity, let us assume that such numbers are proportional to the sizes of their corresponding point sets. Then, the pdf characterizing a given point set is

$$p_c(\mathbf{x}) = \sum_{a=1}^{K^c} \alpha_a^c p(\mathbf{x}|\mathbf{v}_a^c) \text{ with } \sum_{a=1}^{K^c} \alpha_a^c = 1, \ \alpha_a^c \geq 0 \, \forall \alpha_a^c \tag{4.36}$$

Under the Gaussian assumption of the mixtures, we have

$$p(\mathbf{x}|\mathbf{v}_a^c) = G(\mathbf{x}-\mathbf{v}_a^c, \Sigma_a) = \frac{1}{(2\pi)^{d/2}\Sigma_a^{1/2}} \exp\left(-\frac{1}{2}(\mathbf{x}-\mathbf{v}_a^c)^T \Sigma_a^{-1}(\mathbf{x}-\mathbf{v}_a^c)\right) \tag{4.37}$$

denoting Σ_a the ath $d \times d$ covariance matrix as usual. Such matrix is assumed to be diagonal for simplicity. Then, assuming independence between shape points inside a given point set, we obtain the pdf for a given point set that is of the form

$$p(\mathbf{X}^c|\mathbf{V}^c, \boldsymbol{\alpha}^c) = \prod_{i=1}^{n_c} p_c(\mathbf{x}_i) = \prod_{i=1}^{n_c}\sum_{a=1}^{K^c} \alpha_a^c p(\mathbf{x}|\mathbf{v}_a^c) \tag{4.38}$$

being $\boldsymbol{\alpha}^c = (\alpha_1^c, \ldots, \alpha_{K_c}^c)^T$.

Let then $\mathcal{Z}$ denote the so called *average atlas point set*, that is, the point set derived from registrating all the input point sets with respect to a common reference system. Registration means, in this context, the alignment of all point sets $\mathbf{X}^c$ through general (deformable) transformations $\mathbf{f}^c$, each one parameterized by $\boldsymbol{\mu}^c$. The parameters of all transformations must be recovered in order to compute $\mathcal{Z}$. To that end, the pdfs of each of the *deformed shapes* may be expressed in the following terms:

$$p_c = p_c(\mathbf{x}|\mathbf{V}^c, \boldsymbol{\mu}^c) = \sum_{a=1}^{K^c} \alpha_a^c p(\mathbf{x}|\mathbf{f}^c(\mathbf{v}_a^c)) \tag{4.39}$$

In the most general (deformable) case, functions $\mathbf{f}^c$ are usually defined as *thin-plate splines* (TPS) (see Bookstein classic papers [26,40]). The underlying concept of TPS is that a deformable transformation can be decomposed into a linear part and a nonlinear one. All we need is to have a set of n control points $\mathbf{x}_1, \ldots, \mathbf{x}_n \in \mathbb{R}^d$ (remind the B-splines definition of contours in the previous chapter). Then, from the latter points, we may extract a nonrigid function as a mapping $\mathbf{f} : \mathbb{R}^d \to \mathbb{R}^d$:

$$\mathbf{f}(\mathbf{x}) = (\mathbf{A}\mathbf{x} + \mathbf{t}) + \mathbf{W}\mathbf{U}(\mathbf{x}) \tag{4.40}$$

where $\mathbf{A}_{d\times d}$ and $\mathbf{t}_{d\times 1}$ denote the linear part; $\mathbf{W}_{d\times n}$ encodes the nonrigid part; and $\mathbf{U}(\mathbf{x})_{n\times 1}$ encodes n basis functions (as many as control points) $\mathbf{U}_i(\mathbf{x}) = U(||\mathbf{x}-\mathbf{x}_i||)$, each one centered at each $\mathbf{x}_i$, being $U(.)$ a kernel function (for instance $U(r) = r^2\log(r^2)$ when $d = 2$). Thus, what is actually encoding $\mathbf{f}$ is an estimation (interpolation) of the *complete deformation field* from the input points. The estimation process requires the proposed correspondences for each control point: $\mathbf{x}'_1, \ldots, \mathbf{x}'_n$. For instance, in $d = 2$ (2D shapes), we have for each $\mathbf{x}_i = (x_i, y_i)^T$ its corresponding point $\mathbf{x}'_i = (x'_i, y'_i)^T$ (known beforehand or proposed by the matching algorithm). Then, let $\mathbf{K}_{n\times n}$ be a symmetric matrix where the entries are $U(r_{ij}) = U(||\mathbf{x}_i - \mathbf{x}_j||)$ with $i, j = 1, \ldots, n$, and $\mathbf{P}_{n\times 3}$ a matrix where the ith row has the form $(1\ x_i\ y_i)$, that is

$$\mathbf{K} = \begin{pmatrix} 0 & U(r_{12}) & \ldots & U(r_{1n}) \\ U(r_{21}) & 0 & \ldots & U(r_{2n}) \\ \vdots & \vdots & \ddots & \vdots \\ U(r_{n1}) & U(r_{n2}) & \ldots & 0 \end{pmatrix} \quad \mathbf{P} = \begin{pmatrix} 1 & x_1 & y_1 \\ 1 & x_2 & y_2 \\ \vdots & \vdots & \vdots \\ 1 & x_n & y_n \end{pmatrix} \tag{4.41}$$

The latter matrix is the building block of matrix $\mathbf{L}_{(n+3)\times(n+3)}$, and the coordinates of points $\mathbf{x}'_i$ are placed in the columns of matrix $\mathbf{V}_{2\times n}$:

$$\mathbf{L} = \begin{pmatrix} \mathbf{K} & \mathbf{P} \\ \mathbf{P}^T & \mathbf{0}_{3\times 3} \end{pmatrix} \quad \mathbf{V} = \begin{pmatrix} x'_1 & x'_2 & \ldots & x'_n \\ y'_1 & y'_2 & \ldots & y'_n \end{pmatrix} \tag{4.42}$$

Then, the least squares estimation of $\mathbf{A}_1 = (a_1^x, a_1^y)$, and t^x, and $\mathbf{W}^x = (w_1^x, \ldots, w_n^x)$, which are the transformation coefficients for the X dimension, is given by solving

$$\mathbf{L}^{-1}(\mathbf{V}^x|0,0,0)^T = (\mathbf{W}^x|t^x, a_1^x, a_1^y)^T \tag{4.43}$$

being $\mathbf{V}^x =$ the first row of $\mathbf{V}$. A similar rationale is applied for obtaining the coefficients for the Y dimension, and then we obtain the interpolated position for any vector $\mathbf{x} = (x, y)^T$:

$$\mathbf{f}(\mathbf{x}) = \begin{pmatrix} \mathbf{f}_x(\mathbf{x}) \\ \mathbf{f}_y(\mathbf{x}) \end{pmatrix} = \begin{pmatrix} a_1^x & a_1^y \\ a_2^x & a_2^y \end{pmatrix} \begin{pmatrix} x \\ y \end{pmatrix} + \begin{pmatrix} t^x \\ t^y \end{pmatrix} + \begin{pmatrix} \sum_{i=1}^n w_i^x U(||\mathbf{x} - \mathbf{x}_i||) \\ \sum_{i=1}^n w_i^y U(||\mathbf{x} - \mathbf{x}_i||) \end{pmatrix} \tag{4.44}$$

yielding Eq. 4.40. The computation of this interpolation may be speeded up by approximation methods like the Nyström one (see [50]). An important property of $\mathbf{f}(\mathbf{x})$ obtained as described above is that it yields the smoothest interpolation among this type of functions. More precisely, $\mathbf{f}$ minimizes

$$I(\mathbf{f}) = \int\int_{\mathbb{R}^2} \left(\left(\frac{\partial^2 \mathbf{f}}{\partial x^2}\right)^2 + 2\left(\frac{\partial^2 \mathbf{f}}{\partial x \partial y}\right)^2 + \left(\frac{\partial^2 \mathbf{f}}{\partial y^2}\right)^2 \right) dxdy \tag{4.45}$$

which is proportional to $E(\mathbf{f}) = trace(\mathbf{WKW}^T)$, a quantify known as *bending energy* because it quantifies the amount of deformation induced by the obtained mapping. The bending energy is only zero when the nonrigid coefficients are zero. In this latter case, only the rigid (linear) part remains. Furthermore, as a high value of $E(\mathbf{f})$ reflects the existence of outlying matchings when a given correspondence field is recovered, this energy has been recently used for comparing the performance of different matching algorithms [1]. Actually, $E(\mathbf{f})$ is one of the driving forces of the matching algorithm for the distributional shape model because the registration of multiple shapes (*probabilistic atlas construction problem*) may be formulated as finding

$$\mathcal{Z}^* = \arg\min_{\boldsymbol{\mu}^1\cdots\boldsymbol{\mu}^N} \underbrace{H\left(\sum_{c=1}^N \pi_c p_c\right) - \sum_{c=1}^N \pi_c H(p_c)}_{JS_{\boldsymbol{\pi}}(p_1,\ldots,p_N)} + \sum_{c=1}^N E(\mathbf{f}^c) \tag{4.46}$$

where $JS_{\boldsymbol{\pi}}(p_1,\ldots,p_N)$ is the Jensen–Shannon divergence with respect to pdfs $p_1,\ldots,p_N$ and weight vector $\boldsymbol{\pi} = (\pi_1,\ldots,p_N)$, where $\sum_{c=1}^n \pi_c = 1$ and $\pi_c \geq 0\ \forall \pi_c$. Thus, a good alignment must satisfy that the aligned shapes have similar pdfs, that is, similar mixtures. A first way of measuring (quantifying) the similarities between pdfs is to use Jensen–Shannon (JS) divergence, as we did in the previous chapter to drive Active Polygons. The first interesting property of the JS measure in this context is that the pdfs can be weighted, which takes into account the different number of points encoding each shape. The second one is that JS allows the use of different number of cluster centers in each point set.

4.4.2 Multiple Registration and Jensen–Shannon Divergence

Beyond the intuition of the use of JS divergence as a measure of compatibility between multiple pdfs, it is interesting to explore, more formally, the connection between the maximization of the log-likelihood ratio and the minimization of JS divergence. This is not surprising if one considers that the JS can be defined as the Kullback–Leibler divergence between the convex combination of pdfs and the pdfs themselves. Here, we start by taking n_1 i.i.d. samples from p_1, n_2 from p_2, and so on, so that $M = \sum_{c=1}^N n_c$. Then, if the weight of each pdf is defined as $\pi_c = \frac{n_c}{\sum_b n_b}$ (normalizing the number of samples), the likelihood ratio between the *pooled distribution* (the pdf resulting from the convex combination using the latter weights) and the product of all the pdfs is defined as

$$\Lambda = \frac{\prod_{k=1}^M \left(\prod_{c=1}^N \pi_c p_c(\mathbf{X}_k)\right)}{\prod_{c=1}^N \left(\prod_{k_c=1}^{n_c} p_c(\mathbf{X}_{k_c}^c)\right)} \tag{4.47}$$

where $\mathbf{X}_k$ is one of the i.d.d. samples from the pooled distribution which is mapped to the set of pooled samples consisting in $\{\mathbf{X}_{k_c}^c : k_c = 1,\ldots,n_c,\ c = 1,\ldots,N\}$. Then, taking the logarithm of the likelihood ratio, we have

$$\log \Lambda = \sum_{k=1}^M \log\left(\sum_{c=1}^N \pi_c p_c(\mathbf{X}_k)\right) - \sum_{c=1}^N \sum_{k_c=1}^{n_c} \log p_c(\mathbf{X}_{k_c}^c) \tag{4.48}$$

In information theory, the version of the weak law of large numbers (convergence in probability of $\frac{1}{n}\sum_{i=1}^N X_i$ toward the average $E(X)$ when the number of i.i.d. samples n is large enough) is called the *Asymtotic Equipartition Property* (AEP):

$$-\frac{1}{n}p(X_1,\ldots,X_n) = -\frac{1}{n}\sum_{i=1}^{n} p(X_i) \to -E(\log p(x)) = H(X) \tag{4.49}$$

where the arrow ($\to$) indicates *convergence in probability*:

$$\lim_{n\to\infty} Pr\left[\left|-\frac{1}{n}p(X_1,\ldots,X_n) - H(X)\right| > \epsilon\right] = 0,\ \forall\epsilon > 0 \tag{4.50}$$

Therefore, if each n_c is large enough, we have

$$\begin{aligned}\log \Lambda &= -MH\left(\sum_{c=1}^{N}\pi_c p_c\right) + \sum_{c=1}^{N} n_c H(p_c)\\ &= -M\left[H\left(\sum_{c=1}^{N}\pi_c p_c\right) + \sum_{c=1}^{N}\frac{n_c}{N}H(p_c)\right]\\ &= -JS_{\boldsymbol{\pi}}(p_1,\ldots,p_N)\end{aligned} \tag{4.51}$$

Consequently, maximizing the log-likelihood is equivalent to minimizing the JS divergence. In order to proceed to this minimization, we must express each p_c conveniently, adapting Eq. 4.39 we have

$$p_c = p_c(\mathbf{x}|\mathbf{V}^c,\boldsymbol{\mu}^c) = \sum_{a=1}^{K^c}\frac{1}{K^c}p(\mathbf{x}|\mathbf{f}^c(\mathbf{v}_a^c)) = \frac{1}{K^c}\sum_{a=1}^{K^c}G(x-\mathbf{f}^c(\mathbf{v}_a^c),\sigma^2\mathbf{I}) \tag{4.52}$$

that is, we are assuming that $\alpha_a^c = \frac{1}{K^c}$, being K^c the number of clusters of the cth distribution (uniform weighting) and, also for simplicity, a diagonal and identical covariance matrix $\Sigma_a = \sigma^2\mathbf{I}$. In addition, the deformed cluster centers are denoted by $u_a^c = \mathbf{f}^c(\mathbf{v}_a^c)$.

Then, given again one of the i.d.d. samples from the pooled distribution $\mathbf{X}_k$ mapped to one element of the pooled set $\{\mathbf{X}_{k_c}^c : k_c = 1,\ldots,n_c,\ c = 1,\ldots,N\}$ where $\{\mathbf{X}_{k_c}^c\}$ is the set of n_c samples for the cth shape, the entropies of both individual distributions and the pooled one may be estimated by applying again the AEP. For the individual distributions, we have

$$H(p_c) = -\frac{1}{n_c}\sum_{k_c=1}^{n_c}\log p_c(\mathbf{X}_{k_c}^c) = -\frac{1}{n_c}\sum_{k_c=1}^{n_c}\log\left[\frac{1}{K^c}\sum_{a=1}^{K^c}G(\mathbf{X}_{k_c}^c - \mathbf{u}_a^c,\sigma^2\mathbf{I}))\right] \tag{4.53}$$

Regarding the convex combination and choosing $\pi_c = \frac{K^c}{M}$, being $K = \sum_c K^c$ the total number of cluster centers in the N point sets, we have

$$\sum_{c=1}^{N} \pi_c p_c = \sum_{c=1}^{N} \pi_c \left(\frac{1}{K^c} \sum_{a=1}^{K^c} G(x - \mathbf{u}_a^c, \sigma^2 \mathbf{I})) \right) = \frac{1}{K} \sum_{j=1}^{K} G(\mathbf{X}_k - \mathbf{u}_j, \sigma^2 \mathbf{I}) \tag{4.54}$$

being $\mathbf{X}_k$ the pooled samples and $\mathbf{u}_j, j = 1, \ldots, M$ the pooled centers. Consequently, the convex combination can be seen as a Gaussian mixture where its Gaussians are centered on the *deformed cluster centers*. This simplifies the estimation of the Shannon entropy, from AEP, for the convex combination:

$$\begin{aligned} H\left(\sum_{c=1}^{N} \pi_c p_c\right) &= H\left(\frac{1}{K} \sum_{j=1}^{K} G(\mathbf{X}_k - \mathbf{u}_j, \sigma^2 \mathbf{I})\right) \\ &= -\frac{1}{M} \sum_{j=1}^{M} \log\left[\frac{1}{K} \sum_{a=1}^{K} G(\mathbf{X}_k - \mathbf{u}_a, \sigma^2 \mathbf{I})\right] \end{aligned} \tag{4.55}$$

being $M = \sum_c n_c$ the sum of all samples. Therefore, the JS divergence is estimated by

$$\begin{aligned} JS_{\boldsymbol{\pi}}(p_1, \ldots, p_N) &= H\left(\sum_{c=1}^{N} \pi_c p_c\right) - \sum_{c=1}^{N} \pi_c H(p_c) \\ &\approx -\frac{1}{M} \sum_{j=1}^{M} \log\left[\frac{1}{K} \sum_{a=1}^{K} G(\mathbf{X}_j - \mathbf{u}_a, \sigma^2 \mathbf{I})\right] \\ &\quad + \sum_{c=1}^{N} \frac{K_c}{n_c K} \sum_{k_c=1}^{n_c} \log\left[\frac{1}{K^c} \sum_{a=1}^{K^c} G(\mathbf{X}_{k_c}^c - \mathbf{u}_a^c, \sigma^2 \mathbf{I}))\right] \end{aligned} \tag{4.56}$$

Then, after the latter approximation, we may compute the partial derivatives needed to perform a gradient descent of JS divergence with respect to the deformation parameters:

$$\nabla JS = \left[\frac{\partial JS}{\partial \boldsymbol{\mu}^1}, \frac{\partial JS}{\partial \boldsymbol{\mu}^2}, \ldots, \frac{\partial JS}{\partial \boldsymbol{\mu}^N}\right]^T \tag{4.57}$$

and then, for the state vector (current parameters) $\boldsymbol{\Theta} = ({\boldsymbol{\mu}^1}^T, \ldots, {\boldsymbol{\mu}^N}^T)$ and the weighting vector $\boldsymbol{\Gamma} = ({\boldsymbol{\gamma}^1}^T, \ldots, {\boldsymbol{\gamma}^N}^T)$, we have, for instance

$$\boldsymbol{\Theta}_{t+1} = \boldsymbol{\Theta}_t - \boldsymbol{\Gamma}_n \otimes \nabla JS(\boldsymbol{\Theta}_t) \tag{4.58}$$

where $\otimes$ is the Kronecker product. For the sake of modularity, a key element is to compute the derivative of a Gaussian with respect to a $\boldsymbol{\mu}^c$:

$$\begin{aligned}\frac{\partial G(\mathbf{X}^c_{k_c} - \mathbf{u}^c_a, \sigma^2\mathbf{I})}{\partial \boldsymbol{\mu}^c} &= \frac{1}{\sigma^2} G(\mathbf{X}^c_{k_c} - \mathbf{u}^c_a, \sigma^2\mathbf{I})(\mathbf{X}^c_{k_c} - \mathbf{u}^c_a)\frac{\partial \mathbf{u}^c_a}{\partial \boldsymbol{\mu}^c} \\ &= \frac{1}{\sigma^2} G(\mathbf{X}^c_{k_c} - \mathbf{u}^c_a, \sigma^2\mathbf{I})(\mathbf{X}^c_{k_c} - \mathbf{u}^c_a)\frac{\partial \mathbf{f}^c(\mathbf{v}^c_a, \boldsymbol{\mu}^c)}{\partial \boldsymbol{\mu}^c}\end{aligned} \tag{4.59}$$

Therefore

$$\begin{aligned}\frac{\partial JS}{\partial \boldsymbol{\mu}^c} &= -\frac{1}{M}\sum_{j=1}^{M} \frac{1}{\sum_{a=1}^{K} G(\mathbf{X}_j - \mathbf{u}_a, \sigma^2\mathbf{I})} \sum_{a=1}^{K} \frac{\partial G(\mathbf{X}_j - \mathbf{u}_a, \sigma^2\mathbf{I})}{\partial \boldsymbol{\mu}^c} \\ &\quad + \frac{K^c}{n_c K}\sum_{k_c=1}^{n_c} \frac{1}{\sum_{a=1}^{K^c} G(\mathbf{X}^c_{k_c} - \mathbf{u}^c_a, \sigma^2\mathbf{I})} \sum_{a=1}^{K^c} \frac{\partial G(\mathbf{X}^c_{k_c} - \mathbf{u}^c_a, \sigma^2\mathbf{I})}{\partial \boldsymbol{\mu}^c} \\ &= -\frac{1}{M}\sum_{j=1}^{M} \frac{1}{\sum_{a=1}^{K} G(\mathbf{X}_j - \mathbf{u}_a, \sigma^2\mathbf{I})} \sum_{a=1}^{K^c} \frac{\partial G(\mathbf{X}_j - \mathbf{u}^c_a, \sigma^2\mathbf{I})}{\partial \boldsymbol{\mu}^c} \\ &\quad + \frac{K^c}{n_c K}\sum_{k_c=1}^{n_c} \frac{1}{\sum_{a=1}^{K^c} G(\mathbf{X}^c_{k_c} - \mathbf{u}^c_a, \sigma^2\mathbf{I})} \sum_{a=1}^{K^c} \frac{\partial G(\mathbf{X}^c_{k_c} - \mathbf{u}^c_a, \sigma^2\mathbf{I})}{\partial \boldsymbol{\mu}^c}\end{aligned} \tag{4.60}$$

Starting from estimating the cluster centers $\mathbf{V}^c$ for each shape $\mathbf{X}^c$ using a clustering algorithm (see next chapter), next step consists of perform gradient descent (or its conjugate version) setting Θ_0 with zeros. Either numerical or analytical gradient descent can be used, though the analytical version performs better with large deformations. Recall that the expressions for the gradient do not include the regularization term, which only depends on the nonrigid parameters. Anyway, a re-sampling must be done every couple of iterations. After convergence, the deformed point-sets are close to each other, and then it is possible to recover $\mathcal{Z}$ (mean shape) from the optimal deformation parameters. For the sake of computational efficiency, two typical approaches may be taken (even together). The first one is working with *optimization epochs*, that is, find the affine parameters before finding the nonrigid ones together with the re-estimation of the affine parameters. The second one is *hierarchical optimization*, especially proper for large sets of shapes, that is, large N. In this latter case, M is divided into m subsets. The algorithm is applied to each subset separately and then the global atlas is found. It can be proved (see Prob. 4.9) that minimizing the JS divergence with respect to all M is equivalent to minimizing it with respect to the atlases obtained for each of the m subset. The mathematical meaning is that JS divergence is unbiased and, thus, the hierarchical approach is unbiased too. In general terms, the algorithm works pretty well for atlas construction (see Fig. 4.20), the regularization parameter λ being relatively stable in the range $[0.0001, 0.0005]$. With

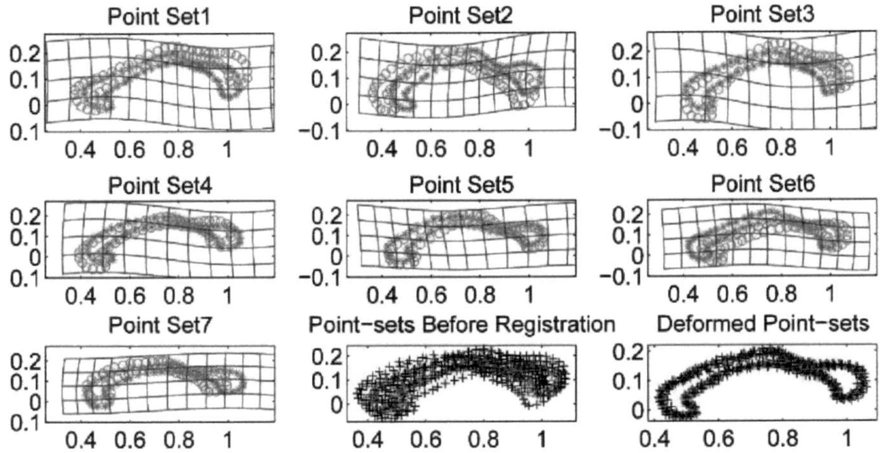

Fig. 4.20. Registering deformations with respect to the atlas and the atlas itself. Point Set1 to Point Set7 show the point-sets in their original state ('o') and after registration with the atlas ('+'). Final registration is showed at the *bottom right*. Figure by F. Wang, B.C. Vemuri, A. Rangarajan and S.J. Eisenschenk (©2008 IEEE).

respect to the robustness (presence of outliers) of the algorithm, it has been proved to be more stable than previous approaches like TRP-RPM [40] or classical registration algorithms like ICP (Iterative Closest Point) [22]. Finally, this method has been extended to 3D, which is a straightforward task.

4.4.3 Information Geometry and Fisher–Rao Information

The distributional shape model described above yields some degree of *bypassing the explicit computation of matching fields*. Thus, when exploiting JS divergence, the problem is posed in a parametric form (find the deformation parameters). Furthermore, a higher degree of bypassing can be reached if we exploit the achievements of *information geometry*. Information geometry deals with mathematical objects in statistics and information theory within the context of *differential geometry* [3,5]. Differential geometry, in turn, studies structures like *differentiable manifolds*, that is, manifolds so smooth that they can be differentiable. This kind of manifolds are studied by Riemannian geometry and, thus, they are dubbed *Riemannian manifolds*. However, it may be a long shot to introduce them formally without any previous intuition. A closer concept to the reader is the one of *statistical manifold*, that is a manifold of probability distributions, like one-dimensional Gaussian distributions. Such manifolds are induced by the parameters characterizing a given family of distributions, and in the case of one-dimensional Gaussians, the manifold is two dimensional $\boldsymbol{\theta} = (\mu, \sigma)$. Then, if a point in the manifold is a probability distribution, what kind of metric is the manifold equipped with? For multi-parameterized distributions $\boldsymbol{\theta} = (\theta_1, \theta_2, \ldots, \theta_N)$, the *Fisher information matrix* is a metric in a Riemannian manifold. The (i, j) component of such matrix is given by

$$\begin{aligned} g_{ij}(\boldsymbol{\theta}) &= \int p(\mathbf{x}|\boldsymbol{\theta}) \left[\frac{\partial}{\partial \theta_i} \log p(\mathbf{x}|\boldsymbol{\theta}) \frac{\partial}{\partial \theta_j} \log p(\mathbf{x}|\boldsymbol{\theta}) \right] d\mathbf{x} \\ &= E\left[\frac{\partial}{\partial \theta_i} \log p(\mathbf{x}|\boldsymbol{\theta}) \frac{\partial}{\partial \theta_j} \log p(\mathbf{x}|\boldsymbol{\theta}) \right] \end{aligned} \tag{4.61}$$

For the trivial case of the one-dimensional Gaussian, we have

$$\begin{aligned} \ln p(x|\mu, \sigma) &= -\ln \sqrt{2\pi} + \ln \frac{1}{\sigma} - \frac{(x-\mu)^2}{2\sigma^2} \\ \frac{\partial}{\partial \mu} \ln p(\mathbf{x}|\boldsymbol{\theta}) &= \frac{(x-\mu)}{\sigma^2} \\ \frac{\partial}{\partial \sigma} \ln p(\mathbf{x}|\boldsymbol{\theta}) &= -\frac{1}{\sigma} + \frac{(x-\mu)^2}{\sigma^3} = \frac{(x-\mu)^2 - \sigma^2}{\sigma^3} \end{aligned} \tag{4.62}$$

Therefore, we obtain

$$g(\mu, \sigma) = \begin{pmatrix} E\left(\frac{(x-\mu)}{\sigma^2}\right)^2 & E\left(\frac{(x-\mu)^3 - (x-\mu)\sigma^2}{\sigma^5}\right) \\ E\left(\frac{(x-\mu)^3 - (x-\mu)\sigma^2}{\sigma^5}\right) & E\left(\frac{(x-\mu)^2 - \sigma^2}{\sigma^3}\right)^2 \end{pmatrix} \tag{4.63}$$

In the general case, $g(\boldsymbol{\theta})$, the so-called *Fisher–Rao metric tensor*, is symmetric and positive definite. But, in what sense $g(\boldsymbol{\theta})$ is a metric? To answer the question we have to look at the differential geometry of the manifold because this manifold is not necessarily flat as in Euclidean geometry. More precisely, we have to focus on the concept of *tangent space*. In differential geometry, the tangent space is defined at any point of the differential manifold and contains the *directions* (vectors) for traveling from that point. For instance, in the case of a sphere, the tangent space at a given point is the plane perpendicular to the sphere radius that touches this and only this point. Consider the sphere with constant curvature (inverse to the radius) as a manifold (which typically has a *smooth* curvature). As we move from $\boldsymbol{\theta}$ to $\boldsymbol{\theta} + \delta\boldsymbol{\theta}$, the tangent spaces change and also the directions. Then, the infinitesimal curve length δs between the latter to points in the manifold is defined as by the equation

$$\delta s^2 = \sum_{ij} g_{ij}(\boldsymbol{\theta}) d\delta\boldsymbol{\theta}_i \delta\boldsymbol{\theta}_j = (\delta\boldsymbol{\theta}^T) g(\boldsymbol{\theta}) \delta\boldsymbol{\theta} \tag{4.64}$$

which is exactly the Euclidean distance if $g(.)$ is the identity matrix. This is why using the Euclidean distance between the pdfs parameters does not generally take into account the geometry of the subspaces where these distributions are embedded. For the case of the Kullback–Leibler divergence, let us consider a one-dimensional variable x depending also on one parameter θ:

$$\begin{aligned} D(p(x|\theta+\delta\theta)||p(x|\theta)) &\equiv D(p(\theta+\epsilon)||p(\theta)) \\ &\equiv D(\epsilon) \approx D(\epsilon)|_{\epsilon=0} + \epsilon D'(\epsilon)|_{\epsilon=0} + \tfrac{1}{2}\epsilon^2 D''(\epsilon)|_{\epsilon=0} \end{aligned} \tag{4.65}$$

which corresponds to the McLaurin expansion. It is obvious that $D(\epsilon)|_{\epsilon=0} = 0$. On the other hand

$$\begin{aligned} D'(\epsilon) = \textstyle\sum_x \Big[&\tfrac{\partial p(x|\theta+\epsilon)}{\partial\theta} \log \tfrac{p(x|\theta+\epsilon)}{p(x|\theta)} \\ &+ p(x|\theta+\epsilon) \left\{ \tfrac{1}{p(x|\theta+\epsilon)} \tfrac{\partial p(x|\theta+\epsilon)}{\partial\theta} + \tfrac{1}{p(x|\theta)} \tfrac{\partial p(x|\theta)}{\partial\theta} \right\} \Big] \end{aligned} \tag{4.66}$$

and consequently obtaining $D'(\epsilon)|_{\epsilon=0} = 0$. Then, for the second derivative, we have

$$\begin{aligned} D''(\epsilon) = \sum_x \Bigg(\Bigg[&\frac{\partial^2 p(x|\theta+\epsilon)}{\partial\theta^2} \log \frac{p(x|\theta+\epsilon)}{p(x|\theta)} \\ &+ \frac{\partial p(x|\theta+\epsilon)}{\partial\theta} \left\{ \frac{1}{p(x|\theta+\epsilon)} \frac{\partial p(x|\theta+\epsilon)}{\partial\theta} + \frac{1}{p(x|\theta)} \frac{\partial p(x|\theta)}{\partial\theta} \right\} \Bigg] \\ &+ \frac{\partial^2 p(x|\theta+\epsilon)}{\partial\theta^2} - \left\{ \frac{\partial p(x|\theta+\epsilon)}{\partial\theta} \frac{1}{p(x|\theta)} - \frac{p(x|\theta+\epsilon)}{p(x|\theta)^2} \frac{\partial p(x|\theta)}{\partial\theta} \right\} \\ &- \frac{p(x|\theta+\epsilon)}{p(x|\theta)} \frac{\partial^2 p(x|\theta+\epsilon)}{\partial\theta^2} \Bigg) \end{aligned} \tag{4.67}$$

Therefore

$$\begin{aligned} D''(\epsilon)|_{\epsilon=0} &= \sum_x \left(\left[\frac{\partial p(x|\theta)}{\partial\theta} \left\{ \frac{1}{p(x|\theta)} \frac{\partial p(x|\theta)}{\partial\theta} + \frac{1}{p(x|\theta)} \frac{\partial p(x|\theta)}{\partial\theta} \right\} \right] \right) \\ &= 2 \sum_x \left(\left[\frac{\partial \log p(x|\theta)}{\partial\theta} \right] \left[\frac{\partial p(x|\theta)}{\partial\theta} \right] \right) \\ &= 2 \sum_x p(x|\theta) \left[\frac{\partial \log p(x|\theta)}{\partial\theta} \right]^2 = 2\, \underbrace{E\left(\left[\frac{\partial \log p(x|\theta)}{\partial\theta} \right]^2 \right)}_{g(\theta)} \end{aligned} \tag{4.68}$$

where $g(\theta)$ is the information that a random variable x has about a given parameter θ. More precisely, the quantity

$$V(x) = \frac{\partial \log p(x|\theta)}{\partial\theta} = \frac{1}{p(x|\theta)} \frac{\partial p(x|\theta)}{\partial\theta} \tag{4.69}$$

is called the *score* of a random variable. It is trivial to prove that its expectation EV is 0 and, thus, $g = EV^2 = var(V)$ is the *variance of the score.* Formally, this is a way of quantifying the amount of information about θ in the data. A trivial example is the fact that the sample average $\bar{x} = \frac{1}{n}\sum_{i=1}^{N} x_i$ is an *estimator* $T(x)$ of the mean θ of a Gaussian distribution. Moreover, $\bar{x}$ is also an *unbiased estimator* because the expected value of the error of the estimator $E_\theta(T(x) - \theta)$ is 0. Then, the Cramér–Rao inequality says that $g(\theta)$ determines a lower bound of the mean-squared error $var(T)$ in estimating θ from the data:

$$var(T) \geq \frac{1}{g(\theta)} \tag{4.70}$$

The proof is trivial if one uses the Cauchy–Schwarz inequality over the product of variances of V and T (see [43], p. 329). Thus, it is quite interesting that $D''(\epsilon)|_{\epsilon=0} \propto g(\theta)$. Moreover, if we return to the McLauring expansion (Eq. 4.65), we have

$$D(p(\theta+\epsilon)||p(\theta)) \approx \epsilon^2 g(\theta) \tag{4.71}$$

whereas in the multi-dimensional case, we would obtain

$$D(p(\theta+\delta\boldsymbol{\theta})||p(\boldsymbol{\theta})) \approx \frac{1}{2}(\delta\boldsymbol{\theta}^T)g(\boldsymbol{\theta})\delta\boldsymbol{\theta} \tag{4.72}$$

which is pretty consistent with the definition of the squared infinitesimal arc-length (see Eq. 4.64). Finally, the Cramér–Rao inequality for the multidimensional case is

$$\Sigma \geq g^{-1}(\boldsymbol{\theta}) \tag{4.73}$$

Σ being the covariance matrix of a set of unbiased estimators for the parameter vector $\boldsymbol{\theta}$.

4.4.4 Dynamics of the Fisher Information Metric

In the latter section, we have presented the Fisher–Rao metric tensor and its connection with the infinitesimal arc-length in a Riemannian manifold. Now, *as we want to travel from a probability distribution to another one through the Riemannian manifold*, we focus on the *dynamics* of the Fisher–Rao metric tensor [34]. The underlying motivation is to connect, as smoothly as possible, two consecutive tangent spaces from the origin distribution toward the final distribution. Such connection is the *affine connection.* In the Euclidean space, as the minimal distance between two points is the straight line, there is no need of considering change of tangents at each point in between because this change of direction does not occur in the Euclidean space. However, and quoting Prof. Amari in a recent videolecture [4]: *These two concepts (minimality of distance, and straightness) coincide in the Riemannian geodesic.* Thus, the *geodesic* is the curve sailing through the surface of the manifold, whose tangent vectors remain parallel during such transportation, and minimizes

$$E = \int_0^1 \sum g(\boldsymbol{\theta}_{ij})\dot{\boldsymbol{\theta}}_i\dot{\boldsymbol{\theta}}_j dt = \int_0^1 \dot{\boldsymbol{\theta}}^T g(\boldsymbol{\theta})\dot{\boldsymbol{\theta}} dt \tag{4.74}$$

being $\dot{\boldsymbol{\theta}}_i = \frac{d\boldsymbol{\theta}_i}{dt}$, and being $t \in [0,1]$ the parameterization of the geodesic path $\boldsymbol{\theta}(t)$. Of course, the geodesic also minimizes $s = \int_0^1 \sqrt{\sum g(\boldsymbol{\theta}_{ij})\dot{\boldsymbol{\theta}}_i\dot{\boldsymbol{\theta}}_j} dt$, the path length. Geodesic computation is invariant to reparameterization and to coordinate transformations. Then, the geodesic can be obtained by minimizing E through the application of the Euler–Lagrange equations characterizing the affine connection[1]

$$\frac{\delta E}{\delta \boldsymbol{\theta}_k} = -2\sum_i g_{ki}\ddot{\boldsymbol{\theta}}_i + \sum_{ij}\left(\left\{\frac{\partial g_{ij}}{\partial \boldsymbol{\theta}_k} - \frac{\partial g_{ik}}{\partial \boldsymbol{\theta}_j} - \frac{\partial g_{kj}}{\partial \boldsymbol{\theta}_i}\right\}\dot{\boldsymbol{\theta}}_i\dot{\boldsymbol{\theta}}_j\right) = 0 \tag{4.75}$$

being $\boldsymbol{\theta}_k$ the kth parameter. The above equations can be rewritten as a function of the *Chirstoffel symbols* of the affine connection

$$\Gamma_{k,ij} = \frac{1}{2}\left\{\frac{\partial g_{ik}}{\partial \boldsymbol{\theta}_j} + \frac{\partial g_{kj}}{\partial \boldsymbol{\theta}_i} - \frac{\partial g_{ij}}{\partial \boldsymbol{\theta}_k}\right\} \tag{4.76}$$

as the following system of second-order differential equations

$$\sum_i g_{ki}\ddot{\boldsymbol{\theta}}_i + \sum_{ij}\left(\Gamma_{k,ij}\dot{\boldsymbol{\theta}}_i\dot{\boldsymbol{\theta}}_j\right) = 0 \tag{4.77}$$

We are assuming a probabilistic shape model with K elements in the mixture, and a common variance σ^2 (free parameter). The only parameters to estimate are the K 2D vectors $\boldsymbol{\mu}_i$ corresponding to the point averages. Thus, each shape is represented by $N = 2K$ parameters $\boldsymbol{\theta} = (\boldsymbol{\mu}_1, \boldsymbol{\mu}_2, \ldots, \boldsymbol{\mu}_N)^T$. This implies that the Fisher tensor is a $N \times N$ matrix, and also that the system of second-order differential equations has N equations. However, it cannot be solved analytically. In these cases, the first tool is the gradient descent:

$$\boldsymbol{\theta}_k^{\tau+1}(t) = \boldsymbol{\theta}_k^{\tau}(t) + \alpha^{\tau}\frac{\delta E}{\delta \boldsymbol{\theta}_k^{\tau}(t)} \tag{4.78}$$

being τ an iteration parameter and α a step one. If we discretize the time from $t = 0$ and $t = 1$ in equally spaced intervals, we initialize $\boldsymbol{\theta}(0)$ with the parameters of one of the shapes and $\boldsymbol{\theta}(1)$ with the parameters of the other. The value of E after finding all parameters yields the distance between both shapes (the length of the geodesic).

Once the geodesics between two shapes are found, it should be interesting to find a way of translating these geodesics to the shape space in such a way that the result is as smoothest as possible, while preserving the

[1] Due to the tight connection with relativity theory, the usual notation for these formulas follow Einstein summation convention. We apply here an *unusual* notation for the sake of clarity.

likelihood. This is called *finding the extrinsic deformations* induced by the intrinsic geodesics. Likelihood preservation is imposed by setting the derivative of the log-likelihood with respect to time. The parameters of the mixture are a sequence of 2D vectors, each one representing a point in the landmark $\boldsymbol{\mu} = (\theta_1, \theta_2)$. The derivative of the log-likelihood referred to that pair of parameters is

$$\frac{d \log L(\mathbf{x}|\boldsymbol{\mu})}{dt} = (\nabla_{\theta_1} \log L)^T \dot{\theta}_1 + (\nabla_{\theta_2} \log L)^T \dot{\theta}_2 + \frac{\partial L}{\partial x_1(t)} u - \frac{\partial L}{\partial x_2(t)} v = 0 \,, \tag{4.79}$$

where $u = \frac{dx_1}{dt}$ and $v = \frac{dx_2}{dt}$ define the vector flow induced by the intrinsic deformation. Therefore, if we enforce smoothing via regularization (thin-plates for instance), the functional to minimize is the following:

$$E = \int \left(\left[\frac{d \log L(\mathbf{x}|\boldsymbol{\mu})}{dt} \right]^2 + \lambda \left[(\nabla^2 u)2 + (\nabla^2 v)2 \right] \right) d\mathbf{x}$$

Example results are shown in Fig. 4.21. When Fisher matrices are based on other estimators like α-order entropy metric tensors, for which closed solutions are available, the results are not so smooth than when the Fisher matrix is used. The α-order metric tensor, for $\alpha = 2$, is defined as

$$g_{ij}^{\alpha}(\boldsymbol{\theta}) = \int \left(\frac{\partial p(\mathbf{x})}{\partial \theta_i} \right) \left(\frac{\partial p(\mathbf{x})}{\partial \theta_j} \right) d\mathbf{x} \tag{4.80}$$

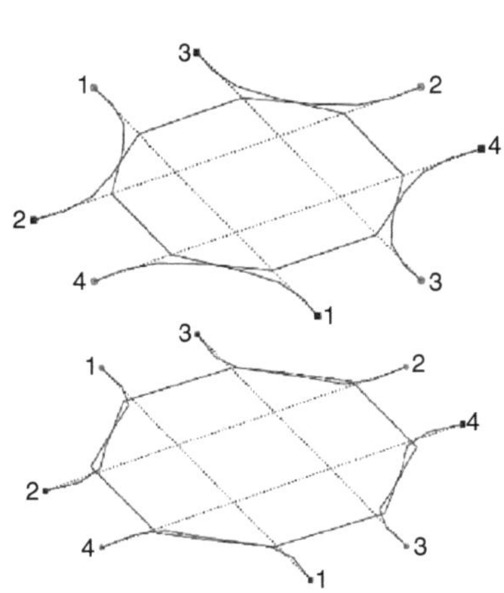

Pairs	Fisher-Rao (10^{-2})	α-Order Entropy (10^{-3})	TPS(10^{-2})
1 vs. 3	14.5905	2.6643	1.4664
1 vs. 4	87.0364	16.2904	10.5958
1 vs. 7	52.8142	10.9610	13.2774
1 vs. 8	3.0751	0.6217	3.0467
2 vs. 4	136.5600	23.8346	17.6227
2 vs. 6	15.8750	3.2368	5.5466
3 vs. 4	153.6000	28.0696	11.1349
3 vs. 6	6.2069	1.1633	2.7389
3 vs. 8	13.3270	2.2415	2.0639
4 vs. 5	59.5560	13.1824	12.8418
4 vs. 7	193.4700	37.7694	34.3238
5 vs. 7	78.6250	12.3373	13.8462
5 vs. 8	20.2640	3.4986	4.0704
6 vs. 7	22.2960	4.0647	5.7042
6 vs. 8	5.9597	1.1400	3.3339
7 vs. 8	40.0820	7.8548	6.8678
7 vs. 9	21.3106	3.9808	5.1912
8 vs. 9	23.9451	5.6155	9.2446

Fig. 4.21. Shape matching results with information geometry. *Left*: Fisher-information (*top*) vs α-entropy (*bottom*). *Dashed lines* are the initializations and solid ones are the final geodesics. Note that the Fisher geodesics are smoother. *Left*: table of distances comparing pairs of shapes 1–9 presented in Fig. 4.19. Figure by A. Peter and A. Rangarajan (©2006 Springer).

An alternative way of solving the above minimization problem is to discretize time in T values and compute the gradients of the following function:

$$E = \sum_{t=1}^{T} \dot{\boldsymbol{\theta}}(t)^T g(\boldsymbol{\theta}(t))\dot{\boldsymbol{\theta}}(t) \tag{4.81}$$

However, in [114] it is noted that the derivatives in $\mathbb{R}^{N\times T}$ are quite unstable, unless at each t, we identify the orthogonal directions that are more likely to contribute more to the energy, and then use only the partial derivatives along these directions to compute the gradient. More precisely, as $\boldsymbol{\theta} \in \mathbb{R}^N$, we diagonalize the $N \times N$ tensor $g(\boldsymbol{\theta})$, which is symmetric, to obtain the orthonormal basis given by the eigenvectors $\{\boldsymbol{\phi}_1, \ldots, \boldsymbol{\phi}_N\}$, assuming that their corresponding eigenvalues are ordered $\lambda_1 > \lambda_2 > \cdots > \lambda_N$. Let then $\boldsymbol{\theta}(t)$ be the unknown point in the geodesic, $\{\boldsymbol{\phi}_j(t)\ j = 1, \ldots, N\}$ its eigenvalues and $\{\lambda_j(t)\ j = 1, \ldots, N\}$ its eigenvectors. The magnitude of each eigenvalue denotes the importance of each of the directions represented by the corresponding eigenvector. Consequently, it is reasonable to truncate the basis to $P \leq N$ eigenvectors attending to retain a given percentage of variability, and thus, of the energy $E(t)$ (at time t). Consequently, if the percentage is fixed, P may vary along time. Then we have the following orthonormal vectors $\{\boldsymbol{\phi}_j(t)\ j = 1, \ldots, P\}$. Let then $\mathbf{V}_j(t) \in \mathbb{R}^{N\times T}$, $j = 1, \ldots, P$ be the vector $(0, \ldots, 0, \boldsymbol{\phi}_j(t), 0, \ldots, 0)^T$. Given $\mathbf{V}_j(t)$, we may define

$$\partial_{jt} E = \frac{E(\boldsymbol{\theta} + \delta \mathbf{V}_j(t)) - E(\boldsymbol{\theta})}{\delta} \tag{4.82}$$

and then, define an approximation of the gradient:

$$\nabla E(\boldsymbol{\theta}) = \frac{\delta E}{\delta \boldsymbol{\theta}} \approx \sum_{j=1}^{P} \sum_{t=1}^{T} (\partial_{jt} E)\mathbf{V}_j(t) \tag{4.83}$$

The gradient search is initialized with the linear path between the shapes and proceeds by deforming the path until convergence.

4.5 Structural Learning with MDL

4.5.1 The Usefulness of Shock Trees

Almost all the data sources used in this book, and particularly in the present chapter, have a vectorial origin. Extending the matching and learning techniques to the domains of graphs is a hard task, which is commencing to intersect with information theory. Let us consider here the simpler case of *trees* where we have a hierarchy which, in addition, is herein assumed to be sampled correctly. Trees have been recognized as good representations for binary

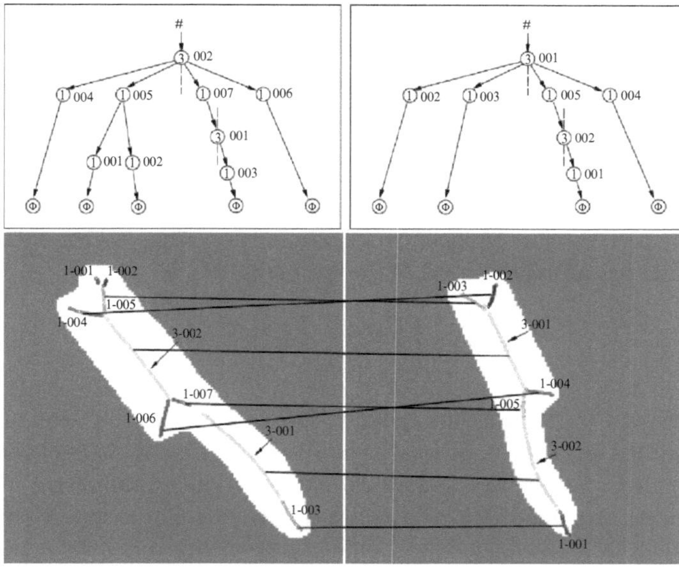

Fig. 4.22. Examples of shock graphs extracted from similar shape and the result of the matching algorithm. Figure by K. Siddiqi, A. Shokoufandeh, S.J. Dickinson and S.W. Zucker (©1999 Springer).

shapes, and tree-versions of the skeleton concept lead to the *shock tree* [145]. A shock tree is rooted in the latest characteristic point (shock) found while computing the skeleton (see Fig. 4.22). Conceptually, the shock graph is built by reversing the grassfire transform, and as a point, is more and more inside the shape, its hierarchy is higher and higher in the tree. Roughly speaking, the tree is build as follows: (i) get the shock point and its branch in the skeleton; (ii) the children of the root are the shocks closer in time of formation and reachable through following a path along a branch; (iii) repeat assuming that each child is a parent; (iv) the leaves of the tree are terminals without a specific shock label (that is, each earlier shock generates a unique child, a terminal node). Using a tree instead of a graph for representing a shape has clear computational advantages because they are easier to match and also to learn than graphs.

4.5.2 A Generative Tree Model Based on Mixtures

The *unsupervised learning of trees* (and consequently of the shapes they represent or encode) is the main topic of this section. Consider that we observe a tree t which is coherent with a model $\mathcal{H}$. We commence by assuming that the probability of observing the tree given the model $\mathcal{M}$ is a mixture k conditional probabilities of observing the tree given models $\mathcal{T}_1, \ldots, \mathcal{T}_k$ [156]:

$$P(t|\mathcal{M}) = \sum_{c=1}^{k} \alpha_c P(t|\mathcal{T}_c) \qquad (4.84)$$

As we have noted above, for the sake of simplicity, sampling errors affect the nodes, but hierarchical relations are preserved. Each tree model $\mathcal{T}_c$ is defined by a set of nodes $\mathcal{N}_c$, a tree order $\mathcal{O}_c$, and the observation probabilities assigned to the nodes of a tree coherent with model $\mathcal{T}_c$ are grouped into the set $\Theta_c = \{\theta_{c,i}\} : i \in \mathcal{N}_c$. Therefore, the probability of observing a node i in the set of trees assigned to the model (class) c is denoted by $\theta_{c,i}$. A simple assumption consists of considering that the nodes of t are independent Bernoulli samples from the model $\mathcal{T}_c$. The independence assumption leads to the factorization of $P(t|\mathcal{T}_c)$. If t is an observation of $\mathcal{T}_c$, we have

$$P(t|\mathcal{T}_c) = \prod_{i \in \mathcal{N}_t} \theta_{c,i} \prod_{j \in \mathcal{N}_c/\mathcal{N}_t} (1 - \theta_{c,i}) \tag{4.85}$$

and otherwise $P(t|\mathcal{T}_c) = 0$. In order to avoid unconnected tree samples, it is also assumed that the root is always observed, that is, it is observed with probability 1. Consequently, structural noise can only eliminate a node (typically a leave), provided that the remaining tree is connected. However, the correspondences between nodes of different trees are not known beforehand and must be inferred. Such correspondences must satisfy the hierarchical constraints (if two nodes belong to the same hierarchy level-order, then their images through the correspondence should also belong to the same order). Then, an interesting question is how to both estimate the models $\mathcal{H}, \mathcal{T}_1, \ldots, \mathcal{T}_k$ that best fit the observed trees $\mathcal{D} = \{t_1, t_2, \ldots, t_N\}$, and the correspondences $\mathcal{C}$ which map nodes of different trees. It is important to note here that the resulting models depend on the correspondences found. The underlying idea is that if a given substructure or structural relation is highly preserved in the data (very frequent), it is reasonable to find it into the model. Regarding information theory, the minimum description length criterion is quite useful here in order to quantify the complexity of a tree model. As we have seen in the previous chapter, the MDL is the sum of the negative of the log-likelihood (cost of describing the data) and the cost of describing the model. The cost, if describing a tree t given the tree model $\mathcal{T}_c$ and the correspondences $\mathcal{C}$, is defined by a finite number when the hierarchy constraints are satisfied:

$$-\log P(t|\mathcal{T}_c, \mathcal{C}) = -\sum_{i \in Im(\mathcal{C})} \log \theta_{c,i} - \sum_{j \in \mathcal{N}/Im(\mathcal{C})} \log(1 - \theta_{c,j}) \tag{4.86}$$

being $Im(.)$ the image of the correspondence, and it is assumed to be ∞ otherwise. Then, the cost of describing a data set $\mathcal{D}$ given $\mathcal{H}, \mathcal{T}_1, \ldots, \mathcal{T}_k$ and $\mathcal{C}$ is defined as

$$L(\mathcal{D}|\mathcal{H}, \mathcal{C}) = -\sum_{t \in \mathcal{D}} \log P(t|\mathcal{H}, \mathcal{C}) = -\sum_{t \in \mathcal{D}} \log \left(\sum_{c=1}^{k} \alpha_c P(t|\mathcal{T}_c, \mathcal{C}) \right) \tag{4.87}$$

Adding another simplifying assumption (a tree t can only be the observation of a unique model), we have hidden binary variables that link trees and modes:

$z_c^t = 1$ if the tree t is observed from model $\mathcal{T}_c$, and $z_c^t = 0$ otherwise. This simplifies the cost of describing the data:

$$L(\mathcal{D}|\mathbf{z}, \mathcal{H}, \mathcal{C}) = -\sum_{t\in\mathcal{D}}\sum_{c=1}^{k} z_c^t \log P(t|\mathcal{T}_c, \mathcal{C}) = \sum_{c=1}^{k}\sum_{t\in\mathcal{D}_c} \log P(t|\mathcal{T}_c, \mathcal{C}) \tag{4.88}$$

being $\mathcal{D}_c = \{t \in \mathcal{D} : z_c^t = 1\}$. On the other hand, the cost of describing the full model $\mathcal{H}$ is built on three components. The first one comes from quantifying the cost of encoding the observation probabilities $\hat{\Theta}_c$:

$$L(\hat{\Theta}_c|\mathbf{z}, \mathcal{H}, \mathcal{C}) = \sum_{c=1}^{k} \frac{n_c}{2}\log(m_c) \tag{4.89}$$

being n_c the number of nodes in the model $\mathcal{T}_c$ and $m_c = \sum_{t\in\mathcal{D}} z_c^t$ the number of trees assigned to the model $\mathcal{T}_c$. The second component comes from quantifying the cost of the mapping implemented by $\mathbf{z}$:

$$L(\mathbf{z}|\mathcal{H}, \mathcal{C}) = -\sum_{c=1}^{k} \log \alpha_c m_c \tag{4.90}$$

The third component concerns the coding of the correspondence $\mathcal{C}$. Assuming equally likely correspondences, the main remaining question is to estimate properly the number of such correspondences. If one takes the number of ordered trees (trees with a number in each node where the hierarchy imposes a partial order) with n nodes, the cost is overestimated because in this context only unordered trees are considered. For this wider class of trees, the Wedderburn–Etherington numbers are commonly used in graph theory. Their asymptotic behavior is exponential: $1, 1, 1, 2, 3, 6, 11, 23, 46, 98, 207, 451, 983, 2{,}179, 4{,}850, 10{,}905, 24{,}631, 56{,}011, 127{,}912, 293{,}547, \ldots$. Thus, the logarithm may be given by $ng + \mathit{const.}$, where $g = (\zeta + \lambda)$, $\zeta \approx 1.31$ and λ is a prior term. Therefore, integrating the three components we have

$$L(\mathcal{H}|\mathcal{C}) = g\sum_{c=1}^{k} n_c + \frac{k-1}{2}\log(m) + \mathit{const.} \tag{4.91}$$

where n_c is the number of nodes in model $\mathcal{T}_c$, and the second term with $m = |\mathcal{D}|$ is the cost of describing the mixing parameters $\boldsymbol{\alpha}$. Dropping the constant, the MDL in this context is given by

$$\begin{aligned} &\mathit{MDL}(\mathcal{D}, \mathcal{H}|\mathcal{C}) \\ &= \sum_{c=1}^{k}\left\{-\sum_{t\in\mathcal{D}_c}\log P(t|\mathcal{T}_c, \mathcal{C}) + n_c\left(\frac{\log(m_c)}{2} + g\right) - m_c \log \alpha_c\right\} \\ &+ \frac{k-1}{2}\log(m)\,. \end{aligned} \tag{4.92}$$

Considering the log-likelihood, let $K_{c,i}^t = \{j \in \mathcal{N}^t | \mathcal{C}(j) = i\}$ be the set of nodes in $\mathcal{D}_c$ for which there is a correspondence with a node $i \in \mathcal{N}^c$. One-to-one matching constrains result in having singletons or empty sets for $K_{c,i}^t$ (its size is either one or zero, respectively). Then, if $l_{c,i} = \sum_{t \in \mathcal{D}_c} |K_{c,i}^t|$ is the number of trees in the data mapping a node to i, and $m_c = |\mathcal{D}_c|$ is the number of trees assigned to $\mathcal{T}_c$, then the maximum likelihood estimation of sampling probability under the Bernouilli model is $\theta_{c,i} = \frac{l_{c,i}}{m_c}$. Then, we have

$$
\begin{aligned}
&\sum_{t \in \mathcal{D}_c} \log P(t|\mathcal{T}_c, \mathcal{C}) \\
&\quad = \sum_{i \in \mathcal{N}_c} m_c \left[\frac{l_{c,i}}{m_c} \log \frac{l_{c,i}}{m_c} + (1 - \frac{l_{c,i}}{m_c}) \log \left(1 - \frac{l_{c,i}}{m_c} \right) \right] \\
&\quad = \sum_{i \in \mathcal{N}_c} m_c \left[\theta_{c,i} \log \theta_{c,i} + (1 - \theta_{c,i}) \log (1 - \theta_{c,i}) \right] \\
&\quad = - \sum_{i \in \mathcal{N}_c} m_c H(\theta_{c,i})
\end{aligned} \tag{4.93}
$$

that is, the individual log-likelihood with respect to each model is given by the weighted entropies of the individual Bernoulli ML estimations corresponding to each node. Furthermore, we estimate the mixing proportions α_c as the observed frequency ratio, that is, $\alpha_c = \frac{m_c}{m}$. Then, defining $H(\boldsymbol{\alpha}) = - \sum_{c=1}^{k} \alpha_c$, we have the following resulting MDL criterion:

$$
\begin{aligned}
&MDL(\mathcal{D}, \mathcal{H}|\mathcal{C}) \\
&\quad = \sum_{c=1}^{k} \sum_{i \in \mathcal{N}_c} \left\{ m_c H(\theta_{i,c}) + \log \frac{m_c}{2} + g + mH(\boldsymbol{\alpha}) + \frac{k-1}{2} \log(m) \right\}
\end{aligned} \tag{4.94}
$$

.

4.5.3 Learning the Mixture

How to learn k tree models from the N input trees in $\mathcal{D}$ so that the MLD cost is minimized? A quite efficient method is to pose the problems in terms of *agglomerative clustering* (see a good method for this in the next chapter). The process is illustrated in Fig. 4.22. Start by having N clusters, that is, one cluster/class per tree. This implies assigning a tree model per sample tree. In each tree, each node has unit sample probability. Then we create one mixture component per sample tree, and compute the MDL cost for the complete model assuming an equiprobable mixing probability. Then, proceed by taking all possible pairs of components and compute tree unionstree unions. The mixing proportions are equal to the sum of proportions of the individual unions. Regarding the sampling probabilities of merged nodes, let $m_i = |\mathcal{D}_i|$

and $m_j = |\mathcal{D}_j|$ the number of tree samples assigned respectively to $\mathcal{T}_i$ and $\mathcal{T}_j$. Let also l_u and l_v be the number of times the nodes in $u \in \mathcal{T}_i$ and $v \in \mathcal{T}_j$ are observed in $\mathcal{D}_i$ and $\mathcal{D}_j$, respectively. Then, if the nodes are not merged, their sampling probabilities are $\theta_u = \frac{l_u}{m_i+m_j}$ and $\theta_v = \frac{l_v}{m_i+m_j}$. However, if the nodes are merged, the sampling probability of the resulting node is $\theta_{uv} = \frac{l_u+l_v}{m_i+m_j}$. For the sake of computational efficiency, in [156] it is defined as the concept of *minimum description length advantage* (MDLA) for two nodes in the following terms:

$$\begin{aligned} MDLA(u,v) &= MDL^u + MDL^v - MDL^{uv} \\ &= (m_i + m_j)[H(\theta_u) + H(\theta_v) - H(\theta_{uv}] \\ &+ \frac{1}{2}\log(m_i + m_j) + g \end{aligned} \tag{4.95}$$

and then, the MDLA for a set of merges $\mathcal{M}$ is

$$MDLA(\mathcal{M}) = \sum_{(u,v)\in\mathcal{M}} MDLA(u,v) \tag{4.96}$$

Given all the possibilities of merging two trees, choose the one which reduces MDL by the greatest amount. If all the proposed merges increase the MDL, then the algorithm stops and does not accept any merging. If one of the mergings succeeds, then we have to compute the costs of merging it with the remaining components and proceed again to choose a new merging. This process continues until no new merging can drop the MDL cost.

4.5.4 Tree Edit-Distance and MDL

Estimating the edit distance, the cost of the optimal sequence of edit operations over nodes and edges is still an open problem in structural pattern matching. In [156], an interesting link between edit distance and MDL is established. Given the optimal correspondence $\mathcal{C}$, the tree-edit distance between trees t and t' is given by

$$D^{edit}(t,t') = \sum_{u\notin dom(\mathcal{C})} r_u + \sum_{v\notin ima(\mathcal{C})} r_v + \sum_{(u,v)\in\mathcal{C}} m_{uv} \tag{4.97}$$

where r_u and r_v are the costs of removing nodes u and v, respectively, whereas m_{uv} is the cost of matching nodes u and v. In terms of sets of nodes of a tree $\mathcal{N}^t$, the edit distance can be rewritten as

$$D^{edit}(t,t') = \sum_{u\in\mathcal{N}^t} r_u + \sum_{v\in\mathcal{N}^{t'}} r_v + \sum_{(u,v)\in\mathcal{C}} (m_{uv} - r_u - rv) \tag{4.98}$$

but what are the costs of r_u, r_v and m_{uv}? Attending to the MDL criterion, these costs are the following ones:

$$
\begin{aligned}
r_z &= (m_t + m_{t'})H(\theta_z) + \frac{1}{2}\log(m_t + m_{t'}) + g, \quad z \in \{u, v\} \\
m_{uv} &= (m_t + m_{t'})H(\theta_{uv}) + \frac{1}{2}\log(m_t + m_{t'})
\end{aligned}
\tag{4.99}
$$

Therefore, edit costs are closely related to the information of the variables associated to the sampling probabilities. Shape clustering examples using these edit costs and spectral clustering are shown in Fig. 4.23 (bottom). In these

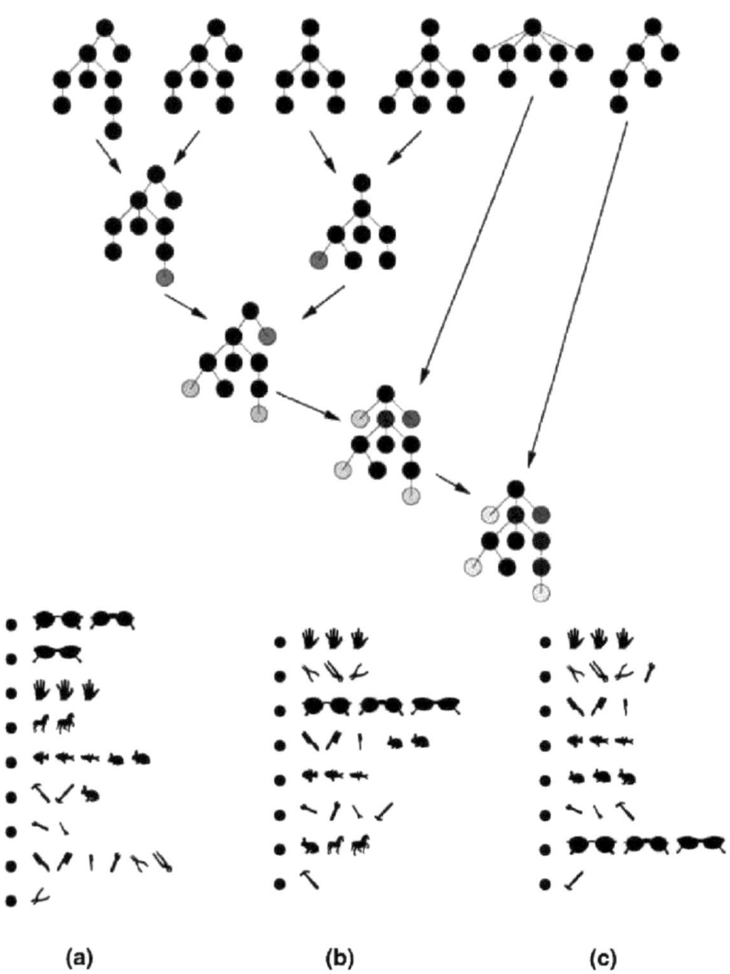

Fig. 4.23. Tree learning. *Top*: dynamic formation of the tree union. The darker the node, the higher its sampling probability. *Bottom*: results of shape clustering from (**a**) mixture of attributed trees; (**b**) weighted edit distance; and (**c**) union of attributed trees. Figure by A. Torsello and E.R. Hancock (©2006 IEEE).

experiments, the *weighted/attribute* versions of both the edit-distance and the union trees come from generalizing the pure structural approach described above to include weights in the trees. These weights are associated to nodes, follow a particular distribution (Gaussian for instance) and influence the probability of sampling them. Finally, it is important to note that other IT principles like *minimum message length* (MML) and other criteria explained along the book can drive this approach [155].

Problems

4.1 Distributions in multidimensional spaces

Perform the following experiment using Octave, Matlab, or some other similar tool. Firstly, generate 100 random points in a 1D space (i.e., 100 numbers). Use integer values between 0 and 255. Calculate the histogram with 256 bins and look at its values. Secondly, generate 100 random points in a 2D space (i.e., 100 pairs of numbers). Calculate the 2D histogram with 256 bins and look at the values. Repeat the experiment with 3D and then with some high dimensions. High dimensions cannot be represented, but you can look at the maximum and minimum values of the bins. What happens to the values of the histogram? If entropy is calculated from the distributions you estimated, what behavior would it present in a high-dimensional space? What would you propose to deal with this problem?

4.2 Parzen window

Look at the formula of the Parzen window's method (Eq. 4.8). Suppose we use in it a Gaussian kernel with some definite covariance matrix ψ and we estimate the distribution of the pixels of the image I (with 256 gray levels). Would the resulting distribution be the same if we previously smooth the image with a Gaussian filter? Would it be possible to resize (reduce) the image in order to estimate a similar distribution? Discuss the differences. You can perform experiments on natural images in order to notice the differences, or you can try to develop analytical proofs.

4.3 Image alignment

Look at Fig. 4.2. It represents the values of two different measures for image alignment. The x and z axes represent horizontal and vertical displacement of the image and the vertical axis represents the value of the measure. It can be seen that mutual information yields smoother results than the normalized cross correlation. Suppose a simpler measure consists in the difference of pixels of the images I and I': $\sum_x \sum_y I_{x,y} - I'_{x,y}$. How would the plot of this measure look like? Use Octave, Matlab or some other tool to perform the experiments on a natural image. Try other simple measures and see which one would be more appropriate for image alignment.

4.4 Joint histograms

In Fig. 4.3, we show the classical way to build a joint histogram. For two similar or equal images (or sets of samples), the joint histogram has high values in its diagonal. Now suppose we have two sets of samples generated by two independent random variables. (If two events are independent, their joint probability is the product of the prior probabilities of each event occurring by itself, $P(A \cap B) = P(A)P(B)$.) How would their joint histogram (see Fig. 4.3) look?

4.5 The histogram-binning problem
Rajwade et al. proposed some methods for dealing with the histogram-binning problem. Their methods present a parameter Q, which quantifies the intensity levels. This parameter must not be confused with the number of bins in classical histograms. Explain their difference. Why setting Q is much more robust than setting the number of histogram bins? Think of cases when the classical histogram of two images would be similar to the area-based histogram. Think of other cases when both kinds of histograms would present a significant difference.

4.6 Alternative metrics and pseudometrics
Figure 4.11 represents the concept of mutual information in a Venn diagram. Draw another one for the $\rho(W, X, Y, Z)$ measure of four variables and shade the areas corresponding to that measure. Think of all the possible values it could take. Make another representation for the conditional mutual information.

4.7 Entropic graphs
Think of a data set formed by two distributions. If we gradually increase the distance of these distributions until they get completely separated, the entropy of the data set would also increase. If we continue separating more and more the distributions, what happens to the entropy value for each one of the following estimators? (a) Classical histograms plugin estimation; (b) Parzen window; (c) minimal spanning tree graph-based estimation; and (d) k-nearest neighbors graph-based estimation, for some fixed k.

4.8 Jensen–Rényi divergence
For the following categorical data sets: $\mathbf{S}_1 = \{1,2,1\}$,$\mathbf{S}_2 = \{1,1,3\}$, and $\mathbf{S}_3 = \{3,3,3,2,3,3\}$, calculate their distributions and then their Jensen–Rényi divergence, giving $\mathbf{S}_1$ and $\mathbf{S}_2$ the same weight and $\mathbf{S}_3$ the double.

4.9 Unbiased JS divergence for multiple registration
Prove that the JS divergence is unbiased for multiple registration of M shapes, that is, the JS divergence with respect to the complete set is equal to the JS divergence with respect to the $m << N$ subsets in which the complete set is partitioned. A good clue for this task is to stress the proof of $JS_{\boldsymbol{\pi}}(p_1, \ldots p_N) - JS_{\boldsymbol{\beta}}(S_1, \ldots S_N) = 0$, where the S_i are the subsets and $\boldsymbol{\beta}$ must be properly defined.

4.10 Fisher matrix of multinomial distributions
The Fisher information matrix of multinomial distributions is the key to specify the Fisher information matrix of a mixture of Gaussians [101]. Actually,

the parameters of the multinomial in this case are the mixing components (which must form a convex combination – sum 1). Derive analytically the Fisher matrix for that kind of distributions.

4.11 The α-order metric tensor of Gaussians and others
An alternative tensor to the Fisher one is the α-order metric one. When $\alpha = 2$, we obtain the expression in Eq. 4.80. Find an analytical expression of this tensor for the Gaussian distribution. Do the same with the multinomial distribution. Compare the obtained tensors with the Fisher–Rao ones.

4.12 Sampling from a tree model
Let $\mathcal{T}_c$ be a tree model consisting of one root node with two children. The individual probability of the root is 1, the probability of the left child is 0.75 and that of the right child is 0.25 (probability is normalized at the same level). Quantify the probability of observing all the possible subgraphs, including the complete tree, by following the independent Bernoulli model (Fig. 4.22)

4.13 Computing the MDL of a tree merging process
Given the definition of MDL for trees, compute the sampling probabilities and MDLs for all steps of the application of the algorithm to the example illustrated in Fig. 4.22. Compute the MDL advantage in each iteration of the algorithm.

4.14 Computing the MDL edit distance
Compute the MDL edit distance for all the matchings (mergings) illustrated in Fig. 4.22.

4.15 Extending MDL for trees with weights
As we have explained along the chapter, it is possible to extend the pure structural framework to an attributed one, where attributes are assigned to nodes. Basically this consists of modeling a weight distribution function, where most relevant nodes have associated larger weights. From the simple assumption that weights are Gaussian distributed, the probability of a given weight can be defined as

$$P_w(w|\mu_{c,i},\sigma_{c,i}) = \begin{cases} \frac{1}{\theta_{c,i}\sigma_{c,i}\sqrt{2\pi}} e^{\left(-\frac{1}{2}\frac{(w-\mu_{c,i})^2}{\sigma_{c,i}^2}\right)} & \text{if } w \geq 0 \\ 0 & \text{otherwise} \end{cases}$$

and the sample probability $\theta_{c,i}$ is the integral of the distribution over positive weights:

$$\theta_{c,i} = \int_0^\infty \frac{1}{\sigma_{c,i}\sqrt{2\pi}} e^{\left(-\frac{1}{2}\frac{(w-\mu_{c,i})^2}{\sigma_{c,i}^2}\right)} dw$$

Given the latter definitions, modify the log-likelihood by multiplying the Bernoulli probability by $P_w(.|.)$ only in the positive case (when we consider $\theta_{c,i}$). Obtain a expression for the MDL criterion under these new conditions. Hints: consider both the change of the observation probability and the addition of two new variables per node (which specify the Gaussian for its weight).

4.6 Key References

- P. Viola and W. M. Wells-III. "Alignment by Maximization of Mutual Information". *5th International Conference on Computer Vision* 24(2): 137–154 (1997)
- J. P. W. Pluim, J. B. A. Maintz, and M. A. Viergever. "Mutual-information-based Registration of Medical Images: A Survey". *IEEE Transactions on Medical Imaging* 22(8): 986–1004 (2003)
- J. Zhang and A. Rangarajan. "Affine Image Registration Using a New Information Metric". *IEEE Conference on Computer Vision and Pattern Recognition* 1: 848–855 (2004)
- A. Rajwade, A. Banerjee and A. Rangarajan. "Probability Density Estimation Using Isocontours and Isosurfaces: Application to Information Theoretic Image Registration". *IEEE Transactions on Pattern Analysis and Machine Intelligence* 31(3): 475–491 (2009)
- A. Ben Hamza and H. Krim. "Image Registration and Segmentation by Maximizing the Jensen-Rényi Divergence". Energy Minimization Methods in Computer Vision and Pattern Recognition – LNCS 2683: 147–163 (2001)
- H. Neemuchwala, A. Hero, and P. Carson. "Image Registration in High Dimensional Space", *International Journal on Imaging Systems and Technology* h16(5): 130–145 (2007)
- A. Peter and A. Rangarajan. "Information Geometry for Landmark Shape Analysis: Unifying Shape Representation and Deformation". *IEEE Transactions on Pattern Analysis and Machine Intelligence* 31(2): 337–350 (2009)
- F. Wang, B. Vemuri, A. Rangarajan, I. M. Schmalfuss, and S.J. Eisenschenck. "Simultaneous Registration of Multiple Point-Sets and Atlas Construction". *IEEE Transactions on Pattern Analysis and Machine Intelligence* 30(11): 2011–2022 (2008)
- A. Torsello and E. R. Hancock. "Learning Shape-Classes Using a Mixture of Tree-Unions". *IEEE Transactions on Pattern Analysis and Machine Intelligence* 28(6): 954–967 (2006)

5

Image and Pattern Clustering

5.1 Introduction

Clustering, or grouping samples which share similar features, is a recurrent problem in computer vision and pattern recognition. The core element of a clustering algorithm is the similarity measure. In this regard information theory offers a wide range of measures (not always metrics) which inspire clustering algorithms through their optimization. In addition, information theory also provides both theoretical frameworks and principles to formulate the clustering problem and provide effective algorithms. Clustering is closely related to the segmentation problem, already presented in Chapter 3. In both problems, finding the optimal number of clusters or regions is a challenging task. In the present chapter we cover this question in depth. To that end we explore several criteria for model order selection.

All the latter concepts are developed through the description and discussion of several information theoretic clustering algorithms: Gaussian mixtures, Information Bottleneck, Robust Information Clustering (RIC) and IT-based Mean Shift. At the end of the chapter we also discuss basic strategies to form clustering ensembles.

5.2 Gaussian Mixtures and Model Selection

5.2.1 Gaussian Mixtures Methods

A probability mixture model is a probability distribution which is the result of the combination of other probability distributions. Mixture models, in particular those formed by Gaussian distributions (or kernels), are widely used in areas involving statistical modeling of data. In statistical pattern recognition, mixture models allow a formal approach to clustering [82]. In traditional clustering methods, different heuristics (e.g. k-means) or agglomerative methods are used [81], while mixture models have a set of parameters which can be adjusted in a formal way. The estimation of these parameters is a clustering, provided that each sample belongs to some kernel of the set of kernels in the mixture. In Bayesian supervised learning the mixture models are used for representation of class-conditional probability distributions [72] and for Bayesian parameter estimation [46].

5.2.2 Defining Gaussian Mixtures

In a mixture model the combination of distributions has to be convex, that is, a linear combination with non-negative weights. Let us define a d-dimensional random variable $\mathbf{y}$ which follows a finite mixture distribution. Its probability density function (pdf) $p(y|\Theta)$ is described by a weighted sum of kernels. These kernels are known distributions, which, in the case of Gaussian mixtures, consist of Gaussian distributions.

$$p(\mathbf{y}|\Theta) = \sum_{i=1}^{K} \pi_i p(\mathbf{y}|\Theta_i) \tag{5.1}$$

$$\text{where } 0 \leq \pi_i \leq 1, \quad i = 1, ..., K, \text{ and } \sum_{i=1}^{K} \pi_i = 1,$$

where K is the number of kernels, $\pi_1, ..., \pi_k$ are the a priori probabilities of each kernel, and Θ_i are the parameters describing the kernel. In Gaussian mixtures these parameters are the means and the covariance matrices, $\Theta_i = \{\mu_i, \Sigma_i\}$. In Fig. 5.1, see an example of 1D data following a Gaussian mixture distribution formed by $K = 3$ kernels of different parameters of each one of them.

Thus, the whole set of parameters of a mixture is $\Theta \equiv \{\Theta_1, ..., \Theta_k, \pi_1, ..., \pi_k\}$. Obtaining the optimal set of parameters Θ^* is usually posed in terms of maximizing the log-likelihood of the pdf to be estimated:

$$\begin{aligned} \ell(Y|\Theta) &= \log p(Y|\Theta) = \log \prod_{n=1}^{N} p(y_n|\Theta) \\ &= \sum_{n=1}^{N} \log \sum_{k=1}^{K} \pi_k p(y_k|\Theta_k). \end{aligned} \tag{5.2}$$

where $Y = \{y_1, ..., y_N\}$ is a set of N independent and identically distributed (i.i.d.) samples of the variable Y. The *Maximum-Likelihood* (ML) estimation

$$\Theta^*_{MT} = \arg\max_{\Theta} \ell(\Theta). \tag{5.3}$$

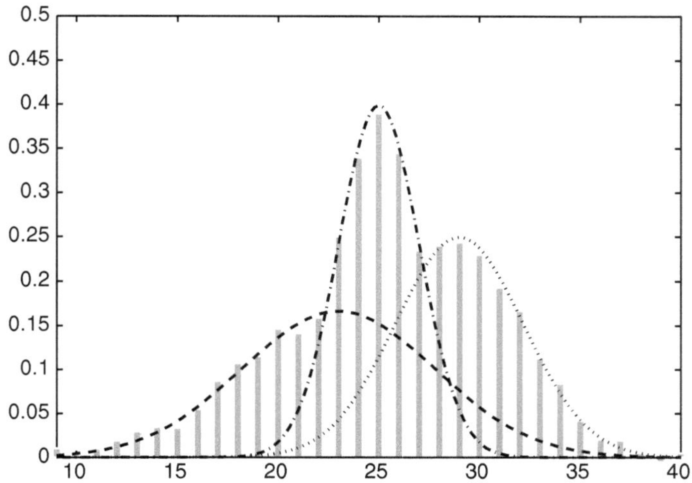

Fig. 5.1. Data of one dimension whose distribution can be approximated by a Gaussian mixture with three kernels of different means and variances.

cannot be determined analytically. The same happens with the *Bayesian Maximum a Posteriori* (MAP) criterion

$$\Theta^*_{MAP} = \arg\max_{\Theta}(\ell(\Theta) + \log p(\Theta)). \tag{5.4}$$

In this case other methods are needed, like expectation maximization (EM) or Markov-chain Monte Carlo algorithms.

5.2.3 EM Algorithm and Its Drawbacks

The expectation maximization algorithm is widely used for fitting mixtures to data [49, 112]. It allows to find maximum-likelihood solutions to problems in which there are *hidden variables*. In the case of Gaussian mixtures, these variables are a set of N labels $Z = \{z^1, ..., z^N\}$ associated to the samples. Each label is a binary vector $z^i = [z_1^{(i)}, ..., z_k^{(i)}]$, with $z_m^{(i)} = 1$ and $z_p^{(i)} = 0$, if $p \neq m$, indicating that $y^{(i)}$ has been generated by the kernel m. The log-likelihood of the complete set of data $X = \{Y, Z\}$ is

$$\log p(Y, Z|\Theta) = \sum_{n=1}^{N}\sum_{k=1}^{K} z_k^n \log[\pi_k p(y_n|\Theta_k)]. \tag{5.5}$$

The EM algorithm is an iterative procedure. It generates a sequence of estimations of the set of parameters $\{\Theta^*(t), t = 1, 2, ...\}$ by alternating two steps: the expectation step and the maximization step, until convergence is achieved. A detailed description of the algorithm is given in [136] and an illustration of its results is given in Fig. 5.2.

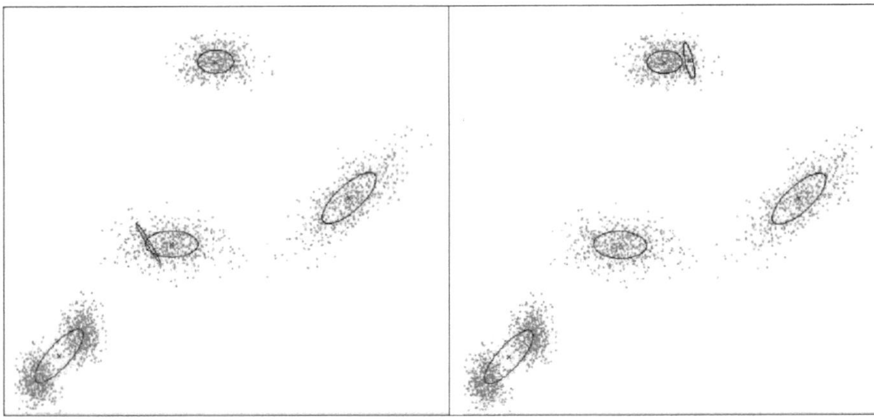

Fig. 5.2. The classical EM is prone to random initialization. The algorithm converges to different local minima in different executions.

E Step

It consists of estimating the expected value of the hidden variables given the visible data Y and current estimation of the parameters $\Theta^*(t)$. Such an expectation can be expressed in the following way:

$$E[z_k^{(n)}|y, \Theta^*(t)] = P[z_k^{(n)} = 1|y, \Theta^*(t)]) = \frac{\pi_k^*(t)p(y^{(n)}|\Theta_k^*(t))}{\Sigma_{j=1}^K \pi_j^*(t)p(y^{(n)}|\Theta_k^*(t))}, \tag{5.6}$$

Thus, the probability of generating $\mathbf{y}_n$ with the kernel k is given by

$$p(k|\mathbf{y}_n) = \frac{\pi_k p(\mathbf{y}^{(n)}|k)}{\Sigma_{j=1}^K \pi_j p(\mathbf{y}^{(n)}|k)} \tag{5.7}$$

M Step

Given the expected Z, the new parameters $\Theta^*(t+1)$ are given by

$$\pi_k = \frac{1}{N}\sum_{n=1}^{N} p(k|\mathbf{y}_n), \tag{5.8}$$

$$\mu_k = \frac{\sum_{n=1}^N p(k|\mathbf{y}_n)\mathbf{y}_n}{\sum_{n=1}^N p(k|\mathbf{y}_n)}, \tag{5.9}$$

$$\Sigma_k = \frac{\sum_{n=1}^N p(k|\mathbf{y}_n)(\mathbf{y}_n - \mu_k)(\mathbf{y}_n - \mu_k)^T}{\sum_{n=1}^N p(k|\mathbf{y}_n)}, \tag{5.10}$$

The EM algorithm results depend very much on the initialization and it usually converges to some local maximum of the log-likelihood function, which does not ensure that the pdfs of the data are properly estimated. In addition, the algorithm requires that the number of elements (kernels) in the mixture is known beforehand. A maximum-likelihood criterion with respect to the number of kernels is not useful because it tends to fit one kernel for each sample.

5.2.4 Model Order Selection

The model order selection problem consists of finding the most appropriate number of clusters in a clustering problem or the number of kernels in the mixture. The number of kernels K is unknown beforehand and cannot be estimated through maximizing the log-likelihood because $\ell(\Theta)$ grows with K. In addition, a wrong K may drive the EM toward an erroneous estimation. In Fig. 5.3 we show the EM result on the parameters of a mixture with $K = 1$ describing two Gaussian distributions.

The model order selection problem has been addressed in different ways. There are algorithms which start with a few number of kernels and add new kernels when necessary. Some measure has to be used in order to detect when some kernel does not fit well the data, and a new kernel has to be added. For example, in [172], a kernel is split or not, depending on the kurtosis measure, used as a measure of non-Gaussianity. Other model order selection methods start with a high number of kernels and fuse some of them. In [55, 56, 59] the EM algorithm is initialized with many kernels randomly placed and then the minimum description length (MDL) principle [140] is used to iteratively remove some of the kernels until the optimal number of them is achieved. Some other approaches are used both to split and fuse kernels. In [176] a general statistical learning framework called de Bayesian Ying-Yang system is proposed and suitable for using for model selection and unsupervised learning of finite mixtures. Other approaches combine EM and genetic algorithms for learning mixtures, using an MDL criterion for finding the best K.

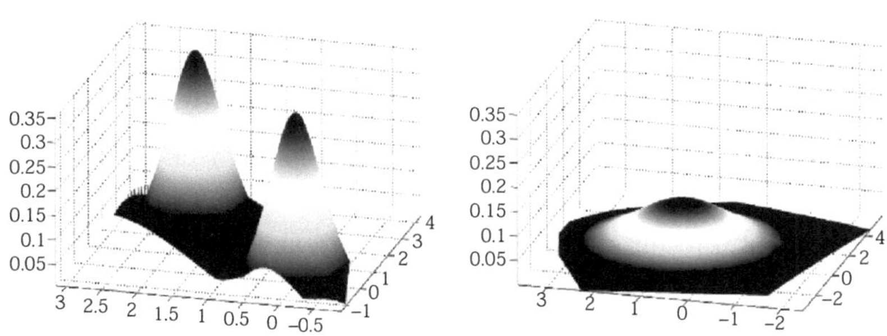

Fig. 5.3. Two Gaussians with averages $\mu_1 = [0, 0]$ y $\mu_2 = [3, 2]$ (*left*) are erroneously described by a unique kernel with $\mu = [1.5, 1]$ (*right*). (Figure by Peñalver et al. [116] (©2009 IEEE)).

In the following sections we present the entropy-based expectation maximization (EBEM) algorithm. This algorithm starts with one kernel and iteratively splits it until achieving the optimal number of kernels. It avoids the initialization problems which EM has, and uses as a criterion the "Gaussianity" of the data which are fit by the kernels of the mixture.

5.3 EBEM Algorithm: Exploiting Entropic Graphs

The EBEM (entropy-based EM) algorithm [116] starts with only one kernel, and finds the maximum-likelihood solution. In each iteration it tests whether the underlying pdf of each kernel is Gaussian. If not, it replaces that kernel with two kernels adequately separated from each other. After the kernel lower Gaussianity has been split into two, new EM steps are performed in order to obtain a new maximum-likelihood solution for the new number K of kernels. The algorithm is initialized with a unique kernel whose parameters of average and covariance are given by the sample. Consequently, the algorithm is not prone to initialization, overcoming the local convergence of the usual EM algorithm.

The EBEM algorithm converges after a few iterations and is suitable for density estimation, pattern recognition and unsupervised color image segmentation. The only parameter which has to be set is the Gaussianity threshold. It is more versatile that just fixing the number of kernels beforehand. We may know how well we want our data to fit the Gaussian distributions of the mixture, but we may not necessarily know how many clusters form the data. For instance, in the color image segmentation context, one may assume that in a image sequence of the same environment, the Gaussianity threshold may be nearly constant whereas the number of kernels will be different for each frame.

5.3.1 The Gaussianity Criterion and Entropy Estimation

Recall that for a discrete variable $Y = \{y_1, ..., y_N\}$ with N values the entropy is defined as

$$H(Y) = -E_y[\log(P(Y))] = -\sum_{i=1}^{N} P(Y = y_i) \log P(Y = y_i). \qquad (5.11)$$

A fundamental result of information theory is the "2nd Gibbs Theorem," which states that Gaussian variables have the maximum entropy among all the variables with equal variance [88]. Then, the entropy of the underlying distribution of a kernel should reach the maximum when the distribution is Gaussian. This theoretical maximum entropy is

$$H_{\max}(Y) = \frac{1}{2} \log[(2\pi e)^d |\Sigma|]. \qquad (5.12)$$

The comparison of the Gaussian entropy with the entropy of the underlying data is the criterion which EBEM uses for deciding whether a kernel is Gaussian or it should be replaced by other kernels which fit better the data.

In order to evaluate the Gaussianity criterion, the Shannon entropy of the data has to be estimated. Several approaches to Shannon entropy estimation have been studied in the past. We can widely classify them as methods which first estimate the density function, and methods which by-pass this and directly estimate the entropy from the set of samples. Among the methods which estimate the density function, also known as "plug-in" entropy estimators [13], there is the well-known "Parzen windows" estimation. Most of the current nonparametric entropy and divergence estimators belong to the "plug-in" methods. They have several limitations. On the one hand the density estimator performance is poor without smoothness conditions. The estimations have high variance and are very sensitive to outliers. On the other hand, the estimation in high-dimensional spaces is difficult, due to the exponentially increasing sparseness of the data. For this reason, entropy has traditionally been evaluated in one (1D) or two dimensional (2D) data. For example, in image analysis, traditionally gray scale images have been used. However, there are datasets whose patterns are defined by thousands of dimensions.

The "nonplugin" estimation methods offer a state-of-the-art [103] alternative for entropy estimation, estimating entropy directly from the data set. This approach allows us to estimate entropy from data sets with arbitrarily high number of dimensions. In image analysis and pattern recognition the work of Hero and Michel [74] is widely used for Rényi entropy estimation. Their methods are based on entropic spanning graphs, for example, minimal spanning trees (MSTs) or k-nearest neighbor graphs. A drawback of these methods for the EBEM algorithm is that the methods based on entropic spanning graphs do not estimate Shannon entropy directly. In the work of Peñalver et al. [116] they develop a method for approximating the value of Shannon entropy from the estimation of Rényi's α-entropy, as explained in the following subsection.

5.3.2 Shannon Entropy from Rényi Entropy Estimation

In Chapter 4 (in Section 4.3.6) we explained how Michel and Hero estimate the Rényi's α-entropy from a minimum spanning tree (MST). Equation 4.32 showed how to obtain the Rényi entropy of order α, that is, $H_\alpha(X_n)$, directly from the samples X_n, by means of the length of the MST. There is a discontinuity at $\alpha = 1$ which does not allow us to use this value of α in the expression of the Rényi entropy: $H_\alpha(f) = 1/(1-\alpha)\log\int_z f^\alpha(z)\,dz$. It is obvious that for $\alpha = 1$ there is a division by zero in $1/(1-\alpha)$. The same happens in the expression of the MST approximation (Eq. 4.32). However, when $\alpha \to 1$, the α-entropy converges to the Shannon entropy:

$$\lim_{\alpha\to 1} H_\alpha(p) = H(p) \tag{5.13}$$

The limit can be calculated using L'Hôpital's rule. Let $f(z)$ be a pdf of z. Its Rényi entropy, in the limit $\alpha \rightarrow 1$ is

$$\lim_{\alpha\rightarrow 1} H_\alpha(f) = \lim_{\alpha\rightarrow 1} \frac{\log \int_z f^\alpha(z)\, dz}{1-\alpha}$$

In $\alpha = 1$ we have that $\log \int_z f^1(z)\, dz = \log 1 = 0$ (note that $f(z)$ is a pdf, then its integral over z is 1). This, divided by $1-\alpha = 0$ is an indetermination of the type $\frac{0}{0}$. By L'Hôpital's rule we have that if

$$\lim_{x\rightarrow c} g(x) = \lim_{x\rightarrow c} h(x) = 0,$$

then

$$\lim_{x\rightarrow c} \frac{g(x)}{h(x)} = \lim_{x\rightarrow c} \frac{g'(x)}{h'(x)}.$$

Substituting the expression of the limit of the Rényi entropy:

$$\lim_{\alpha\rightarrow 1} \frac{\log \int_z f^\alpha(z)\, dz}{1-\alpha} = \lim_{\alpha\rightarrow 1} \frac{\frac{\partial}{\partial\alpha}(\log \int_z f^\alpha(z)\, dz)}{\frac{\partial}{\partial\alpha}(1-\alpha)}$$

The derivate of the divisor is $\frac{\partial}{\partial\alpha}(1-\alpha) = -1$, and the derivate of the dividend is

$$\begin{aligned}
\frac{\partial}{\partial\alpha}\left(\log \int_z f^\alpha(z)\, dz\right) &= \frac{1}{\int_z f^\alpha(z)\, dz} \frac{\partial \int_z f^\alpha(z)\, dz}{\partial\alpha} \\
&= \frac{1}{\int_z f^\alpha(z)\, dz} \int_z f^\alpha(z) \log f(z)\, dz.
\end{aligned}$$

The first term of this expression goes to 1 in the limit because $f(z)$ is a pdf:

$$\begin{aligned}
\lim_{\alpha\rightarrow 1} H_\alpha(f) &= \lim_{\alpha\rightarrow 1} -\frac{\int_z f^\alpha(z) \log f(z)\, dz}{\int_z f^\alpha(z)\, dz} \\
&= -\frac{\int_z f^1(z) \log f(z)\, dz}{1} \\
&= -\int_z f(z) \log f(z)\, dz \equiv H(f)
\end{aligned}$$

Then, in order to obtain a Shannon entropy approximation from α-entropy, α must have a value close to 1. It will be convenient to have a value strictly less than 1, as for $\alpha > 1$ the Rényi entropy is no more concave, as shown in Fig. 4.12. The problem is which value close to 1 is the optimal, given a set of samples.

The experiments presented in [116] show that it is possible to model H_α as a function of α, independently on the on the size and nature of the data. The function is a monotonical decreasing one, as shown in Fig. 5.4 (left)

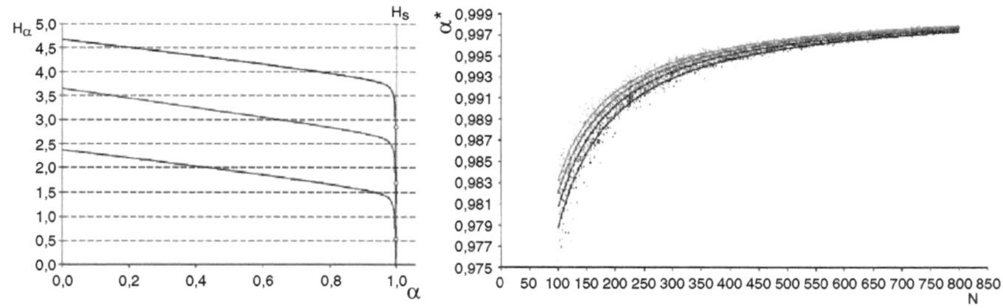

Fig. 5.4. *Left*: H_α for Gaussian distributions with different covariance matrices. *Right*: α^* for dimensions between 2 and 5 and different number of samples.

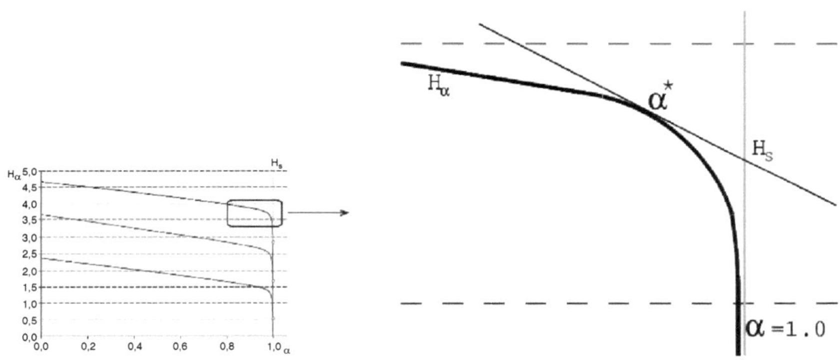

Fig. 5.5. The line tangent to H_α in the point α^* gives the Shannon entropy approximated value at $\alpha = 1$.

(experimental results). For any point of this function, a tangent straight line $y = mx + b$ can be calculated. This tangent is a continuous function and can give us a value at its intersection with $\alpha = 1$, as shown in Fig. 5.5. Only one of all the possible tangent lines is the one which gives us a correct estimation of the Shannon entropy at $\alpha = 1$; let us say that this line is tangent to the function at some point α^*. Then, if we know the correct α^*, we can obtain the Shannon entropy estimation. As H_α is a monotonous decreasing function, the α^* value can be estimated by means of a dichotomic search between two well-separated values of α, for a constant number of samples and dimensions.

It has been experimentally verified that α^* is almost constant for diagonal covariance matrices with variance greater than 0.5, as shown in Fig. 5.4 (right). Figure 5.6 shows the estimation of α^* for pdfs with different covariance matrices and 400 samples.

Then, the optimal α^* depends on the number of dimensions D and samples N. This function can be modeled as

$$\alpha^* = 1 - \frac{a + b \cdot e^{cD}}{N}, \tag{5.14}$$

and its values a, b and c have been calibrated for a set of 1,000 distributions with random $2 \leq d \leq 5$ and number of samples. The resulting function is

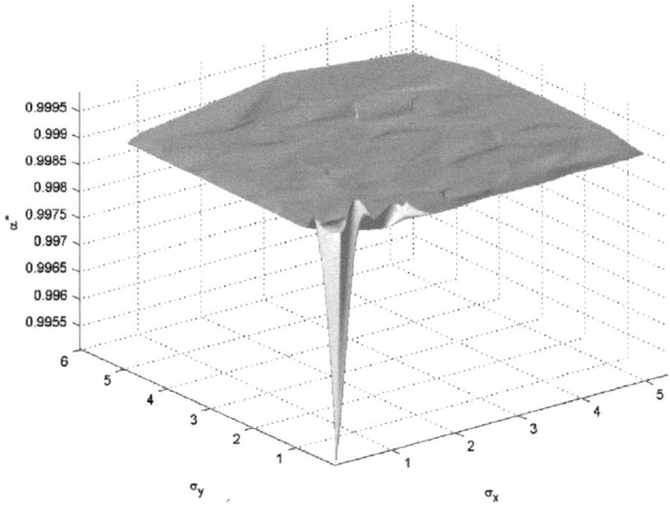

Fig. 5.6. α^* 2D for different covariance values and 400 samples. Value remains almost constant for variances greater than 0.5.

Eq. 5.14 with values $a = 1.271$, $b = 1.3912$ and $c = -0.2488$, as reported in [116]. The function is represented in Fig. 5.4 (right).

5.3.3 Minimum Description Length for EBEM

It can be said that the EBEM algorithm performs a model order selection, even though it has to be tuned with the Gaussianity deficiency threshold. This threshold is a parameter which does not fix the order of the model. It determines the degree of fitness of the model to the data, and with a fixed threshold different model orders can result, depending on the data. However, it might be argued that there is still a parameter which has to be set, and the model order selection is not completely solved. If there is the need not to set any parameter manually, the minimum description length principle can be used.

The minimum description length principle (see Chapter 3) chooses from a set of models the representation which can be expressed with the shortest possible message. The optimal code-length for each parameter is $1/2 \log n$, asymptotically for large n, as shown in [140]. Then, the model order selection criterion is defined as:

$$\mathcal{C}_{MDL}(\Theta_{(k)}, k) = -L(\Theta_{(k)}, y) + \frac{N(k)}{2} \log n, \tag{5.15}$$

where the first term is the log-likelihood and the second term penalizes an excessive number of components, $N(k)$ being the number of parameters required to define a mixture of k kernels.

5.3.4 Kernel-Splitting Equations

When the kernel $K*$, with lowest Gaussianity, has to be split into the K_1 and K_2 kernels (components of the mixture), their parameters $\Theta_{k_1} = (\mu_{k_1}, \Sigma_{k_1})$ and $\Theta_{k_2} = (\mu_{k_2}, \Sigma_{k_2})$ have to be set. The new covariance matrices have two restrictions: they must be definite positive and the overall dispersion must remain almost constant:

$$\begin{aligned} \pi_* &= \pi_1 + \pi_2 \\ \pi_*\mu_* &= \pi_1\mu_1 + \pi_2\mu_2 \\ \pi_*(\Sigma_* + \mu_*\mu_*^T) &= \pi_1(\Sigma_1 + \mu_1\mu_1^T) + \pi_2(\Sigma_2 + \mu_2\mu_2^T) \end{aligned} \tag{5.16}$$

These constraints have more unknown variables than equations. In [48] they perform a spectral decomposition of the actual covariance matrix and they estimate the new eigenvalues and eigenvectors of new covariance matrices.

Let $\sum_* = V_*\Lambda_*V_*^T$ be the spectral decomposition of the covariance matrix $\sum_*$, with $\Lambda_* = diag(\lambda j*^1, ..., \lambda j*^d)$ a diagonal matrix containing the eigenvalues of $\sum_*$ with increasing order, $*$ the component with the lowest entropy ratio, π_*, π_1, π_2 the priors of both original and new components, μ_*, μ_1, μ_2 the means and $\sum_*, \sum_1, \sum_2$ the covariance matrices. Let also be D a $d \times d$ rotation matrix with columns orthonormal unit vectors. D is constructed by generating its lower triangular matrix independently from $d(d-1)/2$ different uniform $U(0,1)$ densities. The proposed split operation is given by

$$\begin{aligned} \pi_1 &= u_1\pi_* \\ \pi_2 &= (1-u_1)\pi_* \\ \mu_1 &= \mu_* - \left(\textstyle\sum_{i=1}^d u_2^i \sqrt{\lambda_*^i} V_*^i\right)\sqrt{\tfrac{\pi_2}{\pi_1}} \\ \mu_2 &= \mu_* - \left(\textstyle\sum_{i=1}^d u_2^i \sqrt{\lambda_*^i} V_*^i\right)\sqrt{\tfrac{\pi_1}{\pi_2}} \\ \Lambda_1 &= diag(u_3)diag(\iota - u_2)diag(\iota + u_2)\Lambda_* \tfrac{\pi_*}{\pi_1} \\ \Lambda_2 &= diag(\iota - u_3)diag(\iota - u_2)diag(\iota + u_2)\Lambda_* \tfrac{\pi_*}{\pi_2} \\ V_1 &= DV_* \\ V_2 &= D^T V_* \end{aligned} \tag{5.17}$$

where, ι is a $d \times 1$ vector of ones, $u_1, u_2 = (u_2^1, u_2^2, ..., u_2^d)^T$ and $u_3 = (u_3^1, u_3^2, ..., u_3^d)^T$ are $2d+1$ random variables needed to construct priors, means and eigenvalues for the new component in the mixture and are calculated as

$$\begin{aligned} u_1 \sim be(2,2), u_2^1 \sim \beta(1, 2d), \\ u_2^j \sim U(-1,1), u_3^1 \sim \beta(1,d), u_3^j \sim U(0,1) \end{aligned} \tag{5.18}$$

with $j = 2, ..., d$. and $\beta()$ a Beta distribution.

The splitting process of a kernel is graphically described in a 2D example, in Fig. 5.7, where it can be seen that the directions and magnitudes of

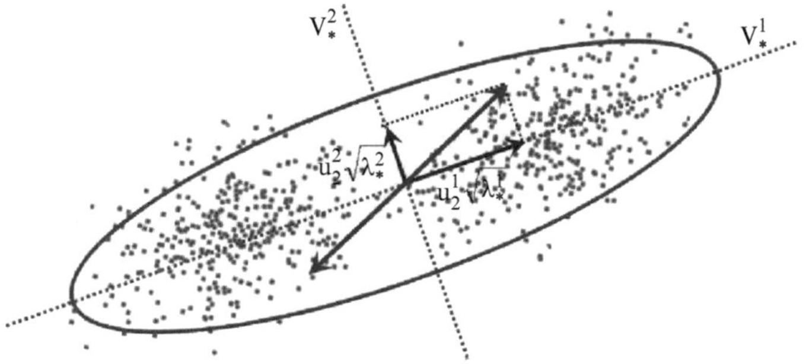

Fig. 5.7. Two-dimensional example of splitting one kernel into two new kernels.

Algorithm 6: Entropy based EM algorithm

EBEM ALGORITHM
Initialization: Start with a unique kernel.
$K \leftarrow 1$.
$\Theta_1 \leftarrow \{\mu_1, \Sigma_1\}$ given by the sample.
repeat:
    repeat
        E Step
        M Step
        Estimate log-likelihood in iteration i: ℓ_i
    until: $|\ell_i - \ell_{i-1}| <$ CONVERGENCE_TH
    Evaluate $H(Y)$ and $H_{max}(Y)$ globally
    if $(H(Y)/H_{max} <$ ENTROPY_TH)
        Select kernel K_* with the lowest ratio and
        decompose into K_1 and K_2
        Initialize parameters Θ_1 and Θ_2(Eq. 5.17)
            Initialize new averages: μ_1, μ_2
            Initialize new eigenvalues matrices: Λ_1, Λ_2
            Initialize new eigenvector matrices: V_1 and V_2
            Set new priors: π_1 and π_2
    else
        Final $\leftarrow$ True
until: Final = True

---

variability are defined by eigenvectors and eigenvalues of the covariance matrix. A description of the whole EBEM algorithm can be seen in Alg. 6.

5.3.5 Experiments

One of the advantages of the EBEM algorithm is that it starts from $K = 1$ components. It keeps on dividing kernels until necessary, and needs no backward steps to make any correction, which means that it does not have the

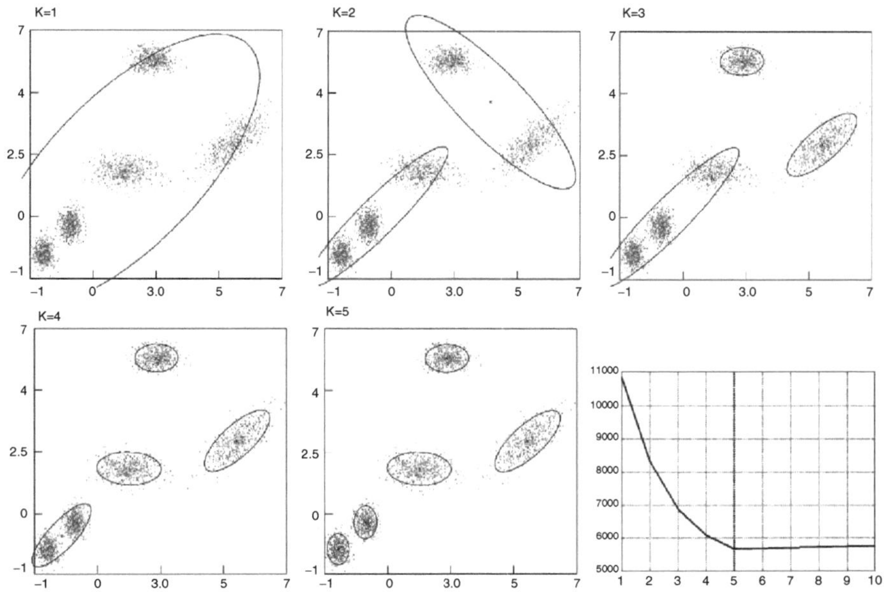

Fig. 5.8. Evolution of the EBEM algorithm from one to five final kernels. After iteration the algorithm finds the correct number of kernels. *Bottom-right*: Evolution of the cost function with MDL criterion.

need to join kernels again. Thus, convergence is achieved in very few iterations. In Fig. 5.8 we can see the evolution of the algorithm on synthetic data. The Gaussianity threshold was set to 0.95 and the convergence threshold of the EM algorithm was set to 0.001.

Another example is shown in Fig. 5.9. It is a difficult case because there are some overlapping distributions. This experiment is a comparison with the MML-based method described in [56]. In [56] the algorithm starts with 20 kernels randomly initialized and finds the correct number of components after 200 iterations. EBEM starts with only one kernel and also finds the correct number of components, with fewer iterations. Figure 5.9 shows the evolution of EBEM in this example. Another advantage of the EBEM algorithm is that it does not need a random initialization.

Finally we also show some color image segmentation results. Color segmentation can be formulated as a clustering problem. In the following experiment the RGB color space information is taken. Each pixel is regarded to as a sample defined by three dimensions, and no position or vicinity information is used. After convergence of the EBEM we obtain the estimated number K of color classes, as well as the membership of the individual pixels to the classes. The data come from natural images with sizes of 189×189 pixels. Some results are represented in Fig. 5.10.

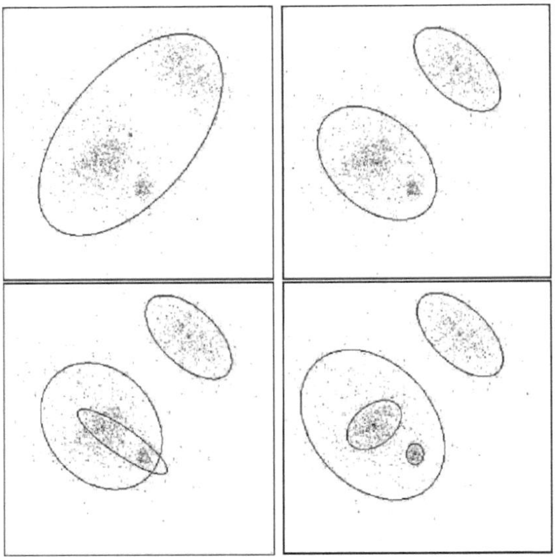

Fig. 5.9. Fitting a Gaussian mixture with overlapping kernels. The algorithm starts with one component and selects the order of the model correctly.

5.4 Information Bottleneck and Rate Distortion Theory

In this section the Information Bottleneck method is discussed. The method may be considered as an improvement over a previous clustering method based on the Rate Distortion Theory, that will be also explained here. The main idea behind Information Bottleneck is that data can be compressed (or assigned to clusters) while preserving important or relevant information, and it has been succesfully applied to computer vision, in the context of object categorization, as long as to other pattern classification problems.

5.4.1 Rate Distortion Theory Based Clustering

When partitioning the data in clusters, a representation for each of these clusters must be chosen; it may be the centroid of the cluster or a random element contained in it. In order to measure the goodness of the representation of a cluster with respect to the data in it, a distortion measure must be defined. Ideal clustering would be achieved in the case of minimal distortion between the data and their corresponding representations. However, as can be seen in Fig. 5.11, distortion is inversely proportional to the complexity of the obtained data model, and as a consequence, data compression (and thus generalization) is lower as distortion decreases. Rate Distortion Theory is a branch of information theory addressed to solve this kind of problems:

Fig. 5.10. Color image segmentation results. Original images (*first column*) and color image segmentation with different Gaussianity deficiency levels (*second and third columns*). See Color Plates. (Courtesy of A. Peñalver.)

given a minimal expected distortion, which is the maximum possible data compression that we can achieve? Or, in other words, how many clusters are needed to represent these data?

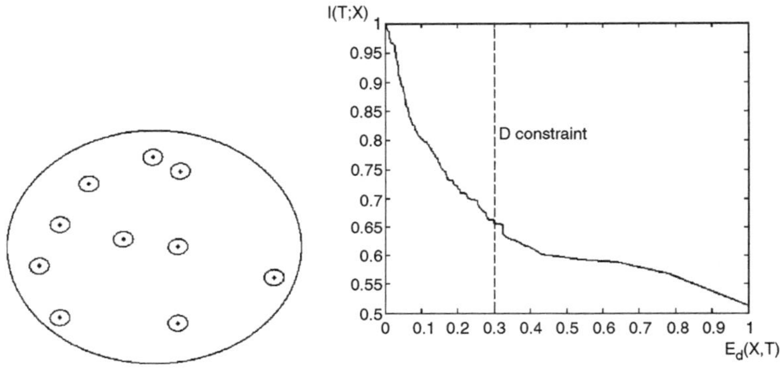

Fig. 5.11. *Left*: the effect of data compression on distortion. If $T = X$ (there is a cluster for each sample), there is zero distortion, but data representation is not compact. In this case, $I(T;X) = H(X)$. By the other hand, if $|T| = 1$ (there is an only cluster), we obtain a very compact representation of the data, yielding $I(T;X) = 0$; however, distortion is also very high. *Right*: plot representing the relation between model complexity $I(T;X)$ and distortion $E_d(X,T)$, just introduced, and an upper bound of D (the *vertical slashed line*).

Given the partition $p(t|x)$ of a dataset $x \in X$ in $t \in T$ clusters, its quality may be measured by

- The complexity of the model $I(T;X)$ or rate; that is, the information that a representative gives of the data in a cluster
- The distortion $E_d(X,T) = \sum ijp(x_i)p(t_j|x_i)d(x_i,t_j)$; that is, the distance from the data to its representation

As Fig. 5.11 shows, if an upper bound constraint D on the expected function is given, lower D values could be used, relaxing the distortion constraint and providing stronger compression. Therefore, Rate Distortion Theory can be expressed as a the problem of searching the most compact model $p(t|x)$ given the distortion constraint D:

$$R(D) = \min_{\{p(t|x):E_d(X,T) \leq D\}} I(T;X) \tag{5.19}$$

and this problem can be solved minimizing the following function, where β is a Lagrange multiplier applied for weighting the distortion term:

$$\mathcal{F}[p(t|x)] = I(T;X) + \beta E_d(X,T) \tag{5.20}$$

Algorithm 7 shows the Blahut–Arimoto algorithm to minimize $\mathcal{F}$, based on alternating steps to refine $p(t|x)$ and $p(t)$, since both are unknown and each depends on the other. The normalization factor $Z(x,\beta)$ ensures that the constraint $\sum_t p(t|x) = 1, \forall x \in X$ is satisfied. Divergence is measured by means of

Algorithm 7: Blahut–Arimoto algorithm

Input: $p(x)$, T, β
Initialize random init $p(t)$
while *no convergence* **do**
  $p(t|x) = \frac{p(t)}{Z(x,\beta)} \exp^{-\beta d(x,t)}$
  $p(t) = \sum_x p(x)p(t|x)$
end
Output: A clustering $p(t|x)$ that minimizes $\mathcal{F}$ for a minimal expected distortion given by β

a distance function $d(x,t)$ between a data sample and its representative. The algorithm runs until convergence is achieved. Several examples are given in Fig. 5.12. This algorithm guarantees that the global minimum is reached and obtains a compact clustering with respect to the minimal expected distortion, defined by parameter β. The main problem is that it requires two input parameters: β and the number of clusters, that must be fixed. However, as can be seen in Fig. 5.12, depending on distortion the algorithm may yield $p(t) = 0$ for one or more clusters; that is, this input parameter actually indicates the maximum number of clusters. Slow convergence for low β values is another drawback of this algorithm. Finally, result will be conditioned by the initial random initialization of $p(t)$.

5.4.2 The Information Bottleneck Principle

The main drawback of the Blahut–Arimoto clustering algorithm, and the Rate Distortion Theory, is the need of defining first a distortion function in order to compare data samples and representatives. In our previous example, this distortion was computed as the euclidean distance between the sample and its corresponding representative. However, in complex problems, the important features of the input signal that define the distortion function may not be explicitly known. However, we may be provided with an additional random variable that helps to identify what is relevant information in the data samples with respect to this new variable in order to avoid loosing too much of this information when clustering data. One example is the text categorization problem. Given a set of words $x \in X$ and a set of text categories $y \in Y$, the objective is to classify a text as belonging to a category y_i depending on the subset of words of X present in the document and their frequency. A common approach to this problem begins with the splitting of the set X into $t \in T$ clusters, and continues with a learning step that builds document category models from these clusters; then, any new document is categorized from the

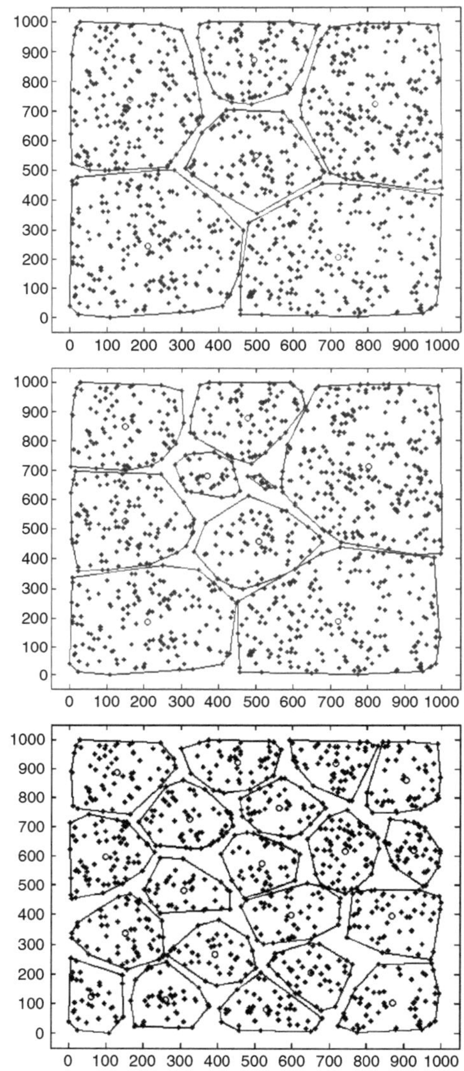

Fig. 5.12. Example of application of the Blahut–Arimoto algorithm to a set X of 1,000 data samples using $\beta = 0.009$, $\beta = 0.01$ and $\beta = 1$, obtained from a uniform distribution, being $|T| = 20$ (in the plots, cluster centroids are represented by *circles*). In all cases, $p(x) = 1/1{,}000, \forall x \in X$. Lower β values decrease the weight of the distortion term in the rate distortion function $\mathcal{F}$; as a consequence, the influence of the distortion term during energy minimization decreases, and the data compression increases, due to the fact that it is partitioned into less data clusters.

frequency histogram that represents the amount of each cluster that is present in this document. In this case X and Y are not independent variables, meaning that $I(X;Y) > 0$.

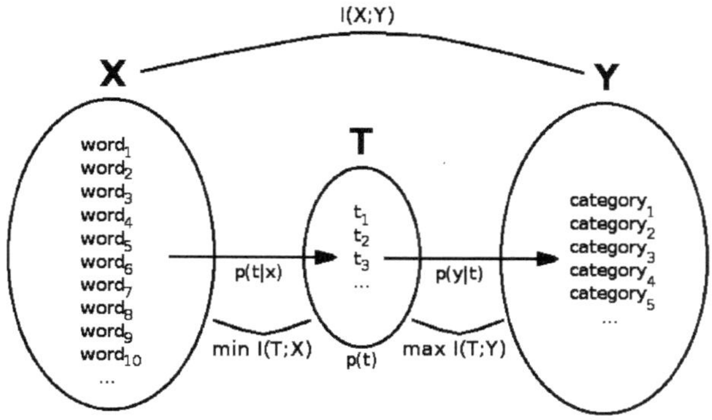

Fig. 5.13. The Information Bottleneck principle applied to text categorization. A text is categorized into a given category $y \in Y$ depending on the words $x \in X$ that are present in it and their frequency. In order to increase efficiency, the words that may be considered during the task are splitted into different clusters $t \in T$. The objective of Information Bottleneck is to achieve a maximal data compression (given by $I(T;X)$) while maximizing the amount of relevant information of X about Y (given by $I(T;Y)$) that the data compression preserves. This $I(T;Y)$ should be as similar to $I(X;Y)$ as possible, $I(X;Y)$ being the amount of information that the original set X gives about Y.

As stated before, a measure of the quality of this clusterization is given by $I(X;T)$. However, another constraint is necessary, due to the fact that if the data compression is indefinitely increased, then the information that X provides about Y is lost. Unlike in Rate Distortion Theory, this additional restriction is not given by distortion for a fixed $|T|$. Information Bottleneck is focused on searching an optimal representation T of X so that predicitions of Y from T are the most similar possible to direct predictions of Y from X, while minimizing $|T|$. That is, the objective is to find an equilibrium between compression, given by $I(X;T)$, and information of clusters about Y, given by $I(T;Y)$. Figure 5.13 illustrates this concept applied to the text categorization problem. The equilibrium is reached by minimizing the following function:

$$\mathcal{L}[p(t|x)] = I(T;X) - \beta I(T;Y) \tag{5.21}$$

Algorithm 8 is a generalization of the Blahut–Arimoto algorithm. The output is a clustering $p(t|x)$ that fulfills the previously indicated constraints. However, due to being based on Blahut–Arimoto, some of the drawbacks of this algorithm are still present: it needs two parameters (the number of clusters and β) and its convergence to the solution is also slow. Furthermore, the algorithm may tend to a local minimum, resulting in a suboptimal solution. It must also be noted that in this algorithm the divergence measure is replaced by the Kullback–Leibler divergence.

In Fig. 5.14 we can see a graph representing $I_x = I(X;T)$ vs. $I_y = I(T;Y)$ for different values of $|T|$. These results were obtained from the input data

---

Algorithm 8: Information Bottleneck based on a generalization of the Blahut–Arimoto algorithm

---

Input: $p(x,y)$, T, β
Initialize random init $p(t|x)$
while *no convergence* **do**
  $p(t|x) = \frac{p(t)}{Z(x,\beta} \exp^{-\beta D_{KL}[p(y|x)||p(y|t)]}$
  $p(y|t) = \frac{1}{p(t)} \sum_x p(y|x)p(t|x)$
  $p(t) = \sum_x p(x)p(t|x)$
end
Output: A clustering $p(t|x)$ that minimizes $\mathcal{L}$ for a given β

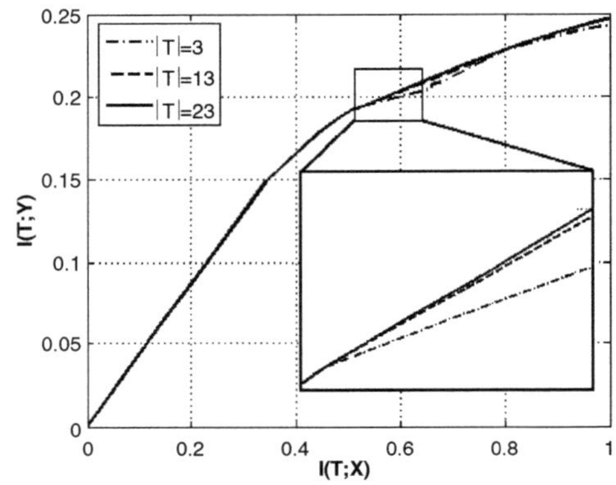

Fig. 5.14. Plot of $I(T;X)$ vs. $I(T;Y)$ for three different values of the cardinality of T: $|T| = 3$, $|T| = 13$ and $|T| = 23$ and for input data in Fig. 5.15. A part of this plot is zoomed in order to show that when plots corresponding to different T cardinalities arrive to a specific value, they diverge. By Altering the β parameter we are displacing through any of these convex curves, depending on the growth rate given in the text. This fact suggests that a deterministic annealing approach could be applied.

in Fig. 5.15. The plot shows that the growth rate of the curve is different depending on the β parameter. Specifically, this ratio is given by

$$\frac{\delta I(T;Y)}{\delta I(X;T)} = \beta^{-1} > 0 \tag{5.22}$$

There is a different convex curve for each different cardinality of the set T. Varying the value of β, we may move through any of these convex curves on the plane $I_x I_y$. This fact suggests that a deterministic annealing (DA) approach could be applied in order to find an optimal clustering.

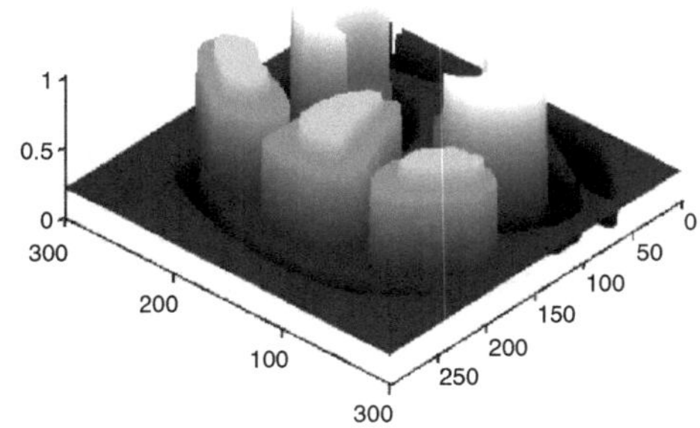

Fig. 5.15. Input data used to compute the plots in Figs. 5.14 and 5.16. Given two classes $Y = \{y_1, y_2\}$, the shown distribution represents $p(X, Y = y_1)$ (from which $p(X, Y)$ can be estimated).

5.5 Agglomerative IB Clustering

The agglomerative Information Bottleneck Clustering (AIB) was designed as an alternative to other methods such as the generalization of the Blahut–Arimoto algorithm explained above. The main feature of this greedy algorithm is that rather than being a soft clustering algorithm, it reduces data by means of a bottom-up hard clustering; that is, each sample is assigned to an only cluster. It is based on relating $I(T;Y)$ with Bayesian error. The algorithm is deterministic, in the sense that it yields hard clustering for any desired number of clusters. Furthermore, it is not parametric. And it is not limited to hard clustering: due to the fact that the results of this method may be considered as the limit of the Information Bottleneck based on deterministic annealing (when zero temperature is reached), using the Information Bottleneck equations shown in previous sections these results must be transformed to soft clustering. However, the main drawbacks of the method are two. First, it is a greedy algorithm, thus its results are optimal in each step but it is globally suboptimal. And second, its bottom up behavior, starting with a cluster for each sample and joining them until only one cluster is achieved, makes this algorithm computationally expensive.

5.5.1 Jensen–Shannon Divergence and Bayesian Classification Error

The Information Bottleneck principle may also be interpreted as a Bayesian classification problem: given a set of samples X, these samples must be classified as members of a set of classes $Y = \{y_1, y_2, \ldots, y_n\}$ with a priori probabilities $\{p(y_i)\}$. In this case, Bayesian error is given by

$$P_{Bayes}(e) = \sum_{x \in X} p(x)(1 - \max_i p(y_i|x)) \tag{5.23}$$

The following equation shows how this Bayesian error is bounded by Jensen–Shannon divergence:

$$\frac{1}{4(M-1)}(H(Y)-JS_{p(y_i)}[p(x|y_i)])^2 \leq P_{Bayes}(e) \leq \frac{1}{2}(H(Y)-JS_{p(y_i)}[p(x|y_i)]) \tag{5.24}$$

where the Jensen–Shannon divergence of M distributions $p_i(x)$, each one having a prior $\Pi_i, 1 \leq i \leq M$, is

$$JS_{\Pi}[p_1, p_2, \ldots, p_M] = H\left[\sum_{i=1}^{M} \Pi_i p_i(x)\right] - \sum_{i=1}^{M} \Pi_i H[p_i(x)] \tag{5.25}$$

The following derivation also relates Jensen–Shannon to Mutual Information:

$$\begin{aligned} JS_{p(y_1,\ldots,p(y_M)}[p(x|y_i),\ldots,p(x|y_M)] &= H\left[\sum_{i=1}^{M} p(y_i)p(x|y_i)\right] - \sum_{i=1}^{M} p(y_i)H[p(x|y_i)] \\ &= H(X) - H(X|Y) = I(X;Y) \end{aligned}$$

We will see later how these properties are applied to AIB.

5.5.2 The AIB Algorithm

Figure 5.12 shows examples of Blahut–Arimoto results. As can be seen, given a fix value for parameter $|T|$, the final number of clusters increases for higher β values. This effect is also produced in the Blahut–Arimoto generalization for Information Bottleneck. In general, given any finite cardinality $|T| \equiv m$, as $\beta \to \infty$ we reach a data partition in which each $x \in X$ is part of an only cluster $t \in T$ for which $D_{KL}[p(y|x)||p(y|t)]$ is minimum and $p(t|x)$ only takes values in $\{0,1\}$. The AIB algorithm starts from an initial state with $|T| = |X|$, and proceeds in each step joining two clusters, selected by means of a greedy criterion, until all samples are compressed in an only cluster. A hard clustering constraint is applied during this process.

Given an optimal partition of X into a set of clusters $T_m = \{t_1, t_2, \ldots, t_m\}$, a union of k of these clusters to create a new union cluster t'_k, and the corresponding new partition T'_m, always yields a loss of information, due to the fact that $I(T_m;Y) \geq I(T'_M;Y)$. Therefore, each greedy step of the algorithm selects the union cluster t'_k that minimizes this loss of information, that is estimated from the following distributions, in order to yield a new partition in which $m' = m - k + 1$:

$$\begin{cases} p(t'_k) = \displaystyle\sum_{i=1}^{k} p(t_i) \\ p(y|t'_k) = \frac{1}{p(t'_k)} \displaystyle\sum_{i=1}^{k} p(t_i, y), \forall y \in y \\ p(t'_k|x) = \begin{cases} 1 \text{ if } x \in t_i \text{ for any } 1 \leq i \leq k \\ 0 \text{ otherwise} \end{cases} \quad \forall x \in X \end{cases} \tag{5.26}$$

Before dealing with the AIB algorithm, some concepts must be introduced:

- The merge prior distribution of t'_k is given by $\Pi_k \equiv (\Pi_1, \Pi_2, \dots, \Pi_k)$, where Π_k is the a priori probability of t_i in t_k, $\Pi_i = \frac{p(t_i)}{p(t'_k)}$.
- The decrease of information in $I_y = I(T;Y)$ due to a merge is

$$\delta I_y(t_1, \dots, t_k) = I(T_M; Y) - I(T'_M; Y) \tag{5.27}$$

- The decrease of information in $I_x = I(T;X)$ due to a merge is

$$\delta I_x(t_1, \dots, t_k) = I(T_M; X) - I(T'_M; X) \tag{5.28}$$

From these definitions, it can be demonstrated that:

$$\delta I_x(t_1, \dots, t_k) = p(t'_k) H[\Pi_k] \geq 0 \tag{5.29}$$

$$\delta I_y(t_1, \dots, t_k) = p(t'_k) JS_{\Pi_k}[p(Y|t_1), \dots, p(Y|t_k)] \geq 0 \tag{5.30}$$

The algorithm is shown in Alg. 9. The cardinality of the initial partition is $m = |X|$, containing each cluster an only sample. In each step, a greedy

---

Algorithm 9: Agglomerative Information Bottleneck

Input: $p(x,y)$, $N = |X|$, $M = |Y|$
Construct $T \equiv X$:
for $i = 1 \dots n$ **do**
  $t_i = \{x_i\}$
  $p(t_i) = p(x_i)$
  $p(y|t_i) = p(y|x_i)$ for every $y \in Y$
  $p(t|x_j) = 1$ if $j = i$ and 0 otherwise
end
$T = \{t_1, \dots, t_j\}$

Information loss precalculation:
for *every* $i, j = 1 \dots N, i < j$ **do**
  $d_{i,j} = (p(t_i) + p(t_j)) JS_{\Pi_2}[p(y|t_i), p(y, t_j)]$
end

Main loop:
for $t = 1 \dots (N-1)$ **do**
  Find $\{\alpha, \beta\} = argmin_{i,j}\{d_{i,j}\}$.
  Merge $\{z_\alpha, z_\beta\} \Rightarrow t'$.
    $p(t') = p(z_\alpha) + p(z_\beta)$
    $p(y|t') = \frac{1}{p(t')}(p(t_\alpha, y) + p(t_\beta, y))$ for every $y \in Y$
    $p(t'|x) = 1$ if $x \in z_\alpha \cup z_\beta$ and 0 otherwise, for every $x \in X$
  Update $T = \{T - \{z_\alpha, z_\beta\}\} \cup \{t'\}$.
  Update $d_{i,j}$ costs w.r.t. t', only for couples that contain z_α or z_β.
end
Output: T_m: m-partition of X into m clusters, for every $1 \leq m \leq N$

---

decision selects the set of clusters to join that minimizes $\delta I_y(t_1, \ldots, t_k)$, always taking $k = 2$ (pairs of clusters). Clusters are joined in pairs due to a property of the information decrease: $\delta I_y(t_1, \ldots, t_k) \leq \delta I_y(t_1, \ldots, t_{k+1})$ and $\delta I_x(t_1, \ldots, t_k) \leq \delta I_x(t_1, \ldots, t_{k+1})$ $\forall k \geq 2$; the meaning of this property is that any cluster union $(t_1, \ldots, t_k) \Rightarrow t'_k$ can be built as $(k-1)$ consecutive unions of cluster pairs; for $1 \leq m \leq |X|$, the optimal partition may be found from $(|X| - m)$ consecutive union of cluster pairs. Therefore, the loss of information δI_y must be computed for each cluster pair in T_m in order to select cluster pairs to join. The algorithm finishes when there is only one cluster left. The result may be expressed as a tree from where cluster sets T_m may be inferred for any $m = |X|, |X| - 1, \ldots, 1$ (see Fig. 5.16).

In each loop iteration the best pair union must be found; thus, complexity is $O(m|Y|)$ for each cluster pair union. However, this complexity may be decreased to $O(|Y|)$ if the mutual information loss due to a pair of clusters merging is estimated directly by means of one of the properties enumerated above: $\delta I_y(t_1, \ldots, t_k) = p(t'_k) JS_{\Pi_k}[p(Y|t_1), \ldots, p(Y|t_k)]$. Another improvement is the precalculation of the cost of merging any pair of clusters. Then, when two clusters t_i and t_j are joined, this cost must only be updated only for pairs that include one of these two clusters.

A plot of $I(T;Y)/I(X;Y)$ vs. $I(T;X)/H(X)$ is shown in Fig. 5.16. As can be seen, the decrease of mutual information $\delta(m)$ when decreasing m, given by the next equation, only can increase

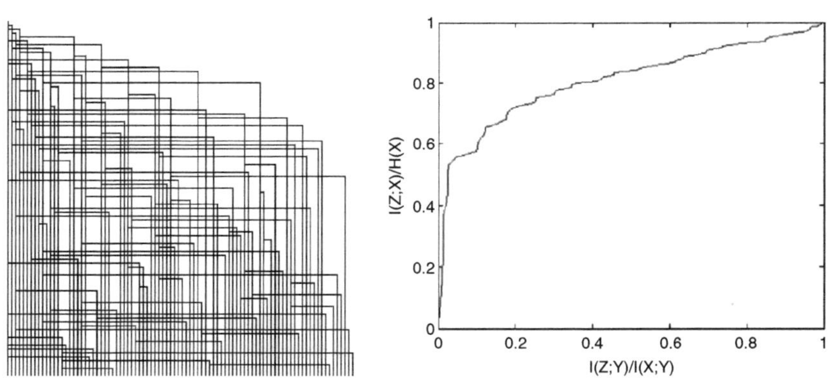

Fig. 5.16. Results of the AIB algorithm applied to a subset of the data in Fig. 5.15. This subset is built from 90 randomly selected elements of X. *Left*: tree showing the pair of clusters joined in each iteration, starting from the bottom (one cluster assigned to each data sample) and finishing at the top (an only cluster containing all the data samples). *Right*: plot of $I(T;Y)/I(X;Y)$ vs. $I(T;X)/H(X)$ (information plane). As the number of clusters decreases, data compression is higher, and as a consequence $I(T;X)$ tends to zero. However, less number of clusters also means that the information that these clusters T give about Y decreases; thus, $I(T;Y)$ also tends to zero. The algorithm yields all the intermediate cluster sets; so a trade-off between data compression and classification efficiency may be searched.

$$\delta(m) = \frac{I(T_m;Y) - I(T_{m-1};Y)}{I(X;Y)} \tag{5.31}$$

This measure provides a method to perform a basic order selection. When $\delta(m)$ reaches a high value, a meaningful partition of clusters has been achieved; further merges will produce a substantial loss of information, and the algorithm can stop at this point. However, this point is reached when m is low, and in this case the complexity of AIB is negligible; thus, usually it is worthy to finish the algorithm and obtain the complete cluster tree, as the one shown in Fig. 5.16.

It must be noted that usually hard clustering from AIB yields worser results than any other Information Bottleneck algorithm based on soft clustering. The cause is that soft clustering is optimal, as it maximizes $I(T;Y)$ while constraining $I(T;X)$. However, any $I(T;X)$ vs. $I(T;Y)$ curve in the information plane, obtained by annealing, intersects with the curve obtained from AIB. Thus, AIB may be considered as a good starting point, from which reverse annealing will guide to an ultimate solution.

5.5.3 Unsupervised Clustering of Images

This section is aimed to show an application of Information Bottleneck to computer vision in the image retrieval field. The objective of an image retrieval system is: given an input query image, answer with a ranked set of images, from most to less similar, extracted from a database. If the images of the database are clustered, then an exhaustive search in the database is not needed, and the process is less computational expensive. Clustering of the database images by means of Information Bottleneck principle may help to maximize the mutual information between clusters and image content, not only improving efficiency, but also performance.

In the work by Goldberger et al. [63], the images are modeled by means of Gaussian Mixtures, and clustered using AIB. However, due to the fact that Gaussian Mixture Models are not discrete, the AIB algorithm must be adapted to continuous distributions. An example of how the images are modeled by means of Gaussian Mixtures is shown in Fig. 5.17. Each pixel is described by a feature vector built from its color values in the Lab color space and its localization in the image (although only these features are used, the model is general and could be extended using other features like texture, for instance). From this representation, a Gaussian Mixture Model $f(x)$ is extracted for each image X, grouping pixels in homogeneous regions defined by Gaussian distributions.

An usual similarity measure in image applications is the Kullback–Leibler divergence, and in this work this is also the measure applied to compare images or clusters. This measure is easily estimated from histograms or discrete distributions, as we have seen during previous chapters. In the case of two distributions f and g representing Gaussian Mixture Models, Kullback–Leibler may be approximated from Monte Carlo simulations:

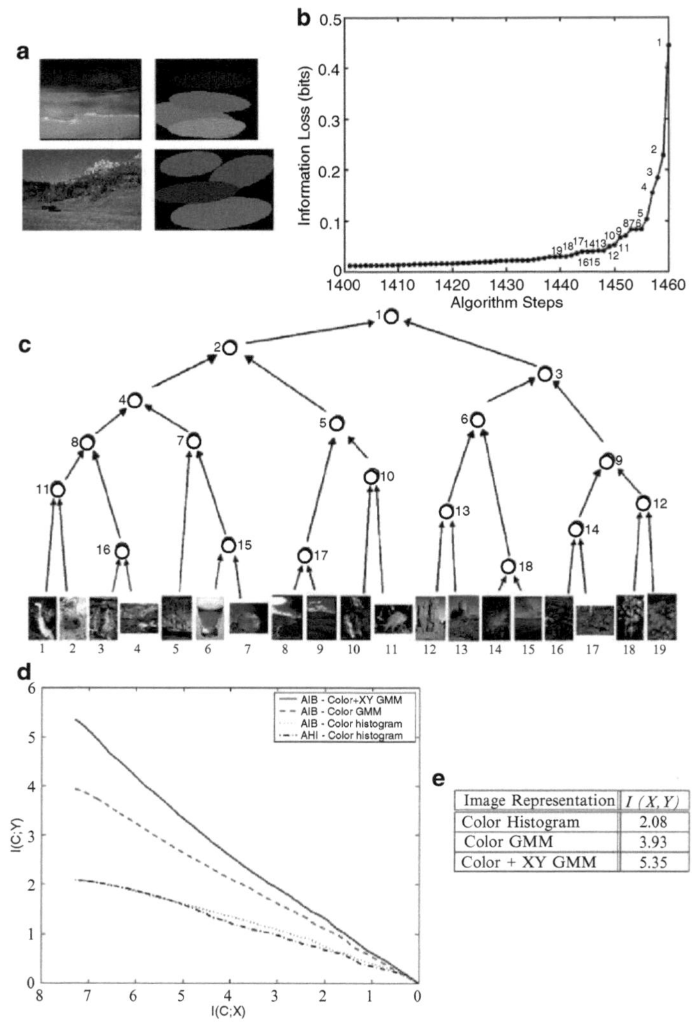

Image Representation	$I(X,Y)$
Color Histogram	2.08
Color GMM	3.93
Color + XY GMM	5.35

Fig. 5.17. From *left* to *right* and from *top* to *bottom*. (**a**) Image representation. Each ellipsoid represents a Gaussian in the Gaussian Mixture Model of the image, with its support region, mean color and spatial layout in the image plane. (**b**) Loss of mutual information during the IAB clustering. The last steps are labeled with the number of clusters in each step. (**c**) Part of the cluster tree formed during AIB, starting from 19 clusters. Each cluster is represented with a representative image. The labeled nodes indicate the order of cluster merging, following the plot in (**b**). (**d**) $I(T;X)$ vs. $I(T;Y)$ plot for four different clustering methods. (**e**) Mutual Information between images and image representations. (Figure by Goldberger et al. (©2006 IEEE)). See Color Plates.

$$D(f||g) = \int f \log \frac{f}{g} \approx \frac{1}{n} \sum_{t=1}^{n} \log \frac{f(x_t)}{g(x_t)} \tag{5.32}$$

where $x_1, x_2, \ldots, x_n$ are sampled from $f(x)$.

As stated before, image clustering is performed by means of AIB, in which X represents the image set to classify, $p(x)$ the priors, that are considered uniform, and Y the random variable associated with the feature vector extracted from a unique pixel. The distribution $f(y|x)$, from which the input parameter $p(x, y)$ may be estimated, is given by

$$f(y|x) = \sum_{j=1}^{k(x)} \alpha_{x,j} N(\mu_{x,j}, \Sigma_{x,j}) \tag{5.33}$$

where $k(x)$ is the number of Gaussian components for image x. There are several equations of Alg. 9 that must be adapted to Gaussian Mixture Model. For instance, given a cluster t, $f(y|t)$ is the mean of all image models that are part of t:

$$f(y|t) = \frac{1}{|t|} \sum_{x \in t} f(y|x) = \frac{1}{|t|} \sum_{j=1}^{k(x)} \alpha_{x,j} N(\mu_{x,j}, \Sigma_{x,j}) \tag{5.34}$$

As can be seen, $f(y|t)$ is also modeled as a Gaussian Mixture. When two clusters t_1 and t_2 are joined, the updating of this joint distribution is given by

$$f(y|t_1 \cup t_2) = \frac{1}{|t_1 \cup t_2|} \sum_{x \in t_1, t_2} f(y|x) = \sum_{i=1,2} \frac{|t_i|}{|t_1 \cup t_2|} f(y|t_i) \tag{5.35}$$

Finally, the cost of merging two clusters is given by

$$d(t_1, t_2) = \sum_{i=1,2} \frac{|t_i|}{|X|} D(f(y|t_i)||f(y|t_1 \cup t_2)) \tag{5.36}$$

where $|X|$ is the size of the image database. From this formulation, AIB can be applied as explained in previous section. In Fig. 5.17, the loss of mutual information for a given database image is shown. In the same figure, part of the cluster tree is also represented.

Mutual information is used in this work as a measure of quality. For instance, the clustering quality may be computed as $I(X;Y)$, X being the unsupervised clustering from AIB and Y a manual labeling. Higher values denote better clustering quality, as the cluster gives more information about the image classes. The quality of the image representation can also be evaluated by means of $I(X;Y)$, in this case X being the set of images and Y the features extracted from their pixels. Due to the fact that a closed-form expression to calculate $I(X;Y)$ for Gaussian Mixtures does not exist, this mutual information is approximated from $I(T;X)$ and $I(T;Y)$. In Fig. 5.17, the $I(T;X)$ vs.

$I(T;Y)$ plot is shown for AIB using Gaussian Mixture Models based on color and localization (AIB – Color+XY GMM), Gaussian Mixture Models based only on color (AIB – Color GMM), and without Gaussian Mixtures for AIB and Agglomerative Histogram Intersection (AIB – Color histogram and AHI – Color histogram). In all cases, $I(X;Y)$ is extracted from the first point of the curves, due to the fact that the sum of all merges cost is exactly $I(X;Y)$. Related to this plot, we show in Fig. 5.17 a table that summarizes these values. The best quality is obtained for color and localization based Gaussian Mixture Models.

Image retrieval from this clustering is straightforward. First the image query is compared to all cluster representatives, and then it is compared to all images contained in the selected cluster. Not only computational efficiency is increased, compared to an exhaustive search, but also previous experiments by Goldberger et al. show that clustering increases retrieval performance. However, this approach has several drawbacks, the main one being that it is still computationally expensive (training being the most time consuming phase). Furthermore, no high level information like shape and texture is used, and as a consequence all images in a category must be similar in shape and appearance.

5.6 Robust Information Clustering

A recent clustering algorithm based on the principles of Rate Distortion and information theory is Robust Information Clustering (RIC) [148]. This algorithm is nonparametric, meaning that it is not based on data pdfs. The idea behind the algorithm is to apply two steps in order to perform a minimax mutual information approach. During the first step, minimization is achieved by means of a rate distortion process which leads to a deterministic annealing (DA) that splits the data into clusters. The annealing process continues until a maximum number of clusters is reached. Maximization step is intended to identify data outliers for each different number of clusters. An optimal number of clusters is estimated in this step based on Structural Risk Minimization.

The Deterministic Annealing step has been described in previous sections (see Blahut–Arimoto algorithm for details). Thus, we focus on the use of the channel capacity, an information theory related measure, to detect outliers and estimate the model order during RIC algorithm. To understand channel capacity, we must think in terms of communications over a noisy channel. Messages $x \in X$ are sent through this channel, being received at the other side as $y \in Y$. In this case, mutual information $I(X;Y)$ represents the quality of the channel when transmitting messages. Thus, channel capacity can be described as the maximum of this mutual information, and represents the maximum information rate that can travel through it:

$$C = \max_{p(x)} I(X;Y) \tag{5.37}$$

When information rate exceeds the channel capacity, then this communication is affected by distortion. Properties of channel capacity are:

1. $C \geq 0$, due to the fact that $I(X|Y) \geq 0$.
2. $C \leq \log|X|$, due to the fact that $C = \max I(X;Y) \leq \max H(X) = \log|X|$.
3. $C \leq \log|Y|$, for the same reason.
4. $I(X;Y)$ is a continuous function of $p(x)$.
5. $I(X;Y)$ is a concave function of $p(x)$ (thus, a maximum value exists).

During Robust Information Clustering, and given the sample priors $p(X)$ and the optimal clustering $\bar{p}(W|X)$ obtained after a Deterministic Annealing process, $p(X)$ is updated in order to maximize the information that samples $x \in X$ yield about their membership to each cluster $w \in W$ (see the last channel capacity property):

$$C(D(p(X))) = \max_{p(X)} C(D(p(X))) = \max_{p(X)} I(p(X); \bar{p}(W|X)) \tag{5.38}$$

This channel capacity is subject to a constraint $D(p(X))$ given by

$$D(p(X)) = \sum_{i=1}^{l} \sum_{k=1}^{K} p(x_i)\bar{p}(w_k|x_i)d(w_k, x_i) \tag{5.39}$$

where $d(w_k|x_i)$ is the dissimilarity between sample $x_i \in X$ and cluster centroid $w_k \in W$. In order to perform the maximization, a Lagrange multiplier $\lambda \geq 0$ may be introduced:

$$C(D(p(X))) = \max_{p(X)} [I(p(X); \bar{p}(W|X)) + \lambda(D(\bar{p}(X)) - D(p(X)))] \tag{5.40}$$

In the latter equation, $\bar{p}(X)$ is a fixed unconditional a priori pmf (usually a equally distributed probability function). Based on a robust density estimation, with $D(p(X)) = D(\bar{p}(X))$ for $p(x_i) \geq 0$, the maximum of the constrained capacity for each x_i is given by

$$p(x_i) = p(x_i) \frac{c_i}{\sum_{i=1}^{l} p(x_i) c_i} \tag{5.41}$$

where:

$$c_i = \exp\left[\sum_{k=1}^{K} (\bar{p}(w_k|x_i) \ln \frac{\bar{p}(w_k|x_i)}{\sum_{j=1}^{l} p(x_j)\bar{p}(w_k|x_i)} - \lambda \bar{p}(w_k|x_i) d(w_k, x_i)\right] \tag{5.42}$$

If $p(x_i) = 0$ for any $x_i \in X$, then x_i is considered an outlier. In this regard, λ may be understood as an outlier control parameter; depending on the value

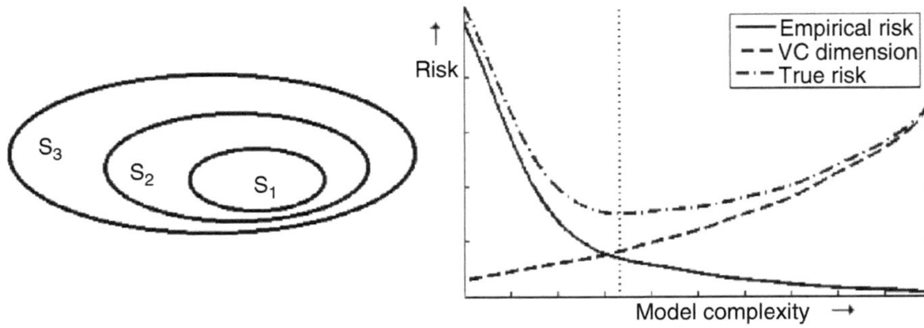

Fig. 5.18. *Left*: after choosing a class of functions to fit the data, they are nested in a hierarchy ordered by increasing complexity. Then, the best parameter configuration of each subset is estimated in order to best generalize the data. *Right*: as complexity of the functions increases, the VC dimension also increases and the empirical error decreases. The *vertical dotted line* represents the complexity for which the sum of VC dimension and empirical error is minimized; thus, that is the model order. Functions over that threshold overfit the data, while functions below that complexity underfit it.

of λ, more samples will be considered as outliers. However, it must be noted that $\lambda = 0$ should not be chosen, as the resultant number of outliers is an essential information during model order selection.

Regarding model order, its estimation is based on Structural Risk Minimization, that involves finding a trade-off between a classifier empirical error (i.e., its fitting quality) and its complexity. Complexity is measured in terms of VC (Vapnik and Chervonenkis) dimension. Figure 5.18 illustrates this process, that may be summarized as

1. From a priori knowledge, choose the class of functions that fit to the data: n degree polynomials, n layer neural networks, or n cluster data labeling, for instance.
2. Classify the functions in this class by means of a hierarchy of nested subsets, ordered by increasing complexity. In the case of polynomials, for instance, they may be ordered in an increasing degree order.
3. Apply an empirical risk minimization for each subset, that is, find for each subset the parameters that better fit the data.
4. Select the model for which the sum of empirical risk (that decreases as complexity increases) and VC dimension (that increases as complexity increases) is minimal.

In the case of RIC algorithm, the set of nested subsets is defined by

$$S_1 \subset S_2 \subset \cdots \subset S_K \subset \cdots \tag{5.43}$$

where $S_K = (Q^K(x_i, W) : w \in \Lambda_K), \forall i$, with a set of functions that indicate the empirical risk for the fitted model during deterministic annealing:

$$Q^K(x_i, W) = \sum_{k=1}^{K} \lim_{T \to 0} p(w_k|x_i) = \sum_{k=1}^{K} \lim_{T \to 0} \frac{p(w_k)\exp(-d(x_i, w_k)/T)}{\sum_{k=1}^{K} p(w_k)\exp(-d(w_k, x_i)/T)} \tag{5.44}$$

When $T \to 0$, then $p(w_k|x_i)$ may be approximated as the complement of a step function, that is linear in parameters and assigns a label to each sample depending on the dissimilarity between sample x_i and cluster w_k. Thus, as stated by Vapnik [165], VC dimension may be estimated from parameter number, being $h_k = (n+1)k$ for each S_k. Then increment in cluster number leads to an increment of complexity: $h_1 \leq h_2 \leq \cdots \leq h_k \leq \cdots$. From this starting point, model order is selected minimizing the following VC bound, similarly to Vapnik application to Support Vector Machines:

$$p_s \leq \eta + \frac{\varepsilon}{2}\left(1 + \left(1 + \eta\frac{4}{\varepsilon}\right)^{1/2}\right) \tag{5.45}$$

where

$$\eta = \frac{m}{l} \tag{5.46}$$

$$\varepsilon = 4\frac{h_k(\ln\frac{2l}{h_k} + 1) - \ln\frac{\xi}{4}}{l} \tag{5.47}$$

l and m being the number of samples and outliers, and $\xi < 1$ a constant.

The RIC algorithm can be seen in Alg. 10. From the dataset $X = \{x_1, \ldots, x_l\}$ and a chosen maximum number of clusters $K_{\max}$, the algorithm returns these data splitted into a set of clusters with centers $W = \{w_1, \ldots, w_k\}$ and identifies the outliers. Song provides an expression to estimate the parameter $K_{\max}$, but depending on data nature this expression may not be valid; thus, we leave it as an open parameter in Alg. 10. Regarding dissimilarity measure, the euclidean distance was the one chosen for this algorithm. Due to this fact, RIC tends to create hyperspherical clusters around each cluster center. Alternative dissimilarity measures may help to adapt the algorithm to kernel based clustering or data that cannot be linearly separable.

An example of application is shown in Fig. 5.19. In this example, RIC is applied to a dataset obtained from four different Gaussian distributions, thus, that is the optimal number of clusters. In this example, we set $T_{\min}$ to 0.1 and α to 0.9. Parameter ξ, used during order selection, is set to 0.2. Finally, the parameters ε and λ, that affect the amount of outliers found during the algorithm, were set to 1 and 0.05, respectively. As can be seen, although clustering based on five clusters from deterministic annealing ($K = 5$) yields no outliers, the algorithm is able to find the optimal number of clusters in the case of $K = 4$, for which several outliers are detected. Note that the model order selection step needs information about noisy data. Therefore, if $\lambda = 0$ is used to solve the same example, the algorithm reaches $K_{\max}$ without discovering the optimal model order of the data.

Algorithm 10: Robust Information Clustering

Input: n-dimensional samples X, $|X| = l$, K_{max}
1. Initialization

- Priors initialization: $\bar{p}(x) = 1/l$
- Temperature initialization: $T > 2\lambda_{max}(V_x)$, where $\lambda_{max}(V_x)$ is the highest eigenvalue of the X covariance matrix V_x
- Cluster initialization: $K = 1$ and $p(w_1)$ (the algorithm starts with an only cluster)

2. Deterministic Annealing: the aim of this step is to obtain the optimal soft clustering $p(w|x)$ for the current K and T
while *no convergence* **do**
  for $i = 1 \dots K$ **do**

$$p(w_i|x) = \frac{p(w_i)\exp^{-(\|x-w_i\|^2/T)}}{\sum_{j=1}^{K} p(w_j)\exp^{-(\|x-w_j\|^2/T)}}$$

$$p(w_i) = \sum_x \bar{p}(x)p(w_i|x)$$

$$w_i = \frac{\sum_x \bar{p}(x)p(w_i|x)}{p(w_i)}$$

  end
end
3. Cooling: $T = \alpha T$, being *alpha* < 1
4. Cluster partition step: calculate $\lambda_{max}(V_{xw})$ for each cluster k, being:

$$V_{xw} = \sum_{i=1}^{l} p(x_i|w_k)(x_i - w_k)(x_i - w_k)^T$$

Given $M = \min_k \lambda_{max}$ for $k = 1 \dots K$, corresponding to cluster $\bar{k}$, which may be splitted during this step.
if $T \leq 2M$ *and* $K < K_{max}$ **then**
  Outlier detection and order selection:
  Initialization: $p(x_i) = 1/l$, *lambda* > 0, *epsilon* > 0, and $\bar{p}(w|x)$ being the optimal clustering partition given by step 2.
  while $\left|\ln \sum_{i=1}^{l} p(x_i)c_i - \ln \max_{i=1\dots l} c_i\right| < \varepsilon$ **do**
    for $i = 1 \dots l$ **do**

$$c_i = \exp[\sum_{k=1}^{K}(\bar{p}(w_k|x_i)\ln\frac{\bar{p}(w_k|x_i)}{\sum_{j=1}^{l} p(x_j)\bar{p}(w_k|x_i)} - \lambda\bar{p}(w_k|x_i)\|w_k - x_i\|^2)]$$

$$p(x_i) = \frac{p(x_i)c_i}{\sum_{i=1}^{l} p(x_i)c_i}$$

    end
  end

(continued)

Algorithm 10: continued

Order selection: Calculate p_s (Eq. 5.45) for the current K. If a minimum p_s is found, the optimal order has been achieved. In this case, STOP the algorithm and return the clustering for $T = 0$

cluster partitioning: cluster $\bar{k}$ is splitted into two new clusters

$w_{K+1} = w_{\bar{k}} + \delta$ and $w_{\bar{k}} = w_{\bar{k}} - \delta$

$p(w_{K+1}) = \frac{p(w_{\bar{k}})}{2}$, $p(w_{\bar{k}}) = \frac{p(w_{\bar{k}})}{2}$

$K = K + 1$

end

5. Temperature check: if $(T > Tmin)$ and $K < Kmax$ return to 2. Otherwise, STOP the algorithm and return the clustering for T=0.

Output: W: k cluster centers, $p(W|X)$

5.7 IT-Based Mean Shift

Information theory not only was included in clustering algorithms during last years, but it also may help to assess or even theoretically study this kind of methods. An example is given in this section, where Mean Shift, a popular clustering and order selection algorithm, is observed from an Information Theoretical point of view [135].

5.7.1 The Mean Shift Algorithm

Mean shift is a nonparametric clustering and order selection algorithm based on pdf estimation from Parzen windows. Given $x \in X = (x_i)_{i=1}^N$, Parzen window is defined as

$$P(x, \sigma) = \frac{1}{N} \sum_{i=1}^{N} G(||x - x_i||^2, \sigma) \tag{5.48}$$

where $G(t, \sigma) = \exp^{-t^2/2\sigma^2}$ is a Gaussian kernel with variance σ. In order to estimate the pdf modes from a set of samples obtained from a multimodal Gaussian density, the mean shift algorithm looks for stationary points where $\nabla P(x, \sigma) = 0$. This problem may be solved by means of an iterative procedure in which x^{t+1} at iteration $t + 1$ is obtained from its value x^t at previous iteration:

$$x^{t+1} = m(x^t) = \frac{\sum_{i=1}^{N} G(||x - x_i||^2, \sigma) x_i}{\sum_{i=1}^{N} G(||x - x_i||^2, \sigma)} \tag{5.49}$$

In the latter equation, $m(x)$ represents the mean of all samples $x_i \in X$ when the Gaussian kernel is centered in x. In order to reach a stationary point, the kernel centered in x must follow the direction of the mean shift vector $m(x) - x$. The effect of normalization in $m(x)$ is that the kernel moves

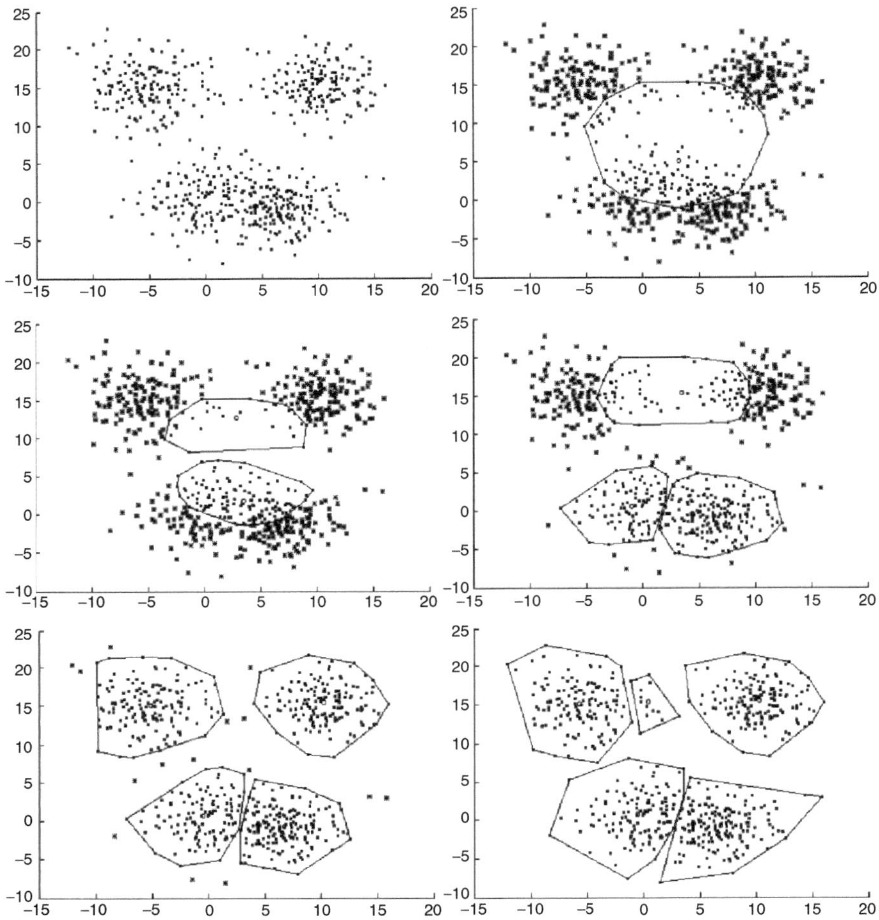

Fig. 5.19. Example of application of RIC to a set of 2D samples gathered from four different Gaussian distributions. From *left* to *right*, and from *top* to *bottom*: data samples, optimal hard clustering for $K = 1$ ($p_s = 0.4393$), optimal hard clustering for $K = 2$ ($p_s = 0.6185$), optimal hard clustering for $K = 3$ ($p_s = 0.6345$), optimal hard clustering for $K = 4$ ($p_s = 0.5140$) and optimal hard clustering for $K = 5$ ($p_s = 0.5615$). In all cases, each cluster is represented by a convex hull containing all its samples, and a *circle* is representing the cluster representative. Outliers are represented by *star* symbols. When $K = 4$ the minimum p_s is found; thus, although in other cases the amount of outliers is lower, $K = 4$ is the optimal number of clusters returned by the algorithm.

with large steps through low density regions and with small steps otherwise; step size estimation is not needed. However, the kernel width σ remains as an important parameter.

Iteratively application of Eq. 5.49 until convergence is known as Gaussian Blurring Mean Shift (GBMS). This process searches the pdf modes while blurring initial dataset. First, samples evolve to the modes of the pdf while mutually approaching. Then, and from a given iteration, data tend to collapse fast, making the algorithm unstable. The blurring effect may be avoided if the pdf estimation is based on the original configuration of the samples. This improvement is known as Gaussian Mean Shift (GMS). In GMS, the pdf estimation is obtained comparing the samples in current iteration with original ones $x_i^o \in X^o$:

$$x^{t+1} = m(x^t) = \frac{\sum_{i=1}^{N} G(||x - x_i^o||^2, \sigma) x_i^o}{\sum_{i=1}^{N} G(||x - x_i^o||^2, \sigma)} \tag{5.50}$$

Samples in X^o are not modified during this algorithm, which is stable and always converges with a linear convergence rate. An additional property of this algorithm is that the trajectory followed by the samples toward local minima is smooth: the angle between two consecutive mean shifts is always in the range $(-\frac{\pi}{2}, \frac{\pi}{2})$. Although GMS algorithm is still sensitive to parameter σ, it may be dynamically adapted using some of the methods reviewed in Chapter 5.

5.7.2 Mean Shift Stop Criterion and Examples

An advantage of GMS over GBMS is that samples are not blurred. Kernels tend to converge to the pdf modes and remain stable. Therefore, a simple stop crietrion for GMS based on a translation threshold may be applied. GMS ends when the sum of mean shift vector magnitudes of the samples is lower than a given threshold, that is

$$\frac{1}{N} \sum_{i=1}^{N} d^t(x_i) < \delta \tag{5.51}$$

$d^t(x_i)$ being the mean shift magnitude corresponding to sample x_i at iteration t:

$$d^t(x_i) = ||x_i^t - x_i^{t-1}||^2 \tag{5.52}$$

Concerning GBMS, there are two main convergence phases. During the first one, all samples rapidly collapse to their pdf modes, while pdf modes are slowly displaced among each others. Then, and depending on Parzen's Window width, all pdf modes also collapse to produce an only sample. Thus, in the case of GBMS, we should apply a stop criterion that stops the algorithm at the end of the first convergence phase. The criterion described in the original work by Rao et al. [135] is a heuristic that relies on information theory.

During the second convergence phase of GBMS, the set $d^t = \{d^t(x_i)\}_{i=1}^N$ contains only k different values, k being the detected number of pdf modes. Splitting d^t into a histogram with sufficient bin size and in the range $[0, \max(d^t)]$, only k of these bins will not be zero valued. The transition from the first convergence phase to the second one can be detected when the difference of Shannon's entropy estimated from this histogram at consecutive iterations is approximately equal to zero:

$$|H(d^{t+1}) - H(d^t)| < 10^{-8} \tag{5.53}$$

Two examples of GMS and GBMS are shown in Figs. 5.20 and 5.21. The data in the first example are extracted from a ring of 16 Gaussian pdfs, with the same variance and different a priori probabilities (i.e. the reason why Gaussians are represented with different heights in that figure). In the case of the second example, the data were extracted from 10 Gaussian pdfs with random mean and variance. As can be seen, in both cases GMS outperforms GBMS, correctly locating the exact number of pdf modes and giving a more accurate prediction of the actual mode location. An effect of GBMS is the collapse of several modes if their separation is lower than the kernel size.

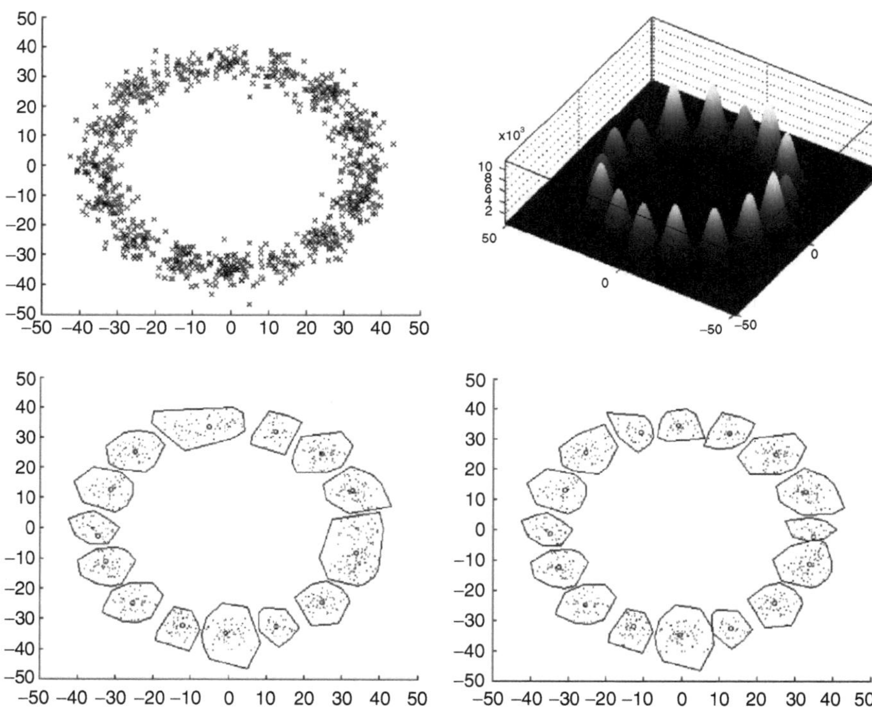

Fig. 5.20. Example of GMS and GBMS application. The samples were obtained from a ring of Gaussian distributions with equal variance and different a priori probabilities. From *left* to *right* and from *top* to *bottom:* input data, pdfs from which data were extracted, GBMS results, and GMS results.

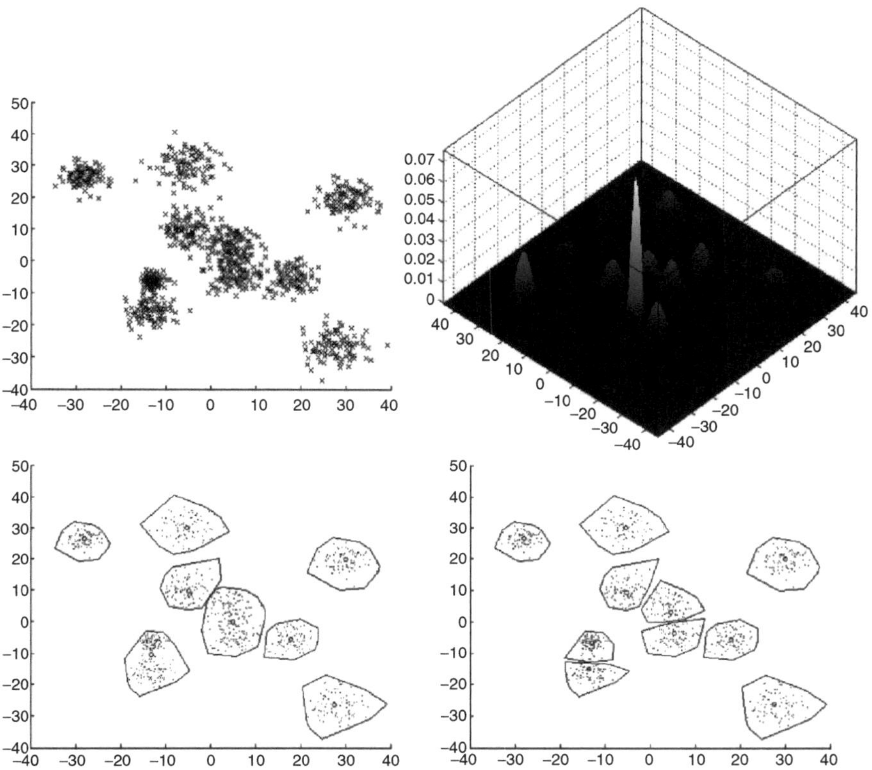

Fig. 5.21. Example of GMS and GBMS application. The samples were obtained from 10 Guassian distributions with random means and variances. From *left* to *right* and from *top* to *bottom:* input data, pdfs from which data were extracted, GBMS results, and GMS results.

5.7.3 Rényi Quadratic and Cross Entropy from Parzen Windows

This section studies the relation between Rényi entropy and pdf estimation based on Parzen windows that will be exploited during the next section. Let us first recall the expression of Rényi quadratic entropy:

$$H(X) = -\log\left(\int P^2(x)\,dx\right) \tag{5.54}$$

Estimating $P(x)$ from Parzen yields a product of two Gaussian kernels. An interesing property is that the integral of the product of two Gaussian kernels is equivalent to a Gaussian kernel which variance is the sum of the original variances. Thus, previous expression can be expressed as

$$H(X) = -\log(V(X)) \tag{5.55}$$

where

$$V(X) = \frac{1}{N^2}\sum_{i=1}^{N}\sum_{j=1}^{N} G(||x_i - x_j||^2, \sigma) \tag{5.56}$$

It must be noted that this last expression considers each pair of samples. The contribution of each sample $x_i \in X$ is given by

$$V(x_i) = \frac{1}{N^2}\sum_{j=1}^{N} G(||x_i - x_j||^2, \sigma) \tag{5.57}$$

In the original paper by Rao et al. [135], samples are described as information particles that interact among them by means of forces similar to the laws of physics. From this supposition emerges the association of $V(x_i)$ to the concept of Information potential. Thus, $V(x_i)$ may be understood as the information potential of x_i over the rest of samples in the dataset (see Fig. 5.22). The derivative is given by

$$\frac{\partial}{\partial x_i} V(x_i) = \frac{1}{N^2}\sum_{j=1}^{N} G(||x_i - x_j||^2, \sigma)\left(\frac{||x_j - x_i||^2}{\sigma^2}\right) \tag{5.58}$$

Following the particle schema, this derivative $F(x_i)$ represents the information net force that all the samples exert over x_i:

$$F(x_i) = \frac{\partial}{\partial x_i} V(x_i) = \sum_{j=1}^{N} F(x_i|x_j) \tag{5.59}$$

where the information force that x_j exerts over x_i is

$$F(x_i|x_j) = \frac{1}{N^2} G(||x_i - x_j||^2, \sigma)\left(\frac{||x_j - x_i||^2}{\sigma^2}\right) \tag{5.60}$$

All this information particle formulation can be extended to the case of interaction between two different datasets $X = (x_i)_{i=1}^{N}$ and $Y = (y_j)_{j=1}^{M}$. Given Parzen window estimation of X and Y pdfs, $P_X(x, \sigma_X)$ and $P_Y(x, \sigma_Y)$, analysis in this case is based on the Rényi cross entropy:

$$H(X;Y) = -\log\left(\int P_X(t)P_Y(t)\,dt\right) \tag{5.61}$$

Then, using the Gaussian kernel property previously introduced:

$$H(X;Y) = -\log(V(X;Y)) \tag{5.62}$$

where

$$V(X;Y) = \frac{1}{NM}\sum_{i=1}^{N}\sum_{j=1}^{M} G(||x_i - y_j||^2, \sigma) \tag{5.63}$$

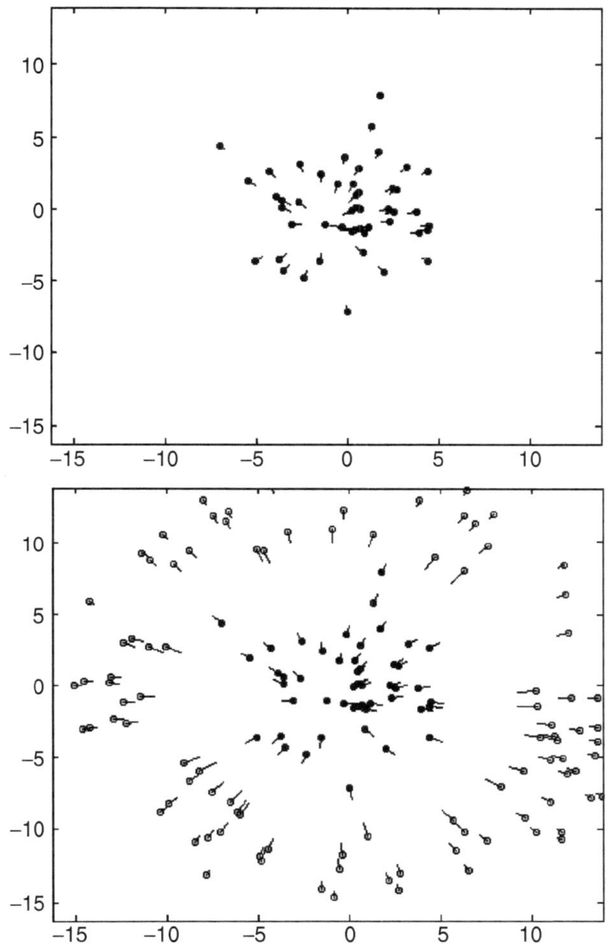

Fig. 5.22. Representation of the information force within a dataset (*top*) and between two different datasets (*bottom*).

and $\sigma^2 = \sigma_X^2 + \sigma_Y^2$. Finally, the information force that Y exerts over a $x_i \in X$ sample (see Fig. 5.22) is

$$
\begin{aligned}
F(x_i; Y) &= \frac{\partial}{\partial x_i} V(x_i; Y) \\
&= \sum F(x_i | x_j) \\
&= \frac{1}{NM} \sum_{j=1}^{M} G(||x_i - y_j||^2, \sigma) \left(\frac{y_j - x_i}{\sigma^2} \right)
\end{aligned}
$$

The opposite force $F(X; y_j)$ is simply derived from $F(x_i; Y)$ swapping $M \leftrightarrow N$ and $X \leftrightarrow Y$.

5.7.4 Mean Shift from an IT Perspective

Studying Mean Shift from an information theory perspective gives insights on what is mean shift actually optimizing, and also provides additional evidence of the stability of GMS over GBMS. From the original dataset X^o, that will not be modified during the process, and an initial $X = X^o$, we may define Mean Shift as an energy minimization problem. The cost function to minimize as X is updated is given by

$$\begin{aligned}
J(X) &= \min_X H(X) \\
&= \min_X - \log(V(X)) \\
&= \max_X V(X) \\
&= \max_X \frac{1}{N^2} \sum_{i=1}^{N} \sum_{j=1}^{N} G(||x_i - x_j||^2, \sigma)
\end{aligned}$$

In the latter expression, the logarithm can be removed. Due to the fact that the logarithm is a monotonic function, its optimization may be translated to an optimization of its parameter, $V(X)$ in this case. As stated before, $x_k = \{1, 2, .., n\} \in X$ are modified in each iteration. In order to search a stable configuration of X, $J(X)$ must be differentiated and equated to zero:

$$2F(x_k) = \frac{2}{N^2} \sum_{j=1}^{N} G(||x_k - x_j||^2, \sigma) \left(\frac{||x_k - x_j||^2}{\sigma^2} \right) = 0 \tag{5.64}$$

After rearranging Eq. 5.64, we obtain exactly the same GBMS iterative sample updating equation in Eq. 5.49. The conclusion extracted after this derivation is that GBMS directly minimizes the dataset's overall Rényi's quadratic entropy. The cause of the unstability of GBMS is the infinite support property of Gaussian kernel, when the only saddle point is given by $H(X) = 0$. In order to avoid GBMS issues, we may choose to minimize Rényi's cross entropy rather than Rényi quadratic entropy. The new cost function is given by

$$J(X) = \max_X V(X; X^o) = \max_X \frac{1}{N^2} \sum_{i=1}^{N} \textstyle\sum_{j=1}^{N} G(||x_i - x_j||^2, \sigma) \tag{5.65}$$

that when differentiated with respect to $x_k = \{1, 2, ..., n\} \in X$ and equaled to zero yields:

$$\frac{\partial}{\partial x_k} J(x) = F(x; X^o) = 0 \tag{5.66}$$

In this case, the update equation is exactly Eq. 5.50. Figure 5.23 shows an example of the evolution of GMS and GBMS cost functions during 59 iterations, from the example in Fig. 5.20. As can be seen, GMS cost function smoothly decreases and converges after a few iterations. Conversely, GBMS

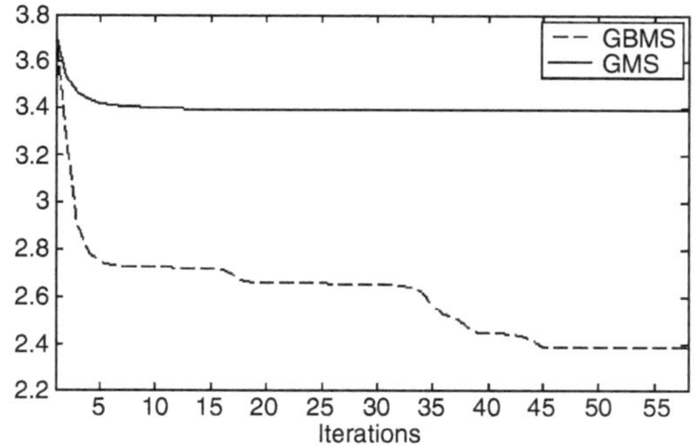

Fig. 5.23. Evolution of GMS and GBMS cost function during 59 mean shift iterations applied to data shown in Fig. 5.20.

cost function decreases during all the process. The GMS stop criterion is more intuitive, due to the fact that is based on its cost function. But GBMS should finish before $H(X) = 0$ is reached, or all samples would have collapsed to an only one. The Rényi cross entropy cost function could have been applied to GBMS. Although a minimum in this graph would be an indicative of the transition from the first convergence phase to the second one, this approach is only valid when pdf modes do not overlap.

5.8 Unsupervised Classification and Clustering Ensembles

In previous sections of this chapter we have explained the problems of clustering and we have presented different algorithms which use IT for performing an unsupervised partition of the data. In this section we address the problem of using different clustering algorithms and combining their results into a single clustering result. Such a strategy is intended to reuse the outputs yielded by different clustering algorithms, producing a new, more robust clustering. Some of the advantages are the possibility to use available clustering solutions and parallelization. One way to generate different clustering results is to perform several clustering procedures by varying the feature sets of the data, as well as by varying the parameters of the clustering algorithms. For example, an EM algorithm can generate different clusterings on the same data, if different initial parameters are used. This is not the case in the EBEM algorithm as it does not depend on initialization; in this case, the Gaussianity deficiency parameter could be varied for producing different outputs. For other algorithms the number of desired clusters in the partition can be changed. Also, very different clustering algorithms could be run in order to combine their results. Therefore, using clustering ensembles can be useful for creating a partition with little hand-tuning or when there is little knowledge about the data.

In clustering ensembles additional problems arise, apart from those inherent to the original clustering problem. The most obvious one is the labeling problem. Suppose we run four different clustering algorithms on the data:

$$\mathcal{X} = \{\mathbf{X}_1, \mathbf{X}_2, \ldots, \mathbf{X}_N\} \tag{5.67}$$

Suppose we have $N = 6$ samples and the algorithms yield four different partitions $\mathcal{C} = \{\mathbf{C}_1, \mathbf{C}_2, \mathbf{C}_3, \mathbf{C}_4\}$ which contain the following labelings:

$$\begin{aligned} \mathbf{C}_1 &= \{1, 1, 2, 2, 3, 3\} \\ \mathbf{C}_2 &= \{2, 2, 3, 3, 1, 1\} \\ \mathbf{C}_3 &= \{2, 1, 1, 4, 3, 3\} \\ \mathbf{C}_4 &= \{2, 1, 4, 1, 3, 3\} \end{aligned} \tag{5.68}$$

These partitions are a result of *hard* clustering, which is to say that the different clusters of a partition are disjoint sets. Contrarily, in *soft* clustering each sample can be assigned to several clusters in different degrees.

How to combine these labels for the clustering ensemble? In the first place, the labels are nominal values and have no relation among different $\mathbf{C}_i$ clusterings. For example $\mathbf{C}_1$ and $\mathbf{C}_2$ correspond to identical partitions despite the values of the labels. Secondly, some clusters may agree on some samples, but disagree on other ones. Also, the fact that some clusterings may yield a different number of clusters is an extra complexity. Therefore some kind of *consensus* is needed.

At first glance, the consensus clustering $\mathbf{C}^*$ should share as much information as possible with the original clusterings $\mathbf{C}_i \in \mathcal{C}$. This could be formulated as a combinatorial optimization problem in terms of mutual information [150]. A different strategy is to summarize the clustering results in a co-association matrix whose values represent the degree of association between objects. Then, some kind of voting strategy can be applied to obtain a final clustering. Another formulation of the problem [131] states that the clusterings $\mathbf{C}_i \in \mathcal{C}$ are, again, the input of a new clustering problem in which the different labels assigned to each sample become its features. In Section 5.8.1 we will explain this formulation.

5.8.1 Representation of Multiple Partitions

In the previous example we showed four different clusterings for the data $\mathcal{X} = \{\mathbf{X}_1, \mathbf{X}_2, \cdots, \mathbf{X}_N\}$ with $N = 6$ samples. Suppose each $\mathbf{X}_j \in \mathcal{X}$ has $D = 2$ features:

$$\begin{aligned} \mathbf{X}_1 &= \{x_{11}, x_{12}\} \\ \mathbf{X}_2 &= \{x_{21}, x_{22}\} \\ \mathbf{X}_3 &= \{x_{31}, x_{32}\} \\ \mathbf{X}_4 &= \{x_{41}, x_{42}\} \\ \mathbf{X}_5 &= \{x_{51}, x_{52}\} \\ \mathbf{X}_6 &= \{x_{61}, x_{62}\} \end{aligned} \tag{5.69}$$

The samples $\mathcal{X}_j$ can be represented together with the labels of their clusterings in $\mathcal{C}$. Also, let us use different labels for each one of the partitions, to make explicit that they have no numerical relation. Now each sample has $H = |\mathcal{C}|$ new features. There are two ways for performing a new clustering: (a) using both original and new features or (b) using only new features. Using only the new features makes the clustering independent of the original features. At this point the problem is transformed into a categorical clustering problem in a new space of features and it can be solved using various statistical and IT-based techniques. The resulting clustering $\mathbf{C}^*$ is known as a *consensus clustering* or a *median partition* and it summarizes the partitions defined by the set of new features.

5.8.2 Consensus Functions

Different consensus functions and heuristics have been designed in the literature. Co-association methods, re-labeling approaches, the mutual information approach, and mixture model of consensus are some relevant approaches [131]. Among the graph-based formulations [150] there are the instance-based, the cluster-based and the hypergraph methods. Some of the characteristics of these approaches are:

- The co-association method evaluates the similarity between objects, based on the number of clusters shared by two objects in all the partitions. It has a quadratic complexity in the number of patterns and it does not provide good co-association value estimation if the number of clusterings is small.
- Re-labeling approaches solve the label correspondence problem heuristically looking for an agreement with some reference partition. Then, a simple voting procedure can associate objects with their clusters and requires a known number of clusters in the target consensus partition.
- The Mutual Information approach has as objective function the mutual information between the empirical probability distribution of labels in the consensus partition and the labels in the ensemble.
- The Mixture Model of consensus has a maximum-likelihood formulation based on a finite mixture model, explained in the next subsection.
- The instance-based graph formulation models pairwise relationships among samples with a fully connected graph, resulting in a partitioning problem of size N^2. Its computational complexity is expensive, and may vary depending on the algorithm used to partition the graph.
- The cluster-based graph formulation constructs a graph to model the similarities among clusters in the ensemble, then partitions this graph to group clusters which correspond to one another.
- Hypergraph methods are heuristics for solving the cluster ensemble problem, by representing the set of clusterings as a hypergraph. In hypergraphs a hyperedge is a generalization of an edge so that it can connect more than

two vertices. In [150] several hypergraph-based approaches are presented, such as a cluster similarity partitioning, a minimum cut objective for hypergraph partitioning, and a cluster correspondence problem.

Next we will explain the mixture model of consensus and two mutual information based objective functions for consensus.

A Mixture Model of Consensus

In this approach for consensus the probabilities of the labels for each pattern are modeled with finite mixtures. Mixture models have already been described in this chapter. Suppose each component of the mixture is described by the parameters $\boldsymbol{\theta}_m$, $1 \leq m \leq M$ where M is the number of components in the mixture. Each component corresponds to a cluster in the consensus clustering, and each cluster also has a prior probability π_m, $1 \leq m \leq M$. Then the parameters to be estimated for the consensus clustering are:

$$\Theta = \{\pi_1, \dots, \pi_M, \boldsymbol{\theta}_1, \dots, \boldsymbol{\theta}_M\} \tag{5.70}$$

In our toy-example we will use $M = 2$ components in the mixture.[1] The data to be described by the mixture are the "new features," which are the labels of all the partitions, obtained from $\mathcal{C}$. We will denote these labels as

$$\begin{aligned} \mathcal{Y} &= \{\mathbf{Y}_1, \dots, \mathbf{Y}_N\} \\ \mathbf{Y}_i &= \{c_{1i}, c_{2i}, \dots, c_{Hi}\} \end{aligned} \tag{5.71}$$

where $H = |\mathcal{C}|$. In other words, using matrices formulation, we have $\mathcal{Y} = \mathcal{C}'$.

It is assumed that all the labels $\mathbf{Y}_i$, $1 \leq i \leq N$ are random and have the same distribution, which is described by the mixture

$$P(\mathbf{Y}_i|\Theta) = \sum_{m=1}^{M} \pi_m P_m(Y_i|\theta_m) \tag{5.72}$$

Labels are also assumed to be independent, so the log-likelihood function for the parameters Θ given $\mathcal{Y}$ is

$$\log \ell(\Theta|\mathcal{Y}) = \log \prod_{i=1}^{N} P(\mathbf{Y}_i|\Theta) \tag{5.73}$$

[1] However, the number of final clusters is not obvious, neither in the example, nor in real experiments. This problem called *model order selection* has already been discussed in this chapter. A minimum description length criterion can be used for selecting the model order.

Now the problem is to estimate the parameters Θ which maximize the likelihood function (Eq. 5.73). For this purpose some densities have to be modeled. In the first place, a model has to be specified for the component-conditional densities which appear in Eq. 5.72. Although the different clusterings of the ensemble are not really independent, it could be assumed that the components of the vector $\mathbf{Y}_i$ (the new features of each sample) are independent, therefore their probabilities will be calculated as a product. In second place, a probability density function (PDF) has to be chosen for the components of $\mathbf{Y}_i$. Since they consist of cluster labels in $\mathbf{C}_i$, the PDF can be modeled as a multinomial distribution.

A distribution of a set of random variates $X_1, X_2, \ldots, X_k$ is multinomial if

$$P(X_1 = x_1, \ldots, X_k = x_k) = \frac{n!}{\prod_{i=1}^{k} x_i!} \prod_{i=1}^{k} \vartheta_i^{x_i} \tag{5.74}$$

where x_i are non-negative integers such that

$$\sum_{i=1}^{k} x_i = n \tag{5.75}$$

and $\theta_i > 0$ are constants such that

$$\sum_{i=1}^{k} \theta_i = 1 \tag{5.76}$$

Putting this together with the product for the conditional probability of $\mathbf{Y}_i$ (for which conditional independence was assumed), the probability density for the components of the vectors $\mathbf{Y}_i$ is expressed as

$$P_m(\mathbf{Y}_i|\Theta_m) = \prod_{j=1}^{H} \prod_{k=1}^{K(j)} \left(\vartheta_{jm}(k)\right)^{\delta(y_{ij},k)} \tag{5.77}$$

where $\vartheta_{jm}(k)$ are the probabilities of each label and $\delta(y_{ij}, k)$ returns 1 when k is the same as the position of the label y_{ij} in the labeling order; otherwise it returns 0. $K(j)$ is the number of different labels existing in the partition j. For example, for the labeling of the partition $\mathbf{C}_2 = \{b, b, c, c, a, a\}$, we have $K(2) = 3$, and the function $\delta(y_{i2}, k)$ would be evaluated to 1 only for the parameters: $\delta(a, 1)$, $\delta(b, 2)$, $\delta(c, 3)$ and it would be 0 for any other parameters. Also note that in Eq. 5.77, for a given mixture the probabilities for each clustering sum 1:

$$\sum_{k=1}^{K(j)} \vartheta_{jm}(k) = 1,\ \forall j \in \{1, \cdots, H\},\ \forall m \in \{1, \cdots, M\}, \tag{5.78}$$

Table 5.1. Original feature space and transformed feature space obtained from the clustering ensemble. Each partition of the ensemble is represented with a different labels in order to emphasize the label correspondence problem.

Sample $\mathbf{X}_i$	Orig. features $\mathcal{X}$	New features $\mathcal{C}$ $\mathbf{C}_1$	$\mathbf{C}_2$	$\mathbf{C}_3$	$\mathbf{C}_4$	Consensus $\mathbf{C}^*$
$\mathbf{X}_1$	$x_{11}\ x_{12}$	1	b	N	β	?
$\mathbf{X}_2$	$x_{21}\ x_{22}$	1	b	M	α	?
$\mathbf{X}_3$	$x_{31}\ x_{32}$	2	c	M	δ	?
$\mathbf{X}_4$	$x_{41}\ x_{42}$	2	c	K	α	?
$\mathbf{X}_5$	$x_{51}\ x_{52}$	3	a	J	γ	?
$\mathbf{X}_6$	$x_{61}\ x_{62}$	3	a	J	γ	?

In our toy-example, taking the values of $\mathcal{Y}$ from Table 5.1 and using Eq. 5.77, the probability for the vector $\mathbf{Y}_1 = \{1, b, N, \beta\}$ to be described by the 2nd mixture component would be:

$$
\begin{aligned}
P_2(\mathbf{Y}_i|\Theta_2) = & \prod_{k=1}^{3}\left(\vartheta_{12}(k)\right)^{\delta(1,k)} \cdot \prod_{k=1}^{3}\left(\vartheta_{22}(k)\right)^{\delta(b,k)} \\
& \cdot \prod_{k=1}^{4}\left(\vartheta_{32}(k)\right)^{\delta(N,k)} \cdot \prod_{k=1}^{4}\left(\vartheta_{42}(k)\right)^{\delta(\beta,k)} \\
= & \ \vartheta_{12}(1)^1\vartheta_{12}(2)^0\vartheta_{12}(3)^0 \\
& \cdot \vartheta_{22}(1)^0\vartheta_{22}(2)^1\vartheta_{22}(3)^0 \\
& \cdot \vartheta_{32}(1)^0\vartheta_{32}(2)^1\vartheta_{32}(3)^0\vartheta_{32}(4)^0 \\
& \cdot \vartheta_{42}(1)^0\vartheta_{42}(2)^1\vartheta_{42}(3)^0\vartheta_{42}(4)^0 \\
= & \ \vartheta_{12}(1)^1\vartheta_{22}(2)^1\vartheta_{32}(2)^1\vartheta_{42}(2)^1
\end{aligned}
\tag{5.79}
$$

This probability refers to the membership of the first sample to the second component of the mixture, corresponding to the second cluster of the final clustering result. The parameters $\vartheta_{jm}(k)$ which describe the mixture are still unknown and have to be estimated.

The likelihood function (Eq. 5.73) can be optimized with the EM algorithm, already explained at the beginning of the chapter. It would not be possible to solve it analytically, as all the parameters in Θ are unknown. The EM is an iterative algorithm for estimation when there are hidden variables. In the present problem we will denote these hidden variables with $\mathcal{Z}$. Given the complete set of data $(\mathcal{Y}, \mathcal{Z})$, their distributions have to be consistent, then the log-likelihood function for this set is given by

$$
\log P(\mathcal{Y}|\Theta) = \log \sum_{z \in \mathcal{Z}} P(\mathcal{Y}, \mathbf{z}|\Theta) \tag{5.80}
$$

The EM algorithm iterates two main steps. In the *E step* the expected values of the hidden variables are estimated, given the data $\mathcal{Y}$ and the current estimation of the parameters Θ. The following equations are the result of the derivation of the equations of the EM algorithm:

$$E[z_{im}] = \frac{\pi'_m \prod_{j=1}^{H} \prod_{k=1}^{K(j)} \left(\vartheta'_{jm}(k)\right)^{\delta(y_{ij},k)}}{\sum_{n=1}^{M} \pi'_n \prod_{j=1}^{H} \prod_{k=1}^{K(j)} \left(\vartheta'_{jn}(k)\right)^{\delta(y_{ij},k)}} \tag{5.81}$$

In the *M step* the parameters of the multinomial distribution are updated in the following way:

$$\pi_m = \frac{\sum_{i=1}^{N} E[z_{im}]}{\sum_{i=1}^{N} \sum_{m=1}^{M} E[z_{im}]} \tag{5.82}$$

$$\vartheta_{jm}(k) = \frac{\sum_{i=1}^{N} \vartheta(y_{ij}, k) E[z_{im}]}{\sum_{i=1}^{N} \sum_{k=1}^{K(j)} \vartheta(y_{ij}, k) E[z_{im}]} \tag{5.83}$$

After the convergence of the EM algorithm, $E[z_{im}]$ will contain the value of the probability that the pattern $\mathbf{Y}_i$ is generated by the mth component of the mixture. The cluster to which the ith sample is assigned is that component of the mixture which has the largest expected value $E[z_{im}]$, $1 \leq i \leq N$

Mutual Information Based Consensus

The mutual information approach to the clustering consensus refers to the maximization of the mutual information between the empirical probability distribution of labels in the consensus partition $\mathbf{C}^*$ and the labels in the ensemble $\{\mathbf{C}_1, \mathbf{C}_2, \mathbf{C}_3, \mathbf{C}_4\}$. One possibility is to use Shannon's definition of mutual information, as do Strehl and Ghosh in [150].

$$I(\mathbf{C}^*; \mathbf{C}_i) = \sum_{k=1}^{K^*} \sum_{j=1}^{K(i)} P(L_k^*, L_j^i) \log \left(\frac{P(L_k^*, L_j^i)}{P(L_k^*) P(L_j^i)} \right) \tag{5.84}$$

where $K(i)$ is the number of different labels in the $\mathbf{C}_i$ partition, K^* is the number of different labels in the consensus partition $\mathbf{C}^*$ (which is the objective partition). The sets L_j^i refer to the samples which are assigned to the cluster j of the clustering i. Thus, the probabilities $P(L_k^*)$, $P(L_j^i)$, and $P(L_k^*, L_j^i)$ are calculated this way:

$$\begin{aligned} P(L_k^*) &= |L_k^*|/N \\ P(L_j^i) &= |L_j^i|/N \\ P(L_k^*, L_j^i) &= |L_k^* \cap L_j^i|/N \end{aligned} \tag{5.85}$$

where N is the total number of samples (6 in the example). This equation calculates the mutual information between only two partitions. For maximizing the amount of information that the hypothetic partition $\mathbf{C}^*$ shares with all partitions in the ensemble $\mathcal{C}$ the sum of all mutual informations has to be maximized:

$$\mathbf{C}^* = \arg\max_{\mathbf{C}} \sum_{i=1}^{H} I(\mathbf{C}; \mathbf{C}_i) \tag{5.86}$$

Let us see an example with the data in Table 5.1. Suppose we want to calculate the mutual information between the fourth clustering $\mathbf{C}_4$ and a consensus clustering $\mathbf{C}^* = \{1, 1, 1, 2, 2, 2\}$. We would have the following cluster labels:

$$\begin{aligned} \mathbf{C}_4 &= \{ \overbrace{\beta,}^{L_2^4} \underbrace{\alpha,}_{L_1^4} \overbrace{\delta,}^{L_3^4} \underbrace{\alpha,}_{L_1^4} \overbrace{\gamma, \gamma}^{L_4^4} \} \\ \mathbf{C}^* &= \{ \underbrace{1,\ 1,\ 1,}_{L_1^*} \underbrace{2,\ 2,\ 2}_{L_2^*} \} \end{aligned} \tag{5.87}$$

Using Eq. 5.84 we calculate the mutual information between the partitions. The logarithm $\log(x)$ is calculated in base 2.

$$\begin{aligned} I(\mathbf{C}^*; \mathbf{C}_4) &= \sum_{k=1}^{2} \sum_{j=1}^{4} \frac{|L_k^* \cap L_j^i|}{N} \log \left(\frac{|L_k^* \cap L_j^i| \cdot 6}{|L_k^*| \cdot |L_j^i|} \right) \\ &= \frac{1}{6} \log \frac{1 \cdot 6}{3 \cdot 2} + \frac{1}{6} \log \frac{1 \cdot 6}{3 \cdot 1} + \frac{1}{6} \log \frac{1 \cdot 6}{3 \cdot 1} + 0 \\ &\quad + \frac{1}{6} \log \frac{1 \cdot 6}{3 \cdot 2} + 0 + 0 + \frac{2}{6} \log \frac{2 \cdot 6}{3 \cdot 2} \\ &= \frac{2}{3} \end{aligned} \tag{5.88}$$

The same calculation should be performed for the other partitions in $\mathcal{C}$ in order to obtain a measure of the amount of information which $\mathbf{C}^*$ shares with the ensemble.

There is another definition of entropy, the *entropy of degree* s, which leads to the formulation of generalized mutual information, used in [131] for clustering consensus evaluation. Entropy of degree s is defined as

$$H^s(\mathbf{P}) = (2^{1-s} - 1)^{-1} \left(\sum_{i=1}^{n} p_i^s - 1 \right) \tag{5.89}$$

where $\mathbf{P} = (p_1, \ldots, p_n)$ is a discrete probability distribution, $s > 0$, and $s \neq 1$. In the limit $s \to 1$ it converges to the Shannon entropy. The entropy of degree s permits simpler characterization than Rényi's entropy of order r. Particularly, in the probabilistic measure of interclass distance, quadratic entropy (entropy of degree $s = 2$) is closely related to classification error. With $s = 2$, generalized mutual information is defined as

$$I^2(\mathbf{C}^*;\mathbf{C}_i) = H^2(\mathbf{C}^*) - H^2(\mathbf{C}^*|\mathbf{C}_i)$$

$$= -2\left(\sum_{j=1}^{K(i)} P(L_j^i)^2 - 1\right) + 2\sum_{r=1}^{K^*} p(L_r^*)\left(\sum_{j=1}^{K(i)} P(L_j^i|L_r^*)^2 - 1\right) \tag{5.90}$$

where the probability $P(L_k^*|L_j^i)$ can be calculated as

$$P(L_k^*|L_j^i) = |L_k^* \cap L_j^i|/|L_j^i| \tag{5.91}$$

The formulation of mutual information presented in Eq. 5.90 corresponds to a clustering consensus criterion which has already been used for clustering, and is known as a "category utility function." Fisher used it in COBWEB [60] for conceptual clustering.

These mutual information functions are objective functions which have to be maximized. For this purpose the generation and evaluation of different hypothetical clusterings $\mathbf{C}^*$ are necessary. Trying all the possible partitions could be impracticable; so some kinds of heuristics or simulated annealing have to be used. For example, if we are performing a hill climbing in the space of possible consensus clusterings, and we are in the state $\mathbf{C}^* = \{1,2,1,2,2,2\}$, there are a set of possible steps to new states. Each one of the six labels could be changed to $l = l \pm 1$; so there are a maximum of $N * 2$ possible states to evaluate. Once all of them are evaluated, the best one is selected and the process is repeated. Such evaluation involves the calculation of the information shared with each one of the partitions in the ensemble. For example, let us evaluate the possible changes to other states from the state $\{1,2,1,2,2,2\}$. The notation L_j^i is used in the same way as illustrated in Eq. 5.87.

$$\begin{aligned}
I^2(\{1,2,1,2,2,2\};\mathcal{C}) &= \sum_{i=1}^{H} I^2(\{1,2,1,2,2,2\};\mathbf{C}_i)\\
&= 0.2222 + 0.2222 + 0.5556 + 0.8889 = 1.8889\\
I^2(\{2,2,1,2,2,2\};\mathcal{C}) &= 0.2222 + 0.2222 + 0.2222 + 0.5556 = 1.2222\\
I^2(\{1,1,1,2,2,2\};\mathcal{C}) &= 0.6667 + 0.6667 + 1.0000 + 0.6667 = 3.0000\\
I^2(\{1,3,1,2,2,2\};\mathcal{C}) &= 0.5556 + 0.5556 + 0.8889 + 0.8889 = 2.8889\\
I^2(\{1,2,2,2,2,2\};\mathcal{C}) &= 0.2222 + 0.2222 + 0.5556 + 0.5556 = 1.5556\\
I^2(\{1,2,1,1,2,2\};\mathcal{C}) &= 0.6667 + 0.6667 + 0.6667 + 0.6667 = 2.6667\\
I^2(\{1,2,1,3,2,2\};\mathcal{C}) &= 0.5556 + 0.5556 + 0.8889 + 0.8889 = 2.8889\\
I^2(\{1,2,1,2,1,2\};\mathcal{C}) &= 0.0000 + 0.0000 + 0.3333 + 0.6667 = 1.0000\\
I^2(\{1,2,1,2,3,2\};\mathcal{C}) &= 0.2222 + 0.2222 + 0.5556 + 0.8889 = 1.8889\\
I^2(\{1,2,1,2,2,1\};\mathcal{C}) &= 0.0000 + 0.0000 + 0.3333 + 0.6667 = 1.0000\\
I^2(\{1,2,1,2,2,3\};\mathcal{C}) &= 0.2222 + 0.2222 + 0.5556 + 0.8889 = 1.8889
\end{aligned} \tag{5.92}$$

The maximal amount of mutual information shared with the partitions of the ensemble is achieved if the next step is to make $\mathbf{C}^* = \{1,1,1,2,2,2\}$. Note that we have allowed to explore the addition of a new class label, "3."

Different constraints regarding the vector space and the possible changes can be imposed. A hill climbing algorithm has been used in this example; however it would stop at the first local maximum. Simulated annealing or a different searching algorithm has to be used to overcome this problem.

Problems

5.1 The entropy based EM algorithm and Gaussianity
The EBEM algorithm performs clustering by fitting Gaussian mixtures to the data; the components of the mixture are the resulting clusters. We presented an experiment with color image segmentation in the RGB color space. Do you think that the colors of the pixels of natural images follow a Gaussian distribution? If the data which we want to clusterize do not follow exactly a mixture of Gaussian distributions, which parameter of EBEM would you tune? Think of the modifications that you would have to perform on the EBEM algorithm in order to make it work for some other model of distribution.

5.2 The entropy based EM algorithm for color image segmentation
In the experiment of EBEM with color image segmentation which we presented, the samples of the data have three dimensions: the dimensions of the RGB color space. However, the positions of the pixels are not taken into account. It would be good to add the criterion of spatial continuity (regions). Think of the modifications that would be necessary.

5.3 The entropy based EM algorithm and MDL
The EBEM algorithm can use the MDL as model order selection criterion. In this case, when the clustering algorithm is applied to image segmentation it has been experimentally observed that the algorithm usually progresses to a high order model. This behavior is due to the fact that MDL underweighs the complexity of the models because it does not consider how many pixels are described by a given component of the mixture. Some authors increase the $1/2$ factor of the complexity term of the MDL formula (see Eq. 5.15). This is done for overweighing complexity and penalizing higher order models. However, the chosen factor depends on the application. Consider the idea of overweighing the complexity term by including not only the number of parameters of the mixture but also the coding efficiency of each component. This is meant to consider both the number of parameters and the amount of pixels encoded by them.

5.4 Information Theory and clustering
Different uses of information theory in the field of data clustering have been introduced through this chapter. Enumerate the information theory concepts applied to Agglomerative Information Bottleneck, Robust Information Clustering and IT Mean Shift. Summarize how these concepts are applied.

5.5 Rate Distortion Theory
Given the samples:

x_1	2	4
x_2	5.1	0
x_3	9.9	1.7
x_4	12.3	5.2

and the representatives of two clusters

t_1	5	2.5
t_2	8.5	3.5

if $p(x_i) = p(x_j) \forall i \neq j$ and $p(t_1|x_i) = p(t_2|x_i) \forall i$, estimate the distortion $E_d(X,T)$ and the model complexity. Think of how to decrease the distortion without varying the number of clusters. How this affects change to the model complexity?

5.6 Information Bottleneck
Recall Alg. 8 (Blahut–Arimoto Information Bottleneck clustering). In which circumstances the inequality $I(T;Y) > I(X;T)$ is true? And the inequality $I(X;T) > I(T;Y)$? Is it possible, in general, to state that any of both alternatives is better than the other one?

5.7 Blahut–Arimoto Information Bottleneck
Given the data given in Prob. 5.5. Let us consider two classes $Y = \{y_1, y_2\}$ for which $p(y_1|x_1) = p(y_2|x_2) = 1$ and $p(y_2|x_3) = p(y_2|x_4)$. Estimate $p(y|t)$ (see Alg. 8). Apply an iteration of Alg. 8 using $\beta = 0.01$. From these results, calculate $\partial I(T;Y)$ and $\partial I(X;T)$, and give an explanation of why these values are obtained.

5.8 Deterministic annealing
Modify Alg. 8 in order to make use of the deterministic annealing principles.

5.9 Agglomerative Information Bottleneck
Let $X = \{x_1, x_2, x_3, x_4\}$ be a set of samples that may be labeled as $Y = \{y_1, y_2\}$. Given the $p(X, Y = y_1)$ distribution:

X	$p(x_i, Y = y_1)$
x_1	0.9218
x_2	0.7382
x_3	0.1763
x_4	0.4057

apply the first iteration of Alg. 5.16, determining the first two samples that are joined and the new posteriors.

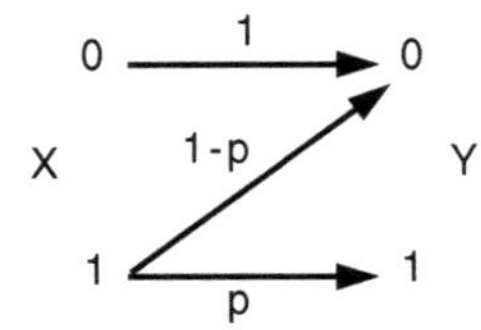

Fig. 5.24. Example channel.

5.10 Model order selection in AIB
The AIB algorithm starts with a high number of clusters (a cluster for each pattern) and proceeds by fusing cluster pairs until there is only one cluster left. Think about the effectiveness of the model order selection criterion described in Eq. 5.31. Hint: explore the information plane (Fig. 5.16).

5.11 Channel capacity
Channel capacity is defined in Eq. 5.37. It is defined over mutual Information, that may be derived as $I(X;Y) = H(Y) - H(Y|X)$ (see Chapter 4). Calculate the capacity of the channel in Fig. 5.24, in which possible inputs (X) and possible outputs (Y) are related by means of a conditional distribution $p(y|x)$ represented by edges. Draw a plot that represents channel capacity vs. p and interpret it.

5.12 RIC and alternative model order selection
The Robust Information Clustering (RIC) algorithm relies on a VC dimension criterion to estimate the optimal order of the model. Suggest an alternative IT-based criterion for this algorithm. Think about expressing the *structural risk* concept in terms of information theory.

5.13 Mean Shift and Information Theory
The interpretation of the Mean Shift algorithm in terms of information theory relies on the Rényi quadratic entropy minimization ($\alpha = 2$). Following a similar rationale, it could be interesting to extend this framework to other α-entropies. Explain the impact of such extension into the modified algorithm.

5.14 X-means clustering and Bayesian Information Criterion
The X-means algorithm is an extension of the classic K-means algorithm. X-means [124] finds automatically the optimal number of clusters by exploiting the Bayesian Information Criterion (BIC) [94, 143]. Such criterion is defined by considering the log-likelihood in the first term and $-K/2 \log n$ as in MDL (K the number of parameters, n the number of samples). Actually, the general MDL converges to BIC when $n \to \infty$. The probability model assumes one Gaussian distribution per cluster. This assumption is plugged into the definition of the log-likelihood. X-means proceeds by running K-means for a given K and deciding whether to split or not each of the K clusters, according to BIC. This process is performed recursively. Discuss the role of the MDL/BIC in the process and its expected behavior in high-dimensional data. Think about the data sparseness and complexity matters.

5.15 Clustering ensembles
The consensus based on mixture models has a maximum-likelihood formulation based on a finite mixture model. In this approach, which kinds of assumptions about the data are made? Compare then with the consensus based on mutual information.

5.9 Key References

- A. Peñalver, F. Escolano, and J.M. Sáez. "EBEM: An Entropy-Based EM Algorithm for Gaussian Mixture Models". *International Conference on Pattern Recognition*, Hong Kong (China) (2006)
- A. Peñalver, F. Escolano, and J.M. Sáez. "Two Entropy-Based Methods for Learning Unsupervised Gaussian Mixture Models". *SSPR/SPR - LNCS* (2006)
- M. Figueiredo and A.K. Jain. "Unsupervised Learning of Finite Mixture Models". *IEEE Transactions on Pattern Analysis and Machine Intelligence* 24(3): 381–396 (2002)
- N. Srebro, G. Shakhnarovich, and S. Roweis. "An Investigation of Computational and Informational Limits in Gaussian Mixture Clustering". *International Conference on Machine Learning* (2006)
- N.Z. Tishby, F. Pereira, and W. Bialek. "The Information Bottleneck method". *37th Allerton Conference on Communication, Control and Computing* (1999)
- N. Slonim, N. Friedman, and N. Tishby. "Multivariate Information Bottleneck". *Neural Computation* 18: 1739–1789 (2006)
- J. Goldberger, S. Gordon, and H. Greenspan. "Unsupervised Image-Set Clustering Using an Information Theoretic Framework". *IEEE Transactions on Image Processing* 15(2): 449–458 (2006)
- N. Slonim and N. Tishby. "Agglomerative Information Bottleneck". In *Proceeding of Neural Information Processing Systems* (1999)
- W. Punch, A. Topchy, and A. Jain. "Clustering Ensembles: Models of Consensus and Weak Partitions". *IEEE Transactions on Pattern Analysis and Machine Intelligence*, 27(12): 1866–1881 (2005)

6

Feature Selection and Transformation

6.1 Introduction

A fundamental problem in pattern classification is to work with a set of features which are appropriate for the classification requirements. The first step is the feature extraction. In image classification, for example, the feature set commonly consists of gradients, salient points, SIFT features, etc. High-level features can also be extracted. For example, the detection of the number of faces and their positions, the detection of walls or surfaces in a structured environment, or text detection are high-level features which also are classification problems in and of themselves.

Once designed the set of features it is convenient to select the most informative of them. The reason for this is that the feature extraction process does not yield the best features for some concrete problems. The original feature set usually contains more features than it is necessary. Some of them could be redundant, and some could introduce noise, or be irrelevant. In some problems the number of features is very high and their dimensionality has to be reduced in order to make the problem tractable. In other problems feature selection provides new knowledge about the data classes. For example, in gene selection [146] a set of genes (features) are sought in order to explain which genes cause some disease. On the other hand, a properly selected feature set significantly improves classification performance. However, feature selection is a challenging task.

There are two major approaches to dimensionality reduction: feature selection and feature transform. Feature selection reduces the feature set by discarding features. A good introduction to feature selection can be found in [69]. Feature transform refers to building a new feature space from the original variables, therefore it is also called feature extraction.

Some well-known feature transform methods are principal component analysis (PCA), linear discriminant analysis (LDA), and independent component analysis (ICA), among others. PCA transform relies on the eigenvalues of the covariance matrix of the data, disregarding the classes. It represents the data optimally in terms of minimal mean-square error between the transformed representation and the original one. Therefore, it is useful for separating noise, but not for finding those features which are discriminative in a classification problem. There is an extension called generalized principal component analysis (GPCA), which is essentially an algebraic geometric approach to clustering. While PCA finds projection directions that maximize the Gaussian variance, GPCA finds subspaces passing through the data. Another feature transform method is LDA, which uses the between-class covariance matrix and the transformed features are discriminative for classification. However, it produces strictly less features than the number of classes of the data. Also, the second-order statistics (covariances) are useful for classes which are represented by well-separated unimodal Gaussian data distributions. The third feature transform mentioned, ICA, finds a maximally clustered projection of the data, maximizing the divergence to the Gaussian density function. Therefore, there is an emphasis on non-Gaussian clusters.

In the feature selection literature, there are three different approaches: *filter* methods [69], *wrapper* methods [24, 96], and *online* methods [127]. Filter feature selection does not take into account the properties of the classifier (it relies on statistical tests over the variables), while wrapper feature selection tests different feature sets by building the classifier. Finally, online feature selection incrementally adds or removes new features during the learning process. There is also a notable difference between the supervised and the unsupervised feature selection methods. In supervised classification the class labels of the training samples are present. Contrarily, in the unsupervised classification problems, the class labels have to be decided by identifying clusters in the training data. This makes the unsupervised feature selection an ill-posed problem, because the clusters depend on the selected features, while the features are selected according to the clusters. An additional problem in clustering is the model order selection, or: which is the optimal number of clusters?

Feature selection (FS) is a combinatorial computational complexity problem. FS methods are usually oriented to find suboptimal solutions in a feasible number of iterations. One problem is how to explore the feature space so that the search is as straightforward as possible. The other problem is the criterion which evaluates the feature subsets. It is a delicate problem, as it has to estimate the usefulness of a subset accurately and inexpensively. Most of the feature selection research is focused on the latter topic, and it is the subject of study of the following sections as well.

6.2 Wrapper and the Cross Validation Criterion

6.2.1 Wrapper for Classifier Evaluation

Wrapper feature selection consists of selecting features according to the classification results that these features yield. Therefore, wrapper feature selection is a classifier-dependent approach. Contrarily, *filter feature selection* is classifier independent, as it is based on statistical analysis on the input variables (features), given the classification labels of the samples. In filter feature selection the classifier itself is built and tested once the features are selected. Wrappers build classifiers each time a feature set has to be evaluated. This makes them more prone to overfitting than filters. It is also worth mentioning that wrappers are usually applied as a multivariate technique, which means that they test whole sets of features.

Let us see a simple example of wrapping for feature selection. For a supervised[1] classification problem with four features and two classes, suppose we have the following data set containing nine samples:

	Features	Class
Sample 1 =	$(x_{11}\ x_{12}\ x_{13}\ x_{14}),$	C1
Sample 2 =	$(x_{21}\ x_{22}\ x_{23}\ x_{24}),$	C1
Sample 3 =	$(x_{31}\ x_{32}\ x_{33}\ x_{34}),$	C1
Sample 4 =	$(x_{41}\ x_{42}\ x_{43}\ x_{44}),$	C1
Sample 5 =	$(x_{51}\ x_{52}\ x_{53}\ x_{54}),$	C2
Sample 6 =	$(x_{61}\ x_{62}\ x_{63}\ x_{64}),$	C2
Sample 7 =	$(x_{71}\ x_{72}\ x_{73}\ x_{74}),$	C2
Sample 8 =	$(x_{81}\ x_{82}\ x_{83}\ x_{84}),$	C2
Sample 9 =	$(x_{91}\ x_{92}\ x_{93}\ x_{94}),$	C2

A wrapper feature selection approach could consist of evaluating different combinations of features. In the previous table each single feature is represented by a column: $F_j = (x_{1j}, x_{2j}, \ldots, x_{9j})$, $j \in \{1, \ldots, 4\}$. The evaluation of a feature set involves building a classifier with the selected features and testing it, so we have to divide the data set into two disjoint sets: the train set for building the classifier and the test set for testing it. For large data sets a good proportion is 75% for the train set and 25% for the test set. Usually it is adequate to perform the partition on randomly ordered samples. On the other hand, for small data sets a strategy which consists of taking only one sample for the test and repeating the process for all the samples exists, as explained later in this section.

It is very important to note that, even if we are provided with a separate test set, we cannot use it for the feature selection process. In other words, during the wrapper feature selection we use the train set which has to be divided into subtrain and subtest sets in order to build classifiers and evaluate them. Once this process is finished, there is a need to test the final results with a data set which has not been used during the feature selection process.

For example, for the wrapper evaluation of the feature sets (F_1, F_2) and (F_1, F_3), two classifiers have to be built and tested. Let us take as train set

[1] In supervised classification, a classifier is built given a set of samples, each one of them labeled with the class to which it belongs. In this section the term classification always refers to supervised classification.

the samples $\{S_1, S_3, S_6, S_8, S_9\}$ and the rest $\{S_2, S_4, S_5, S_7\}$ as test set. Then, the classifiers $\mathcal{C}_1$ and $\mathcal{C}_2$ have to be built with the following data:

$\mathcal{C}_1$:

Features	Class
$(x_{11}\ x_{12})$,	C1
$(x_{31}\ x_{32})$,	C1
$(x_{61}\ x_{62})$,	C2
$(x_{81}\ x_{82})$,	C2
$(x_{91}\ x_{92})$,	C2

$\mathcal{C}_2$:

Features	Class
$(x_{11}\ x_{13})$,	C1
$(x_{31}\ x_{33})$,	C1
$(x_{61}\ x_{63})$,	C2
$(x_{81}\ x_{83})$,	C2
$(x_{91}\ x_{93})$,	C2

and tested with the following data:

$Test_{\mathcal{C}_1}$: Features	$Test_{\mathcal{C}_2}$: Features	$Output$: Class
$(x_{21}\ x_{22})$	$(x_{21}\ x_{23})$	C1
$(x_{41}\ x_{42})$	$(x_{41}\ x_{43})$	C1
$(x_{51}\ x_{52})$	$(x_{51}\ x_{53})$	C2
$(x_{71}\ x_{72})$	$(x_{71}\ x_{73})$	C2

Denoted as *Output* is the set of labels that are expected to be returned by the classifiers for the selected samples. The accuracy of the classifiers is evaluated based on the similarity between its actual output and the desired output. For example, if the classifier $\mathcal{C}_1$ returned $C1, C2, C2, C2$, while the classifier $\mathcal{C}_2$ returned $C1, C2, C1, C1$, then $\mathcal{C}_1$ would be more accurate. The conclusion would be that the feature set (F_1, F_2) works better than (F_1, F_3). This wrapper example is too simple. Actually, drawing such conclusion from just one classification test would be statistically unreliable, and cross validation techniques have to be applied in order to decide which feature set is better than another.

6.2.2 Cross Validation

Cross validation (CV), also known as rotation estimation, is a validation technique used in statistics and, particularly in machine learning, which consists of partitioning a sample of data in several subsets, and performing statistical analysis on different combinations of these subsets. In machine learning and pattern recognition, CV is generally used for estimating the error of a classifier, given a sample of the data. There are two most frequently used CV methods: the *10-fold cross validiation* and the *leave-one-out cross validation* (LOOCV).

The 10-fold cross validation (10-fold CV) method consists of dividing the training set into 10 equally sized partitions and performing 10 classification experiments for calculating their mean error. For each classification, nine of the partitions are put together and used for training (building the classifier), and the other one partition is used for testing. In the next classification, another partition is designed for testing the classifier built with the rest of the partitions. Ten classification experiments are performed so that each partition is used for testing a classifier. Note that the partitioning of the data set is performed only once. It is also important to have a random sample order before partitioning, in order to distribute the samples homogeneously among the partitions.

Leave-one-out cross validation is used in the cases of very reduced training sets. It is equivalent to K-fold CV with K equal to the total number of samples present in the data set. For example, in the well-known NCI60 microarray data set there are 60 samples and 14 classes, where some classes are represented by only two samples. With 10-fold CV there would be cases in which all the samples that represent a class would be in the test set, and the class would not be represented in the training set. Instead, LOOCV would perform 60 experiments, in each one of which, one of the samples would test the classfifier built with the resting 59 samples. The LOOCV error would be the mean error of the 60 classification errors. Also, in the cases when the data set is so reduced that there is no separate test set available, the classification results are usually reported in terms of CV error over the training set.

6.2.3 Image Classification Example

Once explained the basic concept of wrapper feature selection and the cross validation process, let us present an example in the field of computer vision. The problem is the supervised classification of indoor and outdoor images that come from two different sequences taken with a camera mounted on a walking person. Both sequences are obtained along the same path. The first one contains 721 images and is used as a training set. The second sequence, containing 470 images, is used as a test set, so it does not take part in the feature selection process. The images have a 320×240 resolution and they are labeled with six different classes: an office, two corridors, stairs, entrance, and a tree avenue, as shown in Fig. 6.1. The camera used is a stereo camera, so range information (depth) is another feature in the data sets.

One of the most important decisions in a classification problem is how to extract the features from the available data. However, this is not the topic of this chapter and we will just explain a way to extract global low-level features from an image. The technique consists of applying a set of basic filters to each image and taking the histograms of their responses as the features that characterize each image. Then, the feature selection process decides which features are important for the addressed classification problem. The selected feature set depends on the classifier, on the data, and on the labels of the training set.

The filters we use in this example are applied to the whole image and they return a histogram of responses. This means that for each filter we obtain information about the number of pixels in the image which do not respond to it, the number of pixels which completely respond to it, and the intermediate levels of response, depending on the number of bins in the histogram.

There are 18 filters, each one of which has a number of bins in the histogram. The features themselves are given by the bins of the histograms. An example is shown in Fig. 6.2. The filters are the following:

- Nitzberg
- Canny
- Horizontal gradient

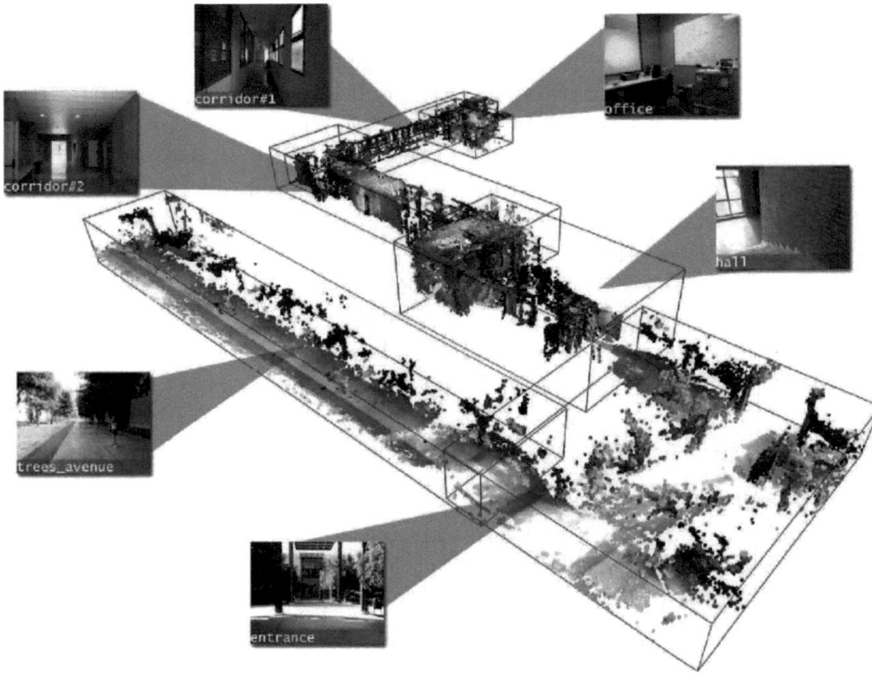

Fig. 6.1. A 3D reconstruction of the route followed during the acquisition of the data set, and examples of each one of the six classes. Image obtained with 6-DOF SLAM. Figure by F. Escolano, B. Bonev, P. Suau, W. Aguilar, Y. Frauel, J.M. Sáez and M.A. Cazorla (©2007 IEEE). See Color Plates.

- Vertical gradient
- Gradient magnitude
- Twelve color filters H_i, $1 \leq i \leq 12$
- Depth information

Some of them are redundant, for example the magnitude and the gradients. Others are similar, like Canny and gradient's magnitude. Finally, some filters may overlap, for example the color filters. The color filters return the probability distribution of some definite color H (from the HSB color space).

The feature selection criterion of the wrapper method is based on the performance of the classifier for a given subset of features, as already explained. For a correct evaluation of the classification error, the cross validation method is used. Tenfold CV is suitable for larger data sets, while LOOCV is useful when the data set is very small. For the following experiments the classification error reported is calculated using 10-fold cross validation. The experiments are performed with a K-Nearest Neighbor (K-NN) classifier with $K = 1$.

There are different strategies for generating feature combinations. The only way to ensure that a feature set is optimum is the exhaustive search

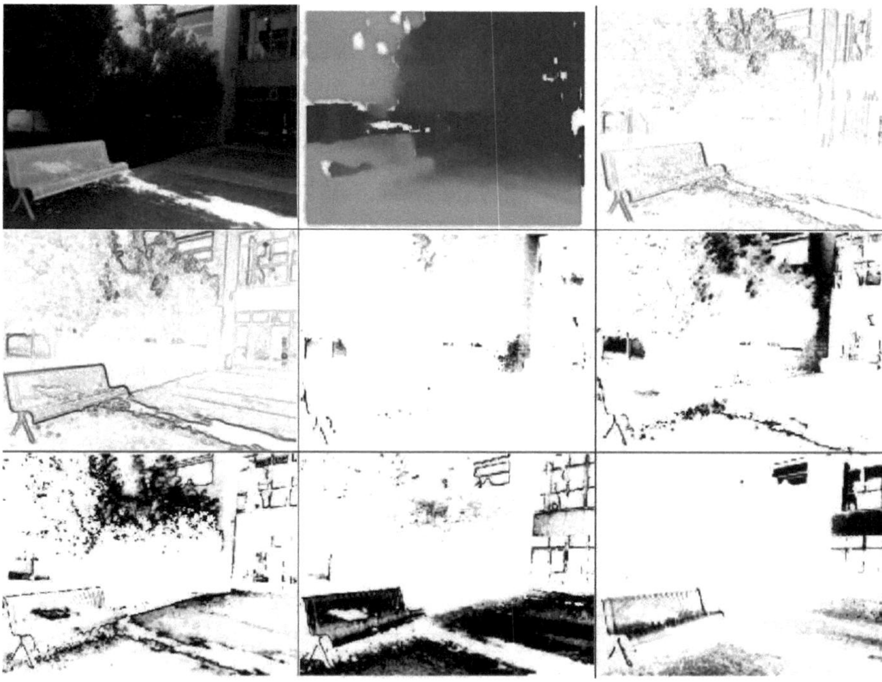

Fig. 6.2. Responses of some filters applied to an image. From *top-bottom* and *left-right*: input image, depth, vertical gradient, gradient magnitude, and four color filters. The rest of the filters are not represented as they yield null output for this input image. Figure by B. Bonev, F. Escolano and M.A. Cazorla (©2008 Springer).

among feature combinations. Exhaustive search is condemned to the *curse of dimensionality*, which refers to the prohibitiveness of the cost when searching in a high-dimensional space. The complexity of an exhaustive search is

$$O(n) = \sum_{i=1}^{n} \binom{n}{i} \tag{6.1}$$

The fastest way to select from a large amount of features is a greedy strategy. Its computational complexity is

$$O(n) = \sum_{i=1}^{n} i \tag{6.2}$$

The algorithm is described in Alg. 11. At the end of each iteration a new feature is selected and its CV error is stored. The process is also outlined in Fig. 6.3.

Algorithm 11: Wrapper Feature Selection

Input: M samples, N_F features
Initialize
$DATA_{M \times N_F} \leftarrow$ vectors of all (M) samples
$F_S = \emptyset$
$F = \{feature_1, feature_2, \dots, features_{N_F}\}$
while $F \neq \emptyset$ **do**
  forall $\forall i \mid feature_i \in F$ **do**
    $D_S = DATA(F_S \cup \{feature_i\})$
    $E_i =$ `10FoldCrossValid`(D_S)
  end
  $selected = \arg\min_i E_i$
  `/* also store` E_i `*/`
  $F_S = F_S \cup \{feature_{selected}\}$
  $F = F \sim \{feature_{selected}\}$
end
Output: F_S

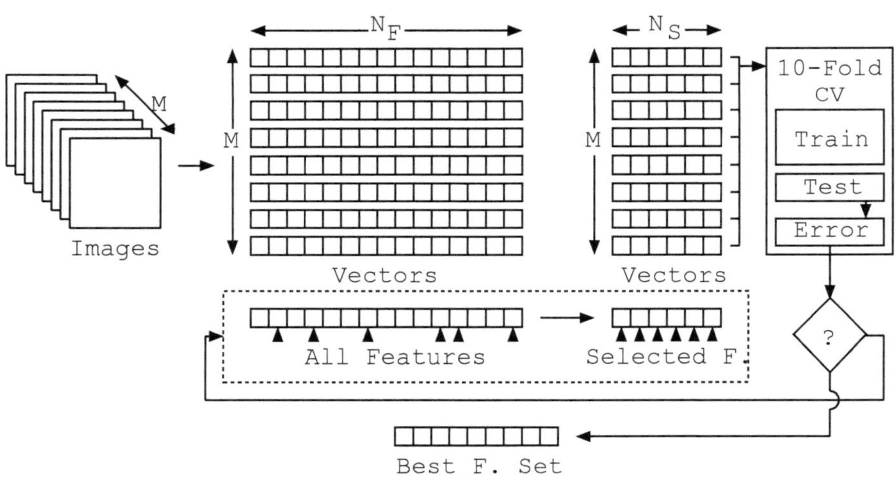

Fig. 6.3. The wrapper feature selection process.

The number of bins used to generate features is an important factor for the feature selection results. A larger number of bins can (but not necessarily) improve the classification (Fig. 6.4). An excessive number of bins overfit the classifier.

Another important factor is the number of classes. As the number of classes increases, the classification performance decays, and more features are needed. In Fig. 6.5 we compare feature selection performance on the same data set, labeled with different numbers of classes.

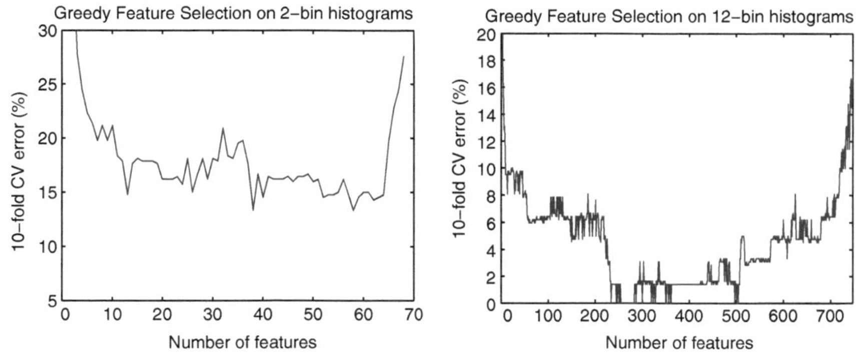

Fig. 6.4. Comparison of feature selection using 2 and 12 bin histograms, on the eight-class indoor experiment.

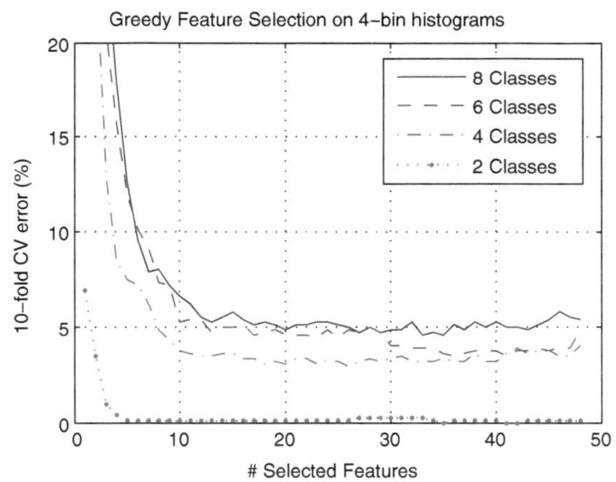

Fig. 6.5. Evolution of the CV error for different number of classes. Figure by B. Bonev, F. Escolano and M.A. Cazorla (©2008 Springer).

6.2.4 Experiments

With the wrapper feature selection process presented, we obtain a sequence of CV classification errors and we can select the feature set with the lowest error. Then, it is time for testing with the separate test set. The test simply consists of querying the classifier for the class of each image from the test set, and comparing this result with the actual class of the image.

The classifier used in this experiment is the K-nearest neighbors (K-NN). An example with test images and their five nearest neighbors from the training set is shown in Fig. 6.6. In Fig. 6.7 we have represented the number of the training image obtained for each one of the test images, ordered in the sequence of acquisition. Ideally the plot would be a straight diagonal line where the first and the last test images correspond to the first and the last train images,

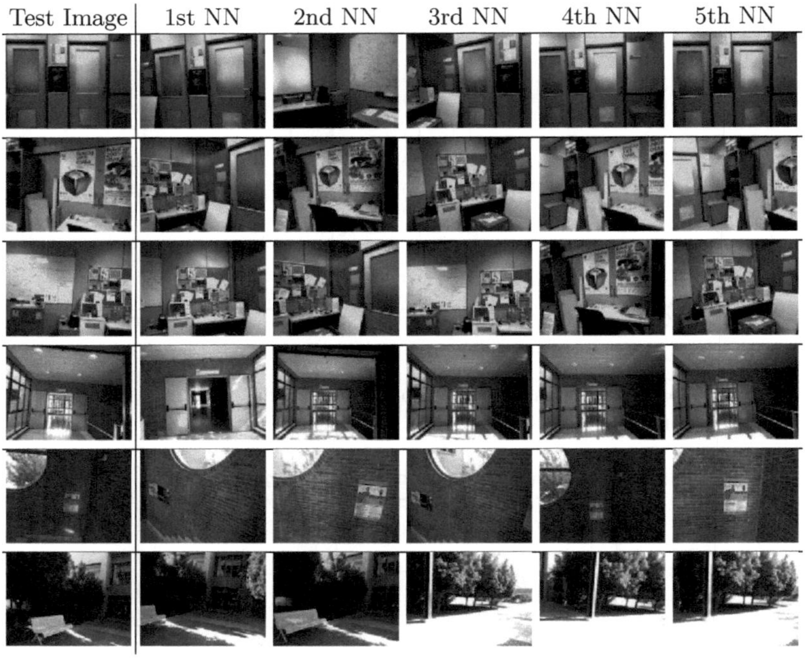

Fig. 6.6. The nearest neighbors of different test images. The training set from which the neighbors are extracted contains 721 images taken during an indoor–outdoor walk. The amount of low-level filters selected for building the classifier is 13, out of 48 in total. Note that the test images of the first column belong to a different set of images. Figure by B. Bonev, F. Escolano and M.A. Cazorla (©2008 Springer).

respectively. This figure illustrates the applicability of this approach to localization tasks, as well as the performance of feature selection on a large data set.

6.3 Filters Based on Mutual Information

6.3.1 Criteria for Filter Feature Selection

In most feature selection approaches there are two well-differentiated issues: the search algorithm and the selection criterion. Another important issue is the stopping criterion used to determine whether an algorithm has achieved a good maximum in the feature space. This section is centered on some IT-based feature selection criteria. A different issue is the way that feature combinations are generated. An exhaustive search among the features set combinations would have a combinatorial complexity with respect to the total number of features. In the following sections we assume the use of a greedy forward feature selection algorithm, which starts from a small feature set, and adds one feature in each iteration.

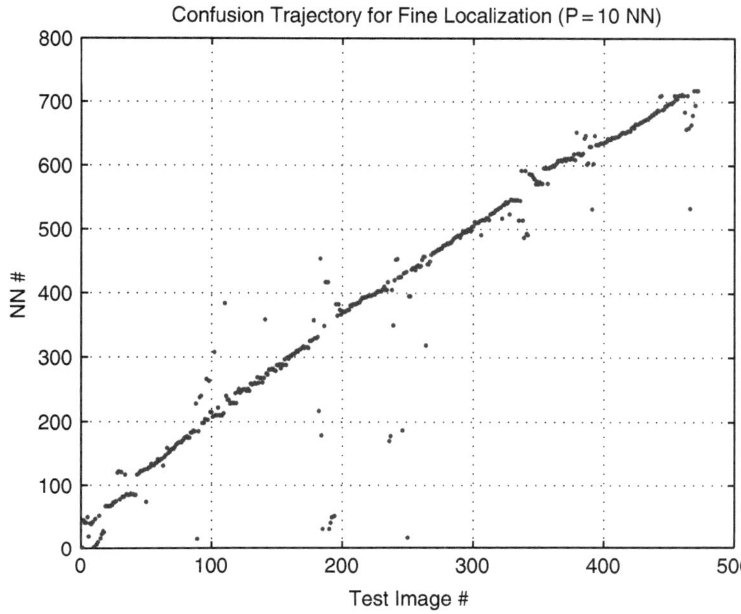

Fig. 6.7. The nearest neighbor (from among the train images, Y axis) of each one of the test images, X axis. In the ideal case the line should be almost straight, as the trajectories of both train and test sets are similar.

In the presence of thousands of features the wrapper approaches are unfeasible because the evaluation of large feature sets is computationally expensive. In filter feature selection the feature subsets are statistically evaluated. Univariate filter methods evaluate a single feature. A way to measure the relevance of a feature for the classification is to evaluate its mutual information (MI) with the classification labels [43]. This is usually suboptimal for building predictors [69] due to the possible redundancy among variables. Peng et al. [125] use not only information about the relevance of a variable but also an estimation of the redundancy among the selected variables. This is the min-Redundancy Max-Relevance (mRMR) feature selection criterion. For their measures they estimate mutual information between pairs of variables. Another feature selection approach estimates the mutual information between a whole set of features and the classes for using the infomax criterion. The idea of maximizing the mutual information between the features and the classes is similar to the example illustrated in Fig. 7.9, where we can see that in the first plot the classifier is the optimal, as well as the mutual information between the two dimensions of the data and the classes (black and white) is maximum. The mutual information is a mathematical measure which captures the dependencies which provide information about the class labels, disregarding those dependencies among features which are irrelevant to the classification.

6.3.2 Mutual Information for Feature Selection

The primary problem of feature selection is the criterion which evaluates a feature set. It must decide whether a feature subset is suitable for the classification problem, or not. The optimal criterion for such purpose would be the Bayesian error rate for the subset of selected features:

$$E(S) = \int_{\mathbf{S}} p(\mathbf{S}) \left(1 - \max_i (p(c_i|\mathbf{S}))\right) d\mathbf{S} \tag{6.3}$$

where $\mathbf{S}$ is the vector of selected features and $c_i \in C$ is a class from all the possible classes C existing in the data.

The Bayesian error rate is the ultimate criterion for discrimination; however, it is not useful as a cost, due to the nonlinearity of the $\max(\cdot)$ function. Then, some alternative cost function has to be used. In the literature there are many bounds on the Bayesian error. An upper bound obtained by Hellman and Raviv (1970) is

$$E(\mathbf{S}) \le \frac{H(C|\mathbf{S})}{2}$$

This bound is related to mutual information, because mutual information can be expressed as

$$I(\mathbf{S}; C) = H(C) - H(C|\mathbf{S})$$

and $H(\mathbf{C})$ is the entropy of the class labels which do not depend on the feature subspace $\mathbf{S}$. Therefore, the mutual information maximization is equivalent to the maximization of the upper bound (Eq. 6.3) of the Bayesian error. There is a Bayesian error lower bound as well, obtained by Fano (1961), and is also related to mutual information.

The relation of mutual information with the Kullback–Leibler (KL) divergence also justifies the use of mutual information for feature selection. The KL divergence is defined as

$$KL(P||Q) = \int_{\mathbf{x}} p(\mathbf{x}) \log \frac{p(\mathbf{x})}{q(\mathbf{x})} d\mathbf{x}$$

for the continuous case and

$$KL(P||Q) = \sum_{\mathbf{x}} p(\mathbf{x}) \log \frac{p(\mathbf{x})}{q(\mathbf{x})}$$

for the discrete case. From the definition of mutual information, and given that the conditional entropy can be expressed as $p(x|y) = p(x, y)/p(y)$, we have that

$$I(X;Y) = \sum_{y \in Y} \sum_{x \in X} p(x,y) \log \frac{p(x,y)}{p(x)p(y)} \tag{6.4}$$

$$= \sum_{y \in Y} p(y) \sum_{x \in X} p(x|y) \log \frac{p(x|y)}{p(x)} \tag{6.5}$$

$$= \sum_{y \in Y} p(y) KL\left(p(x|y)||p(x)\right) \tag{6.6}$$

$$= E_Y(KL\left(p(x|y)||p(x)\right)) \tag{6.7}$$

Then, maximizing mutual information[2] is also equivalent to maximizing the expectation of the KL divergence between the class-conditional densities $P(\mathbf{S}|\mathbf{C})$ and the density of the feature subset $P(\mathbf{S})$. In other words, the density over all classes has to be as distant as possible from the density of each class in the feature subset. Mutual information maximization provides a trade-off between discrimination maximization and redundancy minimization.

There are some practical issues involved into the maximization of mutual information between the features and the classes. Nowadays feature selection problems involve thousands of features, in a continuous feature space. Estimating mutual information between a high-dimensional continuous set of features, and the class labels, is not straightforward, due to the entropy estimation. There exist graph-based methods which do not need the density estimation of the data, thus allowing to estimate the entropy of high-dimensional data with a feasible computational complexity (see Chapter 5).

6.3.3 Individual Features Evaluation, Dependence and Redundancy

Some works on feature selection avoid the multidimensional data entropy estimation by working with single features. This, of course, is not equivalent to the maximization of $I(\mathbf{S};C)$. In the approach of Peng et al. [125] the feature selection criterion takes into account the mutual information of each separate feature and the classes, but also subtract the redundance of each separate feature with the already selected ones. It is explained in the next section.

A simpler approach is to limit the cost function to evaluate only the mutual information between each selected feature $x_i \in \mathbf{S}$ and the classes C:

$$I(\mathbf{S}^*;C) \approx \sum_{x_i \in \mathbf{S}} I(x_i;C) \tag{6.8}$$

Such cost is effective in some concrete cases, as reason Vasconcelos et al. in [166]. The expression of the mutual information of the optimal feature

[2] Some authors refer to the maximization of the mutual information between the features and the classes as *infomax criterion*.

subset $\mathbf{S}^* = \{x_1^*, x_2^*, \ldots, x_{N_{S^*}}^*\}$ of size N_{S^*} can be decomposed into the following sum:

$$I(\mathbf{S}^*; C) = \sum_{i=1}^{N} I(x_i^*; C) - \sum_{i=2}^{N} \left(I(x_i^*; \mathbf{S}_{1,i-1}^*) - I(x_i^*; \mathbf{S}_{1,i-1}^* | C) \right) \tag{6.9}$$

where x_i^* is the ith most important feature and $\mathbf{S}_{1,i-1}^* = \{x_1^*, \ldots, x_{i-1}^*\}$ is the set of the first $i-1$ best features, which have been selected before selecting x_i^*. This expression is obtained by applying the chain rule of mutual information. For the mutual information between N variables $X_1, \ldots, X_N$, and the variable Y, the chain rule is

$$I(X_1, X_2, \ldots, X_N; Y) = \sum_{i=1}^{N} I(X_i; Y | X_{i-1}, X_{i-2}, \ldots, X_1)$$

The property from Eq. 6.9 is helpful for understanding the kind of trade-off between discriminant power maximization and redundancy minimization which is achieved by $I(\mathbf{S}^*; C)$. The first summation measures the individual discriminant power of each feature belonging to the optimal set. The second summation penalizes those features x_i^* which, together with the already selected ones $\mathbf{S}_{1,i-1}^*$, are jointly informative about the class label C. This means that if $\mathbf{S}_{1,i-1}^*$ is already informative about the class label, the informativeness of the feature x_i^* is the kind of redundancy which is penalized. However, those features which are redundant, but do not inform about the class label, are not penalized.

Given this property, Vasconcelos et al. [166] focus the feature selection problem on visual processing with low level features. Several studies report that there exist universal patterns of dependence between the features of biologically plausible image transformations. These universal statistical laws of dependence patterns are independent of the image class. This conjecture implies that the second summation in Eq. 6.9 would probably be close to zero, because of the assumption that the redundancies which carry information about the class are insignificant. In this case, only the first summation would be significant for the feature selection process, and the approximation in Eq. 6.8 would be valid. This is the most relaxed feature selection cost, in which the discriminant power of each feature is individually measured.

An intermediate strategy was introduced by Vasconcelos et al. They sequentially relax the assumption that the dependencies are not informative about the class. By introducing the concept of l-decomposable feature sets they divide the feature set into disjoint subsets of size l. The constraint is that any dependence which is informative about the class label has to be between the features of the same subset, but not between susbsets. If $\mathbf{S}^*$ is the optimal feature subset of size N and it is l-decomposable into the subsets $T_1, \ldots, T_{\lceil N/l \rceil}$, then

$$I(\mathbf{S}^*;C) = \sum_{i=1}^{N} I(x_i^*;C) - \sum_{i=2}^{N} \sum_{j=1}^{\lceil i-1/l \rceil} \left(I(x_i^*;\tilde{\mathbf{T}}_{j,i}) - I(x_i^*;\tilde{\mathbf{T}}_{j,i}|C) \right) \quad (6.10)$$

where $\tilde{\mathbf{T}}_{j,i}$ is the subset of $\mathbf{T}_j$ containing the features of index smaller than k. This cost function makes possible an intermediate strategy which is not as relaxed as Eq. 6.8, and is not as strict as Eq. 6.9. The gradual increase of the size of the subsets $\mathbf{T}_j$ allows to find the l at which the assumption about noninformative dependences between the subsets becomes plausible.

The assumption that the redundancies between features are independent of the image class is not realistic in many feature selection problems, even in the visual processing field. In the following section, we analyze some approaches which do not make the assumption of Eq. 6.8. Instead they take into consideration the interactions between all the features.

6.3.4 The min-Redundancy Max-Relevance Criterion

Peng et al. present in [125] a Filter Feature Selection criterion based on mutual information estimation. Instead of estimating the mutual information $I(\mathbf{S};C)$ between a whole set of features and the class labels (also called prototypes), they estimate it for each one of the selected features separately. On the one hand they maximize the relevance $I(x_j;C)$ of each individual feature $x_j \in \mathbf{F}$. On the other hand they minimize the redundancy between x_j and the rest of selected features $x_i \in \mathbf{S}, i \neq j$. This criterion is known as the min-Redundancy Max-Relevance (mRMR) criterion and its formulation for the selection of the mth feature is

$$\max_{x_j \in \mathbf{F}-\mathbf{S}_{m-1}} \left[I(x_j;C) - \frac{1}{m-1} \sum_{x_i \in \mathbf{S}_{m-1}} I(x_j;x_i) \right] \quad (6.11)$$

This criterion can be used by a greedy algorithm, which in each iteration takes a single feature and decides whether to add it to the selected feature set, or to discard it. This strategy is called forward feature selection. With the mRMR criterion each evaluation of a new feature consists of estimating the mutual information between a feature and the prototypes, as well as the MI between that feature and each one of the already selected ones (Eq. 6.11). An interesting property of this criterion is that it is equivalent to first-order incremental selection using the Max-Dependency (MD) criterion. The MD criterion, presented in the next section, is the maximization of the mutual information between all the selected features (together) and the class, $I(\mathbf{S},C)$.

First-order incremental selection consists of starting with an empty feature set and add, incrementally, a single feature in each subsequent iteration. This implies that by the time the mth feature x_m has to be selected, there already are $m-1$ selected features in the set of selected features $\mathbf{S}_{m-1}$. By defining the following measure for the $x_1, x_2, \ldots, x_n$ scalar variables (i.e., single features):

$$J(x_1, x_2, \ldots, x_n) \\ = \int \cdots \int p(x_1, x_2, \ldots, x_n) \log \frac{p(x_1, x_2, \ldots, x_n)}{p(x_1)p(x_2)\cdots p(x_n)} dx_1 \cdots dx_n,$$

it can be seen that selecting the mth feature with mRMR first-order incremental search is equivalent to maximizing the mutual information between $\mathbf{S}_m$ and the class C. Equations 6.12 and 6.13 represent the simultaneous maximization of their first term and minimization of their second term. We show the equivalence with mutual information in the following equation (Eq. 6.14):

$$\begin{aligned}
I(\mathbf{S}_m; C) &= J(\mathbf{S}_m, C) - J(\mathbf{S}_m) &(6.12)\\
&= J(\mathbf{S}_{m-1}, x_m, C) - J(\mathbf{S}_{m-1}, x_m) &(6.13)\\
&= J(x_1, \ldots, x_{m-1}, x_m, C) - J(x_1, \ldots, x_{m-1}, x_m)\\
&= \int \cdots \int p(x_1, \ldots, x_m, C) \log \frac{p(x_1, \ldots, x_m, C)}{p(x_1)\cdots p(x_m)p(C)} dx_1 \cdots dx_m dC\\
&\quad - \int \cdots \int p(x_1, \ldots, x_m) \log \frac{p(x_1, \ldots, x_m)}{p(x_1)\cdots p(x_m)} dx_1 \cdots dx_m\\
&= \int \cdots \int p(x_1, \ldots, x_m, C) \log \left(\frac{p(x_1, \ldots, x_m, C)}{p(x_1)\cdots p(x_m)p(C)} \right.\\
&\quad \left. \cdot \frac{p(x_1)\cdots p(x_m)}{p(x_1, \ldots, x_m)} \right) dx_1 \cdots dx_m dC\\
&= \int \cdots \int p(x_1, \ldots, x_m, C) \log \frac{p(x_1, \ldots, x_m, C)}{p(x_1, \ldots, x_m)p(C)} dx_1 \cdots dx_m dC\\
&= \int \int p(\mathbf{S}_m, C) \log \frac{p(\mathbf{S}_m, C)}{p(\mathbf{S}_m)p(C)} a^1 d\mathbf{S}_m dC = I(\mathbf{S}_m; C). &(6.14)
\end{aligned}$$

This reasoning can also be denoted in terms of entropy. We can write $J(\cdot)$ as

$$J(x_1, x_2, \ldots, x_n) = H(x_1) + H(x_2) + \cdots + H(x_n) - H(x_1, x_2, \ldots, x_n),$$

therefore

$$J(\mathbf{S}_{m-1}, x_m) = J(\mathbf{S}_m) = \sum_{x_i \in \mathbf{S}_m} H(x_i) - H(\mathbf{S}_m)$$

and

$$J(\mathbf{S}_{m-1}, x_m, C) = J(\mathbf{S}_m, C) = \sum_{x_i \in \mathbf{S}_m} H(x_i) + H(C) - H(\mathbf{S}_m, C)$$

which substituted in Eq. 6.13 results in

$$\begin{aligned}
&J(\mathbf{S}_{m-1}, x_m, C) - J(\mathbf{S}_{m-1}, x_m) \\
&= \sum_{x_i \in \mathbf{S}_m} H(x_i) + H(C) - H(\mathbf{S}_m, C) - \left[\sum_{x_i \in \mathbf{S}_m} H(x_i) - H(\mathbf{S}_m) \right] \\
&= H(C) - H(\mathbf{S}_m, C) + H(\mathbf{S}) = I(\mathbf{S}, C)
\end{aligned}$$

There is a variant of the mRMR criterion. In [130] it is reformulated using a different representation of redundancy. They propose to use a coefficient of uncertainty which consists of dividing the MI between two variables x_j and x_i by the entropy of $H(x_i)$, $x_i \in \mathbf{S}_{m-1}$:

$$\frac{I(x_j; x_i)}{H(x_i)} = \frac{H(x_i) - H(x_i|x_j)}{H(x_i)} = 1 - \frac{H(x_i|x_j)}{H(x_i)}$$

This is a nonsymmetric definition which quantifies the redundancy with a value between 0 and 1. The highest value possible for the negative term $H(x_i|x_j)/H(x_i)$ is 1, which happens when x_i and x_j are independent, then $H(x_i|x_j) = H(x_i)$. The lowest value is 0, when both variables are completely dependent, disregarding their entropy. With this redundancy definition the mRMR criterion expression 6.11 becomes

$$\max_{x_j \in \mathbf{F} - \mathbf{S}_{m-1}} \left[I(x_j; C) - \frac{1}{m-1} \sum_{x_i \in \mathbf{S}_{m-1}} \frac{I(x_j; x_i)}{H(x_i)} \right] \tag{6.15}$$

In the following section we present another mutual-information-based criterion which estimates the MI between the selected features and the class. However, multidimensional data are involved in the estimation, which requires some alternative MI estimation method.

6.3.5 The Max-Dependency Criterion

The Max-Dependency (MD) criterion consists of maximizing the mutual information between the set of selected features $\mathbf{S}$ and the class labels C:

$$\max_{\mathbf{S} \subseteq \mathbf{F}} I(\mathbf{S}; C) \tag{6.16}$$

Then, the mth feature is selected according to

$$\max_{x_j \in \mathbf{F} - \mathbf{S}_{m-1}} I(\mathbf{S}_{m-1}, x_j; C) \tag{6.17}$$

Whilst in mRMR the mutual information is always estimated between two variables of one dimension, in MD the estimation of $I(\mathbf{S}; C)$ is not trivial because $\mathbf{S}$ could consist of a large number of features. In [25] such estimation is performed with the aid of Entropic Spanning Graphs for entropy estimation [74], as explained in Chapter 4. This entropy estimation is suitable for data with a high number of features and a small number of samples, because its complexity depends on the number n_s of samples ($O(n_s \log(n_s))$) but not on the number of dimensions. The MI can be calculated from the entropy estimation in two different ways, with the conditional entropy and with the joint entropy:

$$I(\mathbf{S}; C) = \sum_{x \in \mathbf{S}} \sum_{c \in \mathbf{C}} p(x, c) \log \frac{p(s, c)}{p(x)p(c)} \quad (6.18)$$

$$= H(\mathbf{S}) - H(\mathbf{S}|C) \quad (6.19)$$

$$= H(\mathbf{S}) + H(C) - H(\mathbf{S}, C) \quad (6.20)$$

where x is a feature from the set of selected features $\mathbf{S}$ and c is a class label belonging to the set of prototypes C.

Provided that entropy can be estimated for high-dimensional data sets, different IT-based criteria can be designed, depending on the problem. For example, the Max-min-Dependency (MmD) criterion (Eq. 6.21), in addition to the Max-Dependency maximization, also minimizes the mutual information between the set of discarded features and the classes:

$$\max_{\mathbf{S} \subseteq \mathbf{F}} [I(\mathbf{S}; \mathbf{C}) - I(\mathbf{F} - \mathbf{S}; \mathbf{C})] \quad (6.21)$$

Then, for selecting the mth feature, Eq. 6.22 has to be maximized:

$$\max_{x_j \in \mathbf{F} - \mathbf{S}_{m-1}} [I(\mathbf{S}_{m-1} \cup \{x_j\}; \mathbf{C}) - I(\mathbf{F} - \mathbf{S}_{m-1} - \{x_j\}; \mathbf{C})] \quad (6.22)$$

The aim of the MmD criterion is to avoid leaving out features which have information about the prototypes. In Fig. 6.8 we show the evolution of the criterion as the number of selected features increases, as well as the relative values of the terms $I(\mathbf{S}; \mathbf{C})$ and $I(\mathbf{F} - \mathbf{S}; \mathbf{C})$, together with the 10-fold CV and test errors of the feature sets.

6.3.6 Limitations of the Greedy Search

The term "greedy" refers to the kind of searches in which the decisions cannot be undone. In many problems, the criterion which guides the search does not necessarily lead to the optimal solution and usually falls into a local maximum (minimum). This is the case of forward feature selection. In the previous sections we presented different feature selection criteria. With the following toy problem we show an example of incorrect (or undesirable) feature selection.

Suppose we have a categorical data set. The values of categorical variables are labels and these labels have no order: the comparison of two categorical values can just tell whether they are the same or different. Note that if the data are not categorical but they are ordinal, regardless if they are discrete

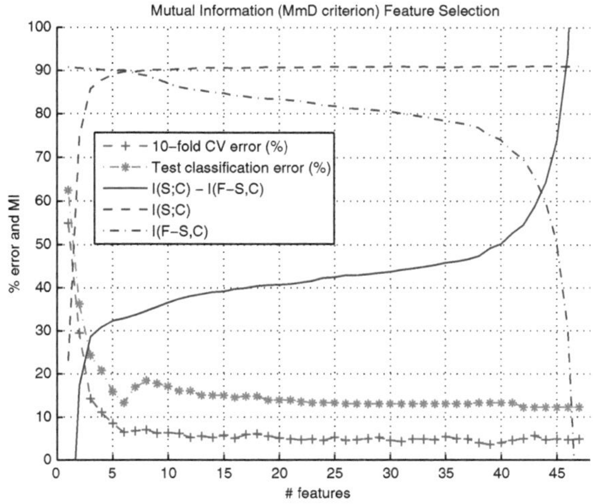

Fig. 6.8. MD and MmD criteria on image data with 48 features. Figure by B. Bonev, F. Escolano and M.A. Cazorla (©2008 Springer).

or continuous, then a histogram has to be built for the estimation of the distribution. For continuous data, a number of histogram bins have to be chosen necessarily, and for some discrete, but ordinal data, it is also convenient. For example, the distribution of the variable $x = \{1, 2, 1{,}002, 1{,}003, 100\}$ could be estimated by a histogram with 1,003 bins (or more) where only five bins would have a value of 1. This kind of histogram is too sparse. A histogram with 10 bins offers a more compact representation, though less precise, and the distribution of x would look like $(\frac{2}{5}, \frac{1}{5}, 0, 0, 0, 0, 0, 0, 0, \frac{2}{5})$. There also are entropy estimation methods which bypass the estimation of the probability distribution, as detailed in Chapter 4. These methods, however, are not suitable for categorical varibles. For simplicity we present an example with categorical data, where the distribution of a variable $x = \{A, B, \Gamma, A, \Gamma\}$ is $Pr(x = A) = \frac{2}{3}$, $Pr(x = B) = \frac{1}{3}$, $Pr(x = \Gamma) = \frac{2}{3}$.

The data set of the toy-example contains five samples defined by three features, and classified into two classes.

x_1	x_2	x_3	C
A	Z	Θ	C_1
B	Δ	Θ	C_1
Γ	E	I	C_1
A	E	I	C_2
Γ	Z	I	C_2

The mutual information between each single feature x_i, $1 \leq i \leq 3$ and the class C is

$$\begin{aligned} I(x_1, C) &= 0.1185 \\ I(x_2, C) &= 0.1185 \\ I(x_3, C) &= 0.2911 \end{aligned} \tag{6.23}$$

Therefore, both mRMR and MD criteria would decide to select x_3 first. For the next feature which could be either x_1 or x_2, mRMR would have to calculate the redundancy of x_3 with each one of them:

$$\begin{aligned} I(x_1, C) - I(x_1, x_3) &= 0.1185 - 0.3958 = -0.2773 \\ I(x_1, C) - I(x_1, x_3) &= 0.1185 - 0.3958 = -0.2773 \end{aligned} \tag{6.24}$$

In this case both values are the same and it does not matter which one to select. The feature sets obtained by mRMR, in order, would be: $\{x_3\}$, $\{x_1, x_3\}$, $\{x_1, x_2, x_3\}$.

To decide the second feature (x_1 or x_2) with the MD criterion, the mutual information between each one of them with x_3, and the class C, has to be estimated. According to the definition of MI, in this discrete case, the formula is

$$I(x_1, x_3; C) = \sum_C \sum_{x_3} \sum_{x_1} \left[p(x_1, x_3, C) \log \frac{p(x_1, x_3, C)}{p(x_1, x_3)p(C)} \right]$$

The joint probability $p(x_1, x_3, C)$ is calculated with a 3D histogram where the first dimension has the values of x_1, i.e., A, B, Γ, the second dimension has the values Θ, I, and the third dimension has the values C_1, C_2. The resulting MI values are

$$\begin{aligned} I(x_1, x_3; C) &= 0.3958 \\ I(x_2, x_3; C) &= 0.3958 \end{aligned} \tag{6.25}$$

Then MD would also select any of them, as first-order forward feature selection with MD and mRMR is equivalent. However, MD can show us that selecting x_3 in first place was not a good decision, given that the combination of x_1 and x_2 has much higher mutual information with the class:

$$I(x_1, x_2; C) = 0.6730 \tag{6.26}$$

Therefore, in this case, we should have used MD with a higher-order forward feature selection, or another search strategy (like backward feature selection or some nongreedy search). A second-order forward selection would have first yielded the set $\{x_1, x_2\}$. Note that the mutual information of all the features and C does not outperform it:

$$I(x_1, x_2, x_3; C) = 0.6730, \tag{6.27}$$

and if two feature sets provide the same information about the class, the preferred is the one with less features: x_3 is not informative about the class, given x_1 and x_2.

The MmD criterion would have selected the features in the right order in this case, because it not only calculates the mutual information about the selected features, but it also calculates it for the nonselected features. Then, in the case of selecting x_3 and leaving unselected x_1 and x_2, MD would prefer not to leave together an unselected pair of features which jointly inform so much about the class. However, MmD faces the same problem in other general cases. Some feature selection criteria could be more suitable for one case or another. However, there is not a criterion which can avoid the local maxima when used in a greedy (forward or backward) feature selection. Greedy searches with a higher-order selection, or algorithms which allow both addition and deletion of features, can alleviate the local minima problem.

6.3.7 Greedy Backward Search

Even though greedy searches can fall into local maxima, it is possible to achieve the highest mutual information possible for a feature set, by means of a greedy backward search. However, the resulting feature set, which provides this maximum mutual information about the class, is usually suboptimal.

There are two kinds of features which can be discarded: irrelevant features and redundant features. If a feature is simply irrelevant to the class label, it can be removed from the feature set and this would have no impact on the mutual information between the rest of the features and the class. It is easy to see that removing other features from the set is not conditioned by the removal of the irrelevant one.

However, when a feature x_i is removed due to its redundancy given other features, it is not so intuitive if we can continue removing from the remaining features, as some subset of them made x_i redundant. By using the mutual information chain rule we can easily see the following. We remove a feature x_i from the set F_n with n features, because that feature provides no additional information about the class, given the rest of the features F_{n-1}. Then we remove another feature $x_{i'}$ because, again, it provides no information about the class given the subset F_{n-2}. In this case, the previously removed one, x_i, will not be necessary anymore, even after the removal of $x_{i'}$. This process can continue until it is not possible to remove any feature because otherwise the mutual information would decrease. Let us illustrate it with the chain rule of mutual information:

$$I(\mathbf{S};C) = I(x_1,\dots,x_n;C) = \sum_{i=1}^{n} I(x_i;C|x_{i-1},x_{i-2},\dots,x_1) \tag{6.28}$$

With this chain rule the mutual information of a multidimensional variable and the class is decomposed into a sum of conditional mutual information. For simplicity, let us see an example with four features:

$$
\begin{aligned}
I(x_1,x_2,x_3,x_4;C) = I(x_1;C) \\
+ I(x_2;C|x_1) \\
+ I(x_3;C|x_1,x_2) \\
+ I(x_4;C|x_1,x_2,x_3)
\end{aligned}
$$

If we decide to remove x_4, it is because it provides no information about C, given the rest of the features, that is: $I(x_4;C|x_1,x_2,x_3) = 0$. Once removed, it can be seen that x_4 does not appear in any other terms, so, x_3, for example, could be removed if $I(x_3;C|x_1,x_2) = 0$, without worrying about the previous removal of x_4.

This backward elimination of features does not usually lead to the minimum feature set. In Fig. 6.9 we have illustrated a sample feature selection problem with four features. The feature x_4 can be removed because $I(x_4;C|x_1,x_2,x_3) = 0$. Actually, this feature is not redundant given other features, but it is completely redundant, because $I(x_4;C) = 0$. The next feature which could be removed is either x_1, x_2, or x_3 because we have that $I(x_1;C|x_2,x_3) = 0$, $I(x_2;C|x_1,x_3) = 0$, and $I(x_3;C|x_1,x_2) = 0$. In such situation greedy searches take their way randomly. See Fig. 6.10 to understand that, if x_3 is taken, the search falls into a local minimum, because neither x_1 nor x_2 can be removed, if we do not want to miss any mutual information with the class. However, if instead of removing x_3, one of the other two features is removed, the final set is $\{x_3\}$, which is the smallest possible one for this example.

The artificial data set "Corral" [87] illustrates well the difference between forward and backward greedy searches with mutual information. In this data

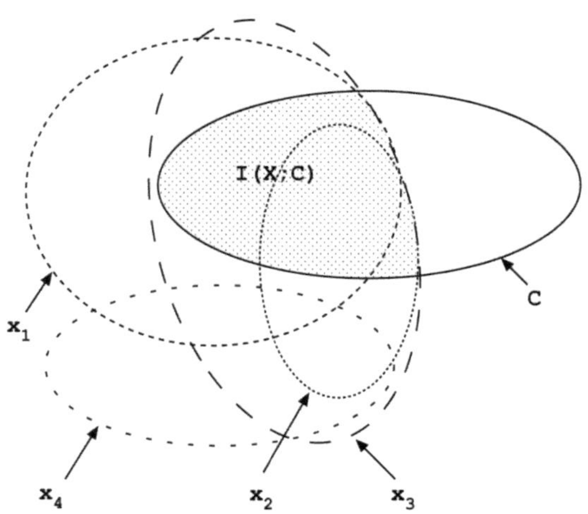

Fig. 6.9. A Venn diagram representation of a simple feature selection problem where C represents the class information, and $X = \{x_1,x_2,x_3,x_4\}$ is the complete feature set. The colored area represents all the mutual information between the features of X and the class information. The feature x_4 does not intersect this area; this means that it is irrelevant.

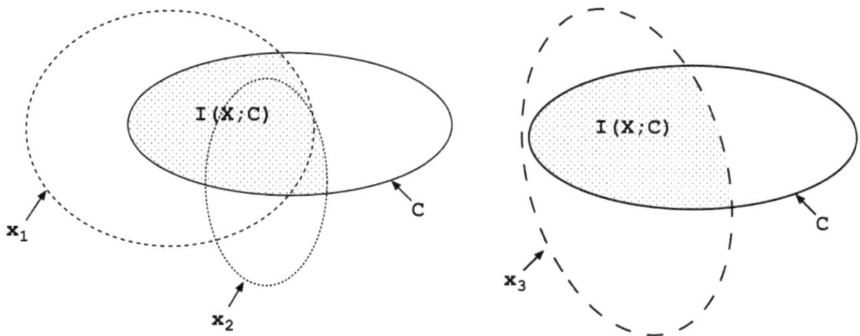

Fig. 6.10. In Fig. 6.9, the features $\{x_1, x_2\}$ (together) do not provide any further class information than x_3 provides by itself, and vice versa: $I(x_1, x_2; C|x_3) = 0$ and $I(x_3; C|x_1, x_2) = 0$. Both feature sets, $\{x_1, x_2\}$ (*left*) and x_3 (*right*), provide the same information about the class as the full feature set.

set there are six binary features, $\{x_1, x_2, x_3, x_4, x_5, x_6\}$. The class label is also binary and it is the result of the operation:

$$C = (x_1 \wedge x_2) \vee (x_3 \wedge x_4)$$

Therefore x_1, x_2, x_3, and x_4 fully determine the class label C. The feature x_5 is irrelevant, and x_6 is a feature highly (75%) correlated with the class label. Some samples could be the following:

x_1	x_2	x_3	x_4	x_5	x_6	C
0	1	1	0	1	0	0
0	1	1	1	1	1	1
1	0	0	0	0	1	0
1	0	1	1	0	1	1

Most feature selection approaches, and in particular those which perform a forward greedy search, first select the highly correlated feature, which is an incorrect decision. Contrarily, when evaluating the mutual information (the MD criterion) in a backward greedy search, the first features to discard are the irrelevant and the correlated ones. Then only the four features defining the class label remain selected.

In practice, the mutual information estimations are not perfect; moreover, the training set usually does not contain enough information to perfectly define the distribution of the features. Then, rather than maintaining a zero decrease of the mutual information when discarding features, the objective is rather to keep it as high as possible, accepting small decreases.

Another way of proving that greedy backward feature elimination is theoretically feasible is presented in the next section.

6.3.8 Markov Blankets for Feature Selection

Markov blankets provide a theoretical framework for proving that some features can be successively discarded (in a greedy way) from the feature set without loosing any information about the class. The Markov blanket of a random variable x_i is a minimal set of variables, such that all other variables conditioned on them are probabilistically independent of the target x_i. (In a Bayesian network, for example, the Markov blanket of a node is represented by the set of its parents, children, and the other parents of the children.)

Before formally defining a Markov blanket, let us define the concept of conditional independence. Two variables $\mathbf{A}$ and $\mathbf{B}$ are conditionally independent, given a set of variables $\mathbf{C}$, if $P(\mathbf{A}|\mathbf{C},\mathbf{B}) = P(\mathbf{A}|\mathbf{C})$. From this definition some properties of conditional independence can be derived. Let us denote the conditional independence between $\mathbf{A}$ and $\mathbf{B}$ given $\mathbf{C}$ as $\mathbf{A} \perp \mathbf{B} \mid \mathbf{C}$. The properties are:

$$\begin{array}{lll}
\text{Symmetry:} & \mathbf{A} \perp \mathbf{B}|\mathbf{C} & \Longrightarrow \mathbf{B} \perp \mathbf{A}|\mathbf{C} \\
\text{Decomposition:} & \mathbf{A},\mathbf{D} \perp \mathbf{B}|\mathbf{C} & \Longrightarrow \mathbf{A} \perp \mathbf{B}|\mathbf{C} \text{ and } \mathbf{D} \perp \mathbf{B}|\mathbf{C} \\
\text{Weak union:} & \mathbf{A} \perp \mathbf{B},\mathbf{D}|\mathbf{C} & \Longrightarrow \mathbf{A} \perp \mathbf{B}|\mathbf{C},\mathbf{D} \\
\text{Contraction:} & \mathbf{A} \perp \mathbf{D}|\mathbf{B},\mathbf{C} \text{ and } \mathbf{A} \perp \mathbf{B}|\mathbf{C} & \Longrightarrow \mathbf{A} \perp \mathbf{D},\mathbf{B}|\mathbf{C} \\
\text{Intersection:} & \mathbf{A} \perp \mathbf{B}|\mathbf{C},\mathbf{D} \text{ and } \mathbf{A} \perp \mathbf{D}|\mathbf{C},\mathbf{B} & \Longrightarrow \mathbf{A} \perp \mathbf{B},\mathbf{D}|\mathbf{C},
\end{array} \tag{6.29}$$

where the Intersection property is only valid for positive probabilities. (Negative probabilities are used in several fields, like quantum mechanics and mathematical finance.)

Markov blankets are defined in terms of conditional independence. The set of variables (or features) $\mathbf{M}$ is a Markov blanket for the variable x_i, if x_i is conditionally independent of the rest of the variables $\mathbf{F} - \mathbf{M} - \{x_i\}$, given $\mathbf{M}$:

$$P(\mathbf{F} - \mathbf{M} - \{x_i\}|\mathbf{M}, x_i) = P(\mathbf{F} - \mathbf{M} - \{x_i\}|\mathbf{M})$$

or

$$x_i \perp \mathbf{F} - \mathbf{M} - \{x_i\} \mid \mathbf{M} \tag{6.30}$$

where $\mathbf{F}$ is the set of features $\{x_1, \dots, x_N\}$. Also, if $\mathbf{M}$ is a Markov blanket of x_i, then the class C is conditionally independent of the feature given the Markov blanket: $x_i \perp C \mid \mathbf{M}$. Given these definitions, if a feature x_i has a Markov blanket among the set of features $\mathbf{F}$ used for classification, then x_i can safely be removed from $\mathbf{F}$ without losing any information for predicting the class.

Once a Markov blanket for x_i is found among $\mathbf{F} = \{x_1, \dots, x_N\}$ and x_i is discarded, the set of selected (still not discarded) features is $\mathbf{S} = \mathbf{F} - \{x_i\}$. In [97] it is proven that, if some other feature x_j has a Markov blanket among $\mathbf{S}$, and x_j is removed, then x_i still has a Markov blanket among $\mathbf{S} - \{x_j\}$. This property of the Markov blankets makes them useful as a criterion for a greedy feature elimination algorithm. The proof is as follows:

Let $\mathbf{M}_i \subseteq \mathbf{S}$ be a Markov blanket for x_i, not necessarily the same blanket which was used to discard the feature. Similarly, let $\mathbf{M}_j \subseteq \mathbf{S}$ be a Markov blanket for x_j. It can happen that $\mathbf{M}_i$ contains x_j, so we have to prove that, after the removal of x_j, the set $\mathbf{M}'_i = \mathbf{M}_i - \{x_j\}$, together with the Markov blanket of $\mathbf{M}_j$, is still a Markov blanket for the initially removed x_i. Intuitively, when we remove x_j, if it forms part of a Markov blanket for some already removed feature x_i, then the Markov blanket of $\mathbf{M}_j$ will still provide the conditional information that x_j provided in $\mathbf{M}_i$. By the definition of Markov blankets in Eq. 6.30, we have to show that, given the blanket $\mathbf{M}'_i \cup \mathbf{M}_j$, the feature x_i is conditionally independent of the rest of the features; let us denote them as $\mathbf{X} = \mathbf{S} - (\mathbf{M}'_i \cup \mathbf{M}_j) - \{x_j\}$. We have to show that

$$x_i \perp \mathbf{X} \mid \mathbf{M}'_i \cup \mathbf{M}_j \tag{6.31}$$

In first place, from the assumption about the Markov blanket of x_j we have that

$$x_j \perp \mathbf{S} - \mathbf{M}_j - \{x_j\} \mid \mathbf{M}_j$$

Using the Decomposition property (Eq. 6.29) we can decompose the set $\mathbf{S} - \mathbf{M}_j - \{x_j\}$ and we obtain

$$x_j \perp \mathbf{X} \cup \mathbf{M}'_i \mid \mathbf{M}_j$$

Using the Weak union property (Eq. 6.29), we can derive from the last statement:

$$x_j \perp \mathbf{X} \mid \mathbf{M}'_i \cup \mathbf{M}_j \tag{6.32}$$

For x_i we follow the same derivations and we have

$$x_i \perp \mathbf{X} \cup (\mathbf{M}_j - \mathbf{M}'_i) \mid \mathbf{M}'_i \cup \{x_j\}$$

and, therefore,

$$x_i \perp \mathbf{X} \mid \mathbf{M}_j \cup \mathbf{M}'_i \cup \{x_j\} \tag{6.33}$$

From Eqs. 6.32 and 6.33, and using the Contraction property (Eq. 6.29) we derive that

$$\{x_i\} \cup \{x_j\} \perp \mathbf{X} \mid \mathbf{M}_j \cup \mathbf{M}'_i$$

which, with the Decomposition property (Eq. 6.29), is equivalent to Eq. 6.31; therefore, it is true that after the removal of x_j, the subset $\mathbf{M}'_i \cup \mathbf{M}_j$ is a Markov blanket for x_i.

In practice it would be very time-consuming to find a Markov blanket for each feature before discarding it. In [97] they propose a heuristic in which they fix a size K for the Markov blankets for which the algorithm searches. The size K depends very much on the nature of the data. If it is too low, it is not possible to find good Markov blankets. If it is too high, the performance is also negatively affected. Among other experiments, the authors of [97] also experiment with the "Corral" data set, already presented in the previous section. With the appropriate K they successfully achieve the correct feature selection on it, similar to the result shown in the previous section with the MD greedy backward elimination.

6.3.9 Applications and Experiments

The applications of filter feature selection are manifold. Usually filter-based techniques are used in problems where the use of a wrapper is not feasible, due to computational time or due to the overfitting which a wrapper may cause with some data sets.

The three filter-based criteria presented are suitable for experiments on data with many features. Particularly, the MD criterion is capable of comparing very high-dimensional data sets. However, it is not suitable for data sets with a large number of samples. The microarray data sets are characterized by a high number of features and a low number of samples, so they are very appropriate for illustrating the performance of the presented criteria. An application of feature selection to microarray data is to identify small sets of genes with good predictive performance for diagnostic purposes. Traditional gene selection methods often select genes according to their individual discriminative power. Such approaches are efficient for high-dimensional data but cannot discover redundancy and basic interactions among genes. Multivariate filters for feature selection overcome this limitation as they evaluate whole sets of features instead of separate features.

Let us see an example with the well-known NCI60 data set. It contains 60 samples (patients), each one containing 6,380 dimensions (features), where each dimension corresponds to the expression level of some gene. The samples are labeled with 14 different classes of human tumor diseases. The purpose of feature selection is to select a set containing those genes which are useful for predicting the disease.

In Fig. 6.11 we have represented the increase of mutual information which leads the greedy selection of new features, and the leave-one-out cross

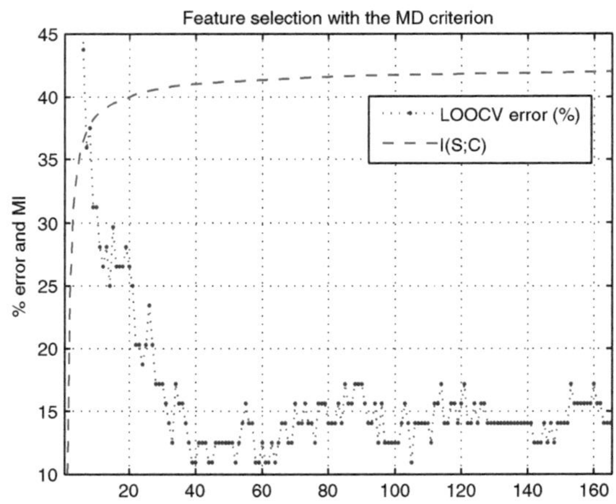

Fig. 6.11. Maximum dependency feature selection performance on the NCI microarray data set with 6,380 features. The mutual information of the selected features is represented.

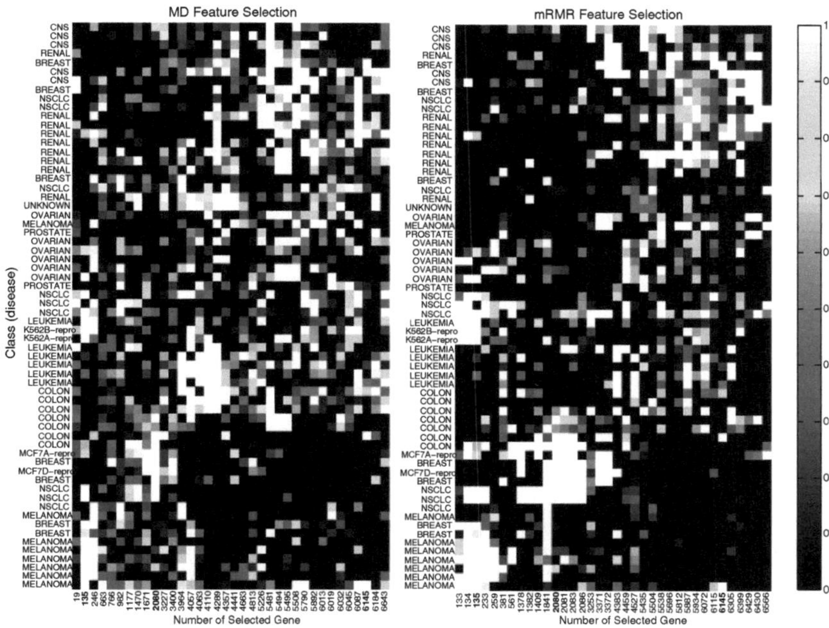

Fig. 6.12. Feature selection on the NCI DNA microarray data. The MD (*left*) and mRMR (*right*) criteria were used. Features (genes) selected by both criteria are marked with an *arrow*. See Color Plates.

validation error.[3] The error keeps on descending until 39 features are selected, then it increases, due to the addition of redundant and noisy features. Although there are 6,380 features in total, only feature sets up to size 165 are represented on the graph.

In Fig. 6.12 we have represented the gene expression matrices of the features selected by MD and mRMR. There are only three genes which were selected by both criteria. This is due to the differences in the mutual information estimation and to the high number of different features in contrast to the small number of samples.

Finally, in Fig. 6.13, we show a comparison of the different criteria in terms of classification errors. The data set used is extracted from image features with a total number of 48 features. Only the errors of the first 20 features sets are represented, as for larger feature sets the error does not decrease. Both the 10-fold CV error and the test error are represented. The latter is calculated with a separate test set which is not used in the feature selection process, that is why the test error is higher than the CV error. The CV errors of the feature

[3] This error measure is used when the number of samples is so small that a test set cannot be built. In filter feature selection the LOOCV error is not used as a selection criterion. It is used after the feature selection process, for evaluating the classification performance achieved.

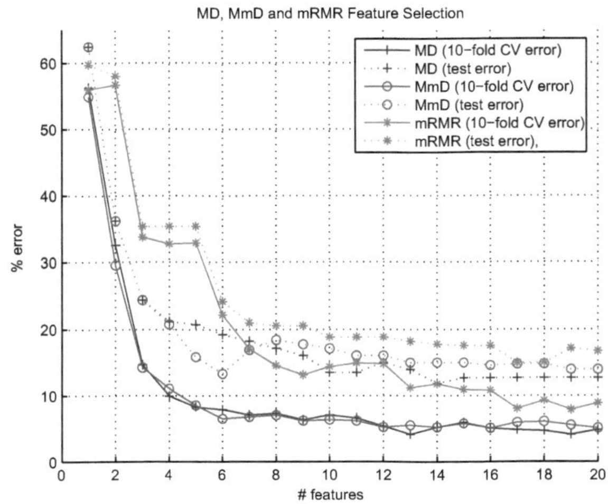

Fig. 6.13. Feature selection performance on image histograms data with 48 features, 700 train samples, and 500 test samples, labeled with six classes. Comparison between the Max-Dependency (MD), Max-min-Dependency (MmD) and the min-Redundancy Max-Relevance (mRMR) criteria. Figure by B. Bonev, F. Escolano and M.A. Cazorla (©2008 Springer).

sets yielded by MmD are very similar to those of MD. Regarding mRMR and MD, in the work of Peng et al. [125], the experimental results given by mRMR outperformed MD, while in the work of Bonev et al. [25] MD outperformed mRMR for high-dimensional feature sets. However, mRMR is theoretically equivalent to first-order incremental MD. The difference in the results is due to the use of different entropy estimators.

6.4 Minimax Feature Selection for Generative Models

6.4.1 Filters and the Maximum Entropy Principle

The maximum entropy principle has been introduced in this book in Chapter 3, and it is one of the IT principles most widely used. We remind the reader the basic idea: when we want to learn a distribution (pdf) from the data and we have expectation constraints (the expectation of several statistical models (features) $G(\cdot)$ must match the samples), the most unbiased (neutral) hypothesis is to take the distribution with maximum entropy which satisfies the constraints:

$$\begin{aligned} p^*(\xi) &= \arg\max_{p(\xi)} - \int p(\xi)\log p(\xi)\,d\xi \\ \text{s.t.} \quad & \int p(\xi)G_j(\xi)\,d\xi = E(G_j(\xi)) = \alpha_j, \quad j = 1,\dots,m \\ & \int p(\xi)\,d\xi = 1 \end{aligned} \tag{6.34}$$

The typical shape of such pdf is an exponential depending on a linear combination of features whose coefficients are the Lagrange multipliers:

$$p^*(\xi) = \frac{1}{Z(\Lambda,\xi)} e^{\sum_{r=1}^{m} \lambda_r G_r(\xi)} \tag{6.35}$$

Herein we consider the problem of learning a probability distribution of a given type of texture from image examples of that texture. Given that pdf we must be able to reproduce or generate images of such textures. Let $f(\mathbf{I})$ be *the true unknown* distribution of images $\mathbf{I}$. In pattern recognition such distribution may represent a set of images corresponding to *similar patterns* (similar texture appearance, for instance). As noted by Field [54], $f(\mathbf{I})$ is highly non-Gaussian. So we have very high-dimensional patterns (images) belonging to a non-Gaussian distribution. A key point here is the selection of the *features* $G(\cdot)$. An interesting approach consists of analyzing the images by applying a certain set of *filters* (Gabor filters, Laplacian of Gaussians, and others like the ones used in Section 6.2) to the images and then *take the histograms of the filtered images as features*. Then $G_j(\mathbf{I})$ denotes here the histogram of the image $\mathbf{I}$ after applying the jth filter (e.g. an instance of a Gabor filter parameterized by a given angle and variance). Such histogram is quantized into say L bins. The use of the filters for extracting the significant information contained in the images is an intelligent method of bypassing the problem of the high dimensionality of the images. However, it is important to *select the most informative filters* which is the other side of the coin of the *minimax* approach described later in this section. Let $\{\mathbf{I}_i^{\text{obs}}\}$ with $i = 1,\ldots,N$ be a set of *observed images* (the training set for learning) and $\{G_j(\mathbf{I})\}$ with $j = 1,\ldots,m$ a set of *histograms, each one derived from the jth filter*. Taking the first-order moment, the average, the statistics of the observations are given by: $\boldsymbol{\alpha}_j = 1/N\sum_{i=1}^{N} G_j(\mathbf{I}_i^{\text{obs}})$ which are vectors of L dimensions (bins). These statistics determine the right hand of the constraints in Eq. 6.34. The Lagrange multipliers contained in the vector $\Lambda = (\boldsymbol{\lambda}_1,\ldots,\boldsymbol{\lambda}_m)$ characterize the log-likelihood of $p(\mathbf{I}^{\text{obs}};\Lambda) = p^*(\mathbf{I})$:

$$L(\Lambda) = \log p(\mathbf{I}^{\text{obs}};\Lambda) = \sum_{i=1}^{N} \log p(\mathbf{I}_i^{\text{obs}};\Lambda) \tag{6.36}$$

and the log-likelihood has two useful properties, related to its derivatives, for finding the optimal Λ:

- *First derivative (Gradient)*

$$\frac{\partial L(\Lambda)}{\partial \boldsymbol{\lambda}_j} = \frac{1}{Z(\Lambda,\mathbf{I})}\frac{\partial Z(\Lambda,\mathbf{I})}{\partial \boldsymbol{\lambda}_j} = E(G_j(\mathbf{I})) - \boldsymbol{\alpha}_j \quad \forall j \tag{6.37}$$

- *Second derivative (Hessian)*

$$\frac{\partial^2 L(\Lambda)}{\partial \boldsymbol{\lambda}_j \partial \boldsymbol{\lambda}_k} = E((G_j(\mathbf{I}) - \boldsymbol{\alpha}_j)(G_j(\mathbf{I}) - \boldsymbol{\alpha}_j)^T) \quad \forall j,k \tag{6.38}$$

The first property provides an iterative method for obtaining the optimal Λ through gradient ascent:

$$\frac{d\boldsymbol{\lambda}_j}{dt} = E(G_j(\mathbf{I})) - \boldsymbol{\alpha}_j, \quad j = 1, \ldots, m \tag{6.39}$$

The convergence of the latter iterative method is ensured by the property associated to the Hessian. It turns out that the Hessian of the log-likelihood is the *covariance matrix* of $(G_1(\mathbf{I}), \ldots, G_m(\mathbf{I}))$, and such a covariance matrix is definite positive under mild conditions. Definite positiveness of the Hessian ensures that $L(\Lambda)$ is concave, and, thus, a unique solution for the optimal Λ exists. However, the main problem of Eq. 6.63 is that the expectations $E(G_j(\mathbf{I}))$ are unknown (only the sample expectations $\boldsymbol{\alpha}_j$ are known). An elegant, though computational intensive way of estimating $E(G_j(\mathbf{I}))$ is to use a Markov chain, because Markov chain Monte Carlo methods, like a Gibbs sampler (see Alg. 13), ensure that in the limit $(M \rightarrow \infty)$ we approximate the expectation:

$$E(G_j(\mathbf{I})) \approx \frac{1}{M}\sum_{i=1}^{M} G_j(\mathbf{I}_i^{\mathrm{syn}}) = \alpha_j^{\mathrm{syn}}(\Lambda), \quad j = 1, \ldots, m \tag{6.40}$$

where $\mathbf{I}_i^{\mathrm{syn}}$ are samples from $p(\mathbf{I}; \Lambda)$, Λ being the current multipliers so far. Such samples can be obtained through a Gibbs sampler (see Alg. 13 where G is the number of intensity values) starting from a pure random image. The jth filter is applied to the ith generated image resulting the $G_j(\mathbf{I}_i^{\mathrm{syn}})$ in the latter equation. It is interesting to remark here that the Λ determine the current, provisional, solution to the maximum entropy problem $p(\mathbf{I}; \Lambda)$, and, thus, the synthesized images match partially the statistics of the observed ones. It is very interesting to remark that the statistics of the observed images are used to generate the synthesized ones. Therefore, if we have a *fixed set of m filters*, the synthesizing algorithm proceeds by computing at iteration $t = 1, 2, \ldots$

$$\frac{d\boldsymbol{\lambda}_j^t}{dt} = \Delta_j^t = \boldsymbol{\alpha}_j^{\mathrm{syn}}(\Lambda^t) - \boldsymbol{\alpha}_j, \quad j = 1, \ldots, m \tag{6.41}$$

Then we obtain $\boldsymbol{\lambda}_j^{t+1} \leftarrow \boldsymbol{\lambda}_j^t + \Delta_j^t$ and consequently Λ^{t+1}. Then, a new iteration begins. Therefore, as we approximate the expectations of each subband and then we integrate all the multipliers in a new Λ^{t+1}, a new model $p(\mathbf{I}; \Lambda^{t+1})$ from which we draw samples with the Markov chains is obtained. As this model matches more and more the statistics of the observations it is not surprising to observe that as the iterations progress, the results resemble more and more the observations, that is, the images from the target class of textures. For a fixed number of filters m, Alg. 12 learns a synthesized image from an observed one. Such an algorithm exploits the Gibbs sampler (Alg. 13) attending to the Markov property (intensity depends on the one of the neighbors). After

Algorithm 12: FRAME

Input: $\mathbf{I}^{obs}$ input image (target), m number of filters
Initialize
Select a group of m filters: $\mathcal{S}_m = \{F_1, F_2, \ldots, F_m\}$
Compute $G_j(\mathbf{I}^{obs})$ for $j = 1, \ldots, m$
Set $\boldsymbol{\lambda}_j \leftarrow \mathbf{0}$ $j = 1, \ldots, m$
Set $\Lambda \leftarrow (\boldsymbol{\lambda}_1, \ldots, \boldsymbol{\lambda}_m)$
Initialize $\mathbf{I}^{syn}$ as a uniform white noise texture
repeat
  Calculate $G_j(\mathbf{I}^{syn})$ for $j = 1, \ldots, m$
  Obtain $\boldsymbol{\alpha}_j^{syn}(\Lambda)$ for $j = 1, \ldots, m$
  Compute $\Delta_j = \boldsymbol{\alpha}_j^{syn}(\Lambda) - \boldsymbol{\alpha}_j$ for $j = 1, \ldots, m$
  Update $\boldsymbol{\lambda}_j \leftarrow \boldsymbol{\lambda}_j + \Delta_j$
  Update $p(\mathbf{I}; \Lambda)$ with the new Λ
  Use a Gibbs sampler to flip $\mathbf{I}^{syn}$ for w weeps under $p(\mathbf{I}; \Lambda)$
until $(d(G_j(\mathbf{I}^{obs}), G_j(\mathbf{I}^{syn})) < \epsilon)$ *for* $j = 1, \ldots, m$;
Output: $\mathbf{I}^{syn}$

Algorithm 13: Gibbs sampler

Input: $\mathbf{I}$ input image, Λ model
Initialize
flips$\leftarrow 0$
repeat
  Randomly pick a location $\mathbf{x} = (x, y)$ under uniform distribution
  forall $v = 0, \ldots, G - 1$ **do**
    Calculate $p(\mathbf{I}(\mathbf{x}) = v | \mathbf{I}(\mathbf{z}) : z \in \mathcal{N}(v))$ by evaluating $p(\mathbf{I}; \Lambda)$ at v
  end
  Randomly flip $\mathbf{I}(\mathbf{x}) \leftarrow v$ under $p(v|z)$
  flips$\leftarrow$ flips + 1
until (flips $= w \times |\mathbf{I}|$);
Output: $\mathbf{I}^{syn}$

applying the sampler and obtaining a new image, the conditional probabilities of having each value must be normalized so that the sum of conditional probabilities sum one. This is key to provide a proper histogram later on. As the FRAME (*Filters, Random Fields and Maximum Entropy*) algorithm depends on a Markov chain it is quite related to simulated annealing in the sense that it starts with a uniform distribution (less structure-hot) and converges to the closest unbiased distribution (target structure-cold) satisfying the expectation constraints. The algorithm converges when the distance between the statistics of the observed image and the ones of the synthesized image does not diverge too much ($d(\cdot)$ can be implemented as the sum of the component-by-component absolute differences of these vectors).

6.4.2 Filter Pursuit through Minimax Entropy

The FRAME algorithm synthesizes the best possible image given the m filters used, but: (i) how many filters do you need for having a good result; (ii) if you constrain the number of filters, which is reasonable for computational reasons, what are the best filters? For instance, in Fig. 6.14, we show how the quality estimation improves as we use more and more filters. A perfect

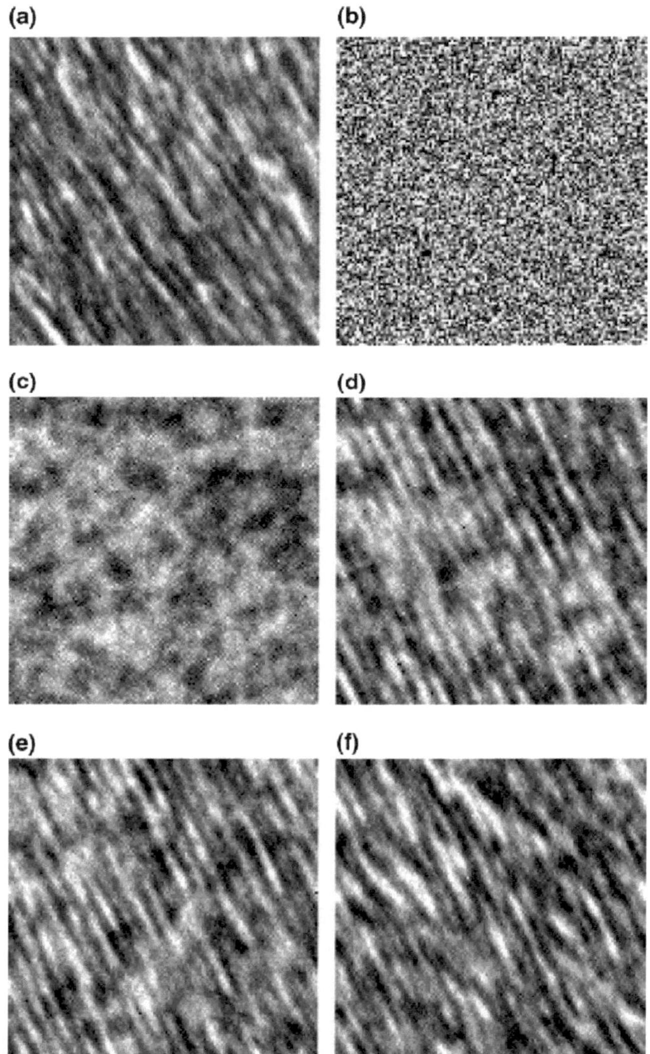

Fig. 6.14. Texture synthesis with FRAME: (**a**) observed image, (**b**) initial random image, (**c**,**d**,**e**,**f**) synthesized images with one, two, three and six filters. Figure by S.C. Zhu, Y.N. Wu and D. Mumford (©1997 MIT Press).

filter selection algorithm should analyze what combinations of the m filters in a potentially large filter bank $\mathcal{B}_{m'} \supseteq \mathcal{S}_m$ yield the best synthesized texture (define $\mathcal{S}$). But, as we have seen along the present chapter, this is not feasible, because different textures need different amounts of filters, and m is unknown beforehand for a given texture type. The usual alternative, which is the one followed in [182, 183], is an *incremental selection of filters* (features). Let $\mathcal{S}_s = \{F_{i_1}, F_{i_2}, \ldots, F_{i_s}\}$, where $i_1, \ldots, i_s$ are indexes identifying filters in $\{1, \ldots, m\}$, be the filters selected at the sth step (obviously $\mathcal{B}_0 = \emptyset$). Then, how to extend $\mathcal{B}_s$ to $\mathcal{B}_{s+1}$? Let $F_{i_{s+1}}$ be the filter maximizing the divergence $d(\beta) = D(\boldsymbol{\alpha}_\beta^{\text{syn}}(\Lambda_s), \boldsymbol{\alpha}_\beta)$ between the sub-band statistics of both the synthesized and the observed image. The arguments of $D(.,.)$ are the marginal distributions of $p(\mathbf{I}; \Lambda_s, \mathcal{S}_s)$ and $f(\mathbf{I})$, respectively. Therefore, maximizing $D(.,.)$ implies maximizing the difference between the latter distributions. Thus, this is why $F_{i_{s+1}} = \arg\max_{F_\beta \in \mathcal{B}/\mathcal{S}_s} d(\beta)$. Usually, the L_p norm is used (for instance with $p = 1$):

$$F_{i_{s+1}} = \arg \max_{F_\beta \in \mathcal{B}/\mathcal{S}_s} \frac{1}{2} |\boldsymbol{\alpha}_\beta - \boldsymbol{\alpha}_\beta^{\text{syn}}(\Lambda_s)|_p \,, \tag{6.42}$$

where the operator $|\cdot|_p$ taking vectorial argument consists of applying it to each pair of coordinates and taking the sum. Then, the *filter pursuit* algorithm is detailed in Alg. 14.

Algorithm 14: Filter Pursuit

Input: $\mathcal{B}_m$ bank of filters, $\mathbf{I}^{obs}$, $\mathbf{I}^{syn}$
Initialize
$s = 0$ $\mathcal{S} \leftarrow \emptyset$
$p(\mathbf{I}) \leftarrow$ uniform distribution
$\mathbf{I}^{syn} \leftarrow$ uniform noise
forall $j = 1, \ldots, m$ **do**
  Compute $\mathbf{I}^{obs,j}$ by applying filter F_j to $\mathbf{I}^{obs}$
  Compute histogram $\boldsymbol{\alpha}_j$ of $\mathbf{I}^{obs,j}$
end
repeat
  forall $F_\beta \in \mathcal{B}/\mathcal{S}$ **do**
    Compute $\mathbf{I}^{syn,j}$ by applying filter F_j to $\mathbf{I}^{syn}$
    Compute histogram $\boldsymbol{\alpha}_j^{syn}$ of $\mathbf{I}^{syn,j}$
    $d(\beta) = \frac{1}{2}|\boldsymbol{\alpha}_j - \boldsymbol{\alpha}_j^{syn}|$
  end
  Choose $F_{i_{s+1}}$ as the filter maximizing $d(\beta)$ among those belonging to $\mathcal{B}/\mathcal{S}$
  $\mathcal{S} \leftarrow \mathcal{S} \cup \{F_{i_{s+1}}\}$
  $s \leftarrow s + 1$
  Given $p(\mathbf{I})$ and $\mathbf{I}^{syn}$ run the FRAME algorithm to obtain $p^*(\mathbf{I})$ and $\mathbf{I}^{syn*}$
  $p(\mathbf{I}) \leftarrow p^*(\mathbf{I})$ and $\mathbf{I}^{syn} \leftarrow \mathbf{I}^{syn*}$
until *($d(\beta) < \epsilon$)*;
Output: $\mathcal{S}$

The rationale of selecting the selecting providing more difference (less redundance) with respect to the yet selected filters is closely connected to the *minimax entropy principle*. This principle can be seen as suggesting that $\mathcal{S}_m$ should minimize the Kullback–Leibler divergence between $p(\mathbf{I}; \Lambda_m, \mathcal{S}_m)$ and $f(\mathbf{I})$ (see Prob. 6.12). Therefore, at each step we should select

$$F_{i_{s+1}} = \arg \max_{F_\beta \in \mathcal{B}/\mathcal{S}_s} H(p(\mathbf{I}; \Lambda_s, \mathcal{S}_s)) - H(p(\mathbf{I}; \Lambda_{s+1}, \mathcal{S}_{s+1})) \tag{6.43}$$

Therefore, Alg. 14 finds

$$\mathcal{S}_m = \arg \min_{\mathcal{S}_m \subset \mathcal{B}} \left\{ \max_{\Omega_m} H(p(\mathbf{I})) \right\} \tag{6.44}$$

where Ω_m is the space of probability distributions satisfying the m expectation constraints imposed by the m filters. In addition, the effectiveness of gain associated to incorporating a given filter to $\mathcal{S}$ is measured by the following decrease of Kullback–Leibler divergence:

$$d(\beta) = D(f(\mathbf{I})||p(\mathbf{I}; \Lambda_s, \mathcal{S}_s)) - D(f(\mathbf{I})||p(\mathbf{I}; \Lambda_{s+1}, \mathcal{S}_{s+1})) \tag{6.45}$$

and, more precisely,

$$d(\beta) = \frac{1}{2}(\boldsymbol{\alpha}_\beta - E_{p(\mathbf{I}; \Lambda_m, \mathcal{S}_m)}(\boldsymbol{\alpha}_\beta))^T \mathbf{C}^{-1} (\boldsymbol{\alpha}_\beta - E_{p(\mathbf{I}; \Lambda_m, \mathcal{S}_m)}(\boldsymbol{\alpha}_\beta)) \tag{6.46}$$

$\mathbf{C}$ being the covariance matrix of the βth histogram. This is a kind of Mahalanobis distance.

6.5 From PCA to gPCA

6.5.1 PCA, FastICA, and Infomax

The search for a proper space where to project the data is more related to the concept of *feature transformation* than to the one of feature selection. However, both concepts are complementary in the sense that both of them point towards optimizing/simplifying the classification and/or clustering problems. It is then no surprising that techniques like PCA (principal component analysis), originally designed to dimensionality reduction, have been widely used for face recognition or classification (see the yet classic papers of Pentland et al. [126, 160]). It is well known that in PCA the N vectorized training images (concatenate rows or columns) $\mathbf{x}_i \in \mathbb{R}^k$ are mapped to a new space (eigenspace) whose origin is the average vector $\hat{\mathbf{x}}$. The differences between input vectors and the average are $\mathbf{d}_i = (\mathbf{x}_i - \hat{\mathbf{x}})$ (centered patterns). Let $\mathbf{X}$ be the $k \times N$ matrix whose colunms are the $\mathbf{d}_i$. Then, the N eigenvectors ϕ_i of $\mathbf{X}^T\mathbf{X}$ are the orthonormal axes of the eigenspace, and the meaning of their respective eigenvalues $\lambda_1 \geq \lambda_2 \geq \cdots \geq \lambda_N$ is the importance of each new dimension of the patterns (the variances of a k-dimensional Gaussian centered on the average). Projecting input patterns on the eigenspace is done by $\mathbf{y} = \mathbf{\Phi}^T(\mathbf{x}_i - \hat{\mathbf{x}})$, $\mathbf{\Phi}^T = \mathbf{\Phi}^{-1}$ being the transpose of $\mathbf{\Phi}$ whose columns are the eigenvectors (we are only changing the base/space of the data). Deprojecting the data without loss of information is straightforward: $\mathbf{\Phi}\mathbf{y} = (\mathbf{x}_i - \hat{\mathbf{x}})$ and $\mathbf{x}_i = \hat{\mathbf{x}} + \mathbf{\Phi}\mathbf{y}$. However, it is possible to obtain deprojections with small loss of information if we remove from $\mathbf{\Phi}$ the eigenvectors (columns) corresponding to the less important eigenvalues. The resulting matrix is $\tilde{\mathbf{\Phi}}$. Then the approximated projection is $\tilde{\mathbf{y}} = \tilde{\mathbf{\Phi}}^T(\mathbf{x}_i - \hat{\mathbf{x}})$ and has as many dimensions as eigenvectors retained, say r. The deprojection is $\tilde{\mathbf{x}}_i = \hat{\mathbf{x}} + \tilde{\mathbf{\Phi}}\tilde{\mathbf{y}}$ where the error $||\tilde{\mathbf{x}}_i - \mathbf{x}_i||$ is given by the sum of the magnitudes of eigenvalues suppressed. The usual result is that with $r << k$ it is possible to deproject without being able for the human being to perceive the difference.

The usual way of using PCA for recognition is to project and give pattern corresponding to the nearest projection as a result. This will be problematic when we have two classes/distributions of patterns (see Fig. 6.15, left) whose projections along the main axes are interleaved. This fact has motivated improvements of PCA in many directions, ICA (*Independent Component Analysis*) being one of them [41]. The underlying rationale of ICA is that relaxing the orthogonality constraint between axes and optimizing their placement so that the projections become statistically independent, the performance of recognition would increase (see Fig. 6.15, center). In other words, given $\mathbf{x}$ it is desirable to find a linear transformation $\mathbf{W}$ where the dimensions of the transformation $\mathbf{y} = \mathbf{W}\mathbf{x}$ are as more independent as possible. It is when we consider the quantification of statistical independence when information theory comes naturally: *statistical independence implies zero mutual information.* That is, given a set of patters $\mathcal{S} = \{\mathbf{x}_1, \ldots, \mathbf{x}_N\}$ when $\mathbf{x}_i \in \mathbb{R}^k$, ICA may be formulated as finding

$$W^* = \arg\min_{\Omega} I(y_1, \ldots, y_k) : \mathbf{y} = \mathbf{W}\mathbf{x}\ \mathbf{x} \in \mathcal{S} \tag{6.47}$$

where Ω is the space of $k \times k$ invertible matrices and I is the mutual information between the transformed variables. However, as multidimensional mutual information cannot be measured for a large k (unless we use a bypass method) initial approximations to the problem assimilated statistical independence with non-Gaussianity (we used a similar trick for defining the

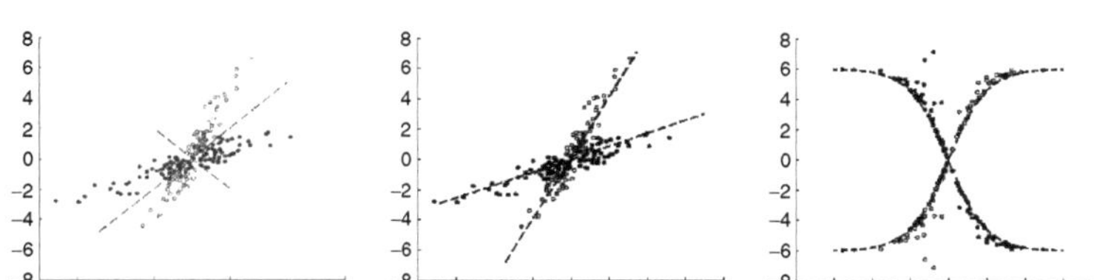

Fig. 6.15. From *left* to *right*: PCA axes, ICA axes, and gPCA axes (polynomials).

concept of *Gaussian deficiency* for mixtures in Chapter 5). Such concept is derived from the one of *neg-entropy*: $J(\mathbf{y}) = H(\mathbf{y}_{Gauss}) - H(\mathbf{y})$, $\mathbf{y}_{Gauss}$ being a Gaussian variable with the same covariance matrix as $\mathbf{y}$. Therefore, as $H(\cdot)$ is the entropy, and the variable with maximum entropy among all with the same covariance is the Gaussian, we have that $J(\mathbf{y}) \geq 0$. It is, thus, interesting to maximize the neg-entropy. However, we have the same problem: the estimation of a multidimensional entropy. However, if y is a one-dimensional standardized variable (zero mean and unit variance) we have that $J(y) \approx [E(G(y)) - E(G(v))]^2$, v being a standardized Gaussian variable, $G(\cdot)$ a non-quadratic function like $G_1(y) = (1/a)\log\cosh(ay)$ or $G_2(y) = e^{-y^2/2}$, and $1 \leq a \leq 2$ is a properly fixed constant [79]. Then, as for a linear invertible transformation $\mathbf{y} = \mathbf{W}\mathbf{x}$ mutual information may be defined as

$$I(y_1, \ldots, y_k) = \sum_{i=1}^{k} H(y_i) - H(\mathbf{y}) = \sum_{i=1}^{k} H(y_i) - H(\mathbf{x}) - \log|\det \mathbf{W}| \quad (6.48)$$

However, we may assume that the y_i are uncorrelated (if two variables are independent then they are uncorrelated but the converse is not true in general) and of unit variance: data are thus *white*. Given original centered $\mathbf{z}$, a *withening transformation*, which decorrelates the data, is done by $\tilde{\mathbf{z}} = \mathbf{\Phi}\mathbf{D}^{-1/2}\mathbf{\Phi}^T$, $\mathbf{\Phi}$ is the matrix of eigenvectors of the covariance matrix of centered data $E(\mathbf{z}\mathbf{z}^T)$ and $\mathbf{D} = \text{diag}\{\lambda_1, \ldots, \lambda_k\}$. White data $\mathbf{y}$ satisfy $E(\mathbf{y}\mathbf{y}^T) = \mathbf{I}$ (unit variances, that is, the covariance is the identity matrix). Then

$$E(\mathbf{y}\mathbf{y}^T) = \mathbf{W}E(\mathbf{x}\mathbf{x}^T)\mathbf{W}^T = \mathbf{I}\,, \quad (6.49)$$

and this implies that

$$1 = \det \mathbf{I} = \det(\mathbf{W}E(\mathbf{x}\mathbf{x}^T)\mathbf{W}^T) = (\det \mathbf{W})E(\mathbf{x}\mathbf{x}^T)(\det \mathbf{W}^T) \quad (6.50)$$

which in turns implies that $\det \mathbf{W}$ must be constant. For withened y_i, entropy and neg-entropy differ by a constant and therefore $I(y_1, \ldots, y_k) = C - \sum_{i=1}^{k} J(y_i)$ and ICA may proceed by maximizing $\sum_{i=1}^{k} J(y_i)$.

One of the earlier algorithms for ICA is based on the *infomax principle* [14]. Consider a nonlinear scalar function $g_i(\cdot)$ like the transfer function in a neural network, but satisfying $g_i' = f_i(y_i)$, f_i being the pdfs (derivatives coincident with the densities). In these networks there are $i = 1, 2, \ldots, k$ neurons, each one with weight $\mathbf{w}_i$. The weights must be chosen to maximize the transferred information. Thus, the network must maximize $H(g_1(\mathbf{w}_1^T\mathbf{x}), \ldots, g_k(\mathbf{w}_k^T\mathbf{x}))$. For a single input, say x, with pdf $f_x(x)$ (see Fig. 6.16, top), and a single neuron with transfer function $g(x)$, the amount of information transferred depends on $f_y(y \equiv g(x))$, and the shape of f_y depends on the matching between the threshold w_0 and the mean $\bar{x}$ and variance of f_x and also on the slope of $g(x)$. The optimal weight is the one maximizing the output information. The purpose is to find w maximizing $I(x, y) = H(y) - H(y|x)$ then, as $H(y|x)$ is sum of entropies of y given different values of x, and it may be assumed to be independent of a specific value. Then, we have

$$\frac{\partial I(x,y)}{\partial w} = \frac{\partial H(y)}{\partial w} \tag{6.51}$$

This is coherent with maximizing the output entropy. We have that

$$f_y(y) = \frac{f_x(x)}{|\partial y/\partial x|} \Rightarrow H(y) = -E(\ln f_y(y)) = E\left(\ln\left|\frac{\partial y}{\partial x}\right|\right) - E(\ln f_x(x)) \tag{6.52}$$

and a maximization algorithm may focus on the first term (output dependent). Then

$$\Delta w \propto \frac{\partial H}{\partial w} = \frac{\partial}{\partial w}\left(\ln\left|\frac{\partial y}{\partial x}\right|\right) = \left(\frac{\partial y}{\partial x}\right)^{-1}\frac{\partial}{\partial w}\left(\frac{\partial y}{\partial x}\right) \tag{6.53}$$

Then, using as $g(\cdot)$ the usual sigmoid transfer: $y = g(x) = 1/(1+e^{-u}) : u = wx + w_0$ whose derivative $\partial y/\partial x$ is $wy(1-y)$ it is straightforward to obtain

$$\Delta w \propto \frac{1}{w} + x(1-2y), \quad \Delta w_0 \propto 1 - 2y \tag{6.54}$$

When extending to many units (neurons) we have a weight matrix $\mathbf{W}$ to estimate, a bias vector $\mathbf{w}_0$ (one bias component per unit), and $\mathbf{y} = g(\mathbf{W}\mathbf{x} + \mathbf{w}_0)$. Here, the connection between the multivariate input–output pdfs depends on the absolute value of the Jacobian $\mathbf{J}$:

$$f_{\mathbf{y}}(\mathbf{y}) = \frac{f_{\mathbf{x}(\mathbf{x})}}{|\mathbf{J}|}, \quad \mathbf{J} = \det\begin{pmatrix} \frac{\partial y_1}{\partial x_1} & \cdots & \frac{\partial y_1}{\partial x_k} \\ \vdots & & \vdots \\ \frac{\partial y_k}{\partial x_1} & \cdots & \frac{\partial y_k}{\partial x_k} \end{pmatrix} \tag{6.55}$$

Extending to multiple unit, we obtain

$$\Delta\mathbf{W} \propto (\mathbf{W}^T)^{-1} + (1-2\mathbf{y})\mathbf{x}^T, \quad \Delta\mathbf{w}_0 \propto \mathbf{1} - 2\mathbf{y} \tag{6.56}$$

which is translated to individual weights w_{ij} as follows:

$$w_{ij} \propto \frac{cof\, w_{ij}}{\det \mathbf{W}} + x_j(1-2y_i) \tag{6.57}$$

$cof\, w_{ij}$ being the co-factor of component w_{ij}, that is, $(-1)^{i+1}$ times the determinant of the matrix resulting from removing the ith row and jth column of $\mathbf{W}$.

The latter rules implement an algorithm which maximizes $I(\mathbf{x};\mathbf{y})$ through maximizing $H(\mathbf{y})$. Considering a two-dimensional case, infomax maximizes $H(y_1, y_2)$, as $H(y_1,y_2) = H(y_1) + H(y_2) - I(y_1,y_2)$ this is equivalent to minimizing $I(y_1,y_2)$ which is the purpose of ICA. However, the latter algorithm does not guarantee the finding of a global minimum unless certain conditions

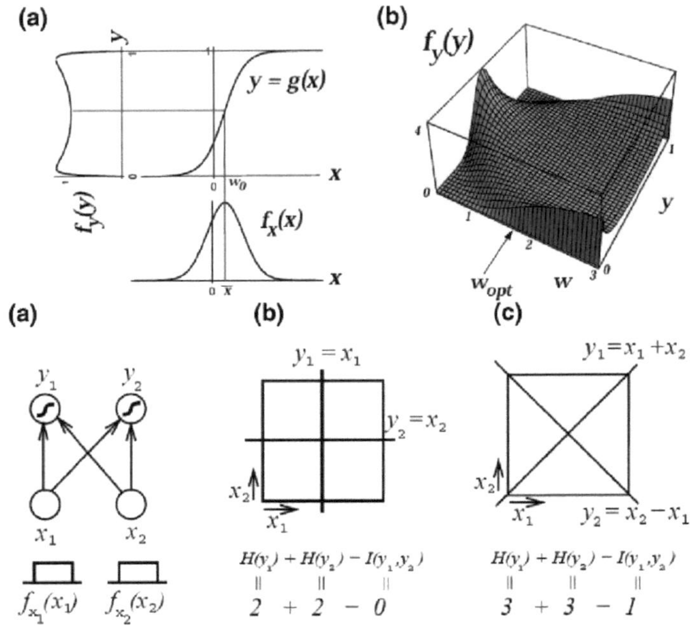

Fig. 6.16. Infomax principle. *Top*: nonlinear transfer functions where the threshold w_0 matches the mean of f_x (**a**); selection of the optimal weight w_{opt} attending to the amount of information transferred (**b**). *Bottom*: maximizing the joint entropy does not always result in minimizing the joint mutual information properly – see details in text. Figures by A.J. Bell and T.J. Sejnowski (©1995 MIT Press).

are satisfied. This is exemplified in Fig. 6.16(bottom). In (a) we have as input two independent variables uniformly distributed, and sigmoidal neurons are used. If the input pdfs are not well matched to the nonlinearity we obtain a solution (c) which has more joint entropy than the correct one (b) but it is clearly worse because it has a larger mutual information (more dependence between the output variables). This kind of configuration appears only when the pdfs of the input are sub-Gaussian (negative kurtosis). However, many natural signals and images are super-Gaussian. Anyway, the transfer functions may be tuned to avoid the latter problem.

Regarding the meaning of the axes in comparison to PCA (where axes retain more and more details of the input patterns as they are less important), in ICA, the axes are more general and represent edge and Gabor filters, and the coding of the images is sparser [15]. These axes are consistent with Field's hypothesis suggesting that cortical neurons with line and edge selectivity form a sparse (distributed) code of natural images. The infomax results shown in Fig. 6.17 are consistent with the ones obtained by the sparseness maximization net presented in [121].

The *FastICA* algorithm, proposed in [78, 80], produces similar axes as the ones obtained by infomax. It has also a design for one unit which is extensible for multiple units. The basic idea for a unit is to find an axis $\mathbf{w}$ maximizing

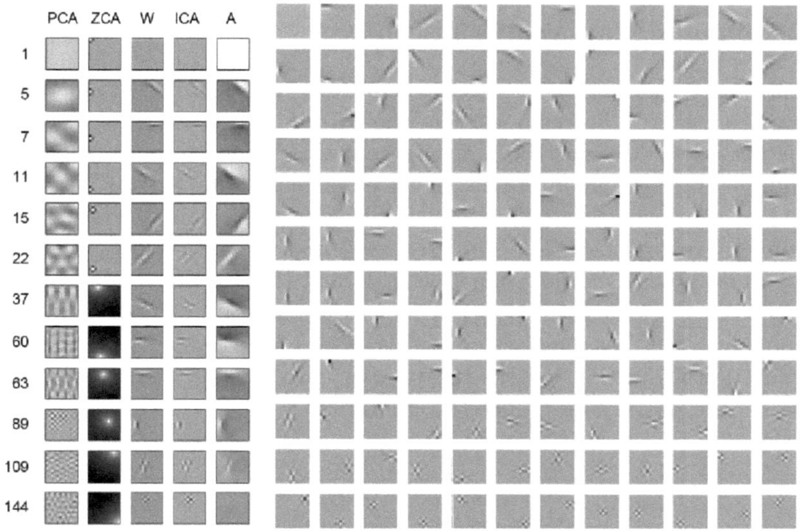

Fig. 6.17. *Left*: axes derived from PCA, ICA, and other variants. *Right*: axes derived from ICA. Figures by A.J. Bell and T.J. Sejnowski (©1995 MIT Press).

$$J(\mathbf{w}^T\mathbf{x}) \approx [E(G(\mathbf{w}^T\mathbf{x})) - E(G(v))]^2 \tag{6.58}$$

subject to $||\mathbf{x}|| = 1$. Input data $\mathbf{x}$ are assumed to be withened, and this implies that $\mathbf{wx}$ has unit variance. Let $g_1(y) = tanh(ay)$ and $g_2(y) = ue^{-u^2/2}$, with the usual setting $a = 1$, be the derivatives of the $G(\cdot)$ functions. Then, FastICA starts with a random $\mathbf{w}$ and computes

$$\mathbf{w}^+ = E(\mathbf{x}g(\mathbf{w}^T\mathbf{x})) - E(g'(\mathbf{w}^T\mathbf{x}))\mathbf{w} \tag{6.59}$$

because of the KKT conditions. Next, $\mathbf{w}^+$ is normalized: $\mathbf{w}^+ \leftarrow \mathbf{w}^+/||\mathbf{w}^+||$, and if a fixed point is not reached then Eq. 6.59 is applied again. Convergence could reach $\mathbf{w}$ or $-\mathbf{w}$ (same direction). Thus the solution to ICA is up to sign. Expectations are approximated by samples means. The extension to more units consists of running the FastICA algorithm for one unit to estimate weight vectors $\mathbf{w}_1, \ldots, \mathbf{w}_k$ and then decorrelate $\mathbf{w}_1^T\mathbf{x}, \ldots, \mathbf{w}_k^T\mathbf{x}$ after every iteration. A good solution to this problem is to estimate the first vector with one-unit FastICA and then exploit the projection of the solution of the second one onto the first. In general, if we have estimated $p < k$ vectors $\mathbf{w}_1, \ldots, \mathbf{w}_p$, then run the one-unit FastICA to obtain $\mathbf{w}_{p+1}$ and after every iteration step subtract from it the sum of its projections over the yet estimated vectors:

$$\mathbf{w}_{p+1} = \mathbf{w}_{p+1} - \sum_{i=1}^{p} \mathbf{w}_{p+1}^T\mathbf{w}_j\mathbf{w}_j \tag{6.60}$$

Then, the obtained vector is renormalized: $\mathbf{w}_{p+1} = \mathbf{w}_{p+1}/\sqrt{\mathbf{w}_{p+1}^T\mathbf{w}_{p+1}}$.

6.5.2 Minimax Mutual Information ICA

Infomax and FastICA are well-known ICA algorithms but information theory has inspired better ones, like the *Minimax Mutual Information ICA Algorithm* [52] or Minimax ICA. In this context, the minimax principle stands for minimizing mutual information within a maximum entropy framework. The algorithm incorporates in this framework the original ideas of Pierre Comon [41] related to *rotate the axes until finding their optimal placement, the one minimizing mutual information.* Then, the starting point is to minimize $I(\mathbf{y}) = \sum_{i=1}^{k} H(y_i) - H(\mathbf{y})$, with $\mathbf{y} = (y_1, \ldots, y_k)^T$. Of course, withening of input data $\mathbf{x}$ is assumed. But, in this case, it is assumed that $\mathbf{y}$ is linked to $\mathbf{x}$ through a rotation matrix $\mathbf{R}$. As joint entropy $H(\mathbf{y})$ is invariant under rotations, the problem is reduced to minimize

$$J(\Theta) = \sum_{i=1}^{k} H(y_i), \tag{6.61}$$

where Θ are the parameters defining the rotation matrix $\mathbf{R}$. Such parameters are the $k(k-1)/2$ *Givens angles* θ_{pq} of a $k \times k$ rotation matrix. The rotation matrix $R^{pq}(\theta_{pq})$ is built by replacing the entries (p,p), (p,q) and (q,p) of the identity matrix by $\cos\theta_{pq}$, $-\sin\theta_{pq}$ and $\cos\theta_{pq}$, respectively. Then, the R is computed as the product of all the 2D rotations:

$$\mathbf{R}(\Theta) = \prod_{p=1}^{n-1} \prod_{q=p+1}^{n} \mathbf{R}^{pq}(\theta_{pq}) \tag{6.62}$$

Then, the method proceeds by estimating the optimal θ_{pq} minimizing Eq. 6.61. A gradient descent method over the given angles would proceed by computing

$$\frac{\partial J(\Theta)}{\partial \theta_{pq}} = \sum_{i=1}^{k} \frac{\partial H(y_i)}{\partial \theta_{pq}} \tag{6.63}$$

which implies computing the derivative of the entropy and thus implies entropy estimation. At this point, the maximum entropy principle tells us to choose the most unbiased distribution (maximum entropy) satisfying the expectation constraints:

$$\begin{aligned} p^*(\xi) &= \arg\max_{p(\xi)} - \int_{-\infty}^{+\infty} p(\xi) \log p(\xi)\, d\xi \\ \text{s.t} \quad & \int_{-\infty}^{+\infty} p(\xi) G_j(\xi)\, d\xi = E(G_j(\xi)) = \alpha_j, \quad j = 1, \ldots, m \\ & \int_{-\infty}^{+\infty} p(\xi)\, d\xi = 1 \end{aligned} \tag{6.64}$$

where the solution has the form

$$p^*(\xi) = \frac{1}{Z(\Lambda, \xi)} e^{\sum_{r=1}^{m} \lambda_r G_r(\xi)} \tag{6.65}$$

Paying attention to the constraints we have

$$\alpha_j = \int_{-\infty}^{+\infty} p(\xi) G_j(\xi)\, d\xi \tag{6.66}$$

Integration by parts states that $\int u dv = uv - \int v\, du$. Considering the form of the solution and applying this rule to

$$u = p(\xi) \quad dv = G_j(\xi)$$

$$du = \left(\sum_{r=1}^{m} \lambda_r G'_r(\xi) \right) p(\xi), \quad v = F_j(\xi) = \int_{-\infty}^{+\infty} G_j(\xi)\, d\xi \tag{6.67}$$

Therefore

$$\alpha_j = p(\xi) F_j(\xi) \,|_{-\infty}^{+\infty} - \int_{-\infty}^{+\infty} F_j(\xi) \left(\sum_{r=1}^{m} \lambda_r G'_r(\xi) \right) p(\xi)\, d\xi \tag{6.68}$$

When the constraint functions are chosen among the moments of the variables, the integrals $F_j(\xi)$ do not diverge faster than the exponential decay of the pdf representing the solution to the maximum entropy problem. Under these conditions, the first term of the latter equation tends to zero and we have

$$\begin{aligned} \alpha_j &= -\sum_{r=1}^{m} \lambda_r \int_{-\infty}^{+\infty} F_j(\xi) G'_r(\xi) p(\xi)\, d\xi \\ &= -\sum_{r=1}^{m} \lambda_r E(F_j(\xi) G'_r(\xi)) = -\sum_{r=1}^{m} \lambda_r \beta_{jr} \end{aligned} \tag{6.69}$$

where the β_{jr} may be obtained from the sample means for approximating $E(F_j(\xi)G'_r(\xi))$. Once we also estimate the α_j from the sample, the vector of Lagrange multipliers $\Lambda = (\lambda_1, \ldots, \lambda_m)^T$ is simply obtained as the solution of a linear system:

$$\Lambda = -\boldsymbol{\beta}^{-1}\boldsymbol{\alpha} \;:\boldsymbol{\alpha} = (\alpha_1, \ldots, \alpha_m)^T, \quad \boldsymbol{\beta} = [\beta_{jr}]_{m\times m} \tag{6.70}$$

In addition to satisfying the standard expectation constraints, the pdf satisfies the constraints derived from obtaining $E(F_j(\xi)G'_r(\xi))$ from the samples (see Prob. 6.8). Then, exploiting $\boldsymbol{\alpha}$ and $\boldsymbol{\beta}$ it is possible to obtain an analytic expression for Eq. 6.63. Firstly, we have

$$
\begin{aligned}
\frac{\partial H(\xi)}{\partial \theta_{pq}} &= -\frac{\partial}{\partial \theta_{pq}} \int_{-\infty}^{+\infty} p(\xi) \log p(\xi)\, d\xi \\
&= -\int_{-\infty}^{+\infty} \left[\frac{\partial p(\xi)}{\partial \theta_{pq}} \log p(\xi) + \frac{\partial p(\xi)}{\partial \theta_{pq}} \right] d\xi \\
&= -\int_{-\infty}^{+\infty} \frac{\partial p(\xi)}{\partial \theta_{pq}} \log p(\xi)\, d\xi - \int_{-\infty}^{+\infty} \frac{\partial p(\xi)}{\partial \theta_{pq}} d\xi \\
&= -\int_{-\infty}^{+\infty} \frac{\partial p(\xi)}{\partial \theta_{pq}} \left(Z(\Lambda, \xi) + \sum_{r=1}^{m} \lambda_r G_r(\xi) \right) d\xi - \int_{-\infty}^{+\infty} \frac{\partial p(\xi)}{\partial \theta_{pq}} d\xi \\
&= -(1 + Z(\Lambda, \xi)) \underbrace{\frac{\partial}{\partial \theta_{pq}} \underbrace{\int_{-\infty}^{+\infty} p(\xi)\, d\xi}_{1}}_{0} - \sum_{r=1}^{m} \lambda_r \int_{-\infty}^{+\infty} G_r(\xi) \frac{\partial p(\xi)}{\partial \theta_{pq}} d\xi \\
&= -\sum_{r=1}^{m} \lambda_r \frac{\partial}{\partial \theta_{pq}} \underbrace{\int_{-\infty}^{+\infty} p(\xi) G_r(\xi)\, d\xi}_{\alpha_r} = -\sum_{r=1}^{m} \lambda_r \frac{\partial \alpha_r}{\partial \theta_{pq}}
\end{aligned}
\tag{6.71}
$$

All the derivations between Eqs. 6.64 and 6.71 are referred to one generic variable ξ. Let $y_i = \xi$ be the random variable associated to the ith dimension of the output (we are solving a maximum entropy problem for each individual output variable). Updating the notation consequently, we obtain

$$
\frac{\partial H(y_i)}{\partial \theta_{pq}} = \sum_{r=1}^{m} \lambda_r^i \frac{\partial \alpha_r^i}{\partial \theta_{pq}}
\tag{6.72}
$$

Therefore, there is a very interesting link between the gradient of the cost function and the expectation constraints related to each output variable. But how to compute $\partial \alpha_r^i / \partial \theta_{pq}$? If we approximate α_r^i by the sample mean, that is, $\alpha_r^i = (1/N) \sum_{j=1}^{N} G_r(y_i^j)$. Now, the latter derivative is expanded using the chain rule:

$$
\begin{aligned}
\frac{\partial H(y_i)}{\partial \theta_{pq}} &= \sum_{j=1}^{N} G_r'(y_i^j) \frac{\partial y_i^j}{\partial \theta_{pq}} \\
&= \sum_{j=1}^{N} G_r'(y_i^j) \left(\frac{\partial y_i^j}{\partial \mathbf{R}_{j:}} \right)^T \left(\frac{\partial \mathbf{R}_{j:}}{\partial \theta_{pq}} \right)^T \\
&= \sum_{j=1}^{N} G_r'(y_i^j) \left(\frac{\partial \mathbf{R}}{\partial \theta_{pq}} \right)_{j:}^T
\end{aligned}
\tag{6.73}
$$

where j constrains the derivative to the jth row of the matrix. Anyway, the derivative of $\mathbf{R}$ with respect to θ_{pq} is given by

$$\frac{\partial \mathbf{R}}{\partial \theta_{pq}} = \left(\prod_{u=1}^{p-1}\prod_{v=u+1}^{k} \mathbf{R}^{uv}(\theta_{uv})\right)\left(\prod_{v=p+1}^{q-1} \mathbf{R}^{pv}(\theta_{pv})\right)$$
$$\times \frac{\partial \mathbf{R}^{pq}(\theta_{pq})}{\partial \theta_{pq}}\left(\prod_{v=q+1}^{k} \mathbf{R}^{pv}(\theta_{pv})\right)\left(\prod_{u=p+1}^{k-1}\prod_{v=u+1}^{k} \mathbf{R}^{uv}(\theta_{uv})\right) \quad (6.74)$$

and the final gradient descent rule minimizes the sum of entropies as follows:

$$\Theta_{t+1} = \Theta_t - \eta \sum_{i=1}^{k} \frac{\partial H(y_i)}{\partial \Theta} \quad (6.75)$$

and the process stabilizes at $\mathbf{R}^*$ (when spurious minima of entropy are avoided). Furthermore, a closer analysis of the partial derivatives in Eq. 6.73, reveals that

$$\frac{\partial J(\Theta)}{\partial \theta_{pq}} = \sum_{i=1}^{k} \frac{\partial H(y_i)}{\partial \theta_{pq}} = \sum_{i=1}^{k} (\boldsymbol{\alpha}^i)^T (\boldsymbol{\beta}^i)^{-T} \left(\frac{\partial \boldsymbol{\alpha}^i}{\partial \theta_{pq}}\right) \quad (6.76)$$

that is, the negative gradient direction depends both on the $\boldsymbol{\alpha}^i$ and the $\boldsymbol{\beta}^i$, which actually depend on $\boldsymbol{\alpha}^i$ and the multipliers. As the $\boldsymbol{\alpha}^i$ depend on the typically, higher-order moments used to define the constraints, the update direction depends both on the gradients of the moments and the gradients of non-Gaussianity: the non-Gaussianity of sub-Gaussian signals is minimized, whereas the one of super-Gaussian signals is maximized. As we show in Fig. 6.18 (left), as the entropy (non-Gaussianity) of the input distribution increases, the same happens with the super-Gaussianity (positive kurtosis) which is a good property as stated when discussing infomax.

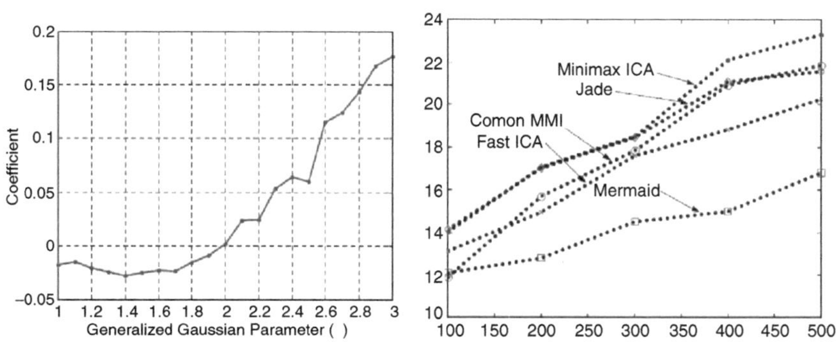

Fig. 6.18. *Left*: increasing of super-Gaussianity for different Generalized Gaussian Distributions parameterized by β:$\beta = 1$ gives the Laplacian, $\beta = 2$, the Gaussian, and $\beta \to \infty$ the uniform distribution. *Right*: SIR comparison between different ICA algorithm as the number of samples increases. Figure by D. Erdogmus, K.E. Hild II, Y.N. Rao and J.C. Príncipe (©2004 MIT Press).

Regarding the performance of minimax ICA, one measure for comparing ICA algorithms is to compute the so-called *signal interference ratio* SIR. Assuming input data resulting from a linear mixing of vectors $\mathbf{s}$ with independent components and then withened with a matrix $\mathbf{W}$, that is, $\mathbf{x} = \mathbf{WHs}$, let $\mathbf{O} = \mathbf{R}^*\mathbf{WH}$ be the argument for computing the SIR

$$\text{SIR}(dB) = \frac{1}{N}\sum_{i=1}^{N} 10\log_{10}\frac{\max_q(O_{iq}^2)}{\mathbf{O}_{i:}\mathbf{O}_{i:}^T - \max_q(O_{iq}^2)} \tag{6.77}$$

where, in each row, the source corresponding to the maximum entry is considered the main signal at that output. Given this measure, it is interesting to analyze how the SIR performance depends on the number of available samples independently of the number of constraints. In Fig. 6.18(right) we show that when generating random mixing matrices, the average SIR is better and better for minimax ICA as the number of samples increases (such samples are needed to estimate high-order moments) being the best algorithm (even than FastICA) and sharing this position with JADE [35] when the number of samples decreases. This is consistent with the fact that JADE uses fourth-order *cumulants* (cumulants are defined by the logarithm of the moment generating function).

The development of ICA algorithms is still in progress. ICA algorithms are also very useful for solving the *blind source separation problem* summarized in the *cocktail party* problem: given several mixed voices demix them. It turns out that the inverse of $\mathbf{W}$ is the mixing matrix [79]. Thus, ICA solves the source separation problem up to the sign. However, regarding pattern classification, recent experimental evidence [167] shows that whitened PCA compares sometimes to ICA: when feature selection is performed in advance, FastICA, withened PCA, and infomax have similar behaviors. However, when feature selection is not performed and there is a high number of components, infomax outperforms the other two methods in recognition rates.

In general, ICA is considered a technique of *projection pursuit* [10], that is, a way of finding the *optimal projections* of the data (here *optimal* means the projection which yields the best clarification of the structure of multidimensional data). Projection pursuit is closely related to feature selection and *Linear Discriminant Analysis* (LDA) [111]. The main difference with respect to ICA and PCA is that LDA looks for projecting the data to maximize the discriminant power. In [66], a technique dubbed the *Adaptive Discriminant Analysis* (ADA) proceeds by iteratively selecting the axis yielding the maximum mutual information between the projection, the class label, and the projection on the yet selected axes. This is, in general, untractable as we have seen along the chapter. In [66], theoretical results are showed for mixtures of two Gaussians.

6.5.3 Generalized PCA (gPCA) and Effective Dimension

Generalized PCA, or *gPCA*, is a proper algebraic approach for solving the chicken-and-egg problem of finding subspaces for the data and finding their

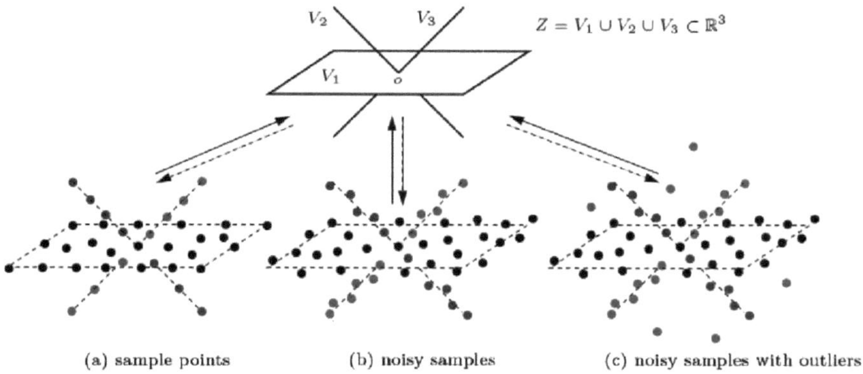

Fig. 6.19. *Top*: arrangement of three subspaces (one plane V_1 and two lines V_2 and V_3). *Bottom*: different instances of the problem with increasing difficulty (from *left* to *right*). Figure by Y. Ma, A.Y. Yang, H. Derksen and R. Fossum (©2008 SIAM).

parametric models [109,169]. Given $\mathcal{A}$ set of vectors in $\mathbb{R}^k$ (the *ambient space*), suppose that such set can be arranged as a set of n subspaces (clusters) $\mathcal{Z} = \mathcal{V}_1 \cup \cdots \cup \mathcal{V}_n \subset \mathbb{R}^k$. Each of the subspaces may have different dimensions $d_1, \ldots, d_n$, and, thus, they may have different bases. Given the two latter elements, it is then possible to associate points to subspaces (clustering, classification, segmentation). For instance, in Fig. 6.19, we have an arrangement of one plane and two lines. Such subspaces are hidden within the data and the task of determining them may be very hard if outlying samples populate the space.

As we have considered above, gPCA is an algebraic approach. More precisely: (i) the number of subspaces is given by the rank of a matrix; (ii) the subspace basis comes from the derivatives of polynomials; and (iii) clustering is equivalent to polynomial fitting. Tasting the flavor of gPCA can be done through a very simple example due to René Vidal for his tutorial in CVPR'08. It consists of finding clusters in $\mathbb{R}$. Assume we have n groups of samples: the first one satisfying $x \approx b_1$ and the second one $x \approx b_2$, and the nth one satisfying $x \approx b_n$. Therefore, a point x may belong either to the first group *or* to the second one, *or* to the nth one. Here, *or* means *product*: $(x - b_1)(x - b_2)\cdots(x - b_n) = 0$. Therefore, we have the polynomial: $p_n(x) = x^n + c_1 x^{n-1} + \cdots + c_n = 0$. Suppose that we have N samples $x_1, \ldots, x_N$, then all these points must satisfy the following system of equations:

$$P_n\mathbf{c} = \underbrace{\begin{pmatrix} x_1^n \ldots x_1 \; 1 \\ x_2^n \ldots x_2 \; 1 \\ \vdots \quad\;\; \vdots \;\; \vdots \\ x_2^n \ldots x_2 \; 1 \end{pmatrix}}_{P_n} \mathbf{c} = \mathbf{0}, \quad \mathbf{c} = \begin{pmatrix} 1 \; c_1 \ldots c_n \end{pmatrix}^T \tag{6.78}$$

Then, the number of clusters is given by analyzing the rank $n = \min\{i : \text{rank}(P_i) = i\}$; the cluster centers are the roots of the polynomial $p_n(x)$. Then the solution to the problem is unique if $N > n$ and has a closed form when $n \leq 4$. Suppose now that we have n *planes*, that is, the samples $\mathbf{x} = (x_1, \ldots, x_k)^T \in \mathbb{R}^k$ are divided into n groups where each group fits a plane. A plane, like $\mathcal{V}_1$, in Fig. 6.19 is defined by $b_1x_1 + b_2x_2 + \cdots + b_kx_k = 0 \equiv \mathbf{b}^T\mathbf{x} = 0$. Therefore, following the *or* rule, two planes are encoded by a polynomial $p_2(\mathbf{x}) = (\mathbf{b}_1^T\mathbf{x})(\mathbf{b}_2^T\mathbf{x})$. An important property of the latter polynomial is that it can be expressed linearly with respect to the polynomial coefficients. For instance, if we have planes in $\mathbb{R}^3$:

$$\begin{aligned}
p_2(\mathbf{x}) &= (\mathbf{b}_1^T\mathbf{x})(\mathbf{b}_2^T\mathbf{x}) = (b_{11}x_1 + b_{21}x_2 + b_{31}x_3)(b_{12}x_1 + b_{22}x_2 + b_{32}x_3) \\
&= (b_{11}x_1)(b_{12}x_1) + (b_{11}x_1)(b_{22}x_2) + (b_{11}x_1)(b_{32}x_3) \\
&\quad +(b_{21}x_2)(b_{12}x_1) + (b_{21}x_2)(b_{22}x_2) + (b_{21}x_2)(b_{32}x_3) \\
&\quad +(b_{31}x_3)(b_{12}x_1) + (b_{31}x_3)(b_{22}x_2) + (b_{31}x_3)(b_{32}x_3) \\
&= (b_{11}b_{12})x_1^2 + (b_{11}b_{22} + b_{21}b_{12})x_1x_2 + (b_{11}b_{32} + b_{31}b_{12})x_1x_3 \\
&\quad +(b_{21}b_{22})x_2^2 + (b_{21}b_{32} + b_{31}b_{22})x_2x_3 + (b_{31}b_{32})x_3^2 \\
&= c_1x_1^2 + c_2x_1x_2 + c_3x_1x_3 + c_4x_2^2 + c_5x_2x_3 + c_6x_3^2 \\
&= \mathbf{c}^T\nu_n(\mathbf{x})
\end{aligned} \tag{6.79}$$

$\nu_n(\mathbf{x})$ being a vector of *monomials* of degree n (in the latter case $n = 2$). In general, we have $M_n^{[k]} = \binom{n+k-1}{n} = (n + k - 1)!/((n + k - 1)(n - 1)!)$ monomials, each one with a coefficient c_h. In the latter example, $M_2^{[3]} = \binom{4}{2} = 4!/(4(1!)) = 6$ monomials and coefficients. Therefore, we have a *mechanism to map or embed a point* $\mathbf{x} \in \mathbb{R}^k$ to a space of $M_n^{[k]}$ dimensions where the basis is given by monomials of degree n and the coordinates are the coefficients c_h for these monomials. Such embedding is known as the *Veronese map* of degree n [71]:

$$\nu_n : \mathbb{R}^k \rightarrow \mathbb{R}^{M_n^{[k]}}, \quad \text{where } \nu_n \begin{pmatrix} x_1 \\ x_2 \\ \vdots \\ x_k \end{pmatrix} = \begin{pmatrix} x_1^n \\ x_1^{n-1}x_2 \\ \vdots \\ x_k^n \end{pmatrix}$$

Given the Veronese maps or mappings for each element in $\mathcal{Z}$ (sample) $\mathbf{x}_1, \ldots, \mathbf{x}_N$ it is then possible to obtain the vector of common coefficients $\mathbf{c}$ as the left null space of the $M_n^{[k]} \times N$ matrix of mappings $\mathbf{L}_n$ (*embedded data matrix*):

$$\mathbf{c}^T\mathbf{L}_n = \mathbf{c}^T(\nu_n(\mathbf{x}_1), \ldots, \nu_n(\mathbf{x}_N)) = \mathbf{0}_N \tag{6.80}$$

This null space may be determined by performing *Singular Value Decomposition* (SVD). The SVD factorization of $\mathbf{L}_n = \mathbf{U}\Sigma\mathbf{V}^T$ decomposes the matrix into: $\mathbf{U}$, whose columns form an orthonormal input basis vectors of

$\mathbf{L}_n$ (left singular vectors), $\mathbf{V}$, whose columns form the output orthonormal basis vectors (right singular vectors), and Σ, a diagonal matrix which contains the *singular values.* Singular values σ are non-negative real numbers. The singular values of $\mathbf{L}_n$ satisfy: $\mathbf{L}_n\mathbf{v} = \sigma\mathbf{u}$ and $\mathbf{L}_n^T\mathbf{u} = \sigma\mathbf{v}$, $\mathbf{v}$ and $\mathbf{u}$ being unit vectors, called left singular vectors for σ in the first case and right ones in the second. The right singular vectors corresponding to zero (or close to zero – vanishing) singular values span the right null space of $\mathbf{L}_n$. The left singular vectors associated to nonzero singular values span the range of $\mathbf{L}_n$. The rank of $\mathbf{L}_n$ is the number of its nonzero singular vectors. Thus, the number of subspaces can be obtained by observing the rank of the embedded data matrix: $m = \min\{i : \text{rank}(L_i) = M_i^{[k]} - 1\}$.

Let $\mathbf{C}$ be the matrix whose columns are the singular vectors associated to vanishing singular values, that is, $\mathbf{C} = (\mathbf{c}_1, \ldots, \mathbf{c}_m)$ forms a basis of the null space of $\mathbf{L}_n$. Let $\mathbf{q}_j(\mathbf{x}) = \mathbf{c}_j^T \nu_n(\mathbf{x})$, and $\mathbf{Q}(\mathbf{x}) = (\mathbf{q}_1(\mathbf{x}), \ldots, \mathbf{q}_m(\mathbf{x}))^T$ a set of polynomials. Then $\mathbf{Q}(\mathbf{x}) = \mathbf{C}^T \nu_n(\mathbf{x})$ is a matrix of dimension $m \times k$, which is key to compute the basis of the subspaces. We can express a polynomial in terms of $\mathbf{q}(\mathbf{x}) = \prod_{j=1}^n (\mathbf{b}_j^T \mathbf{x}) = (\mathbf{b}_1^T \mathbf{x}) \cdots (\mathbf{b}_n^T \mathbf{x})$, and each vector $\mathbf{b}_j$ is orthogonal to the subspace $\mathcal{V}_j$ (in the case of a plane this is obvious because it is the vector defining that plane). For a sample $\mathbf{x}_i \in \mathcal{V}_i$ (*semi-supervised learning*) we have that $\mathbf{b}_i^T \mathbf{x}_i = 0$ by definition.

The derivative of $\mathbf{q}$ with respect to $\mathbf{x}$ is

$$Dp_n(\mathbf{x}) = \frac{\partial \mathbf{q}(\mathbf{x})}{\partial \mathbf{x}} = \sum_j (\mathbf{b}_j) \prod_{l \neq j} (\mathbf{b}_l \mathbf{x}) \tag{6.81}$$

When evaluated at $\mathbf{x}_i \in \mathcal{V}_i$, the derivative vanishes at all addends $j \neq i$ because these addends include $\mathbf{b}_i^T \mathbf{x}_i = 0$ in the factorization. Then, we have

$$Dp_n(\mathbf{x}_i) = \mathbf{b}_i \prod_{l \neq i} (\mathbf{b}_l \mathbf{x}_i) \tag{6.82}$$

and consequently the normal direction to $\mathcal{V}_i$ is given by [168]

$$\mathbf{b}_i = \frac{Dp_n(\mathbf{x}_i)}{||Dp_n(\mathbf{x}_i)||} \tag{6.83}$$

All the derivatives may be grouped in the Jacobian associated to the collection of polynomials. Let

$$\mathcal{J}(\mathbf{Q}(\mathbf{x})) = \begin{pmatrix} \frac{\partial \mathbf{q}_1}{\partial x_1} & \cdots & \frac{\partial \mathbf{q}_1}{\partial x_k} \\ \vdots & \ddots & \vdots \\ \frac{\partial \mathbf{q}_m}{\partial x_1} & \cdots & \frac{\partial \mathbf{q}_m}{\partial x_k} \end{pmatrix} \tag{6.84}$$

be the $m \times k$ *Jacobian* matrix of the collection of $\mathbf{Q}$. Then, it turns out that the rows of $\mathcal{J}(\mathbf{Q}(\mathbf{x}_i))$ evaluated at $\mathbf{x}_i$ span an orthogonal space to $\mathcal{V}_i$. This means that the right null space of the Jacobian yields a basis $\mathbf{B}_i = (\mathbf{b}_1, \ldots, \mathbf{b}_{d_i})$ of $\mathcal{V}_i$, where $d_i = k - \text{rank}(\mathcal{J}(\mathbf{Q}(\mathbf{x}_i))^T)$ and usually $d_i << k$. If we repeat the latter rationale for each $\mathcal{V}_i$ we obtain the basis for all subspaces. Finally, we assign a sample $\mathbf{x}$ to subspace $\mathcal{V}_i$ if $\mathbf{B}_i^T\mathbf{x} = 0$ or we choose the subspace $\mathcal{V}_i$ minimizing $||\mathbf{B}_i^T\mathbf{x}||$ (clustering, segmentation, classification).

Until now we have assumed that we know an example $\mathbf{x}_i$ belonging to $\mathcal{V}_i$, and then it is straightforward to obtain the basis of the corresponding subspace. However, in general, such assumption is unrealistic and we need an *unsupervised learning* version of gPCA. In order to do that we exploit the *Sampson distance.* Assuming that the polynomials in $\mathbf{Q}$ are linearly independent, given a point $\mathbf{x}$ in the arrangement, that is $\mathbf{Q}(\mathbf{x}) = 0$, if we consider the first-order Taylor expansion of $\mathbf{Q}(\mathbf{x})$ at $\mathbf{x}$, the value at $\tilde{\mathbf{x}}$ (the closest point to $\mathbf{x}$) is given by

$$\mathbf{Q}(\tilde{\mathbf{x}}) = \mathbf{Q}(\mathbf{x}) + \mathcal{J}(\mathbf{Q}(\mathbf{x}))(\tilde{\mathbf{x}} - \mathbf{x}) \tag{6.85}$$

which implies

$$||\tilde{\mathbf{x}} - \mathbf{x}||^2 \approx \mathbf{Q}(\mathbf{x})^T(\mathcal{J}(\mathbf{Q}(\mathbf{x}))\mathcal{J}(\mathbf{Q}(\mathbf{x}))^T)^{\dagger}\mathbf{Q}(\tilde{\mathbf{x}}) \tag{6.86}$$

where $(\cdot)^{\dagger}$ denotes the pseudo-inverse. Then, $||\tilde{\mathbf{x}}-\mathbf{x}||^2$ (square of the Euclidean distance) approximates the *Sampson distance.* Then, we will choose to point lying in, say, the nth subspace, $\mathbf{x}_n$ as the one minimizing the latter Sampson distance but having a non-null Jacobian (points having null derivatives lie at the intersection of subspaces and, thus, yield noisy estimations of the normals). This allows us to find the basis $\mathbf{B}_n$. Having this basis for finding the point $\mathbf{x}_{n-1}$ it is useful to exploit the fact that if we have a point $\mathbf{x}_i \in \mathcal{V}_i$, points belonging to $\cup_{l=i}^n \mathbf{V}_l$ satisfy $||\mathbf{B}_i^T\mathbf{x}|| \cdots ||\mathbf{B}_n^T\mathbf{x}|| = 0$. Therefore, given a point $\mathbf{x}_i \in \mathcal{V}_i$, (for instance $i = 1$) a point $\mathbf{x}_{i-1} \in \mathcal{V}_{i-1}$ can be obtained as

$$\mathbf{x}_{i-1} = \arg \min_{\mathbf{x}\in\mathcal{S}:\mathcal{J}(\mathbf{Q}(\mathbf{x}))\neq 0} \frac{\sqrt{\mathbf{Q}(\mathbf{x})^T(\mathcal{J}(\mathbf{Q}(\mathbf{x}))\mathcal{J}(\mathbf{Q}(\mathbf{x}))^T)^{\dagger}\mathbf{Q}(\tilde{\mathbf{x}})} + \delta}{||\mathbf{B}_i^T\mathbf{x}|| \cdots ||\mathbf{B}_n^T\mathbf{x}|| + \delta} \tag{6.87}$$

where $\delta > 0$ is designed to solve the $0/0$ indetermination (perfect data).

The above description is the basic ideal approach of gPCA. It is ideal in the following regard: (i) there are no noisy data which may perturb the result of the SVD decompositions and, consequently, the estimations of both the collection of polynomials and the basis; (ii) the number of subspaces is known beforehand. Therefore, robustness and model-order selection are key elements in gPCA. Both elements (robustness and model-order selection) are considered in the *Minimum Effective Dimension* gPCA algorithm (MED gPCA) [75] where information theory plays a fundamental role. Along this the book we have considered several information-theoretic model-order selection criteria and their applications in *de-grouping problems*: e.g MDL for segmentation/clustering and BIC for clustering. In gPCA we have a mixture of subspaces and we want to identify both these subspaces and the optimal number of them. Furthermore, we need to do that robustly. The application of these criteria to gPCA is not straightforward, mainly due to the fact that there is not necessarily a consensus between information-theoretic criteria and algebraic/geometric structure of data [75]. Thus, the MED gPCA algorithm relies on a new concept known as *Effective Dimension* (ED) and modifies the *Akaike Information Criterion* (AIC) [2] in order to include the ED. This new criterion yields a useful bridge between information theory and geometry for the purposes of model-order selection. Let us start by defining ED. Given an arrangement $\mathcal{Z} = \mathcal{V}_1 \cup \cdots \cup \mathcal{V}_n \subset \mathbb{R}^k$ of n subspaces, each one of dimension $d_i < k$, and N_i sample points $\mathbf{X}_i = \{\mathbf{x}_{ij}\}_{j=1}^{N_i}$ belonging to each subspace $\mathcal{V}_i$, the *effective dimension* of the complete set of sample points $\mathcal{S} = \cup_{i=1}^{n}\mathbf{X}_i$, being $N = \sum_{i=1}^{n} N_i = |\mathcal{S}|$, is defined as

$$\mathrm{ED}(\mathcal{S}, \mathcal{Z}) = \frac{1}{N}\left(\sum_{i=1}^{n} d_i(k-d_i) + \sum_{i=1}^{n} N_i k_i\right) \tag{6.88}$$

Thus the ED is the average of two terms. The first one $d_i(k-d_i)$ is the total number of reals needed to specify a d_i dimensional space in $\mathbb{R}^k$ (the product of the dimension and its margin with respect to the dimension of ambient space). This product is known as the dimension (*Grassman dimension*) of the *Grassmanian manifold* of d_i dimensional subspaces of $\mathbb{R}^k$. On the other hand, the second term $N_i k_i$ is the total number of reals needed to code the N_i samples in $\mathcal{V}_i$. A simple quantitative example may help to understand the concept of ED. Having three subspaces (two lines and a plane in $\mathbb{R}^3$ as shown in Fig. 6.19) with 10 points lying on each line and 30 ones on the plane, we have 50 samples. Then, if we consider having three subspaces, the ED is $\frac{1}{50}(1 \times (3-1) + 1 \times (3-1) + 2 \times (3-2) + 10 \times 1 + 10 \times 1 + 30 \times 2)$ that is $\frac{1}{50}(6+80) = 1.72$. However, if we decide to consider that the two lines define a plane and consider only two subspaces (planes) the ED is given by $\frac{1}{50}(2 \times (3-2) + 2 \times (3-2) + 20 \times 2 + 30 \times 2) = 2.08$ which is higher than the previous choice, and, thus, more complex. Therefore, the better choice is having two lines and one plane (less complexity for the given data). We are minimizing the ED. Actually the *Minimum Effective Dimension* (MED) may be defined in the following terms:

$$\mathrm{MED}(\mathcal{S}) = \min_{\mathcal{V}:\mathcal{S}\subset\mathcal{V}} ED(\mathcal{S}, \mathcal{V}) \tag{6.89}$$

that is, as the minimum ED among the arrangements fitting the samples. Conceptually, the MED is a kind of information theoretic criterion adapted to the context of finding the optimal number of elements (subspaces) of an arrangement, considering that such elements may have different dimensions. It is particularly related to the AIC. Given the data $\mathcal{S} : |\mathcal{S}| = N$, the model parameters $\mathcal{M}$, and the free parameters $k_{\mathcal{M}}$ for the considered class of model, the AIC is defined as

$$\mathrm{AIC} = -\frac{2}{N}\log P(\mathcal{S}|\mathcal{M}) + 2\frac{k_{\mathcal{M}}}{N} \approx E(-\log P(\mathcal{S}|\mathcal{M})) \tag{6.90}$$

when $N \to \infty$. For instance, for the Gaussian noise model with equal variance in all dimensions (isotropy) σ^2, we have that

$$\log P(\mathcal{S}|\mathcal{M}) = -\frac{1}{2\sigma^2}\sum_{i=1}^{N}||\mathbf{x}_i - \tilde{\mathbf{x}}_i||^2 \tag{6.91}$$

$\tilde{\mathbf{x}}_i$ being the best estimate of $\mathbf{x}_i$ given the model $P(\mathcal{S}|\mathcal{M})$. If we want to adapt the latter criterion to a context of models with different dimensions, and we consider the isotropic Gaussian noise model, we obtain that AIC minimizes

$$\text{AIC}(d_{\mathcal{M}}) = \frac{1}{N}\sum_{i=1}^{N}||\mathbf{x}_i - \tilde{\mathbf{x}}_i||^2 + 2\frac{d_{\mathcal{M}}}{N}\sigma^2 \tag{6.92}$$

selecting the model class $\mathcal{M}^*$, $d_{\mathcal{M}}$ being the dimension of a given model class. AIC is related to BIC (used in the XMeans algorithm, see Prob. 5.14) in the sense that in the latter one, one must simply replace the factor 2 in the second summand by $\log(N)$ which results in a higher amount of penalization for the same data and model. In the context of gPCA, we are considering Grassmanian subspaces of $\mathbb{R}^k$ with dimension $d(k-d)$ which implies that AIC minimizes

$$\text{AIC}(d) = \frac{1}{N}\sum_{i=1}^{N}||\mathbf{x}_i - \tilde{\mathbf{x}}_i||^2 + 2\frac{d(k-d)}{N}\sigma^2 \tag{6.93}$$

The AIC is extended in [92] in order to penalize subspaces of higher dimensions for encoding the data. This can be easily done by adding Nd to the Grassmannian dimensions. This is the *Geometric AIC* (GAIC):

$$\text{GAIC}(d) = \frac{1}{N}\sum_{i=1}^{N}||\mathbf{x}_i - \tilde{\mathbf{x}}_i||^2 + 2\frac{d(k-d)+Nd}{N}\sigma^2 \tag{6.94}$$

When applied to encoding the samples $\mathcal{S}$ with a given arrangement of, say n, subspaces $\mathcal{Z}$ the GAIC is formulated as

$$\begin{aligned}\text{GAIC}(\mathcal{S},\mathcal{Z}) &= \frac{1}{N}\sum_{i=1}^{N}||\mathbf{x}_i - \tilde{\mathbf{x}}_i||^2 + 2\sum_{j=1}^{n}\frac{d_j(k-d_j)+Nd_j}{N}\sigma^2\\ &= \frac{1}{N}\sum_{i=1}^{N}||\mathbf{x}_i - \tilde{\mathbf{x}}_i||^2 + 2N\sigma^2 ED(\mathcal{S},\mathcal{Z})\end{aligned} \tag{6.95}$$

which reveals a formal link between GAIC and the ED (and the MED when GAIC is minimized). However, there are several limitations that preclude the direct use of GAIC in the context of gPCA. First, having subspaces of different dimensions makes almost impossible to choose a prior model. Second, the latter rationale can be easily extended to assuming a distribution for the sample data. And third, even when the variance is known, the GAIC minimizes the *average residual* $||\mathbf{x}_i - \tilde{\mathbf{x}}_i||^2$ which is not an effective measure against outliers. The presence of outliers is characterized by a very large residual.

This a very important practical problem where one can find two extremes in context of high noise rates: samples are considered either as belonging to a unique subspace or to N one-dimensional spaces, each one defined by a sample. Thus, it seems very convenient in this context to find the optimal trade-off between dealing noise and obtaining a good model fitness. For instance, in the SVD version of PCA, the $k \times N$ data matrix $\mathbf{X} = (\mathbf{x}_1, \mathbf{x}_2, \ldots, \mathbf{x}_N)$ is factorized through SVD $\mathbf{X} = \mathbf{U}\Sigma\mathbf{V}^T$, the columns of $\mathbf{U}$ give a basis and the rank provides the dimension of the subspace. As in the eigenvectors/eigenvalues formulation of PCA, the singular values left represent the squared error of the representation (*residual error*). Representing the ordered singular values vs. the dimension of the subspace (see Fig. 6.20, left) it is interesting to note the existence of a *knee point* after the optimal dimension. The remaining singular values represent the sum of square errors, and weights $w_1, w_2 > 0$, derived from the knee points, represent the solution to an optimization problem consisting on minimizing

$$J_{\mathrm{PCA}}(\mathcal{Z}) = w_1 \sum_{i=1}^{N} ||\mathbf{x}_i - \tilde{\mathbf{x}}_i||^2 + w_2 dim(\mathcal{Z}) \tag{6.96}$$

$\mathcal{Z}$ being the subspace, $\tilde{\mathbf{x}}_i$ the closest point to $\mathbf{x}_i$ in that subspace and $dim(\mathcal{Z})$ the dimension of the subspace. The latter objective function is closely related to the MLD-AIC like principles: the first term is data likelihood (*residual error* in the terminology of least squares) and the second one is complexity. In order to translate this idea into the gPCA language it is essential to firstly redefine the MED properly. In this regard, noisy data imply that the ED is highly dependent on the maximum allowable residual error, known as *error tolerance* τ. The higher τ the lower optimal dimension of the subspace (even zero dimensional). However, when noisy data arise, and we set a τ lower to the horizontal coordinate of the knee point, samples must be either considered as independent subspaces or as unique one-dimensional subspace. The consequence is

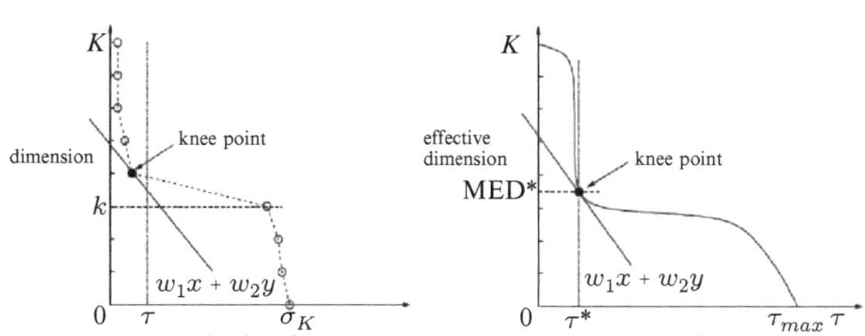

Fig. 6.20. Optimal dimensions. In PCA (*left*) and in gPCA (*right*). In both cases K is the dimension of the ambient space and k is the optimal dimension. Figures by K. Huang, Y. Ma and R. Vidal (©2004 IEEE).

that the optimal dimension (MED in gPCA) approaches the dimension of the ambient space. Thus, a redefinition of the MED must consider the error tolerance:

$$\text{MED}(\mathcal{X}, \tau) = \min_{\mathcal{Z}:||\mathbf{X}-\tilde{\mathbf{X}}||_\infty \leq \tau} ED(\tilde{\mathbf{X}}, \mathcal{Z}) \tag{6.97}$$

being $||\mathbf{X} - \tilde{\mathbf{X}}||_\infty = \sum_{i=1}^{N} ||\mathbf{x}_i - \tilde{\mathbf{x}}_i||_\infty$ and $||\mathbf{x} - \tilde{\mathbf{x}}||_\infty = \max_{\mathbf{x}\in\mathbf{X}} ||\mathbf{x} - \tilde{\mathbf{x}}||$. In Fig. 6.20 we plot the MED vs. τ. With the new definition of MED we must optimize functions of the type:

$$J_{\text{gPCA}}(\mathcal{Z}) = w_1 \sum_{i=1}^{N} ||\mathbf{x}_i - \tilde{\mathbf{x}}_i||_\infty + w_2 \text{MED}(\mathcal{X}, \tau) \tag{6.98}$$

Adopting a criterion like the one in Eq. 6.93 in gPCA by preserving the geometric structure of the arrangement can be done by embedding this criterion properly in the algebraic original formulation of gPCA. As we have explained above, an arrangement of n subspaces can be described by a set of polynomials of degree n. But when subspaces with different dimensions arise (e.g. lines and planes) also polynomials with less dimensions than n can fit the data. In the classical example of two lines and one plane (Fig. 6.19) a polynomial of second degree may also fit the data. The intriguing and decisive question is whether this plane may be then partitioned into two lines. If so we will reduce the ED, and this will happen until no subdivision is possible. In an ideal noise-free setting, as $M_n^{[k]}$ decides the rank of matrix $\mathbf{L}_n$ and then bounds the number of subspaces, it is interesting to start by $n = 1$, and then increase n until we find a polynomial of degree n which fits all the data ($\mathbf{L}_n$ has lower rank). Then, it is possible to separate all the n subspaces following the gPCA steps described above. Then, we can try to apply the process recursively to each of the n until there are no lower dimensional subspaces in each group or there are too much groups. In the real case of noisy data, we must fix n and find the subspaces (one at a time) with a given error tolerance. For a fixed n: (i) find the fist subspace and assign to it the points with an error less than τ; (ii) repeat the process to fit the remaining subspaces to the data points (find points associated to subspaces and then the orthogonal directions of the subspace). However, the value of τ is key to know whether the rank of $\mathbf{L}_n$ can be identified correctly. If we under-estimate this rank, the number of subspaces is not enough to characterize all the data. If we over-estimate the rank we cannot identify all the subspaces because all points have been assigned to a subspace in advance. We can define a range of ranks (between r_{min} and r_{max}) and determine whether the rank is under or over-estimated in this range. If none of the ranks provides a segmentation in n subspaces within a tolerance of τ, it is quite clear that we must increase the number of subspaces.

In Alg. 15 $\alpha =< .,. >$ denotes the *subspace angle* which is an estimation of the amount of dependence between two subspaces (low angle indicates high dependence and vice versa). Orthogonal subspaces have a $\pi/2$ angle. If the

Algorithm 15: Robust-gPCA

Input: $\mathbf{X}$ samples matrix, τ tolerance
Initialize
$n = 1$, success=false
repeat
  $\mathbf{L}_n(\mathbf{X}) \leftarrow (\nu_n(\mathbf{x}_1), \ldots, \nu_n(\mathbf{x}_N))^T \in \mathbb{R}^{M_n^{[k]} \times N}$
  $r_{max} = M_n^{[k]} - 1$, $r_{min} = arg\,\min_i \left\{ \frac{\sigma_i(\mathbf{L}_n)}{\sum_{j=1}^{i-1} \sigma_j(\mathbf{L}_n)} \leq 0.02 \right\}$
  while $(r_{min} \leq r_{max})$**and**(**not** *(success)*) **do**
    $r = \lfloor (r_{min} + r_{max})/2 \rfloor$;
    over-estimated=false under-estimated=false
    Compute the last $M_n^{[k]} - r$ eigenvectors $\{\mathbf{b}_i\}$ of $\mathbf{L}_n$
    Obtain the polynomials $\{\mathbf{q}_j(\mathbf{x}) = \mathbf{b}_j^T \nu_n(\mathbf{x})\} : \mathbf{q}(\mathbf{x}) = \prod_{j=1}^{n} (\mathbf{b}_j^T \mathbf{x})$
    Find n samples $\mathbf{x}_j, \mathbf{x}_k$ where
    $\alpha_{jk} =< \mathbf{B}_j, \mathbf{B}_k >: \mathbf{B}_z = span\{\mathcal{J}(\mathbf{Q}(\mathbf{x}_z))\} > 2\tau$
    Otherwise (no samples satisfying constraint) over-estimated=true
    if over-estimated **then** $r_{max} \leftarrow r - 1$
    else
      Assign each point in $\mathbf{X}$ to its closest subspace considering error tolerance τ and obtain n groups $\mathbf{X}_k$
      if fail **then**
        under-estimated = true
      end
      if under-estimated **then**
        $r_{min} \leftarrow r + 1$
      else
        success = true
      end
    end
  end
  if success **then**
    forall $k = 1 : n$ **do**
      Robust-gPCA($\mathbf{X}_k$,τ)
  end
  else
    $n \leftarrow n + 1$
  end
until (success;
or$(n \geq n_{max})$) **Output**: MED gPCA

highest angle we can find is too small, this indicates that the classic examples of two lines and one plane are illustrated in Fig. 6.21. The algorithm starts with $n = 1$, enters the *while* loop and there is no way to find a suitable rank for that group (always over-estimated). Then, we increase n to $n = 2$, and it is possible to find two independent subspaces because we have two groups (the plane and the two lines). Then, for each group a new recursion level

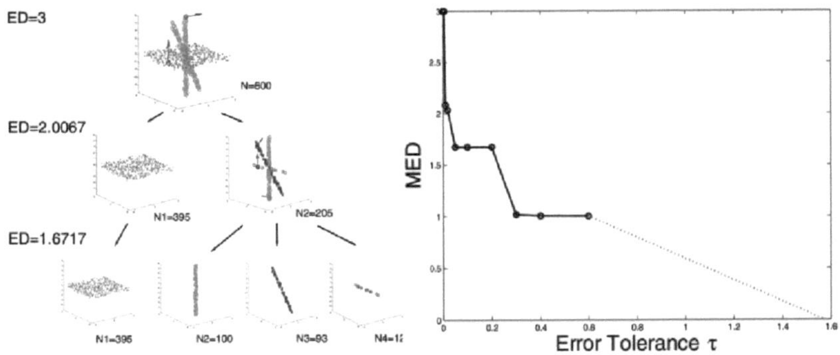

Fig. 6.21. *Right*: result of the iterative robust algorithm for performing gPCA recursively while minimizing the effective dimension. Such minimization is derived from the recursive partition. *Left*: decreasing of ED with τ. Figure by Y. Ma, A.Y. Yang, H. Derksen and R. Fossum (©2008 SIAM).

Fig. 6.22. Unsupervised segmentation with gPCA. Figure by K. Huang, Y. Ma and R. Vidal (©2004 IEEE). See Color Plates.

is started, and for each of them we start by setting $n = 1$. The recursion level associated to the plane stops after finding a suitable rank. This implies succeeding, enter a second level of recursion, started without succeeding, and having $r_{\min} > r_{\max}$. Then n increases until $n_{\max}$ is found and the algorithm stops. On the other hand, the first level of recursion for the two lines increases until $n = 2$ where it is possible to find independent subspaces and a third level of recursion starts. For each line, the rationale for the plane is applicable. As it is expected, entering new levels of recursion decreases the ED. Finally, in Fig. 6.22 we show the results of unsupervised segmentation. Each pixel is associated to a 16×16 window. As each pixel has color information (RGB) we have patterns of $16 \times 16 \times 3 = 768$ dimensions for performing PCA. The first 12 eigenvectors are used for gPCA segmentation. Only one level of recursion is needed in this case, because for each of the groups there is no need re-partition the group again.

Problems

6.1 Filters and wrappers
There are two major approaches to feature selection: filter and wrapper. Which one is more prone to overfitting? Does filter feature selection always improve the classification results?

6.2 Filter based on mutual information – Estimation of mutual information
The mutual information $I(\mathbf{S};C)$ can be calculated in two different ways, with the conditional entropy 6.19 and with the joint entropy 6.19. Do you think that the conditional form is suitable for the feature selection problem? Note that C is discrete. Would the conditional form be suitable for a problem with a continuous C?

6.3 Mutual information calculation
When using mutual information as a criterion for feature selection, if the features are continuous, some estimation technique has to be used. However, if the data are categorical, the mutual information formula for the discrete case can be applied. In the following toy example the data are categorical. How many feature sets are possible? Calculate the mutual information with the class, for each one of them.

Data set with four samples defined by three features and classified into two classes:

x_1	x_2	x_3	C
A	Δ	Θ	C_1
B	Z	I	C_1
A	E	I	C_2
Γ	Z	I	C_2

6.4 Markov blankets for feature selection
The Makrov blanket criterion for feature selection removes those features for which a Markov blanket is found among the set of remaining features. What happens in the case of completely irrelevant features? Which Markov blanket is found for them?

6.5 Conditional dependence in feature selection
Markov blankets and the MD criteria do not assume independence between features, and can have into account complex dependence patterns. Remember that both of them are filters for feature selection, so they are independent of the classifier. Are they of benefit to the classification rates of any kind of classifier? A hint: the "Corral" data set (already explained in previous sections), when used with the Naive Bayes classifier, performs better with all the features, than with just the first four of them (which completely determines the class label). Why?

6.6 Mutual information and conditional dependence
Venn diagrams are useful for the intuitive understanding mutual information. Draw a Venn diagram for the already described "Corral" data set.

6.7 Filter based on mutual information – complexity
The mRMR and the MD criteria have different complexity issues. The estimation of entropy is an important complexity factor. The following classic data sets have different dimensionalities and numbers of samples. State your reasons for using one criterion or another on each one of them. Apart from complexity, consider some implementation issues, for example, storing the results of calcula which would be repeated many times.

- NCI data set. Contains 60 samples (patients), each one of them with 6,380 features (genes). The samples are labeled with 14 different classes of human tumor diseases.
- Census Income, also known as Adult data set. Contains 48,842 samples (people from the census), described by 14 features. The labels indicate whether income exceeds \$50,000/year based on census data.

The MD criterion can be used both in forward and backward searches. Which one would you use for each one of the previous data sets? Which data set would be more feasible for an exhaustive search?

6.8 Satisfying additional constraints in maximum entropy
Prove that, when computing the Lagrange multipliers by exploiting Eq. 6.69, the resulting maximum entropy pdf satisfies also the extra constraints:

$$E(F_j(x)G'_r(x)) = \frac{1}{N}\sum_{i=1}^{N} F_j(x_i)G'_r(x_i)$$

for $j = 1, \ldots, m$ and $k = 1, \ldots, m$, N being the number of samples x_i, and $G_r(x) = x^r$. Obtain the β variables and the Lagrange multipliers for that case.

6.9 MML for feature selection and Gaussian clusters
In [101], the authors propose a method for selecting features while estimating clusters through Gaussian mixtures. The basic IT criterion used in such work is the *Minimum Message Length* (MML) [173] to measure the saliency of the data dimensions (features). Such criterion can be formulated in the following terms:

$$\mathrm{MML}(\Theta) = -\log p(\Theta) - \log(\mathcal{D}|\Theta) + \frac{1}{2}|\mathbf{I}(\Theta)| + \frac{c}{2}\left(1 + \log\frac{1}{12}\right) \tag{6.99}$$

where Θ are the parameters of the model (means, variances), $\mathcal{D}$ are the data (then $\log(\mathcal{D}|\Theta)$ is the log-likelihood), and $\mathbf{I}(\Theta)$ is the Fisher information matrix.

Irrelevant features have a near-zero saliency (MML). Consider, for instance, the paradigmatic case in the positive quadrant of $\mathbb{R}^2$ (samples as defined as (x_i, y_i)) where two different *vertical oriented* Gaussians have the same covariance matrix Σ with horizontal variance clearly smaller than the vertical one: $\sigma_x^2 \ll \sigma_y^2$. The mean in y of the first Gaussians is clearly greater than the second (by an amount of d_y), the same happens in the x-dimension with d_x. This means that the Gaussians may be considered *independent*. Show then, in this case, that both x and y dimensions are *relevant*. However, when both d_y tends to 0, and d_x is small but not zero, and we replace both Gaussians by a unique one using PCA, it turns out that only the y dimension is relevant. Quantify such relevances through MML. Study the change of relevance as d_x grows (always using the PCA approximation for the sake of simplicity).

6.10 Complexity of the FRAME algorithm
What is the computational complexity of Alg. 12? Set a given image size, intensity range, sweeps, and filters and give an estimation of its execution time in a current laptop.

6.11 FRAME and one-dimensional patterns
The method described in Alg. 12 can be used also for synthesizing/reproducing one-dimensional patterns (like sounds coming from speech or spike trains). Conjecture what should be the role of filters like edge-detectors (contrast), averaging filters (smoothing), or simply intensity ones (use directly the histogram of the signal intensity). Think about using an adequate number of filters depending on the number of examples available.

6.12 Minimax entropy and filter pursuit
When describing Alg. 14 we have presented a connection between selecting the optimal filter and minimizing the Kullback–Leibler divergence between $f(\mathbf{I})$ and $p(\mathbf{I}; \Lambda_m, \mathcal{S}_m)$. Prove that such divergence can be expressed in terms of the differences of the entropies of the latter distributions. If so, as $H(f(\mathbf{I}))$ is fixed, in order to minimize such Kullback–Leibler divergence we only need to minimize $H(p(\mathbf{I}; \Lambda_m, \mathcal{S}_m))$. Hint: use the fact that

$$E_{p(\mathbf{I};\Lambda_m,\mathcal{S}_m)}(\boldsymbol{\alpha}_j) = E_{f(\mathbf{I})}(\boldsymbol{\alpha}_j), \quad j = 1, \ldots, m$$

6.13 Kullback–Leibler gradient
In Alg. 14 prove that if we have two alternatives of Lagrange multipliers Λ and the optimal Λ^*

$$D(f(\mathbf{I})||p(\mathbf{I};\Lambda)) = D(f(\mathbf{I})||p(\mathbf{I};\Lambda^*)) + D(p(\mathbf{I};\Lambda^*)||p(\mathbf{I};\Lambda))$$

which means that, given the positiveness of the Kullback–Leibler divergence, the optimal choice Λ^* has always the minimal divergence.

6.14 Arrangements, Veronese maps, and gPCA

In gPCA, an arrangement of n subspaces can be expressed by a set of polynomials of degree n. Consider the following configuration arrangement in $\mathbb{R}^3$ (each sample is of the form $\mathbf{x} = (x_1, x_2, x_3)^T$: $\mathcal{Z} = \mathcal{V}_1 \cup \mathcal{V}_2$), where $\mathcal{V}_1 = \{\mathbf{x} : x_1 = x_2 = 0\}$ (a vertical line passing through the origin) and $\mathcal{V}_2 = \{\mathbf{x} : x_3 = 0\}$ (the horizontal plane passing through the origin). The union may be found as following:

$$\begin{aligned}\mathcal{V}_1 \cup \mathcal{V}_2 &= \{\mathbf{x} : (x_1 = x_2 = 0) \vee (x_3 = 0)\}\\ &= \{\mathbf{x} : (x_1 = 0 \vee x_3 = 0) \wedge (x_2 = 0 \vee x_3 = 0)\}\\ &= \{\mathbf{x} : (x_1 x_3 = 0) \wedge (x_2 x_3 = 0)\}\end{aligned}$$

and then the union is represented by $q_1(\mathbf{x}) = (x_1 x_3)$ and $q_2(\mathbf{x}) = (x_2 x_3)$. Then, the generic Jacobian matrix is given by

$$\mathcal{J}(\mathbf{Q}(\mathbf{x})) = \begin{pmatrix} \frac{\partial q_1}{\partial x_1} & \frac{\partial q_1}{\partial x_2} & \frac{\partial q_1}{\partial x_3} \\ \frac{\partial q_2}{\partial x_1} & \frac{\partial q_2}{\partial x_2} & \frac{\partial q_2}{\partial x_3} \end{pmatrix} = \begin{pmatrix} x_3 & 0 & x_1 \\ 0 & x_3 & x_2 \end{pmatrix}$$

where the first rows $Dp_1(\mathbf{x})$ and the second are $Dp_2(\mathbf{x})$. Then choosing a $\mathbf{z}_1 = (0\ 0\ 1)^T$ point with $x_3 = 1$ for the line, and choosing a point $\mathbf{z}_2 = (1\ 1\ 0)^T$ with $x_1 = x_2 = 1$ for the plane. We have

$$\mathcal{J}(\mathbf{Q}(\mathbf{z}_1)) = \begin{pmatrix} 1\,0\,0 \\ 0\,1\,0 \end{pmatrix}, \quad \mathcal{J}(\mathbf{Q}(\mathbf{z}_2)) = \begin{pmatrix} 0\,0\,1 \\ 0\,0\,1 \end{pmatrix}$$

where it is clear that the mull of the transpose of the Jacobian yields the basis of the line (a unique vector $\mathbf{B}_1 = \{(0\ 0\ 1)^T\}$ because the Jacobian has rank 2) and of the second (with rank 1) $\mathbf{B}_1 = \{(0\ 1\ 0)^T, (-1\ 0\ 0)^T\}$. Check that the obtained basis vectors are all orthogonal to their corresponding $Dp_1(\mathbf{z}_i)$ and $Dp_2(\mathbf{z}_i)$. Given the latter explanations, compute these polynomials from the Veronese map. Hint: for $n = 2$ in $\mathbb{R}^3$ ($k = 3$) we have $M_n^{[k]} = 6$ monomials and coefficients. To that end suggest 10 samples for each subspace. Then compute the $\mathbf{Q}$ collection of vectors and then the Jacobian assuming that we have a known sample per subspace. Show that the results are coherent with the ones described above. Find also the subspaces associated to each sample (segmentation). Repeat the problem unsupervisedly (using the Sampson distance). Introduce noise samples and reproduce the estimation and the segmentation to test the possible change of rank and the deficiencies in the segmentation.

6.15 gPCA and Minimum Effective Dimension

Consider the following configuration in $\mathbb{R}^3$: two orthogonal planes, one fitting 500 samples and the other plane fitting 100, and two lines, each one associated to 200 samples. Compute the MED and the optimal arrangement.

6.6 Key References

- I. Guyon and A. Elisseeff. "An Introduction to Variable and Feature Selection". *Journal of Machine Learning Research* 3:1157–1182 (2003)
- K. Torkkola. "Feature Extraction by Non-Parametric Mutual Information Maximization". *Journal of Machine Learning Research* 3:1415–1438 (2003)
- H. Peng, F. Long, and C. Ding. "Feature Selection Based on Mutual Information: Criteria of Max-Dependency, Max-Relevance, and Min-Redundancy". *IEEE Transactions on Pattern Analysis and Machine Intelligence* 27(8):1226–1238 (2005)
- B. Bonev, F. Escolano, and M. Cazorla. "Feature Selection, Mutual Information, and the Classification of High-Dimensional Patterns". *Pattern Analysis and Applications* 1433–7541 (2008)
- A. Vicente, P.O. Hoyer, and A. Hyvärinen. "Equivalence of Some Common Linear Feature Extraction Techniques for Appearence-Based Object Recognition Tasks". *IEEE Transactions on Pattern Analysis and Machine Intelligence* 29(5):896–900 (2007)
- N. Vasconcelos and M. Vasconcelos. "Scalable Discriminant Feature Selection for Image Retrieval and Recognition". *Computer Vision and Pattern Recognition Conference*, Washington, DC (USA) (2004)
- D. Koller and M. Sahami. "Toward Optimal Feature Selection". ICML-96: *Proceedings of the Thirteenth International Conference on Machine Learning*, pp. 284–292, San Francisco, CA: Morgan Kaufmann, Bari (Italy) (1996)
- M. Law, M. Figueiredo, and A.K. Jain. "Simultaneous Feature Selection and Clustering Using a Mixture Model". *IEEE Transactions on Pattern Analysis and Machine Intelligence* 26(9):1154–1166 (2004)
- S.C. Zhu, Y.N. Wu, and D.B. Mumford. "FRAME: Filters, Random field And Maximum Entropy: Towards a Unified Theory for Texture Modeling". *International Journal of Computer Vision* 27(2):1–20 (1998)
- A. Hyvärinen and E. Oja. "Independent Component Analysis: Algorithms and Applications". *Neural Networks* 13(4–5):411–430 (2000)
- T. Bell and T. Sejnowski. "An Information-Maximization Approach to Blind Separation and Blind Deconvolution". *Neural Computation* 7:1129–1159 (1995)
- D. Erdogmus, K.E. Hild II, Y.N. Rao, and J.C. Príncipe. "Minimax Mutual Information Approach for Independent Component Analysis". *Neural Computation* 16:1235–1252 (2004)
- Y. Ma, A.Y. Yang, H. Derksen, and R. Fossum. "Estimation of Subspace Arrangements with Applications in Modeling and Segmenting Mixed data". *SIAM Review* 50(3):413–458 (2008)

7

Classifier Design

7.1 Introduction

The classic information-theoretic classifier is the decision tree. It is well known that one of its drawbacks is that it tends to overfitting, that is, it yields large classification errors with test data. This is why the typical optimization of this kind of classifiers is some type of pruning. In this chapter, however, we introduce other alternatives. After reminding the basics of the incremental (local) approach to grow trees, we introduce a global method which is applicable when a probability model is available. The process is carried out by a dynamic programming algorithm. Next, we present an algorithmic framework which adapts decision trees for classifying images. Here it is interesting to note how the tests are built, and the fundamental lesson is that the large amount of possible tests (even when considering only binary relationships between parts of the image) recommends avoiding building unique but deep trees, in favor of a bunch of shallow trees. This is the keypoint of the chapter, the emergence of *ensemble classifying methods*, complex classifiers built in the aggregation/combination of simpler ones, and the role of IT in their design. In this regard, the method adapted to images is particularly interesting because it yields experimental results in the domain of OCRs showing that tree-averaging is useful. This work inspired the yet classical *random forests* approach where we analyze their generalization error and show applications in different domains like bioinformatics, and present links with Boosting. Following the *ensemble* focus of this chapter, next section introduces two approaches to improve Boosting. After introducing the Adaboost algorithm we show how boosting can be driven both by mutual information maximization (*infomax*) and by maximizing Jensen–Shannon divergence (*JBoost*). The main difference between the two latter IT-boosting approaches relies on feature selection (linear or nonlinear features). Then, we introduce to the reader the world of *maximum entropy classifiers*, where we present the basic iterative-scaling algorithm for finding the Lagrange multipliers. Such algorithm is quite different from the one described in Chapter 3, where we estimate the multipliers in a different way (coupled with a generative model). We also estabish connections among iterative scaling, the family of exponential distributions and information projection. Finally we cannot close this chapter and the book itself without paying special attention to the extension of information projection (Bregman divergences), which has recently inspired new methods for building linear classifiers.

7.2 Model-Based Decision Trees

7.2.1 Reviewing Information Gain

Let $\mathcal{X} = \{\mathbf{x}\}$ be the training set (example patterns) and $\mathcal{Y} = \{y\}$ the class labels which are assigned to each element of $\mathcal{X}$ in a supervised manner. As it is well known, a *classification tree*, like CART [31] or C4.5 [132], is a model extracted from the latter associations which tries to correctly predict the most likely class for unseen patterns. In order to build such model, one must consider that each pattern $\mathbf{x} \in \mathcal{X}$ is a feature vector $\mathbf{x} = (x_1, x_2, \ldots, x_N)$, and also that each feature x_t has associated a *test* $I(x_i > c_i) \in \{0, 1\}$, whose value depends on whether the value x_i is above (1) or below (0) a given threshold c_i. Thus, a tree $\mathcal{T}$ (binary when assuming the latter type of test) consists of a set of *internal* nodes $\dot{\mathcal{T}}$, each one associated to a test, and a set of *terminal* nodes (leaves) $\partial\mathcal{T}$, each one associated to a class label. The outcome $h_{\mathcal{T}}(\mathbf{x}) \in \mathcal{Y}$ relies on the sequence of outcomes of the tests/questions followed for reaching a terminal.

Building $\mathcal{T}$ implies establishing a correspondence $\pi(t) = i$, with $i \in \{1, 2, \ldots, N\}$, between each $t \in \dot{\mathcal{T}}$ and each test $X_t = I(x_i > c_i)$. Such correspondence induces a partial order between the tests (nodes) in the tree (what tests should be performed first and what should be performed later), and the first challenging task here is *how to choose the proper order*. Let Y a random variable encoding the *true class of the examples* and defined over $\mathcal{Y}$. The uncertainty about such true class is encoded, as usual, by the entropy

$$H(Y) = \sum_{y\in\mathcal{Y}} P(Y = y) \log_2 P(Y = y) \tag{7.1}$$

$P(Y = y)$ being the proportion of examples in $\mathcal{X}$ labeled as y. In the classical *forward greedy method* the test associated to the root of the tree, that is, $\pi(t) = i$, with $t = 1$, is the one *gaining* more information about Y, that is, the one maximizing

$$H(Y) - H_t(Y|X_t) \tag{7.2}$$

being a sort of conditional entropy $H_t(Y|X_t)$ based on the possible outcomes of the test:

$$H_t(Y|X_t) \equiv P(X_t = 0)H_{t_0}(Y) + P(X_t = 1)H_{t_1}(Y) \tag{7.3}$$

where $P(X_t = 1) = \frac{|\mathcal{X}_t|}{|\mathcal{X}|}$ is the fraction of examples in $\mathcal{X}$ satisfying the test X_t and $P(X_t = 0) = 1 - P(X_t = 1)$. Moreover, H_{t_0} and H_{t_1} are the entropies associated to the two descents of t:

$$H_{t_0}(Y) \equiv H(Y|X_t = 0) = -\sum_{y\in\mathcal{Y}} P(Y = y|X_t = 0)\log_2 P(Y = y|X_t = 0)$$

$$H_{t_1}(Y) \equiv H(Y|X_t = 1) = -\sum_{y\in\mathcal{Y}} P(Y = y|X_t = 1)\log_2 P(Y = y|X_t = 1)$$

where $P(Y = y|X_t = 1)$ is the fraction of $\mathcal{X}_t$ of class y, and $P(Y = y|X_t = 0)$ is the fraction of $\mathcal{X} \sim \mathcal{X}_t$ of class y. Consequently, $H_t(Y|X_t)$ is consistent with the definition of conditional entropy:

$$\begin{aligned} H_t(Y|X_t) &= P(X_t = 0)H(Y|X_t = 0) + P(X_t = 1)H(Y|X_t = 1) \\ &= \sum_{k\in\{0,1\}} P(X_t = k)H(Y|X_t = k) \end{aligned}$$

Once $i = \pi(t)$ is selected for $t = 1$ it proceeds to select recursively $\pi(t_0)$ and $\pi(t_1)$. Let us assume, for simplicity, that $\mathcal{X} \sim \mathcal{X}_t$ is so homogeneous that $H_{t_0}(Y) < \epsilon$ and $\epsilon \approx 0$. This means that, after renaming $t_0 = l$, we have $l \in \partial\mathcal{T}$, that is, a leaf, and all examples in $\mathcal{X} \sim \mathcal{X}_t$ will be labeled with the most frequent class, say y_l. On the other hand, if $H_{t_1}(Y) > \epsilon$, after renaming $t = t_1$, we should select a feature $j = \pi(t)$ maximizing

$$H_t(Y) - H_t(Y|X_1 = 1, X_t) = H(Y|X_1 = 1) - H_t(Y|X_1 = 1, X_t)\ . \tag{7.4}$$

$$H_t(Y|X_1 = 1, X_t) = \sum_{k\in\{0,1\}} P(X_1 = 1, X_t = k)H(Y|X_1 = 1, X_t = k)$$

where, for instance, $P(X_1 = 1, X_t = 1) = \frac{|\mathcal{X}_1 \bigcap \mathcal{X}_t|}{|\mathcal{X}|}$ is the fraction of examples satisfying tests X_1 and X_t. In addition, if we denote by Q_t the set of outcomes (test results) preceding X_t, the feature $j = \pi(t)$ should minimize

$$H_t(Y) - H_t(Y|Q_t, X_t) = H(Y|Q_t) - H_t(Y|Q_t, X_t) \tag{7.5}$$

Such a process should be applied recursively until it is not possible to refine any node, that is, until all leaves are reached. As the nature of the process is greedy, it is very probable to obtain a suboptimal tree in terms of classification performance.

7.2.2 The Global Criterion

Alternatively to the greedy method, consider that the path for reaching a leaf $l \in \partial\mathcal{T}$ has length P and let $Q_l = X_{\pi(1)} \bigcap X_{\pi(2)} \bigcap \ldots \bigcap X_{\pi(P-1)}$. Then $P(Q_t) = P(X_{\pi(1)} = k_{\pi(1)}, \ldots, X_{\pi(P-1)} = k_{\pi(P-1)})$ is the probability of reaching that leaf. A global error is partially given by minimizing the *average terminal entropy*

$$H(Y|\mathcal{T}) = \sum_{l\in\partial\mathcal{T}} P(Q_l)H_l(Y) = \sum_{l\in\partial\mathcal{T}} P(Q_l)H(Y|Q_l) \tag{7.6}$$

which means that it is desirable to either have small entropies at the leaves, when the $P(Q_l)$ are significant, or to have small $P(Q_l)$ when $H(Y|Q_l)$ becomes significant.

The latter criterion is complemented by a *complexity-based* one, which is compatible with the MDL principle: the minimization of the *expected depth*

$$Ed(\mathcal{T}) = \sum_{l\in\partial\mathcal{T}} P(Q_l)d(l) \tag{7.7}$$

$d(l)$ being the depth of the leaf (the depth of the root node is zero).

When combining the two latter criteria into a single one it is key to highlight that the optimal tree depends on the joint distribution of the set of possible tests $\mathbf{X} = \{X_1, \ldots, X_N\}$ and the *true class* Y. Such joint distribution $(\mathbf{X}, Y)$ is the model $\mathcal{M}$. It has been pointed out [8, 62] that a model is available in certain computer vision applications like face detection where there is a "rare" class a, corresponding to the object (face) to be detected, and a "common" one b corresponding to the background. In this latter case we may estimate the prior density $p_0(y)$, $y \in \mathcal{Y} = \{a, b\}$ and exploit this knowledge to speed up the building of the classification tree.

Including the latter considerations, the *global optimization problem* can be posed in terms of finding

$$\mathcal{T}^* = \arg\min_{\mathcal{T}\in\Omega} C(\mathcal{T}, \mathcal{M}) = H(Y|\mathcal{T}) + \lambda Ed(\mathcal{T}) \tag{7.8}$$

where Ω is the set of possible trees (partial orders) depending on the available features (tests) and $\lambda > 0$ is a control parameter. Moreover, the maximal depth of $\mathcal{T}^*$ is bounded by $D = \max_{l\in\partial\mathcal{T}} d(l)$.

What is interesting in the latter formulation is that the cost $C(\mathcal{T}, \mathcal{M})$ can be computed recursively. If we assume that the test X_t is associated to the root of the tree we have

$$\begin{aligned} C(\mathcal{T}, \mathcal{M}) &= \lambda + P(X_t{=}0)C(\mathcal{T}_0, \{\mathcal{M}|X_t{=}0\}) + P(X_t = 1)C(\mathcal{T}_1, \{\mathcal{M}|X_t{=}1\}) \\ &= \lambda + \sum_{k\in\{0,1\}} P(X_t = k)C(\mathcal{T}_k, \{\mathcal{M}|X_t = k\}) \end{aligned}$$

and therefore

$$C(\mathcal{T}_k, \{\mathcal{M}|X_t = k\}) = H(Y|\mathcal{T}_k, X_t = k) + \lambda Ed(\mathcal{T}_k)$$

$\mathcal{T}_k$ being the k-th subtree and $\{\mathcal{M}|X_t = k\}$ the resulting model after fixing $X_t = k$. This means that, once X_t is selected for the root, we have only $N-1$ tests to consider. Anyway, finding $C^*(\mathcal{M}, D)$, the minimum value of $C(\mathcal{T}, \mathcal{M})$ over all trees with maximal depth D requires a near $O(N!)$ effort, and thus is only practical for a small number of tests and small depths.

In addition, considering that $C^*(\mathcal{M}, D)$ is the minimum value of $C(\mathcal{T}, \mathcal{M})$ over all trees with maximal depth D, we have that for $D > 0$:

$$C^*(\mathcal{M}, D) = \min \begin{cases} H(p_0) \\ \lambda + \min_{t \in \mathbf{X}} \sum_{k \in \{0,1\}} P(X_t = k) C^*(\{\mathcal{M}|X_t = k\}, D-1) \end{cases} \tag{7.9}$$

and obviously $C^*(\mathcal{M}, 0) = H(p_0)$. As stated above, finding such a minimum is not practical in the general case, unless some assumptions were introduced.

7.2.3 Rare Classes with the Greedy Approach

For instance, in a two-class context $\mathcal{Y} = \{a, b\}$, consider that we have a "rare" class a and a "common" one b, and we assume the prior $p_0(a) \approx 10^{-1}$ and $p_0(b) = 1 - p_0(a)$. Consider for instance the following test: always $X_1 = 1$ when the true class is the rare one, and $X_1 = 1$ randomly when the true class is the common one; that is $P(X_1 = 1|Y = a) = 1$ and $P(X_1 = 1|Y = b) = 0.5$. If the rare class corresponds to unfrequent elements to detect, such test never yields false negatives (always fires when such elements appear) but the rate of false positives (firing when such elements do not appear) is 0.5. Suppose also that we have a second test X_2 complementary to the first one: $X_2 = 1$ randomly when the true class is the rare one, and never $X_2 = 1$ when the true class is the common one, that is $P(X_2 = 1|Y = a) = 0.5$ and $P(X_2 = 1|Y = b) = 0$. Suppose that we have three versions of X_2, namely X_2', X_2'', and X_2''' which is not a rare case in visual detection where we have many similar tests (features) with similar lack of importance.

Given the latter specifications (see Table 7.1), a local approach has to decide between X_1 and any version of X_2 for the root r, based on $H_r(Y|X_1)$ and $H_r(Y|X_2)$. The initial entropy is

$$\begin{aligned} H(Y) &= -\underbrace{P(Y=a)}_{p_0(a)} \log_2 P(Y=a) - \underbrace{P(Y=b)}_{p_0(b)} \log_2 P(Y=b) \\ &= -\underbrace{\frac{5}{10^4} \log_2 \left(\frac{5}{10^4}\right)}_{\approx 0} - \frac{9,995}{10^4} \log_2 \left(\frac{9,995}{10^4}\right) = 0.0007 \end{aligned}$$

Table 7.1. Simple example.

Example	X_1	$X_2^{'}$	$X_2^{''}$	$X_2^{'''}$	Class
$\mathbf{x}_1$	1	0	0	0	b
$\mathbf{x}_2$	1	0	0	0	b
...	...	...	...	...	...
$\mathbf{x}_{5,000}$	1	0	0	0	b
$\mathbf{x}_{5,001}$	0	0	0	0	b
$\mathbf{x}_{5,002}$	0	0	0	0	b
...	...	...	...	...	...
$\mathbf{x}_{9,995}$	0	0	0	0	b
$\mathbf{x}_{9,996}$	1	0	0	1	a
$\mathbf{x}_{9,997}$	1	0	1	0	a
$\mathbf{x}_{9,998}$	1	1	0	0	a
$\mathbf{x}_{9,999}$	1	1	0	1	a
$\mathbf{x}_{10,000}$	1	1	1	0	a
Test	X_1	$X_2^{'}$	$X_2^{''}$	$X_2^{'''}$	Class
$\mathbf{x}_1^{'}$	1	0	0	0	a
$\mathbf{x}_2^{'}$	1	0	1	1	a
$\mathbf{x}_3^{'}$	1	1	1	1	a

When computing conditional entropies for selecting X_1 or X_2 for the root we have

$$\begin{aligned} H(Y|X_1) &= P(X_1=0)\underbrace{H(Y|X_1=0)}_{0} + P(X_1=1)H(Y|X_1=1) \\ &= \frac{(5,000+5)}{10^4}\left\{-\frac{5}{5,005}\underbrace{\log_2\left(\frac{5}{5,005}\right)}_{\approx 0} - \frac{5,000}{5,005}\log_2\left(\frac{5,000}{5,005}\right)\right\} \\ &= 0.0007 = H(Y) \end{aligned}$$

whereas

$$\begin{aligned} H(Y|X_2^{'}) &= P(X_2^{'}=0)H(Y|X_2^{'}=0) + P(X_2^{'}=1)\underbrace{H(Y|X_2^{'}=1)}_{0} \\ &= \frac{(9,995+2)}{10^4}\left\{-\frac{2}{9,997}\log_2\left(\frac{2}{9,997}\right) - \frac{9,995}{9,997}\log_2\left(\frac{9,995}{9,997}\right)\right\} \\ &= 0.0003 \approx \frac{H(Y)}{2} \end{aligned}$$

and

$$H(Y|X_2^{''}) = P(X_2^{''}=0)H(Y|X_2^{''}=0) + P(X_2^{''}=1)\underbrace{H(Y|X_2^{''}=1)}_{0}$$
$$= \frac{(9,995+3)}{10^4}\left\{-\frac{3}{9,998}\log_2\left(\frac{3}{9,998}\right) - \frac{9,995}{9,998}\log_2\left(\frac{9,995}{9,998}\right)\right\}$$
$$= 0.0004.$$

and $H(Y|X_2^{'''})$

$$H(Y|X_2^{'''}) = P(X_2^{'''}=0)H(Y|X_2^{'''}=0) + P(X_2^{'''}=1)\underbrace{H(Y|X_2^{'''}=1)}_{0}$$
$$= \frac{(9,995+3)}{10^4}\left\{-\frac{3}{9,998}\log_2\left(\frac{3}{9,998}\right) - \frac{9,995}{9,998}\log_2\left(\frac{9,995}{9,998}\right)\right\}$$
$$= 0.0004$$

The latter results evidence that $H(Y|X_2)$, for all versions of the X_2 test, will be always lower than $H(Y|X_1)$ because $H(Y|X_1)$ is dominated by the case $X_1 = 1$ because in this case the fraction of examples of class b is reduced to $1/2$ and also all patterns with class a are considered. This increases the entropy approaching $H(Y)$. However, when considering X_2, the dominating option is $X_2 = 0$ and in case almost all patterns considered are of class b except few patterns of class a which reduces the entropy.

Therefore, in the classical local approach, $X_2^{'}$ will be chosen as the test for the root node. After such a selection we have in $\mathcal{X} \sim \mathcal{X}_2^{'}$ almost all of them of class b but two of class a, whereas in $\mathcal{X}_2^{'}$ there are three examples of class a. This means that the child of the root node for $X_2^{'} = 0$ should be analyzed more in depth whereas child for $X_2^{'} = 1$ results in a zero entropy and does not require further analysis (labeled with class a). In these conditions, what is the best following test to refine the root? Let us reevaluate the chance of X_1:

$$\begin{aligned} H(Y|X_2^{'}=0,X_1) &= P(X_2^{'}=0,X_1=0)\overbrace{H(Y|X_2^{'}=0,X_1=0)}^{0} \\ &\quad + P(X_2^{'}=0,X_1=1)H(Y|X_2^{'}=0,X_1=1) \\ &= \frac{(5,000+2)}{10^4}H(Y|X_2^{'}=0,X_1=1) \\ &= \frac{(5,000+2)}{10^4}\left\{-\frac{2}{5,002}\log_2\left(\frac{2}{5,002}\right)\right. \\ &\quad \left. - \frac{5,000}{5,002}\log_2\left(\frac{5,000}{5,002}\right)\right\} = 0.0003 \end{aligned}$$

and consider also $X_2^{''}$:

$$\begin{aligned}
H(Y|X_2^{'}=0,X_2^{''}) &= P(X_2^{'}=0,X_2^{''}=0)H(Y|X_2^{'}=0,X_2^{''}=0)\\
&\quad +P(X_2^{'}=0,X_2^{''}=1)\underbrace{H(Y|X_2^{'}=0,X_2^{''}=1)}_{0}\\
&= \frac{(9,995+1)}{10^4}H(Y|X_2^{'}=0,X_2^{''}=0)\\
&= \frac{(9,995+1)}{10^4}\left\{-\frac{1}{9,996}\log_2\left(\frac{1}{9,996}\right)\right.\\
&\quad \left.-\frac{9,995}{9,996}\log_2\left(\frac{9,995}{9,996}\right)\right\} = 0.0002
\end{aligned}$$

and $X_2^{'''}$:

$$\begin{aligned}
H(Y|X_2^{'}=0,X_2^{'''}) &= P(X_2^{'}=0,X_2^{'''}=0)H(Y|X_2^{'}=0,X_2^{'''}=0)\\
&\quad +\underbrace{P(X_2^{'}=0,X_2^{'''}=1)}_{0}H(Y|X_2^{'}=0,X_2^{'''}=1)\\
&= \frac{(9,995+2)}{10^4}H(Y|X_2^{'}=0,X_2^{'''}=0)\\
&= \frac{(9,995+2)}{10^4}\left\{-\frac{2}{9,997}\log_2\left(\frac{2}{9,997}\right)\right.\\
&\quad \left.-\frac{9,995}{9,997}\log_2\left(\frac{9,995}{9,997}\right)\right\} = 0.0003
\end{aligned}$$

Again, X_1 is discarded. What happens now is that the child for $X_2^{''}=1$ has an example of class a and it is declared a leaf. On the other hand, the branch for $X_2^{''}=0$ has associated many examples of class b and only one of class a, and further analysis is needed. Should this be the time for X_1? Again

$$\begin{aligned}
H(Y|X_2^{'}=0,X_2^{''}=0,X_1) &= P(X_2^{'}=0,X_2^{''}=0,X_1=0)\\
&\quad \overbrace{H(Y|X_2^{'}=0,X_2^{''}=0,X_1=0)}^{0}\\
&\quad +P(X_2^{'}=0,X_2^{''}=0,X_1=1)\\
&\quad H(Y|X_2^{'}=0,X_2^{''}=0,X_1=1)\\
&= \frac{(9,995+1)}{10^4}H(Y|X_2^{'}=0,X_2^{''}=0,X_1=1)\\
&= \frac{(9,995+1)}{10^4}\left\{-\frac{1}{9,996}\log_2\left(\frac{1}{9,996}\right)\right.\\
&\quad \left.-\frac{9,995}{9,996}\log_2\left(\frac{9,995}{9,996}\right)\right\} = 0.0002
\end{aligned}$$

However, if we consider $X_2^{'''}$ what we obtain is

$$
\begin{aligned}
H(Y|X_2^{'}=0,X_2^{''}=0,X_2^{'''}) = P(X_2^{'}=0,X_2^{''}=0,X_2^{'''}=0)&\overbrace{H(Y|X_2^{'}=0,X_2^{''}=0,X_2^{'''}=0)}^{0}\\
&+\frac{(9,995+1)}{10^4}H(Y|X_2^{'}=0,X_2^{''}=0,X_2^{'''}=1)\\
&=\frac{(9,995+1)}{10^4}\left\{-\frac{1}{9,996}\log_2\left(\frac{1}{9,996}\right)\right.\\
&\left.-\frac{9,995}{9,996}\log_2\left(\frac{9,995}{9,996}\right)\right\}=0.0002
\end{aligned}
$$

Then we may select $X_2^{'''}$ and X_1. This means that with a lower value of $p_0(a)$, X_1 would not be selected with high probability. After selecting, for instance $X_2^{'''}$, we may have that the child for $X_2^{'''}=0$ is always of class b and the child for $X_2^{'''}=1$ is a unique example of class a. Consequently, all leaves are reached without selecting test X_1. Given the latter greedy tree, let us call it $\mathcal{T}_{local}=(X_2^{'},X_2^{''},X_2^{'''})$ attending to its levels, we have that the rate of misclassification is 0 when the true class $Y=b$ although to discover it we should reach the deepest leaf. When we test the tree with a examples not contemplated in the training we have that only one of them is misclassified (the misclassification error when $Y=a$ is $\frac{1}{8}$ (12.5%)). In order to compute the mean depth of the tree we must calculate the probabilities of reaching each of the leaves. Following an inorder traversal we have the following indexes for the four leaves $l=3,5,6,7$:

$$
\begin{aligned}
P(Q_3) &= P(X_2^{'}=1)=3\times 10^{-4}\\
P(Q_5) &= P(X_2^{'}=0,X_2^{''}=1)=10^{-4}\\
P(Q_6) &= P(X_2^{'}=0,X_2^{''}=0,X_2^{'''}=0)=9,995\times 10^{-4}\\
P(Q_7) &= P(X_2^{'}=0,X_2^{''}=0,X_2^{'''}=1)=10^{-4}
\end{aligned}
$$

and then

$$
\begin{aligned}
Ed(\mathcal{T}_{local}) &= \sum_{l\in\partial\mathcal{T}_{local}} P(Q_l)d(l)\\
&= P(Q_3)d(3)+P(Q_5)d(5)+P(Q_6)d(6)+P(Q_7)d(7)\\
&= 3\times 10^{-4}\times 1+10^{-4}\times 2+9,995\times 10^{-4}\times 3+10^{-4}\times 3\\
&= 2.9993
\end{aligned}
$$

which is highly conditioned by $P(Q_6)d(6)$, that is, it is very probable to reach that leaf, the unique labeled with b. In order to reduce such average depth it should be desirable to put b leaves at the lowest depth as possible, but this implies changing the relative order between the tests. The fundamental question is that in $\mathcal{T}_{local}$ uneffective tests (features) are chosen so systematically by the greedy method, and tests which work perfectly, but for a rare class are relegated or even not selected.

On the other hand, the evaluation $H(Y|\mathcal{T}_{local}) = 0$ results from a typical tree where all leaves have null entropy. Therefore the cost $C(\mathcal{T}_{local}, \mathcal{M}) = \lambda Ed(\mathcal{T}_{local})$, and if we set $\lambda = 10^{-4}$ we have a cost of 0.0003.

7.2.4 Rare Classes with Global Optimization

In order to perform global optimization following the latter example, we are going to assume that we have roughly two classes of tests $\mathbf{X} = \{X_1, X_2\}$ with similar probability distributions, say $X_1 = \{X_1^{'}, X_1^{''}, X_1^{'''}\}$ and $X_2 = \{X_2^{'}, X_2^{''}, X_2^{'''}\}$. In computer vision arena, each of these versions could be understood as filters $\phi()$ with similar statistical behavior. In order to make the problem tractable it seems reasonable to consider the context of building trees with instances of X_1 and X_2, and the results of the test for 10^4 data are indicated in Table 7.2. In this latter case, the probabilities of each type of filter are:

$$
\begin{aligned}
P(X_1 = 0) &= 4,995 \times 10^{-4} = 0.4995 \\
P(X_1 = 1) &= 5,005 \times 10^{-4} = 0.5005 \\
P(X_2 = 0) &= 9,997 \times 10^{-4} = 0.9997 \\
P(X_2 = 1) &= 3 \times 10^{-4} \qquad = 0.0003
\end{aligned}
$$

The latter probabilities are needed for computing $C^*(\mathcal{M}, D)$ in Eq. 7.9, and, of course, its associated tree. Setting for instance $D = 3$, we should compute $C^*(\mathcal{M}, D)$, being $\mathcal{M} = p_0$ because the model is mainly conditioned by our knowledge of such prior. However, for doing that we need to compute $C^*(\mathcal{M}_1, D-1)$ where $\mathcal{M}_1 = \{\mathcal{M}|X_t = k\}$ for $t = 1 \ldots N$ (in this $N = 2$) and $k = 0, 1$ is the set of distributions:

$$
\mathcal{M}_1 = \{p(\cdot|X_t = k), \quad 1 \le t \le N, \quad k \in \{0,1\}\}
$$

and in turn compute $C^*(\mathcal{M}_2, D-2)$ being

$$
\mathcal{M}_1 = \{p(\cdot|X_t = k, X_r = l), \quad 1 \le t, r \le N, \quad k, l \in \{0,1\}\}
$$

and finally $C^*(\mathcal{M}_3, D-3) = C^*(\mathcal{M}_3, 0)$ from

$$
\mathcal{M}_3 = \{p(\cdot|X_t = k, X_r = l, X_s = m), \quad 1 \le t, r, s \le N, \quad k, l, m \in \{0,1\}\}
$$

The latter equations are consistent with the key idea that the evolution of $(\mathbf{X}, Y)$ depends only on the evolution of the posterior distribution as the tests are performed. Assuming conditional independence between the tests (X_1 and

Table 7.2. Complete set of tests.

Example	$X_1^{'}$	$X_1^{''}$	$X_1^{'''}$	$X_2^{'}$	$X_2^{''}$	$X_2^{'''}$	Class
$\mathbf{x}_1$	1	1	1	0	0	0	b
...	...	...	...	...	...	...	...
$\mathbf{x}_{1,250}$	1	1	1	0	0	0	b
$\mathbf{x}_{1,251}$	1	1	0	0	0	0	b
...	...	...	...	...	...	...	...
$\mathbf{x}_{2,499}$	1	1	0	0	0	0	b
$\mathbf{x}_{2,500}$	1	0	0	0	0	0	b
...	...	...	...	...	...	...	...
$\mathbf{x}_{3,750}$	1	0	0	0	0	0	b
$\mathbf{x}_{3,751}$	1	0	1	0	0	0	b
...	...	...	...	...	...	...	...
$\mathbf{x}_{5,000}$	1	0	1	0	0	0	b
$\mathbf{x}_{5,001}$	0	1	1	0	0	0	b
...	...	...	...	...	...	...	...
$\mathbf{x}_{6,250}$	0	1	1	0	0	0	b
$\mathbf{x}_{6,251}$	0	1	0	0	0	0	b
...	...	...	...	...	...	...	...
$\mathbf{x}_{7,500}$	0	1	0	0	0	0	b
$\mathbf{x}_{7,501}$	0	0	0	0	0	0	b
...	...	...	...	...	...	...	...
$\mathbf{x}_{8,750}$	0	0	0	0	0	0	b
$\mathbf{x}_{8,751}$	0	0	1	0	0	0	b
...	...	...	...	...	...	...	...
$\mathbf{x}_{9,995}$	0	0	1	0	0	0	b
$\mathbf{x}_{9,996}$	1	1	1	0	0	1	a
$\mathbf{x}_{9,997}$	1	1	1	0	1	0	a
$\mathbf{x}_{9,998}$	1	1	1	1	0	0	a
$\mathbf{x}_{9,999}$	1	1	1	1	0	1	a
$\mathbf{x}_{10,000}$	1	1	1	1	1	0	a
Test	$X_1^{'}$	$X_1^{''}$	$X_1^{'''}$	$X_2^{'}$	$X_2^{''}$	$X_2^{'''}$	**Class**
$\mathbf{x}_1^{'}$	1	1	1	0	0	0	a
$\mathbf{x}_2^{'}$	1	1	1	0	1	1	a
$\mathbf{x}_3^{'}$	1	1	1	1	1	1	a

X_2 in this case) it is possible to select the best test at each level and this will be done by a sequence of functions:

$$\Psi_d : \mathcal{M}_d \rightarrow \{1 \dots N\}$$

In addition, from the latter equations, it is obvious that we have only two tests in our example: X_1 and X_2, and consequently we should allow to use them *repeatedly*, that is, allow $p(\cdot|X_1 = 0, X_1 = 1)$ and so on. This can be achieved in practice by having *versions* or the tests with similar statistical behavior as we argued above. Thus, the real thing for taking examples and computing the posteriors (not really them by their entropy) will be the underlying assumption of $p(\cdot|X_1^{'} = 0, X_1^{''} = 1)$, that is, the first time X_1 is named we use its first version, the second time we use the second one, and so on. However, although we use this trick, the algorithm needs to compute $\sum_{0=1}^{D} |\mathcal{M}_d|$ which means, in our case, $\sum_{0=1}^{D}(2 \times N)^d = 1 + 4 + 16 + 64$ posteriors. However, if we do not take into account the order in which the tests are performed along a branch the complexity is

$$|\mathcal{M}_d| = \binom{d + 2M - 1}{2M - 1}$$

which reduces the latter numbers to $1 + 4 + 10 + 20$ (see Fig. 7.1, top) and makes the problem more tractable.

Taking into account the latter considerations it is possible to elicit a dynamic programming like solution following a *bottom-up* path, that is, starting from computing the entropies of all posteriors after $D = 3$. The top-down collection of values states that the tests selected at the first and second levels are the same X_2 (Fig. 7.1, bottom). However, due to ambiguity, it is possible to take X_2 as optimal test for the third level which yields $\mathcal{T}_{local} \equiv \mathcal{T}_{global_1}$

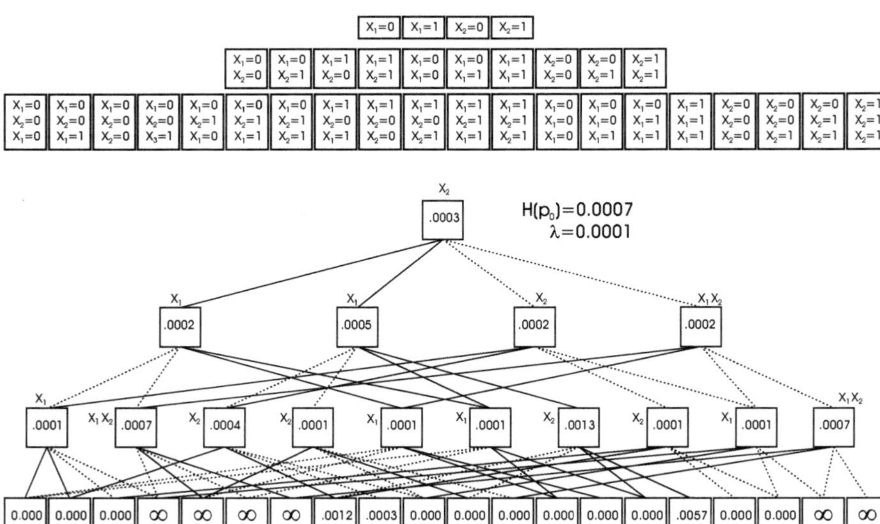

Fig. 7.1. Bottom-up method for computing the tree with minimal cost. *Top:* cells needed to be filled during the dynamic programming process. *Bottom:* bottom-up process indicating the cost cells and the provisional winner test at each level. *Dashed lines* correspond to information coming from X_1 and *solid lines* to information coming from X_2.

having exactly the optimal cost. However, it seems that it is also possible to take X_1 for the third level and obtaining for instance $\mathcal{T}_{global_2} = (X_2^{'}, X_2^{''}, X_1^{'})$. The probabilities of the four leaves are:

$$\begin{aligned}
P(Q_3) &= P(X_2^{'} = 1) = 3 \times 10^{-4},\\
P(Q_5) &= P(X_2^{'} = 0, X_2^{''} = 1) = 10^{-4},\\
P(Q_6) &= P(X_2^{'} = 0, X_2^{''} = 0, X_1^{'} = 0) = 4,995 \times 10^{-4},\\
P(Q_7) &= P(X_2^{'} = 0, X_2^{''} = 0, X_1^{'} = 1) = 5,001 \times 10^{-4},
\end{aligned}$$

and the mean depth is

$$\begin{aligned}
Ed(\mathcal{T}_{global_2}) &= \sum_{l \in \partial \mathcal{T}_{global_2}} P(Q_l)d(l)\\
&= P(Q_3)d(3) + P(Q_5)d(5) + P(Q_6)d(6) + P(Q_7)d(7)\\
&= 3 \times 10^{-4} \times 1 + 10^{-4} \times 2 + 4,995 \times 10^{-4} \times 3\\
&\quad + 5{,}001 \times 10^{-4} \times 3\\
&= 2.9993
\end{aligned}$$

and

$$\begin{aligned}
H(Y|\mathcal{T}_{global_2}) &= \sum_{l \in \partial \mathcal{T}_{global_2}} P(Q_l)d(l)\\
&= P(Q_3)d(3) + P(Q_5)d(5) + P(Q_6)d(6) + P(Q_7)d(7)\\
&= 0 + 0 + 0 + H(Y|Q_7)\\
&= 0 + 0 + 0 + \frac{5,000}{5,0001} \log_2 \left(\frac{5,000}{5,001}\right)\\
&= 0.0003
\end{aligned}$$

Let us evaluate the *global cost* $C(\mathcal{T}, \mathcal{M})$ of the obtained tree

$$\begin{aligned}
C(\mathcal{T}_{global_2}, \mathcal{M}) &= H(Y|\mathcal{T}_{global_2}) + \lambda Ed(\mathcal{T}_{global_2})\\
&= 0.0003 + 0.0001 \times 2.9993\\
&= 0.0005999 > \underbrace{C^*(\mathcal{M}, D)}_{0.0003}
\end{aligned}$$

which indicates that this is a suboptimal choice, having, at least, the same misclassification error than $\mathcal{T}_{local}$. However, in deeper trees what is typically observed is that the global method improves the misclassification rate of the local one and produces more balanced trees including X_1-type nodes (Fig. 7.2).

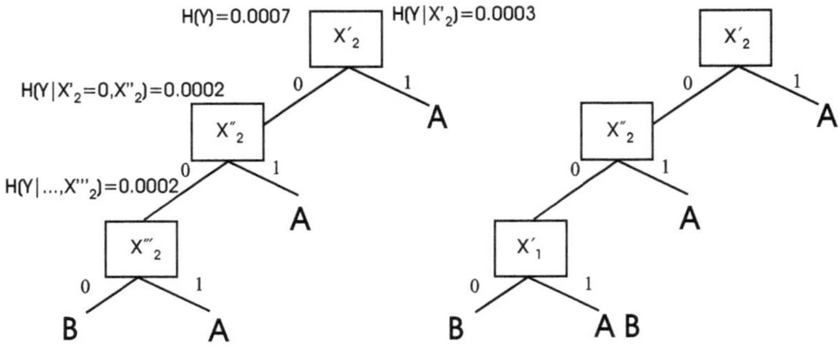

Fig. 7.2. Classification trees. *Left:* using the classical greedy approach. Ambiguity: nonoptimal tree with greater cost than the optimal one.

7.3 Shape Quantization and Multiple Randomized Trees

7.3.1 Simple Tags and Their Arrangements

Classical decision trees, reviewed and optimized in the latter section, are designed to classify vectorial data $\mathbf{x} = (x_1, x_2, \ldots, x_N)$. Thus, when one wants to classify images (for instance bitmaps of written characters) with these methods, it is important to extract significant features, and even reduce N as possible (see Chapter 5). Alternatively it is possible to build trees where the tests X_t operate over small windows and yield 0 or 1 depending on whether the window corresponds to a *special configuration*, called *tag*. Consider for instance the case of binary bitmaps. Let us, for instance, consider 16 examples of the 4 simple arithmetic symbols: $+$, $-$, $\div$, and $\times$. All of them are 7×7 binary bitmaps (see Fig. 7.3). Considering the low resolution of the examples we may retain all the 16 tags of dimension 2×2 although the tag corresponding to number 0 should be deleted because it is the least informative (we will require that at least one of the pixels is 1 (black in the figure)).

In the process of building a decision tree exploiting the tags, we may assign a test to each tag, that is $X_1, \ldots, X_{15}$ which answer 1 when the associated tag is present in the bitmap and 0 otherwise. However, for the sake of invariance and higher discriminative power, it is better to associate the tests to the satisfaction of "binary" spatial relationships between tags, although the complexity of the learning process increases significantly with the number of tags. Consider for instance four types of binary relations: "north," "south," "west" and "east." Then, $X_{5\uparrow 13}$ will test whether tag 5 is north of tag 13, whereas $X_{3\rightarrow 8}$ means that tag 5 is at west of tag 8. When analyzing the latter relationships, "north," for instance, means that the second row of the tag must be greater than the first row of the second tag.

Analyzing only the four first initial (ideal) bitmaps we have extracted 38 *canonic* binary relationships (see Prob. 7.2). Canonic means that many relations are equivalent (e.g. $X_{1\uparrow 3}$ is the same as $X_{1\uparrow 12}$). In our example the

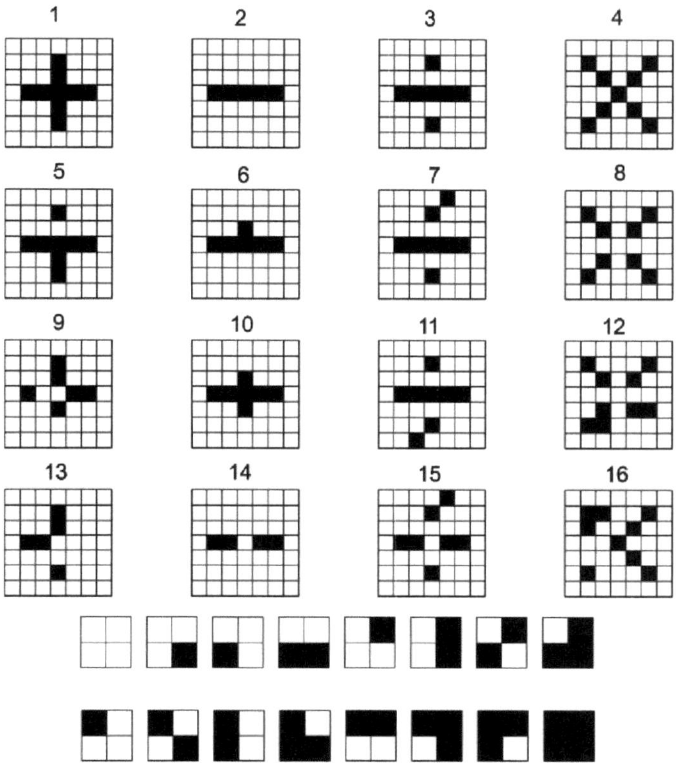

Fig. 7.3. *Top:* Sixteen example bitmaps. They are numbered 1 to 16 from top-left to bottom-right. Each column has examples of a different class. Bitmaps 1 to 4 represent the ideal prototypes for each of the four classes. *Bottom:* The 16 tags coding the binary numbers from 0000 to 1111. In all these images 0 is colored white and 1 is black.

initial entropy is $H(Y) = -4\left(\frac{1}{4}\right)\log_2\left(\frac{1}{4}\right) = 2$ as there are four classes and four examples per class.

7.3.2 Algorithm for the Simple Tree

Given $\mathbf{B}$, the set of canonic binary relations, called *binary arrangements*, the basic process for finding a decision tree consistent with the latter training set, could be initiated by finding the relation minimizing the conditional entropy. The tree is shown in Fig. 7.4 and $X_{3\uparrow 8}$ is the best local choice because it allows to discriminate between two superclasses: $(-, \times)$ and $(+, \div)$ yielding the minimal relative entropy $H(Y|X_{3\uparrow 8}) = 1.0$. This choice is almost obvious because when $X_{3\uparrow 8} = 0$ class $-$ without "north" relations is grouped with class $\times$ without the tag 3. In this latter case, it is then easy to discriminate between classes $-$ and $\times$ using $X_{3\to 3}$ which answers "no" in $\times$. Thus, the leftmost path of the tree (in general: "no","no",...) is built obeying to the

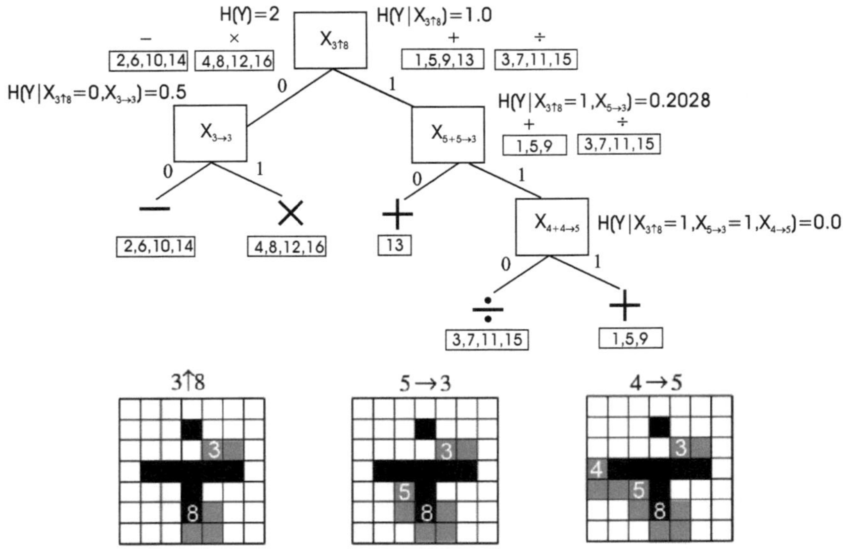

Fig. 7.4. *Top:* tree inferred exploiting binary arrangements **B** and the minimal extensions $\mathbf{A}_t$ found by the classical greedy algorithm. *Bottom:* minimal extensions selected by the algorithm for example 5. Gray masks indicate the tag and the number of tag is in the anchoring pixel (*upper-left*).

rule to find the binary arrangement which is satisfied by almost one example and best reduces the conditional entropy. So, the code of this branch will have the prefix 00.... But, what happens when the prefix 00...1 appears?

The rightmost branch of the tree illustrates the opposite case, the prefix is 11.... Once a binary arrangement has been satisfied, it proceeds to *complete* it by adding a *minimal extension*, that is, a new relation between existing tags or by adding a new tag and a relation between this tag and one of the existing ones. We denote by $\mathbf{A}_t$ the set of minimal extensions for node t in the tree. In this case, tags are anchored to the coordinates of the upper-leftmost pixel (the *anchoring pixel*) and overlapping between new tags (of course of different types) is allowed. Once a 1 appears in the prefix, we must find the *pending arrangement*, that is, the one minimizing the conditional entropy. For instance, our first *pending arrangement* is $X_{5+5\to 3}$, that is, we add tag 5 and the relation $5 \to 3$ yielding a conditional entropy of $H(Y|X_{3\uparrow 8}=1, X_{5\to 3}) = 0.2028$. After that, it is easy to discriminate between + and ÷ with the pending arrangement $X_{4+4\to 5}$ yielding a 0.0 conditional entropy. This result is close to the ideal Twenty Questions (TQ): The mean number of queries EQ to determine the true class is the expected length of the codes associated to the terminal leaves: $C_1 = 00$, $C_2 = 01$, $C_3 = 10$, $C_4 = 110$, and $C_5 = 111$:

$$EQ = \sum_{l \in \mathcal{T}} P(C_l)L(C_l) = \frac{1}{4} \times 2 + \frac{1}{4} \times 2 + \frac{1}{16} \times 1 + \frac{1}{4} \times 3 + \frac{3}{16} \times 3 = 2.3750$$

that is $H(Y) \leq EQ < H(Y) + 1$, being $H(Y) = 2$. This is derived from the Huffman code which determines the optimal sequence of questions to classify an object. This becomes clear from observing that in the tree, each test divides (not always) the masses of examples in subsets with almost equal size.

7.3.3 More Complex Tags and Arrangements

In real experiments for handwritten character recognition, the example images have larger resolutions (for instance 70×70) and the size of the, usually squared, window should be increased in order to capture relevant details of the shapes. Even for a $N = 4$ height, $2^{N \times N} = 65,536$ masks should be considered. This implies the need of discovering clusters of masks. We may take a large sample $\mathbf{U}$ of 4×4 windows from the training bitmaps. Then, we build a binary classification tree $\mathbf{U}$, where *the tags will be the nodes of the tree*. If the window size is 4×4, we may perform 16 binary questions of the type *"Is pixel (i,j)=1?"*. Of course, the root node contains no questions (this is indicated by a gray tone in Fig. 7.4 (top)). In order to select the following question (pixel) at node t, it must divide the subpopulation of samples $\mathbf{U}_t$ into two groups with similar mass as possible. This is the underlying principle of Twenty Questions and it is addressed to reduce the entropy of the empirical distribution of configurations: the Huffman code is the one with minimal expected length and this means that following this approach, the upper bound of the entropy will be minimal and also the entropy itself.

The first level has two tags and their children inherit the questions of the parent before selecting a new one. The second level has four tags and so on. The number of questions in a tag corresponds to the level (root is level zero). And for five questions we have $2 + 4 + 8 + 16 + 32 = 62$ tags. Furthermore, taking into account also the fact that eight binary relations must be considered (including "north-west", and so on). As we have seen above, exploiting a sequence of binary arrangements (let us denote such sequence as $\mathbf{A} = \{2 \searrow, 2 \nearrow 7\}$) has an interesting impact in reducing the entropy of the conditional distribution $P(Y = y|X_{\mathbf{A}} = 1)$ (see Fig. 7.5). Higher order relations may be considered. All of these relations may be metric ones, provided that scale translational invariances are preserved. For instance, an example of ternary relations between three anchoring pixels x, y and z is $||x - y|| < ||x - z||$, and if we add a fourth point t we have the quaternary relation $||x - y|| < ||z - t||$. As the number of tags, relationships, and their order increase, it is more and more probable to reach leaves of zero conditional entropy, but the shape class is correctly determined. However, the complexity of the learning process is significantly increased, and this is why in practice the number of tags and the number of relations are bounded (for instance 20 tags and 20 relations). Furthermore, it is more practical to complete minimal arrangements (for instance when binary relationships are considered) without trying to yield a connected graph, that is, with a set of binary graphs.

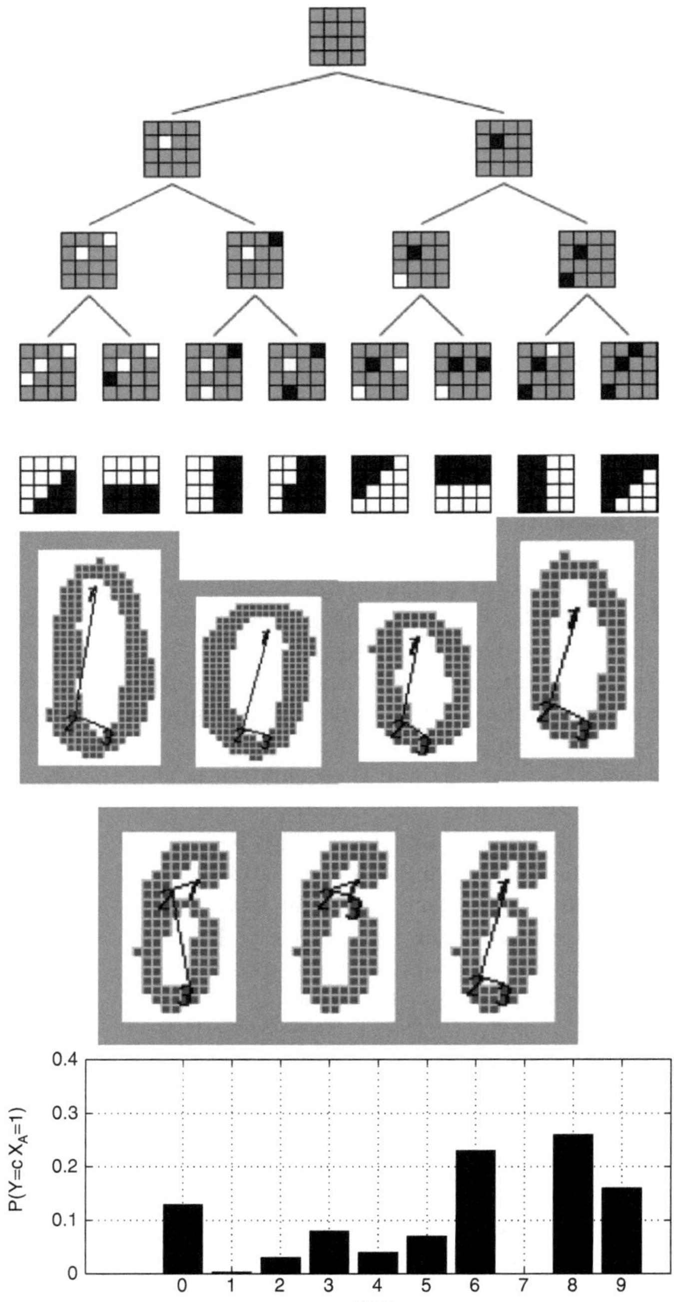

Fig. 7.5. *Top*: first three tag levels and the most common configuration below each leaf. *Center*: instances of geometric arrangements of 0 and 6. *Bottom*: conditional distribution for the 10 digits and the arrangement satisfied by the shapes in the middle. Top and Center figures by Y. Amit and D. Geman (©1997 MIT Press).

7.3.4 Randomizing and Multiple Trees

Let us then suppose that we have a set $\mathbf{X}$ of M possible tests and also that these tests are not too complex to be viable in practice. The purpose of a classification tree $\mathcal{T}$ is to minimize $H(Y|\mathbf{X})$, which is equivalent to maximizing $P(Y|\mathbf{X})$. However, during the construction of $\mathcal{T}$ not all tests are selected, we maximize instead $P(Y|\mathcal{T})$ which is only a good approximation of maximizing $P(Y|\mathbf{X})$ when M, and hence the depth of the tree, is large.

Instead of learning a single deep tree, suppose that we replace it by a collection of K shallow trees $\mathcal{T}_1, \mathcal{T}_2, \ldots, \mathcal{T}_K$. Then we will have to estimate

$$P(Y = c|\mathcal{T}_k) \;\; k = 1, \ldots, K.$$

Considering that the number M of possible tests $\mathbf{X}$ is huge, and considering also that a random selection from the set $\mathbf{B}$ of binary tests is likely to increase the information about the Y random variable[1] it seems reasonable to select a random subset of $\mathbf{B}$ for each node for growing a particular tree instead of reviewing "all" the possible arrangements. Besides reducing the complexity of the process, randomization has three additional good effects. Firstly, it allows to characterize the same shape from different points of views (one per tree) as we show in Fig. 7.6 (top). Secondly, it reduces the statistical dependency between different trees. And thirdly, it introduces more robustness during classification.

The latter second and third advantages are closely related and emerge from the way that the different trees are combined. These trees may be aggregated as follows. Suppose that we have $|\mathcal{Y}| = C$ classes. Thus, the leaves of each tree will have only values $c = 1, \ldots, C$. However, as we bound the depth of each tree, say to D, it is expected to have some high entropic leaves. Generally speaking we are interested in estimating $P(Y = c|\mathcal{T}_k = l)$, where $l \in \partial\mathcal{T}_k$ is the posterior probability that the correct class is c at leaf l. So we must estimate $C \times 2^D$ parameters per tree ($K \times C \times 2^D$ contemplating all trees). This means that we will have $|\mathcal{X}|/(C2^D)$ examples available per parameter and consequently large training sets are needed. However, this need may be attenuated considering that most of the parameters will be zero and estimating only, for instance, the five largest elements (bins) of a given posterior.

Once all posteriors are estimated, and following the notation

$$\mu_{\mathcal{T}_k}(c) = P(Y = c|\mathcal{T}_k = l)$$

we have that $\mu_{\mathcal{T}_k}(\mathbf{x})$ denotes the posterior for the leaf reached by a given input $\mathbf{x}$. *Aggregation by averaging* consists of computing:

$$\bar{\mu}(\mathbf{x}) = \frac{1}{K}\sum_{k=1}^{K} \mu_{\mathcal{T}_k}(\mathbf{x}) \tag{7.10}$$

[1] In the example in Fig. 7.4 we have first selected a binary arrangement.

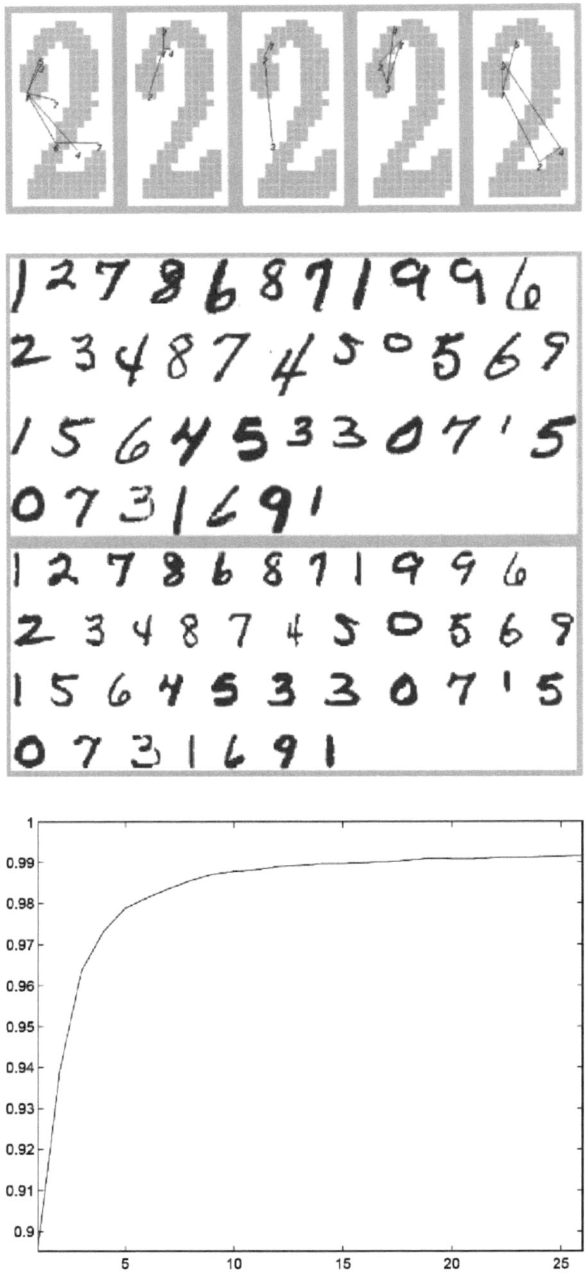

Fig. 7.6. Randomization and multiple trees. *Top:* inferred graphs in the leaves of five differently grown trees. *Center:* examples of handwritten digit images, before (*up*) and after (*down*) preprocessing. *Bottom:* conditional distribution for the ten digits and the arrangement satisfied by the shapes in the middle. Figure by Y. Amit and D. Geman (©1997 MIT Press).

Finally, the class assigned to a given input $\mathbf{x}$ is the mode of the averaged distribution:

$$\hat{Y} = \arg\max_{c} \bar{\mu}_c \tag{7.11}$$

This approach has been successfully applied to the OCR domain . Considering the NIST database with 223, 000 binary digits written by more than two thousand writers, and using 100, 000 for training and 50, 000 for testing (see Fig. 7.6, center) the classification rate significantly increases with the number of trees (see Fig. 7.6, bottom). In these experiments the average depth of the trees was 8.8 and the average number of terminal nodes was 600.

7.4 Random Forests

7.4.1 The Basic Concept

As we have seen in the latter section, the combination of multiple trees increases significantly the classification rate. The formal model, developed years later by Breiman [30], is termed "Random Forests" (RFs). Given a labeled training set $(\mathcal{X}, \mathcal{Y})$, an RF is a collection of tree classifiers $\mathcal{F} = \{h_k(\mathbf{x}),\ k = 1, \ldots, K\}$. Probably the best way of understanding RFs is to sketch the way they are usually built.

First of all, we must consider the dimension of the training set $|\mathcal{X}|$ and the maximum number of variables (dimensions) of their elements (examples), which is N. For each tree. Then, the kth tree will be built as follows: (i) Select randomly and with replacement $|\mathcal{X}|$ samples from $\mathcal{X}$, a procedure usually denoted *bootstrapping*[2]; this is its training set $\mathcal{X}_k$. (ii) Each tree is grown by selecting randomly at each node $n << N$ variables (tests) and finding the best split with them. (iii) Let each tree grow as usual, that is, without pruning (reduce the conditional entropy as much as possible). After building the K forest, an $\mathbf{x}$ input is classified by determining the most voted class among all the trees (the class representing the majority of individual choices), namely

$$h_{\mathcal{F}}(\mathbf{x}) = MAJ\{h_1(\mathbf{x}), h_2(\mathbf{x}), \ldots, h_K(\mathbf{x})\}$$

a procedure usually denoted as *bagging*.

7.4.2 The Generalization Error of the RF Ensemble

The analysis of multiple randomized trees (Section 7.3.4) reveals that from a potentially large set of features (and their relationships) building multiple trees from different sets of features results in a significant increment of the recognition rate. However, beyond the intuition that many trees improve

[2] Strictly speaking, bootstrapping is the complete procedure of extracting several training sets by sampling with replacement.

the performance, one must measure or bound, if possible, the error rate of the forest, and check in which conditions it is low enough. Such error may be quantified by the *margin*: the difference between the probability of predicting the correct class and the maximum probability of predicting the wrong class. If $(\mathbf{x}, y)$ (an example of the training and its correct class) is a sample of the random vector $\mathcal{X}, \mathcal{Y}$ which represents the corresponding training set, the margin is formally defined as

$$mar(\mathcal{X}, \mathcal{Y}) = P(h_{\mathcal{F}}(\mathbf{x}) = y) - \max_{z \neq y} P(h_{\mathcal{F}}(\mathbf{x}) = z) \tag{7.12}$$

The margin will be in the range $[-1, 1]$ and it is obvious that the largest the margin the more confidence in the classification. Therefore, the *generalization error* of the forest is

$$GE = P_{\mathcal{X},\mathcal{Y}}(mar(\mathcal{X}, \mathcal{Y}) < 0)\ . \tag{7.13}$$

that is, the probability of having a negative margin over the distribution $\mathcal{X}, \mathcal{Y}$ of training sets. Therefore, for the sake of the classification performance, it is interesting that the GE, the latter probability, has a higher bound as low as possible. In the case of RFs such a bound is related to the correlation between the trees in the forest. Let us for instance define the *strength of the forest* as

$$s = E_{\mathcal{X},\mathcal{Y}} mar(\mathcal{X}, \mathcal{Y}) \tag{7.14}$$

that is, the expectation of the margin over the distribution $\mathcal{X}, \mathcal{Y}$, and let

$$z^* = \arg\max_{z \neq y} P(h_{\mathcal{F}}(\mathbf{x}) = z) \tag{7.15}$$

Then, we may redefine the margin in the following terms:

$$\begin{aligned} mar(\mathcal{X}, \mathcal{Y}) &= P(h_{\mathcal{F}}(\mathbf{x}) = y) - P(h_{\mathcal{F}}(\mathbf{x}) = z^*) \\ &= E_{\Theta}\left[\underbrace{I(h_{\mathcal{F}}(\mathbf{x}) = y) - I(h_{\mathcal{F}}(\mathbf{x}) = z^*)}_{rmar(\mathcal{X},\mathcal{Y},\Theta)}\right] \\ &= E_{\Theta}\left[rmar(\mathcal{X}, \mathcal{Y}, \Theta)\right] \end{aligned} \tag{7.16}$$

$I(\cdot)$ being an indicator (counting) function, which returns the number of times the argument is satisfied, $rmar(.,.,.)$ the so-called *raw margin*, and $\Theta = \{\Theta_k\}$ a bag of parameters sets (one set per tree). For instance, in the case of bagging we have $\Theta_k = \mathcal{X}_k$. Consequently, the margin is the expectation of the raw margin with respect to Θ, that is, with respect to all the possible ways of bootstraping the training set.

Let us now consider two different parameter sets Θ and Θ'. Assuming i.i.d., the property $(E_{\Theta} f(\Theta))^2 = E_{\Theta} f(\Theta) \times E_{\Theta'} f(\Theta')$ is satisfied. Consequently, and applying Eq. 7.16, it is verified that

$$mar(\mathcal{X},\mathcal{Y})^2 = E_{\Theta,\Theta'}\left[rmar(\mathcal{X},\mathcal{Y},\Theta) \times rmar(\mathcal{X},\mathcal{Y},\Theta')\right] \tag{7.17}$$

and consequently, defining

$$\begin{aligned} v &= Var_{\mathcal{X},\mathcal{Y}}(mar(\mathcal{X},\mathcal{Y})) \\ &= E_{\mathcal{X},\mathcal{Y}}[mar(\mathcal{X},\mathcal{Y})^2] - s^2 \\ &= E_{\mathcal{X},\mathcal{Y}}[E_{\Theta,\Theta'}[rmar(\mathcal{X},\mathcal{Y},\Theta) \times rmar(\mathcal{X},\mathcal{Y},\Theta')]]] - s^2 \end{aligned}$$

then, as $s^2 = (E_{\mathcal{X},\mathcal{Y}}[E_{\Theta} rmar(\mathcal{X},\mathcal{Y},\Theta)])^2$, interchanging $E_{\mathcal{X},\mathcal{Y}}[E_{\Theta,\Theta'}][\cdot]$ and $E_{\Theta,\Theta'}[E_{\mathcal{X},\mathcal{Y}}][\cdot]$ and the same holds for $E_{\mathcal{X},\mathcal{Y}}[E_{\Theta}][.]$, and applying the definition of covariance $Cov(A,B) = E(A \times B) - E(A) \times E(B)$ we have that

$$\begin{aligned} v &= E_{\Theta,\Theta'}\{Cov_{\mathcal{X},\mathcal{Y}}\left[rmar(\mathcal{X},\mathcal{Y},\Theta), rmar(\mathcal{X},\mathcal{Y},\Theta')\right]\} \\ &= E_{\Theta,\Theta'}\{\rho(\Theta,\Theta') \times Std(\Theta) \times Std(\Theta')\} \end{aligned} \tag{7.18}$$

$\rho(.,.)$ being the correlation between $rmar(\mathcal{X},\mathcal{Y},\Theta)$ and $rmar(\mathcal{X},\mathcal{Y},\Theta')$ holding both parameter sets fixed and $Std(.)$ being the standard deviation of the corresponding raw margin holding the argument fixed. The latter definition establishes an interesting link between the training set and the bootstrapped sets.

Let us now define $\bar{\rho}(\Theta,\Theta')$ as the *mean value of the correlation*:

$$\bar{\rho}(\Theta,\Theta') = \frac{\overbrace{E_{\Theta,\Theta'}\{\rho(\Theta,\Theta') \times Std(\Theta) \times Std(\Theta')\}}^{v}}{E_{\Theta,\Theta'}[Std(\Theta) \times Std(\Theta')]} \tag{7.19}$$

As Θ and Θ' are independent with the same distribution the following relation holds:

$$\bar{\rho}(\Theta,\Theta') = \frac{v}{(E_{\Theta}[Std(\Theta)])^2} \tag{7.20}$$

which is equivalent to

$$v = \bar{\rho}(\Theta,\Theta') \times (E_{\Theta}[Std(\Theta)])^2 \tag{7.21}$$

And we have

$$v \leq \bar{\rho}(E_{\Theta}[Std(\Theta)])^2 \equiv \bar{\rho} E_{\Theta}[Var(\Theta)] \tag{7.22}$$

Furthermore, as $Var_{\Theta}(Std(\Theta)) = E_{\Theta}[Std(\Theta)^2] - (E_{\Theta}[Std(\Theta)])^2$, we have that

$$v \leq (E_{\Theta}[Std(\Theta)^2] \equiv E_{\Theta}[Var(\Theta)]) \tag{7.23}$$

As $Var(\Theta) = E_{\mathcal{X},\mathcal{Y}}[rmar(\mathcal{X},\mathcal{Y},\Theta)^2] - (E_{\mathcal{X},\mathcal{Y}}[rmar(\mathcal{X},\mathcal{Y},\Theta)])^2$, consequently:

$$\begin{aligned} E_{\Theta}[Var(\Theta)] &= E_{\Theta}[E_{\mathcal{X},\mathcal{Y}}[rmar(\mathcal{X},\mathcal{Y},\Theta)^2]] - E_{\Theta}[(E_{\mathcal{X},\mathcal{Y}}[rmar(\mathcal{X},\mathcal{Y},\Theta)])^2] \\ &\leq 1 - s^2 \end{aligned} \tag{7.24}$$

Thus, we have two interesting connections: (i) an upper bound for the variance of the margin depending on the average correlation; and (ii) another upper bound between the expectation of the variance with respect to Θ, which depends on the strength of the forest. The coupling of the latter connections is given by the *Chebishev inequality*. Such inequality gives an upper bound for rare events in the following sense: $P(|x-\mu| \geq \alpha) \leq \sigma^2/\alpha^2$. In the case of random forests we have the following setting:

$$GE = P_{\mathcal{X},\mathcal{Y}}(mar(\mathcal{X},\mathcal{Y}) < 0) \leq \bar{\rho}\frac{(1-s^2)}{s^2} \tag{7.25}$$

which means that the generalization error is bounded by a function depending on both the correlation and the strength of the set classifiers. High correlation between trees in the forest results in poor generalization and vice versa. Simultaneously, low strength of the set of classifiers results in poor generalization.

7.4.3 Out-of-the-Bag Estimates of the Error Bound

When building the tree $\mathcal{T}_k \in \mathcal{F}$ and selecting $\mathcal{X}_k = \Theta_k \subset \mathcal{X}$ through bootstrapping, we have seen that independent selection has a deep impact on the bounds of the generalization error. Thus, it is interesting to increase the strength and reduce the correlation as much as possible. A useful mechanism for quantifying and controlling these factors is the so called *out-of-the-bag* (oob) estimate of GE. This mechanism starts by discarding 1/3 of the $|\mathcal{X}|$ samples for building each tree (oob samples). Let

$$Q(\mathbf{x},z) = \frac{\sum_{k=1}^{K} I(h_k(\mathbf{x}) = z : (\mathbf{x},y) \notin (\mathcal{X}_k,\mathcal{Y}))}{\sum_{k=1}^{K} I((\mathbf{x},y) \notin (\mathcal{X}_k,\mathcal{Y}))} \tag{7.26}$$

be the oob proportions of votes for a wrong class z. Such votes come from the test set, and this means that y$Q(\mathbf{x},z)$ is formally connected with GE. First of all, it is an estimate for $p(h_{\mathcal{F}}(\mathbf{x}) = z)$. From the definition of the strength in Eq. 7.14, we have

$$s = E_{\mathcal{X},\mathcal{Y}}(P(h_{\mathcal{F}}(\mathbf{x}) = y) - \max_{z \neq y} P(h_{\mathcal{F}}(\mathbf{x}) = z)) \tag{7.27}$$

Thus, $\hat{s}$, the approximated strength, is defined as

$$\hat{s} = \frac{1}{|\mathcal{X}|}\sum_{\mathbf{x}\in\mathcal{X}}(Q(\mathbf{x},y) - \max_{z\neq y} Q(\mathbf{x},z)) \tag{7.28}$$

With respect to the correlation, it is defined in Eq. 7.20 as a function of the variance and the expectation of the standard deviation:

$$\bar{\rho} = \frac{v}{(E_\Theta[Std(\Theta)])^2} \approx \frac{\frac{1}{|\mathcal{X}|}\sum_{\mathbf{x}\in\mathcal{X}}(Q(\mathbf{x},y) - \max_{z\neq y} Q(\mathbf{x},z))^2 - \hat{s}^2}{(E_\Theta[Std(\Theta)])^2} \quad (7.29)$$

The standard deviation is defined as $Std(\Theta) = \sqrt{p_y + p_{z^*} + (p_y - p_{z^*})^2}$, $p_y = E_{\mathcal{X},\mathcal{Y}}(h_{\mathcal{F}}(\mathbf{x}) = y)$, $p_{z^*} = E_{\mathcal{X},\mathcal{Y}}(h_{\mathcal{F}}(\mathbf{x}) = z^*)$. Then, after building the kth classifier from Θ_k we use $Q(\mathbf{x}, z)$ in order to compute z^* for every example in $\mathcal{X}$. Consequently, we have

$$\hat{p}_y(\Theta_k) = \frac{1}{\frac{1}{3}|\mathcal{X}|}\sum_{\mathbf{x}\in\mathcal{X}\sim\mathcal{X}_k} I(h_k(\mathbf{x}) = y) \; .$$

$$\hat{p}_{z^*}(\Theta_k) = \frac{1}{\frac{1}{3}|\mathcal{X}|}\sum_{\mathbf{x}\in\mathcal{X}\sim\mathcal{X}_k} I(h_k(\mathbf{x}) = z^*) \quad (7.30)$$

Therefore, $Std(\Theta_k)$ can be approximated from the latter expressions. Finally, $Std(\Theta)$ is approximated by the average of $Std(\Theta_k)$ over all the classifiers forming the forest. Thus, having estimates both for the variance and the standard deviation we have the estimate for the correlation.

7.4.4 Variable Selection: Forest RI vs. Forest-RC

Besides the bootstrapping and its consequences in the GE bound, another key element of random forests is the way of selecting $n << N$ features (tests) for building each individual classifier. Random selection of features, with a fixed size n, is argued to be a good mechanism. This method is known as *Forest-RI.* A typical choice of n is $n = \lfloor \log_2 N + 1 \rfloor$. It is interesting not to include too much variables, because the higher n the higher the correlation. However, the higher n, the higher the strength. Consequently, it is important to find a value for n which represents a good trade-off between incrementing correlation and incrementing the strength. In order to do that, a good mechanism is to exploit the oob estimates. In this regard, when there are few input variables, Forest-RI may lead to a high generalization error because an important percentage of the variables will be selected. In these cases, it may be interesting to build new variables by performing linear combinations of exisiting ones. At each node, l variables are selected and combined linearly with coefficients belonging to $[-1, 1]$. Then, n linear combinations are generated. Anyway, the final choice of n will depend on the trade-off described above. This latter mechanism is known as *Forest-RC.* This mechanism has been recently applied to *image classification* under the approach of *bag-of-visual-words*(BoW) [28, 53, 147, 178, 181]. Broadly speaking, images are represented by several invariant (scale and affine) features like improved versions of the Kadir–Brady one described in Chapter 2. Such features are encoded with suitable descriptors like the popular SIFT one [107], a 128-feature vector with good orientation invariance properties. The descriptors (once clustered) define a *visual vocabulary*. The more extended classification mechanism is to obtain a histogram with the frequencies of each *visual word* for each image. Such histogram is compared with stored histograms in the database and the closest one is chosen as output. The two main weaknesses of the BoW approach are: (i) to deal with clutter; and (ii) to represent the geometric structure properly and exploit it in classification. Spatial pyramids [102] seek a solution for both problems. The underlying idea of this representation is to partition the image into increasingly fine subregions and compute histograms of the features inside each subregion. A refinement of this idea is given in [27]. Besides histograms of descriptors, also orientation histograms are included (Fig. 7.7, top). With respect to robustness against clutter, a method for the automatic selection of the Region of Interest (ROI) for the object (discarding clutter) is also proposed in the latter paper. A rectangular ROI is learnt by maximizing the area of similarity between objects from the same class (see Fig. 7.7, bottom). The idea is to compute histograms only in these ROIs. Thus, for each level in the pyramid, constrained to the ROI, two types of histograms are computed: appearance ones and shape ones. This is quite flexible, allowing to give more weight to shape or to appearance depending on the class. When building a tree in the forest, the type of descriptor is randomly selected, as well as the pyramid level from which it is chosen. If $\mathbf{x}$ is the descriptor (histogram) of a given learning example, each bin is a feature, considering that the number of bins, say N, depends on the selected pyramid level. Anyway, the *corresponding test* for such example is $\mathbf{n}^T\mathbf{x} + b \leq 0$ for the right child, and >0 for the left one. Thus all trees are binary. The components of $\mathbf{n}$ are chosen randomly from $[-1, 1]$. The variable b is chosen randomly between 0 and the distance of $\mathbf{x}$ from the origin. The number of zeros n_z imposed for that test is also chosen randomly and yields a number $n = N - n_z$ of *effective features* used. For each node, r tests, with $r = 100D$ (increasing linearly with the node depth D), are performed and the best one in terms of conditional entropy reduction is selected. The purpose of this selection is to reduce the correlation between different trees, accordingly with the trade-off described above. The number of effective features is not critical for the performance, and the method is comparable with state-of-the-art approaches to BoW [178] (around 80% with 101 categories and only 45% with 256 ones).

Another interesting aspect to cover within random forest is not the number of features used but the possibility of selecting variables on behalf of their *importance* (Chapter 6 is devoted to feature selection, and some techniques presented there could be useful in this context). In the context of random forests, important variables are the ones which, when noised, maximize *the increasing of misclassification error with respect to the oob error with all variables intact*. This definition makes equal importance to sensitiveness. However, this criterion has been proved to be inadequate in some settings: for instance in scenarios with heterogeneous variables [151], where subsampling

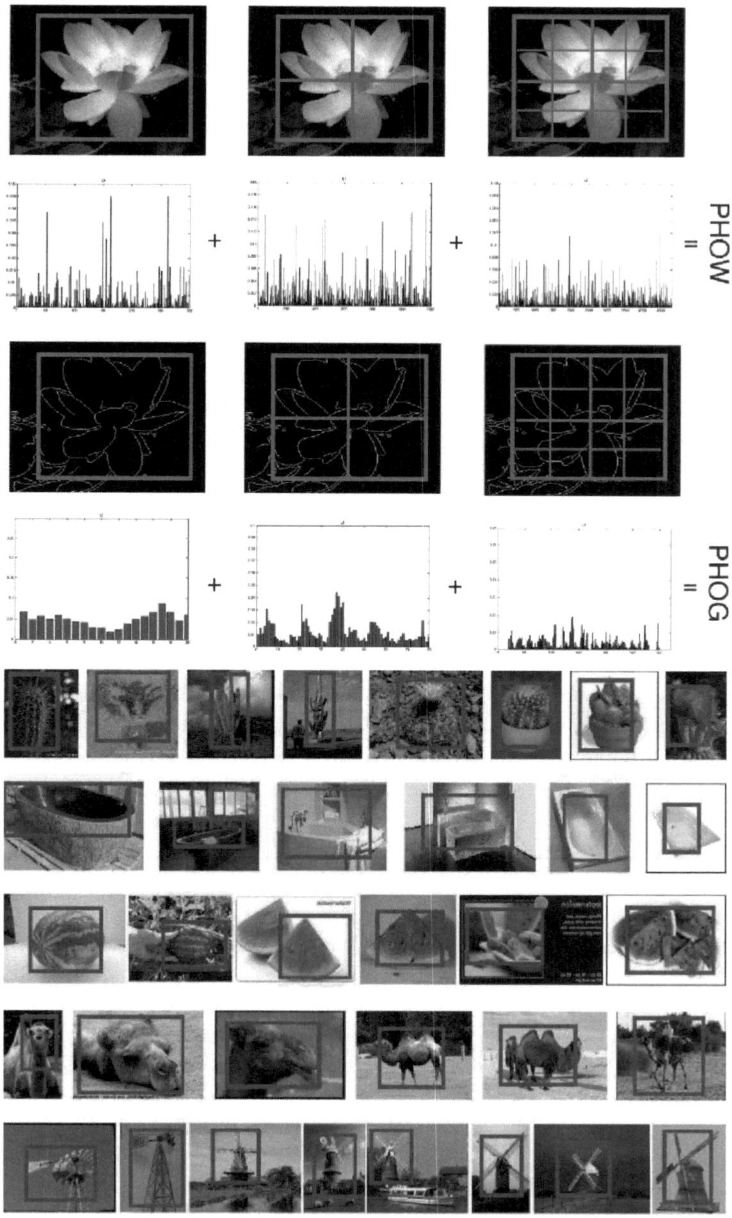

Fig. 7.7. *Top:* appearance and shape histograms computed at different levels of the pyramid (the number of bins of each histogram depends on the level of the pyramid where it is computed). *Bottom:* several ROIs learn for different categories in the Caltech-256 Database (256 categories). Figure by A. Bosch, A. Zisserman and X. Muñoz (©2007 IEEE). See Color Plates.

without replacement is proposed as an alternative mechanism for driving random forests. With respect to a definition of importance, in this context, in terms of the *information content* about the class to which the example belongs, it seems an interesting open problem (see also [164]).

7.5 Infomax and Jensen–Shannon Boosting

Boosting algorithms are becoming popular among research community. The underlying idea of Boosting is to combine a set of weak learners to improve them and build a strong classifier. This idea is supported by the work of Kearns and Valiant [95], who proved that, when enough training data is available, learners whose performance is slightly better than random guessers could be joined to form a good classifier. Later, Schapire presented a polynomial time Boosting algorithm [142]. Adaboost algorithm [61] is considered the first practical approximation of Boosting to real-world problems.

Given a set of N labeled samples $(\mathbf{x_1}, y_1), ..., (\mathbf{x_N}, y_N), \mathbf{x_i} \in \mathcal{X}, y_i \in \{0,1\}$, AdaBoost algorithm builds a strong classifier $h_f(x)$ from a linear combination of T simple weak classifiers or hypothesis $h_t(x)$. Examples of weak learners are mononeural perceptrons, decision trees, and in general, any classifier that works at least slightly better than a random guesser, with error $\epsilon_t < 1/2$. Simpler weak learners will yield better results than using multilayer perceptrons, support vector machines, and other complexer methods. The main loop of the algorithm, that is shown at Alg. 16, is repeated T times, in order to learn T weak classifiers from weighted samples; after each iteration, wrongly

---

Algorithm 16: Freund and Schapire Adaboost algorithm

Input: set of N labeled examples $(\mathbf{x_1}, y_1), ..., (\mathbf{x_N}, y_N)$
Initialize the samples weight vector $D_1(i) = \frac{1}{N}, i = 1, .., N$
for *t=1 to T* **do**
  Train a new weak classifier h_t providing it with the distribution D_t
  Calculate the error of h_t as $\epsilon_t = \sum_{i=1}^{N} D(i)|h_t(x_i) - y_i|$
  if $\epsilon_t > \frac{1}{2}$ **then**
    Stop algorithm
  end
  Set $\alpha_t = \frac{\epsilon_t}{(1-\epsilon_t)}$
  Update weights vector $D_{t+1}(i) = D_t(i)\alpha_t^{1-|h_t(x_i)-y_i|}$
  Normalize weights vector so D_{t+1} is a distribution
end
Output: Strong classifier:

$$h_f(x) = \begin{cases} 1, \sum_{t=1}^{T} (log\frac{1}{\alpha_t})h_t(x) \geq \frac{1}{2}\sum_{t=1}^{T} log\frac{1}{\alpha_t} \\ 0, otherwise \end{cases}$$

---

classified samples increase their weights, and as a consequence, the algorithm will focus on them at next iteration. Classification error of each weak learner is also stored in order to build the strong classifier.

An example can be seen in Fig. 7.8. As stated before, samples are weighted during algorithm, allowing weak learners to focus on previously misclassified data. Some learning algorithms can handle this reweighting in a natural way, but in other cases it is not possible. Another possibility is to use weights in order to perform a resampling of the training samples.

This section will show how information theory can be applied jointly with Boosting through two algorithms. The first one, Infomax Boosting algorithm [108], is based on the Infomax principle introduced by Linsker for neural networks [105]. The main idea is that neurons are trained in order to maximize the mutual information between their inputs and output. Translation of this idea to Adaboost is simple: the algorithm will try to select in each iteration the weak learner that maximizes the mutual information between input samples and class labels.

On the other hand, we may find several algorithms based on divergence measures like Kullback–Leibler to select at each iteration the best weak classifier. In the case of most of these methods, the only difference is the divergence measure applied. At present, best results are achieved by JSBoost learning [76], based on Jensen–Shannon divergence, which yields several advantages over other dissimilarity measures.

Both boosting algorithms described in this section were first applied to face detection. The objective of these algorithms is to detect the exact positions of all faces in an image. The image is splitted into image patches, at different scales, and the classifier classifies each image patch as being *face* or *not face*. In order to achieve this, during training process two classes are used: a set of *positive samples* (image patches corresponding to faces) and a set of *negative samples* (image patches that do not contain any face), having all these samples the same size. Depending on the boosting algorithm, different features are learnt as weak learners to build the strong classifier. Benefits of using features are clear: computational cost is lower, due to the fact that image patches are transformed into a lower level representation, and specific domain knowledge may be incorporated to learning process. An example of features as basic classifiers in a boosting algorithm may be found in Viola's face detection algorithm based on Adaboost [170]. In this case, rectangle features based on Haar basis functions are applied to 24×24 image regions, as Fig. 7.10 shows. The set of all possible features is large, therefore Adaboost is applied to extract the most discriminative ones.

7.5.1 The Infomax Boosting Algorithm

The main idea of this algorithm is that information theory may be used during Boosting in order to select the most informative weak learner in each

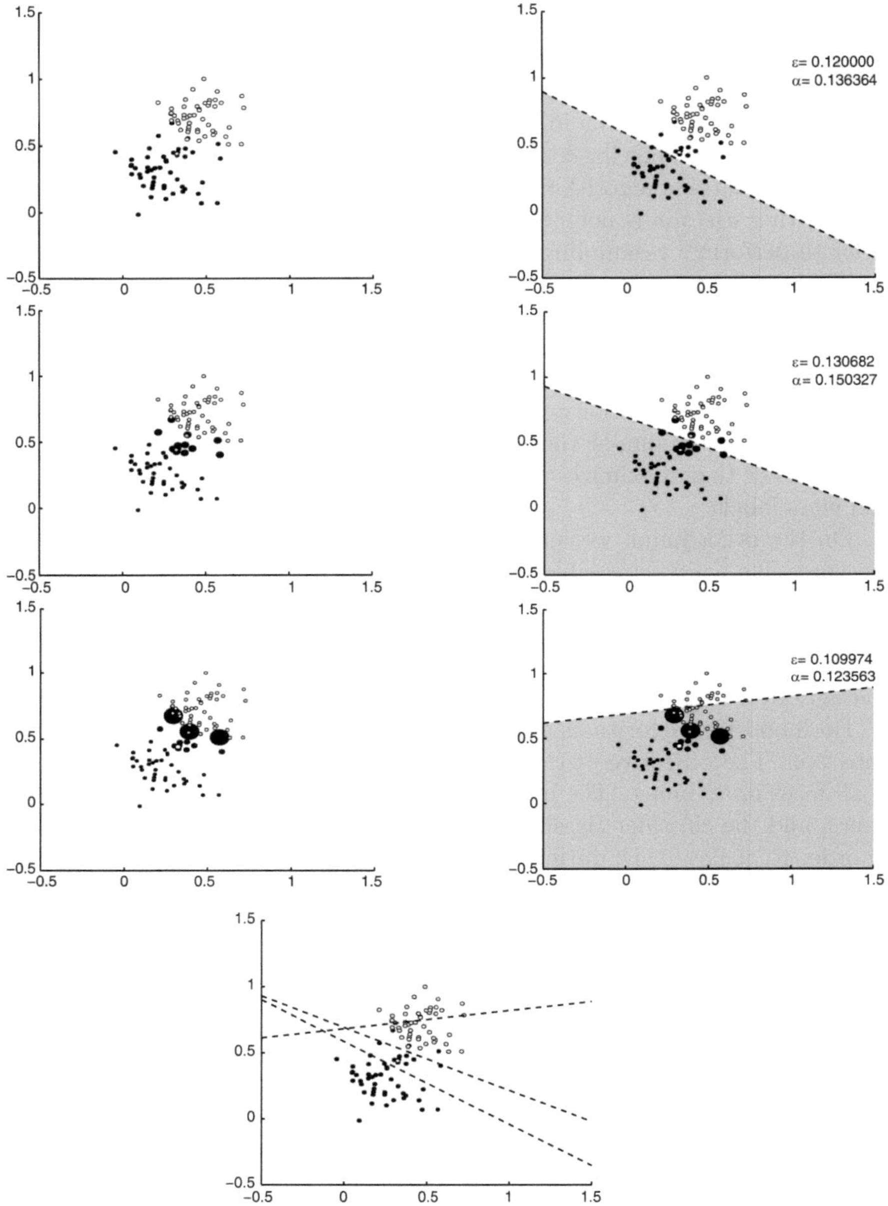

Fig. 7.8. Adaboost applied to samples extracted from two different Gaussian distributions, using three iterations (three weak classifiers are trained). Different weights are represented as different sample sizes. *First row:* initial samples and first weak classifier. *Second row:* reweighted samples after first iteration and second classifier. *Third row:* reweighted samples after second iteration and third classifier. *Fourth row:* final classifier.

iteration, thus discarding classification error as weak learner selection criterion. The Infomax principle states that this most informative learner is the one that maximizes the mutual information between input (training sample) and output (class label). However, computing mutual information requires numerical integration and knowing data densities; in the case of feature based boosting, both requirements result in inefficient approaches. An efficient implementation in the case of boosting based on high-dimensional samples and features may be achieved by means of the method explained in this section, based on kernel density estimation and quadratic mutual information.

Features may be described as low-dimensional representation of high dimensional data $\mathbf{x} \in \mathbb{R}^d$. Let us define a linear projection feature as the output of a function $\phi : \mathbb{R}^d \rightarrow \mathbb{R}^{d'}$ with $d' \ll d$:

$$\phi(x) = \phi^T x \tag{7.31}$$

with $\phi \in \mathbb{R}^d$ and $\phi^T \phi = 1$.

Infomax feature selection

The most informative feature at each iteration is called *Infomax feature*, and the objective of the algorithm is to find this Infomax feature at each iteration of the boosting process. In order to measure how informative a feature is, mutual dependence between input samples and class label may be computed. A high dependence will mean that input samples will provide more information about which class can it be labeled as. A natural measure of mutual dependence between mapped feature $\phi^T x$ and class label c is mutual information:

$$I(\phi^T x; c) = \sum_{c=1}^{C} \int_x p(\phi^T x, c) \log \frac{p(\phi^T x, c)}{p(\phi^T x)p(c)} d\phi^T x \tag{7.32}$$

Therefore, Infomax feature ϕ^* yields the maximum mutual information with respect to all other features:

$$\phi^* = \arg\max_{\phi} I(\phi^T x; c) \tag{7.33}$$

Figure 7.9 shows a simple example of mutual information as a measure to detect the best of three classifiers, from the one that perfectly splits two classes to one that cannot distinguish between classes. First classifier yields the highest mutual dependence between samples and class labels; in the case of the third classifier, input samples are not informative at all. As can be seen, mutual information and classification error are related.

When trying to find the Infomax feature, two problems arise. The first one is that the exact probability density function $p(\phi^T x, c)$ is not known.

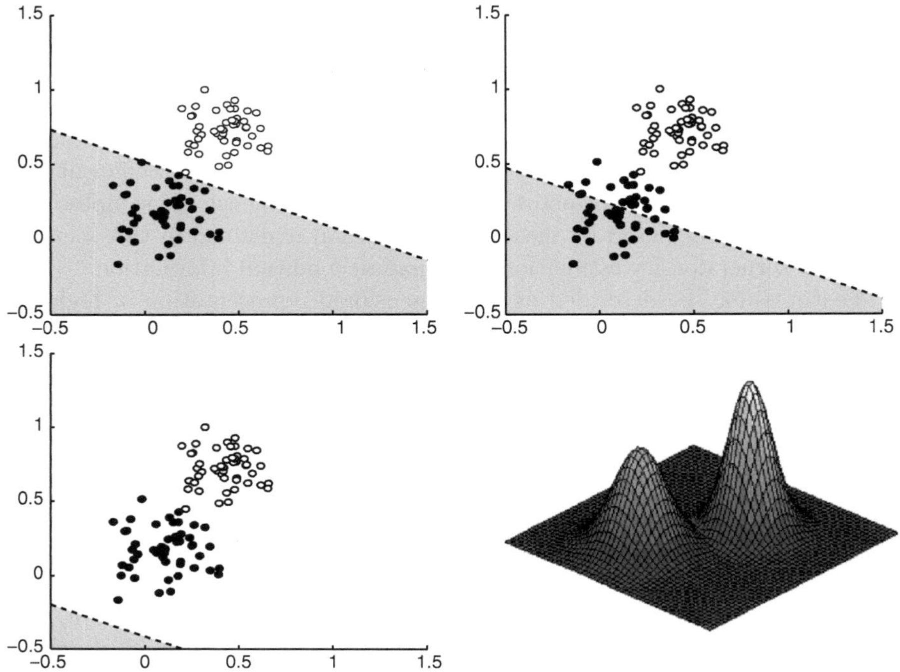

Fig. 7.9. Three different classifiers applied to the same data, obtained from two different Gaussian distributions (represented at *bottom right* of the figure). Mutual information between inputs and class labels, obtained by means of entropic graphs (as explained in Chapters 3 and 4), are 6.2437, 3.5066, and 0, respectively. The highest value is achieved in the first case; it is the *Infomax* classifier.

However, it can be estimated by means of a kernel density estimation method like Parzen's windows (explained in Chapter 5). Given a multivariate Gaussian with mean μ and variance σ^2:

$$G_\sigma(x-\mu) = \frac{1}{(2\pi)^{d/2\pi^d}} \exp\left(-\frac{(x-\mu)^T)(x-\mu)}{2\sigma^2}\right) \tag{7.34}$$

the probability density function can be approximated combining several of these Gaussians kernels:

$$p(\phi^T x|c) = \sum_{i=1}^{N_c} w_i^c G_\sigma(\phi^T x - \phi^T x_i^c) \tag{7.35}$$

where N_c is the number of classes, $\mathbf{x}_i^c$ represents input sample i from class c, and w_i^c is a non-negative weight applied to each training sample of class c, satistying that $\sum_{i=1}^{N_c} w_i^c = 1$.

Although solving this first problem, the derivation of the numerical integration of Eq. 7.32 is not simple. However, knowing that Mutual Information between projected data $\phi^T x$ and class label c may be expressed in terms of Kullback–Leibler divergence:

$$I(\phi^T x; c) = D(p(\phi^T x, c)||p(\phi^T x)p(c)) \tag{7.36}$$

this second problem may be solved using a different divergence measure of densities, like *quadratic divergence*:

$$Q(p||q) = \int_x (p(x) - q(x))^2 \, dx \tag{7.37}$$

Thus, initial mutual information expression may be reformulated as quadratic mutual information between projected data and class label:

$$I_Q(\phi^T x; c) = \sum_{c=1}^{C} \int_{\phi^T x} (p(\phi^T x, c) - p(\phi^T x)p(c))^2 \, d\phi^T x \tag{7.38}$$

and after joining this expression with the probability density functions approximated by means of Eq. 1.5:

$$\begin{aligned} I_Q(\phi^T x; c) = &\sum_{c=1}^{C} P_c^2 \sum_{i=1}^{N_c} \sum_{j=1}^{N_c} w_i^c w_j^c G_{\sqrt{2\sigma}}(y_i^c - y_j^c) \\ &+ \sum_{c_1=1}^{C} \sum_{c_2=1}^{C} \sum_{i=1}^{N_c} \sum_{j=1}^{N_{c'}} u_i^c v_j^{c'} G_{\sqrt{2\sigma}}(y_i^c - y_j^{c'}) \end{aligned} \tag{7.39}$$

where w_i^c is the associated weight of x_i^c in the approximated probability density function and

$$P_c = p(c) \tag{7.40}$$

$$y_i^c = \phi^T x_i^c \tag{7.41}$$

$$u_i^c = P_c \left(\sum_{c'=1}^{C} P_{c'}^2 - 2P_c \right) w_i^c \tag{7.42}$$

$$v_i^c = P_c w_i^c \tag{7.43}$$

Thus, the Infomax feature is searched in feature space by means of a quadratic mutual information gradient ascent. This gradient is computed from previous equation:

$$\frac{\partial I_Q}{\partial \phi} = \sum_{c=1}^{C} \sum_{i=1}^{N_c} \frac{\partial I_Q}{\partial y_i^c} \frac{\partial y_i^c}{\partial \phi} = \sum_{c=1}^{C} \sum_{i=1}^{N_c} \frac{\partial I_Q}{\partial y_i^c} x_i^c \tag{7.44}$$

where:

$$\frac{\partial I_Q}{\partial y_i^c} = P_c^2 \sum_{j=1}^{N_c} w_i^c w_j^c H_1 \left(\frac{y_i^c - y_j^c}{2\sigma} \right) + \sum_{c'=1}^{C} \sum_{j=1}^{N_{c'}} u_i^c v_j^{c'} H_1 \left(\frac{y_i^c - y_j^{c'}}{2\sigma} \right) \quad (7.45)$$

$H_1(x) = -2xe^{-x^2}$ being the first degree Hermite function. Therefore, if an initial estimate ϕ_0 is chosen, by gradient ascent next values can be computed as $\phi_{k+1} = \phi_k + \nu \frac{\partial \mathcal{I}_Q}{\partial \phi}$, with $\nu > 0$, until convergence is reached.

Infomax Boosting vs. Adaboost

Infomax Boosting modifies the original Adaboost algorithm by using different criteria to select the weak learner at each iteration: rather than selecting the classifier having less error, the most informative one is chosen. The new algorithm, as shown in Alg. 17, is straightforward. The main change is in

---

Algorithm 17: Infomax Boosting algorithm

Input: set of N_+ positive and N_- negative labeled examples $\{x_1^-, .., x_{N_-}^-, x_1^+, .., x_{N_+}^+\}$, class priors P_+ and P_- and width parameter of Gaussian kernels σ^2.

Initialize $w_{i,1}^+ = \frac{1}{N_+}, i = 1, .., N_+$ and $w_{i,1}^- = \frac{1}{N_-}, i = 1, .., N_-$

for *t=1 to T* **do**

Choose the Infomax feature ϕ_t with $w_{i,t}^+$,$w_{i,t}^-$ and σ^2.

Construct kernel density estimation of class-dependent densities using:

$$p_+^{(t)}(\phi_t^T x) = \sum_{i=1}^{N_+} w_{i,t}^+ G_\sigma(\phi_t^T x - \phi_t^T x_i^+)$$

$$p_-^{(t)}(\phi_t^T x) = \sum_{i=1}^{N_-} w_{i,t}^- G_\sigma(\phi_t^T x - \phi_t^T x_i^-)$$

Build weak classifier $f_t(x, c) = \log \frac{p_+^{(t)}(\phi_t^T x) P_+}{p_-^{(t)}(\phi_t^T x) P_-}$

Update weights:

$w_{i,t+1}^+ = w_{i,t}^+ \exp(-f_t(x_i^+))$

$w_{i,t+1}^- = w_{i,t}^- \exp(-f_t(x_i^-))$

Normalize weight vectors so w_{t+1}^+ and w_{t+1}^- are distributions

end

Output: Strong classifier: $f(x) = \sum_{t=1}^{T} f_t(x)$

---

the inner loop, where all concepts explained during this section are incorporated. A common modification of Adaboost is also included at initialization, where weight values are calculated separately for samples belonging to different classes, in order to improve efficiency. It must be noted that strong classifier output is not a discrete class label, but a real number, due to the fact that Infomax Boosting is based on *real Adaboost*, and that class labels are in $\{-1, 1\}$.

A question that may arise is if Infomax performance is really better than in the case of Adaboost algorithm. Infomax boosting was originally applied to face detection, like Viola's face detection based on Adaboost. As can be seen in Fig. 7.10, in this case features were obtained from oriented and scaled Gaussians and Gaussian derivatives of first and second order, by means of convolutions. Figure 7.10 shows results of comparing both algorithms. Regarding false positive rate, its fall as more features are added is similar; however, at first iterations, error for Infomax is quite lower, needing a lower number of features to decrease this error below 1%. In the case of ROC curves, it can be clearly seen that Infomax detection rate is higher (96.3–85.1%) than Adaboost value. The conclusion is that not only Infomax performance is better, but that low error rates can also be achieved faster than in the case of Adaboost, thanks to the Infomax principle, which allows algorithm to focus on the most informative features.

7.5.2 Jensen–Shannon Boosting

Another way to apply information theory to boosting is using a symmetric divergence measure in order to select the most discriminative weak learner at each iteration, that is, the feature that optimally differentiates between the two classes. One example is KLBoost [106], based on a symmetric Kullback–Leibler divergence measure. However, Kullback–Leibler has limited numerical stability. Jensen–Shannon improves this stability and introduces other advantages. JSBoost algorithm, based on Jensen–Shannon divergence measure, is quite similar to boosting algorithms explained above, thus we focus only on how weak learners are selected and differences between it and other boosting techniques.

Jensen–Shannon Feature Pursuit

The objective in each iteration is to find the JS Feature, the feature $\phi : \mathbb{R}^d \to \mathbb{R}^d$ that better discriminates between positive and negative classes. Unlike Infomax algorithm, designed only for linear features, JSBoost algorithm can rely also on not linear features. Once feature ϕ_i at iteration i is learned, two histograms h_i^+ and h_i^- are calculated from positive and negative samples mapped with this feature. Finally, from these two distributions, a weak classifier $\varphi_i()$: $\mathbb{R} \to \mathbb{R}$ is built in a way that $\varphi_i(\phi_i(x)) > 0$ for

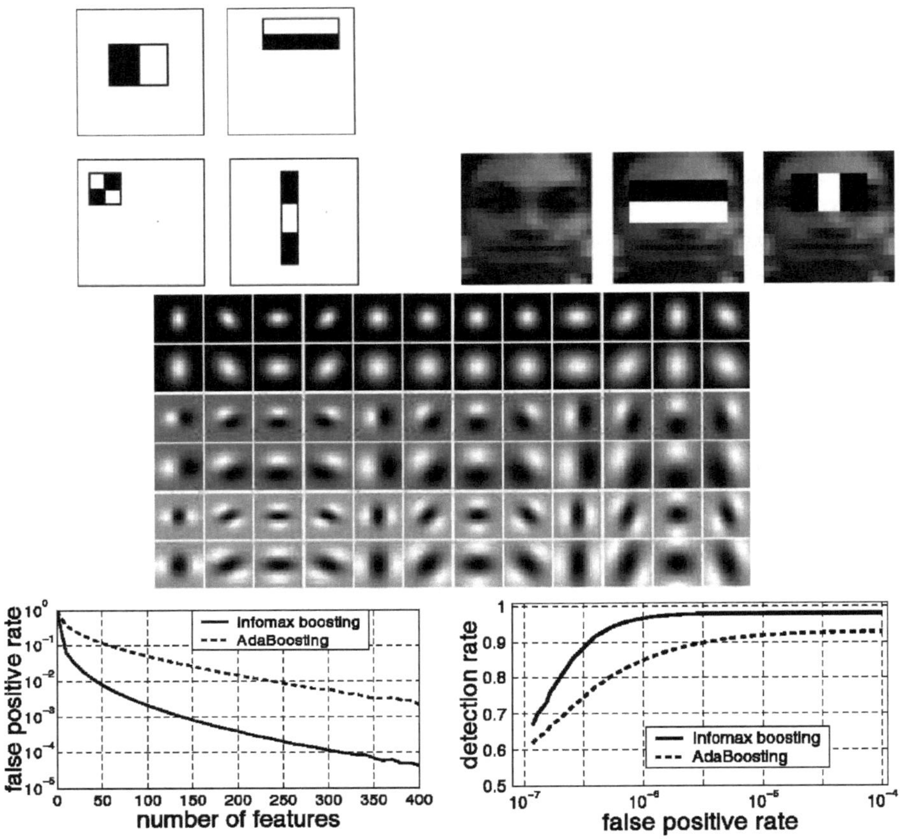

Fig. 7.10. Comparison between Adaboost and Infomax. *Top row left:* example of Haar features, relative to detection window, used in Viola's face detection based on Adaboost. Pixel values from *white rectangles* are subtracted from pixel values in *black rectangles*. *Top row right:* example of application of two Haar features to the same face image. *Center row:* feature bank used in Lyu's face detection based on Infomax. (Figure by S. Lyu (©2005 IEEE)). *Bottom row:* Lyu's experiments comparing Infomax and Adaboost false positive rate and ROC curves. (Figure by S. Lyu (©2005 IEEE)).

positive samples $(h_i^+(\phi_i(x)) > h_i^-(\phi_i(x)))$ and $\varphi_i(\phi_i(x)) < 0$ for negative ones $(h_i^-(\phi_i(x)) > h_i^+(\phi_i(x)))$:

$$\varphi_i(x) = \frac{1}{2}\log\frac{h_i^+(\phi_i(x))}{h_i^-(\phi_i(x))} \tag{7.46}$$

In order to select the most discriminative feature at each iteration, Jensen–Shannon divergence is used, rather than Kullback–Leibler, due to the fact that this last measure is undefined for $h_i^+(\phi_i(x)) = 0$ and $h_i^-(\phi_i(x)) = 0$:

$$JS(\phi_i) = \int \left\{ h_i^+(\phi_i(x)) \cdot \log \frac{h_i^+(\phi_i(x))}{\frac{1}{2}[h_i^+(\phi_i(x)) + h_i^-(\phi_i(x))]} \right\} d\phi_i(x) \quad (7.47)$$

Jensen–Shannon, as Kullback–Leibler, is not a symmetric measure. This symmetry is achieved redefining it as

$$SJS(\phi_i) = \int \left\{ h_i^+(\phi_i(x)) \log \frac{h_i^+(\phi_i(x))}{\frac{1}{2}[h_i^+(\phi_i(x)) + h_i^-(\phi_i(x))]} + h_i^-(\phi_i(x)) \log \frac{h_i^-(\phi_i(x))}{\frac{1}{2}[h_i^+(\phi_i(x)) + h_i^-(\phi_i(x))]} \right\} d\phi_i(x) \quad (7.48)$$

At iteration k, JS feature is calculated as

$$\phi_i^* = \arg\max_{\phi_i} JS(\phi_i) \quad (7.49)$$

Figure 7.11 shows an example of application of Jensen–Shannon divergence, used in this case to discriminate between three different Haar based features in the face detection problem. These Haar features are exactly the same than the ones used in Viola's Adaboost face detection algorithm. The divergence between feature based classifier output between positive and negative samples is highest in the case of the first feature; it is the JS Feature. As in the case of Infomax Boosting, Jensen–Shannon divergence and error classification are related measures.

JSBoost vs. other boosting algorithms

Algorithm 18 shows JSBoosting. The main difference with previous methods is how the final classifier is built. The weak learners are not weighted by a coefficient, like for example in Infomax. Another small difference is how weights are initialized; also, a coefficient β_k is applied at iteration k to weight updating, in order to control how fast the weight is updated. The value of β_k is defined as

$$\beta_k = \log \frac{1 - \epsilon_k}{\epsilon_k} \quad (7.50)$$

where ϵ_k is the training error of weak learner at iteration k. However, as stated before, SJS value is used to select a weak classifier at each iteration, rather than this classification error.

Like in other cases, JSBoost was first applied to face detection problem; the operator or feature from where weak learners were built is called Local Binary Pattern (LBP), that is a not linear feature. Experiments comparing face detection performance using this kind of features demonstrate that

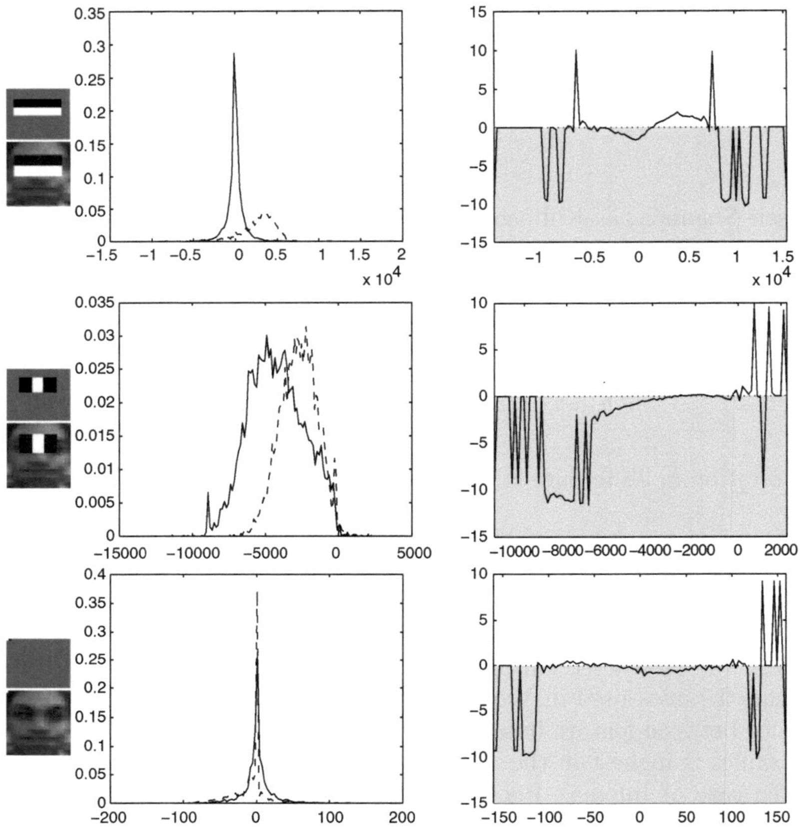

Fig. 7.11. Example of JS Feature pursuit. Three Haar based features and an example of application to a positive sample are shown, from best to worst. At the center of each row, the corresponding $h^+(x)$ (*dash line*) and $h^-(x)$ (*solid line*) distributions are represented, and at the right, $\phi(x)$ is shown. SJS values for each feature, from top to bottom, are 6360.6, 2812.9, and 1837.7, respectively.

JSBoost (98.4% detection rate) improves KLBoost results (98.1%) and outperforms Real Adaboost ones (97.9%). Furthermore, JSBoost achieves higher detection rates with a lower number of iterations than other methods.

7.6 Maximum Entropy Principle for Classification

7.6.1 Improved Iterative Scaling

Maximum entropy is a general technique for estimating probability distributions from data. It has been successfully applied in pattern recognition tasks of different fields, including natural language processing, computer vision, and bioinformatics. The maximum entropy classifier is based on the idea that the

Algorithm 18: JSBoosting algorithm

Input: set of N_+ positive and N_- negative labeled examples $\{x_1^-, ..., x_{N_-}^-, x_1^+, ..., x_{N+}^+\}$

Initialize $w_i = \frac{1}{2N_+}$ for positive samples and $w_i = \frac{1}{2N_-}$ for negative samples

for *k=1 to K* **do**

  Select JS feature ϕ_k by Jensen–Shannon divergence using weights w_i

  $f_k(x) = \frac{1}{2}\log\frac{h_k^+(\phi_k(x))}{h_k^-(\phi_k(x))}$

  Update weights $w_i = w_i \cdot \exp(-\beta_k) \cdot y_i \cdot f_k(x_i), i = 1, \ldots, N$, and normalize weights so that $\sum_i w_i = 1$

end

Output: Strong classifier:

$$h_f(x) = \begin{cases} 1, \sum_{1=1}^{K} \frac{1}{2}\log\frac{h_k^+(\phi_k(x))}{h_k^-(\phi_k(x))} \geq 0 \\ 0, otherwise \end{cases}$$

Table 7.3. A toy example of a classification problem with two classes $\mathcal{C} = \{motorbike, car\}$ and an original feature space of $D = 3$ features. There are five samples in this training set.

	Original features		
c	1: "has gears"	2: "# wheels"	3: "# seats"
Motorbike	yes	3	3
Motorbike	no	2	1
Motorbike	yes	2	2
Car	yes	4	5
Car	yes	4	2

most suitable model for classification is the most uniform one, given some constraints which we call features. The maximum entropy classifier has to learn a conditional distribution from labeled training data. Let $x \in \mathcal{X}$ be a sample and $c \in \mathcal{C}$ be a label, then the distribution to be learnt is $P(c|x)$, which is to say, we want to know the class given a sample. The sample is characterized by a set of D features (dimensions). For the maximum entropy classifier we will formulate the features as $f_i(x, c)$, and $1 \leq i \leq N_F$, and $N_F = D|\mathcal{C}|$. Each sample has D features for each one of the existing classes, as shown in the following example.

Table 7.3 contains the original feature space of a classification problem where two classes of vehicles are described by a set of features. In Table 7.4 the features are represented according to the formulation of the maximum entropy classifier.

Table 7.4. The values of the feature function $f_i(x, c)$ for the training data.

	$f_i(x,c)$					
Class	$Class(i) = 1 = motorbike$			$Class(i) = 2 = car$		
feature	"has gears"	"# wheels"	"# seats"	"has gears"	"# wheels"	"# seats"
$x, c \setminus i$	$i = 1$	$i = 2$	$i = 3$	$i = 4$	$i = 5$	$i = 6$
$x = 1, c = 1$	1	3	3	0	0	0
$x = 2, c = 1$	0	2	1	0	0	0
$x = 3, c = 1$	1	2	2	0	0	0
$x = 4, c = 2$	0	0	0	1	4	5
$x = 5, c = 2$	0	0	0	1	4	2

The feature function is defined as

$$f_i(x, c) = \begin{cases} \text{mod}(i, c) & \text{if class}(i) = c \\ 0 & \text{otherwise} \end{cases} \tag{7.51}$$

In the following equations f will form part of a product, multiplying a probability, so when the value of the feature is 0, the probability will also be null. Originally the features f are defined as a binary function; however, a generalization is made to allow real positive values. If a feature has a natural value h, for the model this is the same as declaring the existence of the same feature h times. Generalizing in the same way real positive values are allowed for the features. The maximum entropy classifier has been widely used in Natural Language Processing [19] where many problems are easily formulated in terms of binary features which may be present in a text zero times, once, or more than once.

The maximum entropy principle is used to model a conditional distribution $P(c|x)$ which is restricted to have as expected values for the features $f_i(x, c)$, the values shown in 7.4. These expected values with respect to the data distribution $P(x, c)$ are denoted as

$$\begin{aligned} P(f) &\equiv \sum_{x \in \mathcal{X}} \sum_{c \in \mathcal{C}} P(x, c) f(x, c) \\ &= \sum_{x \in \mathcal{X}} P(x) \sum_{c \in \mathcal{C}} P(c|x) f(x, c) \end{aligned} \tag{7.52}$$

Here $P(x, c)$ and $P(x)$ are the expected values in the training sample. The equation is a constraint of the model, it forces the expected value of f to be the same as the expected value of f in the training data. In other words, the model has to agree with the training set on how often to output a feature.

The empirically consistent classifier that maximizes entropy is known as the *conditional exponential model*. It can be expressed as

$$P(c|x) = \frac{1}{Z(x)} \exp\left(\sum_{i=1}^{N_F} \lambda_i f_i(x, c)\right) \tag{7.53}$$

where λ_i are the weights to be estimated. In the model there are N_F weights, one for each feature function f_i. If a weight is zero, the corresponding feature has no effect on classification decisions. If a weight is positive then the corresponding feature will increase the probability estimates for labels where this feature is present, and decrease them if the weight is negative. $Z(x)$ is a normalizing factor for ensuring a correct probability and it does not depend on $c \in \mathcal{C}$:

$$Z(x) = \sum_{c \in \mathcal{C}} \exp\left(\sum_{i=1}^{N_F} \lambda_i f_i(x, c)\right) \tag{7.54}$$

A detailed derivation is presented in [18]. In his work it is proved that this model is the one that is closest to the expected probability of the features in the training data in the sense of Kullback–Leibler divergence. The estimation of the parameters (weights) is a very important point and is not trivial.

In the following equation the log-likelihood function of the empirical distribution of the training data is expressed:

$$\begin{aligned} L(p) &\equiv \log \prod_{x \in \mathcal{X}} \prod_{c \in \mathcal{C}} P(c|x)^{P(x,c)} \\ &= \sum_{x \in \mathcal{X}} \sum_{c \in \mathcal{C}} P(x, c) \log P(c|x) \end{aligned} \tag{7.55}$$

This is a dual maximum-likelihood problem. It could be solved with the Expectation Maximization (EM) algorithm. However, when the optimization is subjected to the feature constraints, the solution to the primal maximum entropy problem is also the solution to the dual maximum-likelihood problem of Eq. 7.55, provided that the model has the same exponential form. In [18] it is also explained that due to the convex likelihood surface of the model, and due to the fact that there are no local maxima, it is possible to find the solution using a hill climbing algorithm. A widely used hill climbing method is the *Generalized Iterative Scaling*. The *Improved Iterative Scaling* performs better and will be outlined in this subsection. Also, there are other optimization techniques that can be used, such as gradient descent, variable metric, and so on.

The *Improved Iterative Scaling* algorithm (see Alg. 19) performs a hill-climbing search among the log-likelihood space formed by the parameters λ_i. It is guaranteed to converge on the parameters that define the maximum entropy classifier for the features of a given data set.

Equation 7.56 can be solved in closed-form or with other root-finding procedure, such as Newton's numerical method.

Once the parameters λ_i are calculated, the conditional probability distribution $P(c|x)$ is modeled and classification of new examples can be performed using Eq. 7.53.

Let us see a classification toy-example with the data from Table 7.4. The Improved Iterative Scaling algorithm converges to the weights:

$$(\lambda_1, \ldots, \lambda_6) = (0.588, -0.098, -0.090, -0.588, 0.098, 0.090)$$

Algorithm 19: Improved Iterative Scaling

Input: A distribution of classified samples $p(x,c)$ and a set of N_F features f_i.
Initialize $\lambda_i = 0$, $\forall i \in \{1,2,\ldots,N_F\}$
repeat
  foreach $i \in \{1,2,\ldots,N_F\}$ **do**
    Solve for $\Delta\lambda_i$ the equation:

$$\sum_{x\in\mathcal{X}} P(x) \sum_{c\in\mathcal{C}} P(c|x) f_i(x,c) \exp(\Delta\lambda_i \sum_{j=1}^{N_F} f_j(x,c)) \\ = \sum_{x\in\mathcal{X}} \sum_{c\in\mathcal{C}} P(x,c) f_i(x,c) \tag{7.56}$$

    Update $\lambda_i \leftarrow \lambda_i + \Delta\lambda_i$
  end
until *convergence of all* λ_i ;
Output: The λ_i parameters of the *conditional exponential model* (Eq. 7.53).

Table 7.5. The values of the feature function $f_i(x,c)$ for the new sample supposing that its class is *motorbike* ($x=6, c=1$) and another entry for the supposition that the class is *car* ($x=6, c=2$).

	$f_i(x,c)$					
Class	$Class(i) = 1 = motorbike$			$Class(i) = 2 = car$		
feature	"has gears"	"# wheels"	"# seats"	"has gears"	"# wheels"	"# seats"
$x, c \setminus i$	$i=1$	$i=2$	$i=3$	$i=4$	$i=5$	$i=6$
$x=6, c=1$	1	4	4	0	0	0
$x=6, c=2$	0	0	0	1	4	4

We want to classify a new unlabeled sample which has the following features:

`has gears, 4 wheels, 4 seats`

We have to calculate the probabilities for each one of the existing classes, in this case $\mathcal{C} = motorbike, car$. Let us put the features using the notation of Table 7.4. We have two new samples, one of them is supposed to belong to the first class, and the other one to the second class (see Table 7.5).

According to the estimated model the conditional probability for the sample ($x=6, c=1$) is

$$P(c=1|x=6) = \frac{1}{Z(6)} \exp\left(\sum_{i=1}^{N_F} \lambda_i f_i(6,1)\right) \tag{7.57}$$

where the exponent is equal to

$$\begin{aligned}
&\sum_{i=1}^{N_F} \lambda_i f_i(6,1) \\
&= 0.588 \cdot 1 - 0.098 \cdot 4 - 0.090 \cdot 4 - 0.588 \cdot 0 + 0.098 \cdot 0 + 0.090 \cdot 0 \\
&= -0.164
\end{aligned} \tag{7.58}$$

and the normalizing factor is

$$\begin{aligned}
Z(6) &= \exp\left(\sum_{i=1}^{N_F} \lambda_i f_i(6,1)\right) + \exp\left(\sum_{i=1}^{N_F} \lambda_i f_i(6,2)\right) \\
&= \exp(0.588 \cdot 1 - 0.098 \cdot 4 - 0.090 \cdot 4 - 0.588 \cdot 0 + 0.098 \cdot 0 + 0.090 \cdot 0) \\
&\quad + \exp(0.588 \cdot 0 - 0.098 \cdot 0 - 0.090 \cdot 0 - 0.588 \cdot 1 + 0.098 \cdot 4 + 0.090 \cdot 4) \\
&= 0.8487 + 1.1782 = 2.027
\end{aligned} \tag{7.59}$$

so the probability is

$$P(c=1|x=6) = \frac{1}{2.027} \exp(-0.164) = 0.4187 \tag{7.60}$$

Performing the same calculations for the other class ($x = 6, c = 2$) we have

$$P(c=2|x=6) = \frac{1}{Z(6)} \exp\left(\sum_{i=1}^{N_F} \lambda_i f_i(6,2)\right) \tag{7.61}$$

where the exponent is equal to

$$\begin{aligned}
&\sum_{i=1}^{N_F} \lambda_i f_i(6,2) \\
&= 0.588 \cdot 0 - 0.098 \cdot 0 - 0.090 \cdot 0 - 0.588 \cdot 1 + 0.098 \cdot 4 + 0.090 \cdot 4 \\
&= 0.164
\end{aligned} \tag{7.62}$$

and the probability is

$$P(c=2|x=6) = \frac{1}{2.027} \exp(0.164) = 0.5813 \tag{7.63}$$

Therefore the maximum entropy classifier would label the sample "`has gears, 4 wheels, 4 seats`" as a *car*, with a 0.5813 probability.

7.6.2 Maximum Entropy and Information Projection

Besides classifier design, along this book the ME principle has been applied to several problems (it appears in almost all chapters). In all cases, but when presenting *iterative scaling*, we have outlined solutions based on the *primal* formulation of the problem (that is, go ahead to estimate the Lagrange multipliers related to the expectation constraints). In the previous subsection we have referred to iterative scaling as an algorithm solving the *dual maximum-likelihood problem*. Actually, iterative scaling and its improvements are good methods for working in the dual space, but, what is the origin of this *dual framework*? Let us start by the classical discrete formulation of the ME problem:

$$
\begin{aligned}
p^*(x) &= \arg\max_{p(x)} \left\{ \sum_x p(x) \frac{1}{\log p(x)} \right\} \\
\text{s.t} \quad & \sum_x p(x) F_j(x) = a_j, \; j = 1, \ldots, m \\
& \sum_x p(x) = 1 \\
& p(x) \geq 0 \; \forall x \; ,
\end{aligned} \tag{7.64}
$$

where a_j are empirical estimations of $E(G_j(x))$. It is particularly interesting here to remind the fact that the objective function is concave and, also, that all constraints are linear. These properties are also conserved in the following *generalization* of the ME problem:

$$
\begin{aligned}
p^*(x) &= \arg\min_{p(x)} \left\{ D(p||\pi) = \sum_x p(x) \log \frac{p(x)}{\pi(x)} \right\} \\
\text{s.t} \quad & \sum_x p(x) F_j(x) = a_j, \; j = 1, \ldots, m \\
& \sum_x p(x) = 1 \\
& p(x) \geq 0 \; \forall x
\end{aligned} \tag{7.65}
$$

where $\pi(x)$ is a *prior* distribution, and thus, when the prior is uniform we have the ME problem. This generalization is known as the *Kullback's minimum cross-entropy principle* [100]. The primal solution is obtained through Lagrange multipliers, and has the form of the following *exponential function* (see Section 3.2.2):

$$
p^*(x) = \frac{1}{Z(\Lambda)} e^{\sum_{j=1}^m \lambda_j F_j(x)} \pi(x) \tag{7.66}
$$

where $\Lambda = (\lambda_1, \ldots, \lambda_m)$, and the main difference with respect to the solution to the ME problem is the factor corresponding to the prior $\pi(x)$. Thinking of the definition of $p^*(x)$ in Eq. 7.66 as a *family of exponential functions* (discrete or continuous) characterized by: an *input space* $\mathcal{S} \subseteq \mathbb{R}^k$, where the x are defined, $\Lambda \in \mathbb{R}^m$ (*parameters*), $F(x) = (F_1(x), \ldots, F_m)$ (*feature set*), and $\pi(x)$: $\mathbb{R}^k \to \mathbb{R}$ (*base measure*, not necessarily a probability measure). Thus, the compact form of Eq. 7.66 is

$$
p_\Lambda(x) = \frac{1}{Z(\Lambda)} e^{\Lambda \cdot F(x)} \pi(x) \tag{7.67}
$$

and considering that $Z(\Lambda) = \sum_x e^{\Lambda \cdot F(x)} \pi(x)$, the *log-partition function* is denoted $G(\Lambda) = \ln \sum_x e^{\Lambda \cdot F(x)} \pi(x)$. Thus it is almost obvious that

$$p_\Lambda(x) = \frac{1}{e^{G(\Lambda)}} e^{\Lambda\cdot F(x)}\pi(x) = e^{\Lambda\cdot F(x)-G(\Lambda)}\pi(x) \tag{7.68}$$

where the right-hand side of the latter equation is the usual definition of an exponential family:

$$p_\Lambda(x) = e^{\Lambda\cdot F(x)-G(\Lambda)}\pi(x) \tag{7.69}$$

The simplest member of this family is the Bernoulli distribution defined over $\mathcal{S} = \{0,1\}$ (discrete distribution) by the well-known parameter $\theta \in [0,1]$ (the success probability) so that $p(x) = \theta^x \times (1-\theta)^{1-x}$, $x \in S$. It turns out that $E(x) = \theta$. Beyond the input space $\mathcal{S}$, this distribution is posed in exponential form by setting: $T(x) = x$, $\pi(x) = 1$ and choosing

$$\begin{aligned}
p_\lambda(x) &\propto e^{\lambda x}\\
p_\lambda(x) &= e^{\lambda x - G(\lambda)}\\
G(\lambda) &= \ln\sum_x e^{\lambda x} = \ln\left(e^{\lambda\cdot 0} + e^{\lambda\cdot 1}\right) = \ln(1+e^\lambda)\\
p_\lambda(x) &= \frac{e^{\lambda x}}{1+e^\lambda}\\
p_\lambda(1) &= \frac{e^\lambda}{1+e^\lambda} = \theta\ .\\
p_\lambda(0) &= 1 - p_\lambda(1) = 1 - \frac{e^\lambda}{1+e^\lambda} = \frac{1}{1+e^\lambda} = 1-\theta
\end{aligned} \tag{7.70}$$

The latter equations reveal a *correspondence* between the *natural space* λ and the usual parameter $\theta = e^\lambda/(1+e^\lambda)$. It is also interesting to note here that the usual parameter is related to the expectation of the distribution. Strictly speaking, in the general case the *natural space* has as many dimensions as the number of features, and it is defined as $\mathcal{N} = \{\Lambda \in R^m : -1 < G(\Lambda) < 1\}$. Furthermore, many properties of the exponential distributions, including the correspondence between parameter spaces, emerge from the analysis of the log-partition. First of all, $G(\Lambda) = \ln\sum_x e^{\Lambda\cdot F(x)}\pi(x)$ is strictly convex with respect to Λ. The partial derivatives of this function are:

$$\begin{aligned}
\frac{\partial G}{\partial \lambda_i} &= \frac{1}{G(\Lambda)}\frac{\partial}{\partial \lambda_i}\sum_x e^{\Lambda\cdot F(x)}\pi(x)\\
&= \frac{1}{G(\Lambda)}\sum_x\left(\frac{\partial}{\partial\lambda_i}e^{\sum_{j=1}^m \lambda_j F_j(x)}\right)\pi(x)\\
&= \frac{1}{G(\Lambda)}\sum_x e^{\sum_{j=1}^m \lambda_j F_j(x)}\pi(x)F_i(x)\\
&= \frac{1}{G(\Lambda)}\sum_x e^{\Lambda\cdot F(x)}\pi(x)F_i(x)\\
&= \sum_x p_\Lambda(x)F_i(x)\\
&= E_\Lambda(F_i(x)) = a_i
\end{aligned} \tag{7.71}$$

This result is not surprising if one considers that $G'(\Lambda) = E_\Lambda(F(x)) = \mathbf{a}$, which is obvious to derive from above. Furthermore, the convexity of the log-partition function makes this derivative unique and thus it is possible to make a *one-to-one* correspondence between the *natural parameters* $\Lambda \in \mathbb{R}^m$ and the *expectation* parameters $\mathbf{a}$ lying also in $\mathbb{R}^m$.

Beyond the latter correspondence, the deepest impact of the connection between the derivative of the log-partition and the expectations in learning is the finding of the distribution maximizing the log-likelihood of the data. Let $\mathbf{X} = \{x_1, \ldots, x_N\}$ be a set of i.i.d. samples of a distribution $p_\Lambda(x)$ belonging to the exponential family. Then, the log-likelihood is

$$\begin{aligned}\ell(\mathbf{X}|\Lambda) &= \log \prod_{i=1}^{N} p_\Lambda(x_i) = \sum_{i=1}^{N} \log(p_\Lambda(x_i)) \\ &= \sum_{i=1}^{N} (\Lambda \cdot F(x_i) - G(\Lambda) + \log \pi(x_i)) \\ &= \Lambda \left(\sum_{i=1}^{N} \cdot F(x_i) \right) - NG(\Lambda) + \sum_{i=1}^{N} \log \pi(x_i) \end{aligned} \tag{7.72}$$

Then, for finding the distribution maximizing the log-likelihood we must set to zero its derivative with respect to Λ:

$$\begin{aligned}\ell'(\mathbf{X}|\Lambda) = 0 &\Rightarrow \left(\sum_{i=1}^{N} \cdot F(x_i) \right) - NG'(\Lambda) = 0 \\ &\Rightarrow G'(\Lambda) \equiv \mathbf{a} = \frac{1}{N} \sum_{i=1}^{N} \cdot F(x_i) \end{aligned} \tag{7.73}$$

which implies that Λ corresponds to the unique possible distribution fitting the average vector $\mathbf{a}$. Thus, if we consider the one-to-one correspondence, between the natural space and the expectation space established through $G'(\Lambda)$, the maximum-likelihood distribution is determined up to the prior $\pi(x)$. If we know the prior beforehand, there is a unique member of the exponential distribution family, $p^*(x)$ satisfying $E_\Lambda(F(x)) = \mathbf{a}$ which is the closest to $\pi(x)$, that is, which minimizes $D(p||\pi)$. That is, any other distribution $p(x)$ satisfying the constraints is farther from the prior:

$$\begin{aligned} &D(p||\pi) - D(p^*||\pi) \\ &= \sum_x p(x) \log \frac{p(x)}{\pi(x)} - \sum_x p^*(x) \log \frac{p^*(x)}{\pi(x)} \end{aligned}$$

$$= \sum_x p(x)\frac{p(x)}{\pi(x)} - \sum_x p^*(x)(\Lambda^* \cdot F(x) - G(\Lambda^*))$$

$$= \sum_x p(x)\frac{p(x)}{\pi(x)} - \sum_x p(x)(\Lambda^* \cdot F(x)) - \sum_x p^*(x)G(\Lambda^*)$$

$$= \sum_x p(x)\frac{p(x)}{\pi(x)} - \sum_x p(x)(\Lambda^* \cdot F(x)) - \sum_x p^*(x)G(\Lambda^*)$$

$$= \sum_x p(x)\frac{p(x)}{\pi(x)} - \sum_x p(x)(\Lambda^* \cdot F(x) \sum_x p(x)G(\Lambda^*)$$

$$= \sum_x p(x)\frac{p(x)}{\pi(x)} - \sum_x p(x)(\Lambda^* \cdot F(x) - G(\Lambda^*))$$

$$= \sum_x p(x)\frac{p(x)}{\pi(x)} - \sum_x p(x)\log\frac{p^*(x)}{\pi(x)}$$

$$= \sum_x p(x)\log\frac{p(x)}{p^*(x)} = D(p||p^*) \tag{7.74}$$

where the change of variable from $p^*(x)$ to $p(x)$ in the right term of the fourth line is justified by the fact that both pdfs satisfy the expectation constraints. Thus: $\sum_x p^*(x)(\Lambda^* \cdot F(x)) = \sum_x p(x)(\Lambda^* \cdot F(x))$ because $\sum_x p(x)^* F(x) = E_{\Lambda^*}(F(x)) = \sum_x p(x)F(x) = E_{\Lambda}(F(x)) = \mathbf{a}$. This is necessary so that the Kullback–Leibler divergence satisfies a *triangular equality* (see Fig. 7.12, top):

$$D(p||\pi) = D(p||p^*) + D(p^*||\pi) \tag{7.75}$$

Otherwise (in the general case that $p^*(x)$ to $p(x)$ satisfy, for instance, the same set of *inequality constraints*) the constrained minimization of cross-entropy (Kullback–Leibler divergence with respect to a prior) satisfies [45, 144]:

$$D(p||\pi) \geq D(p||p^*) + D(p^*||\pi) \tag{7.76}$$

In order to prove the latter inequality, we firstly remind that $D(p^*||\pi) = min_{p \in \Omega} D(p||\pi)$, Ω being a convex space of probability distributions (for instance, the space of members of the exponential family). Then, let $p^*_\alpha(x) = (1-\alpha)p^*(x) + \alpha p(x)$, with $\alpha \in [0,1]$, be pdfs derived from a convex combination of $p^*(x)$ and $p(x)$. It is obvious that

$$D(p^*_\alpha||p) \geq D(p^*||p) \equiv D(p^*_0||p) \geq 0 \tag{7.77}$$

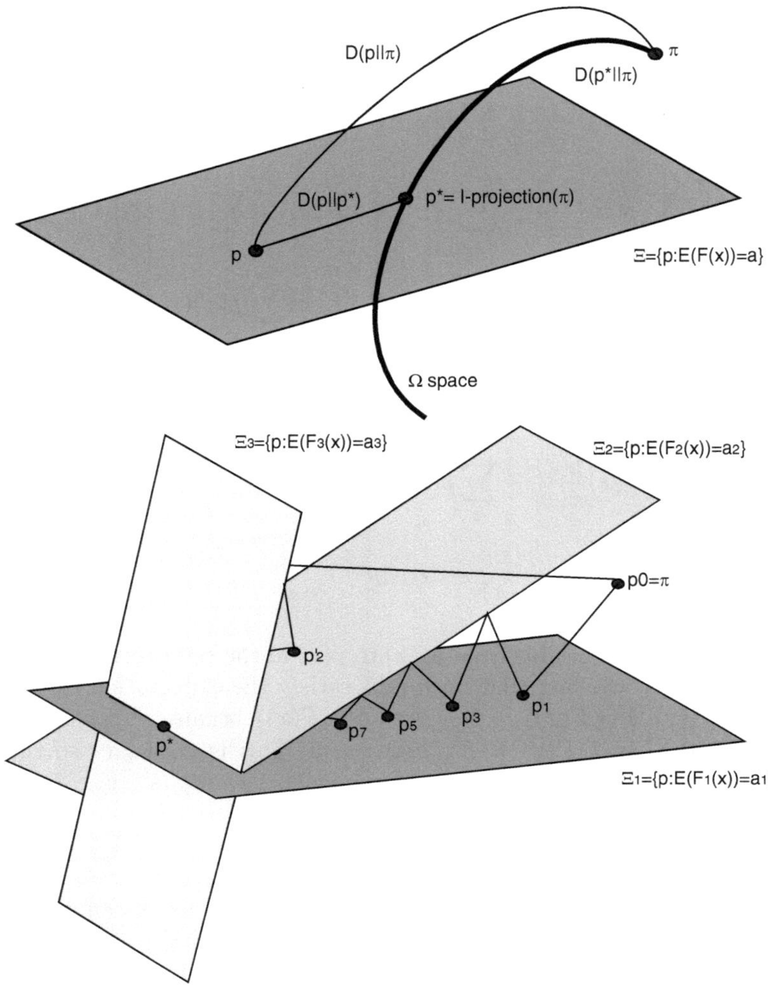

Fig. 7.12. *Top:* triangular equality. *Bottom:* solving by alternating projections in different subspaces until convergence.

and, thus, $D'(p_0^*||p) \geq 0$. Now, taking the derivative of $D(p_\alpha^*||p)$ with respect to α:

$$
\begin{aligned}
&0 \leq D'(p_\alpha^*||p) \\
&= \frac{d}{d\alpha}\left\{\sum_x ((1-\alpha)p^*(x) + \alpha p(x)) \log \frac{(1-\alpha)p^*(x) + \alpha p(x)}{\pi(x)}\right\}\Bigg|_{\alpha=0} \\
&= \sum_x (p(x) - p^*(x))\left\{1 + \log \frac{p^*(x)}{\pi(x)}\right\} \\
&= \sum_x \left\{p(x) + p(x)\log \frac{p^*(x)}{\pi(x)}\right\} - \sum_x \left\{p^*(x) + p^*(x)\log \frac{p^*(x)}{\pi(x)}\right\}
\end{aligned}
$$

$$= \sum_x \left\{ p(x) \log \frac{p^*(x)}{\pi(x)} \right\} - \sum_x \left\{ p^*(x) \log \frac{p^*(x)}{\pi(x)} \right\}$$
$$= \sum_x \left\{ p(x) \log \frac{p(x)}{\pi(x)} \frac{p^*(x)}{p(x)} \right\} - \sum_x \left\{ p^*(x) \log \frac{p^*(x)}{\pi(x)} \right\}$$
$$= \sum_x \left\{ p(x) \log \frac{p(x)}{\pi(x)} + p(x) \log \frac{p^*(x)}{p(x)} \right\} - \sum_x \left\{ p^*(x) \log \frac{p^*(x)}{\pi(x)} \right\}$$
$$= D(p||\pi) - D(p^*||p) - D(p^*||\pi) \tag{7.78}$$

Therefore,

$$D(p||\pi) - D(p^*||p) - D(p^*||\pi) \geq 0 \equiv D(p||\pi) \geq D(p^*||p) + D(p^*||\pi) \tag{7.79}$$

Thus, for the equality case (the case where the equality constraints are satisfied) it is timely to highlight the *geometric interpretation* of such *triangular equality*. In geometric terms, all the pdfs of the exponential family which satisfy the same set of equality constraints lie in an affine subspace Ξ (here is the connection with information geometry, introduced in Chapter 4). Therefore, p^* can be seen as the *projection* of π onto such subspace. That is the definition of *I-projection* or *information projection*. Consequently, as p is also the I-projection of π in the same subspace, we have a sort of Pythagorean theorem from the latter equality. Thus, the Kullback–Leibler divergences play the role of the Euclidean distance in the Pythagorean theorem. In addition, pi and p^* belong to Ω, the convex space. Actually, the facts that Ω is a convex space, and also the convexity of $D(p||\pi)$, guarantee that p^* is unique as we have seen above. Anyway, the p^* distribution has the three following equivalent properties: (i) it is the I-projection of the prior; (ii) it is the maximum likelihood distribution in Ω; and (iii) $p^* \in \Xi \cap \Omega$.

In algorithmic terms, the latter geometric interpretation suggests an iterative approach to find p^* from an initial p, for instance π. A very intuitive approach was suggested by Burr 20 years ago [32]. For the sake of simplicity, consider that we have only two expectation constraints to satisfy, that is $F(x) = (F_1(x), F_2(x))$, and $E(F_1(x)) = a_1$, $E(F_2(x)) = a_2$, thus being $\mathbf{a} = (a_1, a_2)$, and $\Lambda = (\lambda_1, \lambda_2)$ (here, for simplicity, we do not include λ_0, the multiplier ensuring up-to-one sum). Each of the constraints has associated a different convex space of probability distributions satisfying them: Ξ_1 and Ξ_2, respectively. Obviously, p^* lies in $\Xi_1 \cap \Xi_2$. However, starting by $p_{t=0} = \pi$ it is possible to alternatively generate $p_t \in \Xi_1$ (t even) and $p_t \in \Xi_2$ (t odd) until convergence to p^*. More precisely, p_1 has the form

$$p_1(x) = e^{\lambda_1 F_1(x) - G(\lambda_1)} \pi(x) \tag{7.80}$$

as $G'(\lambda_1) = E_{\lambda_1}(F_1(x)) = a_1$, then λ_1 is uniquely determined. Let us then project p_1 onto Ξ_2 using p_1 as prior. Then we have

$$\begin{aligned} p_2(x) &= e^{\lambda_2 F_2(x) - G(\lambda_2)} p_1(x) \\ &= e^{\lambda_2 F_2(x) - G(\lambda_2)} e^{\lambda_1 F_1(x) - G(\lambda_1)} \pi(x) \\ &= e^{\lambda_1 F_1(x) + \lambda_2 F_2(x) - (G(\lambda_1) + G(\lambda_2))} \pi(x) \end{aligned} \tag{7.81}$$

where we must find λ_2, given λ_1, which, in general, does not satisfy the constraint generating Ξ_2. However, a key element in this approach is that the triangular equality (Eq. 7.75) ensures that: (i) p_1 is closer to p^* than p_0; and (ii) p_2 is closer to p^* than p_1. In order to prove that, the cited triangular equality is rewritten as

$$D(p||\pi) = D(p^*||\pi) - D(p^*||p) \equiv D(p^*||p) = D(p^*||\pi) - D(p||\pi) \tag{7.82}$$

because $D(p||p^*) = -D(p^*||p)$. This new form of the triangular equality is more suitable for giving the intuition of the convergence of alternate projections:

$$\begin{aligned} D(p^*||p_1) &= D(p^*||p_0) - D(p_1||p_0) \\ D(p^*||p_2) &= D(p^*||p_1) - D(p_2||p_1) \\ &\dots \\ D(p^*||p_t) &= D(p^*||p_{t-1}) - D(p_t||p_{t-1}) \end{aligned} \tag{7.83}$$

The latter iteration alternates projecting onto subspace Ξ_1 and Ξ_2, using the previous probability as a prior, and finding the multipliers satisfying the corresponding constraint, until a $p^* \in \Xi_1 \cap \Xi_2$ is found. As we have seen above, such minimizer is unique. The convergence speed of this process was experimentally compared with a Newton–Raphson iterative scheme and the speed of the alternated method is faster or equal, but never slower, than the one of the iterative schemes.

A generalization of the latter idea, when having m constraints, is to project the current prior (result of iteration t) onto the subspace $\Xi_{t(\mod m)}$ (*iterative scaling*). This scheme assumes an arbitrary order in the selection of the next subspace to project onto. As we illustrate in Fig. 7.12 (bottom), alternating between near parallel subspaces slows down the search, with respect to alternating between near orthogonal subspaces. However, the search for the constraint farthest from the current point could add an inadmissible overload to the search process. It seems more convenient to *take into account all the constraints at each step*. This is the motivation of *generalized iterative scaling* [47], whose geometric interpretation is due to Csziszár [44]. Let Ξ be the space of probability distributions satisfying m expectation constraints:

$$\Xi = \left\{ p : \sum_x p(x) F_j(x) = a_j \;\; j = 1, \dots, m \right\} \tag{7.84}$$

The latter space is called a *linear family*. It is assumed, without loss of generality, that $F_j(x) \geq 0$ and $\sum_{j=1}^m F_j(x) = 1$. If the latter conditions do not apply for our problem, the original features F_j^o are affine scaled properly ($F_j = aF_j^o + b$) so that $\sum_{j=1}^m F_j(x) \leq 1$, and if the latter inequality is strict, an additional function ($F_{m+1} = 1 - \sum_{j=1}^m F_j$) and, thus, a new constraint, must be added. Let us also define the families

$$\Psi = \{\tilde{\pi}(x,j) = \pi(x)F_j(x) \;\; j = 1,\ldots,m\}$$

$$\tilde{\Psi}_1 = \left\{\tilde{\pi}(x,j) = \pi(x)F_j(x) : \sum_{j=1}^{m} \tilde{\pi}(x,j) = a_j \;\; j = 1,\ldots,m\right\}$$

$$\tilde{\Psi}_2 = \left\{\tilde{p}(x,j) = p(x)F_j(x) : \sum_{j=1}^{m} \tilde{p}(x,j) = a_j \;\; j = 1,\ldots,m\right\} \tag{7.85}$$

The latter definitions ensure that $\tilde{p}^*(x,j) = p^*(x)\tilde{\pi}(x,j)$. Furthermore, as $\tilde{\Psi}_1$ and $\tilde{\Psi}_2$ are linear families whose marginals with respect to j yield $\mathbf{a}$, then $\Psi = \tilde{\Psi}_1 \cap \tilde{\Psi}_2$. This implies that we may iterate alternating projections onto the latter spaces until convergence to the optimal p^* lying in the intersection. We define $\tilde{p}_{2t}(x,j) = p_t(x)F_j$, $t = 0,1,\ldots$, with $p_0 = \pi$. Then, let $\tilde{p}_{2t+1}$ be the I-projection of $\tilde{p}_{2t}$ on $\tilde{\Psi}_1$, and, $\tilde{p}_{2t+2}$ the I-projection of $\tilde{p}_{2t+1}$ on $\tilde{\Psi}_2$. The projection of $\tilde{p}_{2t}$ on $\tilde{\Psi}_1$ can be obtained by exploiting the fact that we are projecting on a family of distribution defined by a marginal. Consequently the projection can be obtained by *scaling the projecting pdf with respect to the value of the corresponding marginal*:

$$\begin{aligned}\tilde{p}_{2t+1}(x,j) &= \tilde{p}_{2t}(x,j)\frac{\sum_j \tilde{p}_{2t}(x,j)}{\sum_x \tilde{p}_{2t}(x,j)} \\ &= p_t(x)F_j(x)\frac{a_j}{a_{j,t}}, \quad a_{j,t} = \sum_x p_t(x)F_j(x)\end{aligned} \tag{7.86}$$

where $0/0 = 0$. Then, the projection of $\tilde{p}_{2t+2}$ is obtained by minimizing the cross entropy of pdfs in $\tilde{\Psi}_2$ from $\tilde{p}_{2t+1}$. That is, we must minimize:

$$\begin{aligned}D(\tilde{p}(x)||\tilde{p}_{2t+1}(x)) &= \sum_x \sum_{j=1}^{m}\left\{\tilde{p}(x,j)\log\frac{\tilde{p}(x,j)}{\tilde{p}_{2t+1}(x,j)}\right\} \\ &= \sum_x \sum_{j=1}^{m}\left\{p(x)F_j(x)\log\frac{p(x)F_j(x)}{p_t(x)F_j(x)\frac{a_j}{a_{j,t}}}\right\} \\ &= \sum_x p(x)\sum_{j=1}^{m} F_j(x)\left\{\log\frac{p(x)}{p_t(x)\frac{a_j}{a_{j,t}}}\right\} \\ &= \sum_x p(x)\left\{\log\frac{p(x)}{p_t(x)} + \sum_{j=1}^{m} F_j(x)\log\frac{a_{j,n}}{a_j}\right\} \\ &= \sum_x p(x)\left\{\log\frac{p(x)}{p_t(x)} + \log\prod_{j=1}^{m}\left(\frac{a_{j,n}}{a_j}\right)^{F_j(x)}\right\}\end{aligned} \tag{7.87}$$

In order to find a minimizer for the latter equation, the definition of the following quantity:

$$R_{t+1}(x) = p_t(x)\prod_{j=1}^{m}\left(\frac{a_j}{a_{j,n}}\right)^{F_j(x)} \quad \text{that is} \quad p_t(x) = \frac{R_{t+1}(x)}{\prod_{j=1}^{m}\left(\frac{a_j}{a_{j,n}}\right)^{F_j(x)}} \tag{7.88}$$

and the inclusion of a normalizing factor Z_{t+1} so that $R_{t+1}(x)/Z_{t+1}$ is a probability distribution, imply

$$\begin{aligned}
D(\tilde{p}(x)||\tilde{p}_{2t+1}(x)) &= \sum_x p(x)\left\{\log\frac{p(x)}{p_t(x)} + \log\prod_{j=1}^{m}\left(\frac{a_{j,t}}{a_j}\right)^{F_j(x)}\right\}\\
&= \sum_x p(x)\left\{\log\frac{p(x)}{p_t(x)\prod_{j=1}^{m}\left(\frac{a_j}{a_{j,t}}\right)^{F_j(x)}}\right\}\\
&= \sum_x p(x)\left\{\log\frac{p(x)}{R_{t+1}(x)} + \log\frac{1}{Z_{t+1}} - \log\frac{1}{Z_{t+1}}\right\}\\
&= \sum_x p(x)\left\{\log\frac{p(x)}{\frac{1}{Z_{t+1}}R_{t+1}(x)} + \log\frac{1}{Z_{t+1}}\right\}\\
&= D\left(p(x)||\frac{1}{Z_{t+1}}R_{t+1}(x)\right) + \log\frac{1}{Z_{t+1}}
\end{aligned} \tag{7.89}$$

Thus, the minimizing distribution is $p_{t+1}(x) = R_{t+1}(x)/Z_{t+1}$, the minimum being $1/Z_{t+1}$. Consequently we obtain

$$\tilde{p}_{2t+1}(x,j) = p_{t+1}(x)F_j(x), \quad p_{t+1}(x) = \frac{1}{Z_{t+1}}p_t(x)\prod_{j=1}^{m}\left(\frac{a_j}{a_{j,t}}\right)^{F_j(x)} \tag{7.90}$$

and the following recurrence relation, defining the *generalized iterative scaling* [47], is satisfied:

$$R_{t+1}(x) = R_t(x)\prod_{j=1}^{m}\left(\frac{a_j}{b_{j,t}}\right)^{F_j(x)}, \quad b_{j,t} = \sum_x R_t(x)F_j(x) \tag{7.91}$$

with $R_0 = p_0 = \pi$. Furthermore, the so-called *improved* generalized iterative scaling, described in the previous section, has the latter recurrence relation as starting point. However, this algorithm is designed from the maximization of the log-likelihood. More precisely, such maximization is performed by deriving the log-likelihood with respect to the multipliers. As we have shown in Eq. 7.72, the log-likelihood for a pdf belonging to the exponential family is given by

$$\ell(\mathbf{X}|\Lambda) = \sum_{i=1}^{N} \log(p_\Lambda(x_i)) = \sum_{i=1}^{N} (\Lambda \cdot F(x_i) - G(\Lambda) + \log \pi(x_i)) \tag{7.92}$$

If we are seeking for an iterative gradient-ascent approach $\Lambda' \leftarrow \Lambda + \Delta$, we need to choose, at each step Δ satisfying the following monoticity inequality $\ell(\mathbf{X}|\Lambda + \Delta) - \ell(\mathbf{X}|\Lambda) \geq 0$. Then, obtaining the following equivalence:

$$\ell(\mathbf{X}|\Lambda + \Delta) - \ell(\mathbf{X}|\Lambda) = \sum_{i=1}^{N} (\Delta \cdot F(x_i)) - G(\Lambda + \Delta) + G(\Lambda) \tag{7.93}$$

is straightforward. Then, from the definitions of $G(\Lambda)$ and $G(\Lambda + \Delta)$ in terms of the logarithm of the partition function we have

$$\begin{aligned}
&\sum_{i=1}^{N} (\Delta \cdot F(x_i)) - G(\Lambda + \Delta) + G(\Lambda) \\
&= \sum_{i=1}^{N} (\Delta \cdot F(x_i)) - \log\left(\sum_{x_i} e^{(\Lambda+\Delta)\cdot F(x_i)} \pi(x_i)\right) + \log\left(\sum_{x_i} e^{\Lambda\cdot F(x_i)} \pi(x_i)\right) \\
&= \sum_{i=1}^{N} (\Delta \cdot F(x_i)) - \log\left(\frac{\sum_{x_i} e^{(\Lambda+\Delta)\cdot F(x_i)} \pi(x_i)}{\sum_{x_i} e^{\Lambda\cdot F(x_i)} \pi(x_i)}\right) \\
&= \sum_{i=1}^{N} (\Delta \cdot F(x_i)) - \log\left(\frac{\sum_{x_i} p(x_i) Z(\Lambda) e^{\Delta\cdot F(x_i)}}{\sum_{x_i} p(x_i) Z(\Lambda)}\right) \\
&= \sum_{i=1}^{N} (\Delta \cdot F(x_i)) - \log\left(\frac{\sum_{x_i} p(x_i) e^{\Delta\cdot F(x_i)}}{\sum_{x_i} p(x_i)}\right) \\
&= \sum_{i=1}^{N} (\Delta \cdot F(x_i)) - \log\left(\sum_{x_i} p(x_i) e^{\Delta\cdot F(x_i)}\right) = \ell(\mathbf{X}|\Lambda + \Delta) - \ell(\mathbf{X}|\Lambda).
\end{aligned} \tag{7.94}$$

Then, as $-\log \alpha \geq 1 - \alpha$ for $\alpha > 0$, we have

$$\ell(\mathbf{X}|\Lambda+\Delta) - \ell(\mathbf{X}|\Lambda) \geq \underbrace{\sum_{i=1}^{N} (\Delta \cdot F(x_i)) + 1 - \sum_{x_i} \left(p(x_i) e^{\Delta\cdot F(x_i)}\right)}_{\mathcal{A}(\Delta|\Lambda)} \geq 0 \tag{7.95}$$

Then, the right side of the latter equation is a lower bound of the log-likelihood increment. Next formal task consists of posing that lower bound $\mathcal{A}(\Delta|\Lambda)$ in terms of the $\delta_j : \Delta = (\delta_1, \ldots, \delta_j, \ldots, \delta_m)$ so that setting to zero the partial derivatives with respect to each δ_j allows to obtain it. To that end, it is key to realize that the derivatives of $\mathcal{A}(\Delta|\Lambda)$ will leave each λ_j as a function of all the other ones due to the exponential (coupled variables). In order to decouple each λ_j so that each partial derivative depends only on it, a useful trick (see [17]) is to perform the following transformation on the exponential:

$$e^{\Delta \cdot F(x_i)} = e^{\sum_{j=1}^{m} \delta_j F_j(x_i)} = e^{\left(\sum_{j=1}^{m} F_j(x_i)\right) \sum_{j=1}^{m} \frac{\delta_j F_j(x_i)}{\sum_{j=1}^{m} F_j(x_i)}} \tag{7.96}$$

Then, let us define $p_j(x) = F_j(x)/(\sum_j F_j(x))$ the pdf resulting from the introduction of $\sum_j F_j(x)$ in the exponentiation. The real trick comes when Jensen's inequality for $p_j(x)$ is exploited. As we have seen in other chapters in the book (see for instance Chapter 3 on segmentation), given a convex function $\varphi(x)$ we have that $\varphi(E(x)) \leq E(\varphi(x))$. Therefore, as the exponential is convex we have that $e^{\sum_x p_j(x) q(x)} \leq \sum_x p(x) e^{q(x)}$. Then setting $q(x) = \sum_j F_j(x)$ we have that

$$\mathcal{A}(\Delta|\Lambda) \geq \underbrace{\sum_{i=1}^{N} \left(\sum_{j=1}^{m} \delta_j F_j(x_i) \right) + 1 - \sum_{x_i} p(x_i) \left(\sum_{j=1}^{m} p_j(x_i) e^{\delta_j \sum_{j=1}^{m} F_j(x_i)} \right)}_{\mathcal{B}(\Delta|\Lambda)}$$

$$= \sum_{i=1}^{N} \left(\sum_{j=1}^{m} \delta_j F_j(x_i) \right) + 1 - \sum_{x_i} p(x_i) \left(e^{\delta_j \sum_{j=1}^{m} F_j(x_i)} \right). \tag{7.97}$$

Then we have

$$\frac{\partial \mathcal{B}(\Delta|\Lambda)}{\partial \delta_j} = \sum_{i=1}^{N} F_j(x_i) - \sum_{x_i} p(x_i) \left(F_j(x_i) e^{\delta_j \sum_{k=1}^{m} F_k(x_i)} \right) \tag{7.98}$$

and setting each partial derivative to zero we obtain the updating equations. The difference between the expression above and the ones used in Alg. 19 is the inclusion of the conditional model used for formulating the classifier. The verification of these equations is left as an exercise (see Prob. 7.12). A faster version of the Improved Iterative Scaling algorithm is proposed in [86]. The underlying idea of this latter algorithm is to exploit tighter bounds by decoupling only part of the variables.

7.7 Bregman Divergences and Classification

7.7.1 Concept and Properties

If exponential families, described in the previous section, are a general formal way of describing different probability distributions, Bregman divergences [29] generalize the concept of divergence associated to each member of the family. We will reach this point later in this section. Let us first define formally a Bregman divergence. Such divergence $D_\phi(\mathbf{x}, \mathbf{y})$ between the arguments, which may be real-valued vectors or probability distributions, is characterized by a convex differentiable function $\phi(\cdot)$ called the *generator* so that

$$D_\phi(\mathbf{x}, \mathbf{y}) = \phi(\mathbf{x}) - \phi(\mathbf{y}) - (\mathbf{x} - \mathbf{y})^T \nabla \phi(\mathbf{y}) \tag{7.99}$$

When considering one-dimensional variables x and y, the first-order Taylor expansion of $\phi(x)$ at y, with $x \geq y$, is $\phi(x) = \phi(y) + \phi'(y)(x - y)$. Therefore, the Bregman divergence is usually seen as the *tail of the Taylor expansion* (the terms remaining after subtracting the first-order approximation). Thus, Bregman divergence is defined as the difference between the true value of $\phi(x)$ and the value of the tangent to $\phi(y)$ at x. Of course, the magnitude of this difference, the tail, for fixed x and y, depends on the generator. Some yet classical generators and their divergences are:

$$\begin{aligned}
\phi(\mathbf{x}) &= ||\mathbf{x}||^2 \;\Rightarrow D_\phi(\mathbf{x},\mathbf{y}) = ||\mathbf{x}-\mathbf{y}||^2 \text{ (Euclidean)},\\
\phi(\mathbf{x}) &= \sum_i x_i \log x_i \;\Rightarrow D_\phi(\mathbf{x},\mathbf{y}) = \sum_i x_i \log\frac{x_i}{y_i} \text{ (Kullback–Leibler)},\\
\phi(\mathbf{x}) &= -\sum_i \log x_i \;\Rightarrow D_\phi(\mathbf{x},\mathbf{y}) = \sum_i \left(\frac{x_i}{y_i} - \log\frac{x_i}{y_i} - 1\right) \text{ (Itakura–Saito)}
\end{aligned} \tag{7.100}$$

In general, Bregman divergences satisfy $D_\phi(\mathbf{x},\mathbf{y}) \geq 0$ with equality to 0 if and only if $\mathbf{x} = \mathbf{y}$. However, they are not *metrics* because neither the symmetry nor the triangle equality hold in general. However, we have the following pythagorean inequality:

$$D_\phi(\mathbf{x},\mathbf{y}) \geq D_\phi(\mathbf{z},\mathbf{y}) + D_\phi(\mathbf{x},\mathbf{z}) \tag{7.101}$$

where the equality holds when $\mathbf{z} = \min D_\phi(\mathbf{w},\mathbf{y})$ with $\mathbf{w} \in \Omega$ the so-called *Bregman* projection of $\mathbf{y}$ onto the convex set Ω (this is the generalization of information projection illustrated in Fig. 7.12). Thus, given the connection between Bregman divergences and information projection, it is not surprising that when considering distributions belonging to the exponential family, each distribution has associated a *natural* divergence. Actually, there is an interesting *bijection* between exponential pdfs and Bregman divergences [11]. Such bijection is established through the concept of *Legendre duality.* As we have seen in the previous section, the expressions

$$p_\Lambda(x) = e^{\Lambda\cdot F(x) - G(\Lambda)}\pi(x) \;\text{ and }\; G'(\Lambda) = E_\Lambda(F(x)) = \mathbf{a} \tag{7.102}$$

establish an interesting bijection between natural (Λ) and expectation $\mathbf{a}$ parameters, due to the strict convexity of $G(\cdot)$. More precisely, given a real-valued function $G(\cdot)$ in R^m. Then, its *conjugate function* $G^*(\cdot)$ is defined as

$$G^*(\mathbf{a}) = \sup_{\Lambda\in\Omega}\{\mathbf{a}\cdot\Lambda - G(\Lambda)\} \tag{7.103}$$

being $\Omega = \text{dom}(G)$. The latter equation is also known as the *Legendre transformation* of $G(\cdot)$. When $G(\cdot)$ is strictly convex and differentiable over $\Theta = \text{int}(\Omega)$, there is a unique Λ^* corresponding to the sup$(\cdot)$ in the latter equation. Then, by setting the gradient at this point to zero we obtain

$$\nabla(\mathbf{a}\cdot\Lambda - G(\Lambda))|_{\Lambda=\Lambda^*} = 0 \;\Rightarrow\; \mathbf{a} = \nabla G(\Lambda^*) \tag{7.104}$$

The inverse $\nabla G^{-1} : \Theta^* \to \Theta$, with $\Theta^* = \text{int}(\text{dom}(G^*))$, exists because ∇G is monotonic given the convexity of G. As G^* is a Legendre transform, we have that the first derivatives of G and G^* are inverse of each other: $\nabla G = (\nabla G^*)^{-1}$ and $(\nabla G)^{-1} = \nabla G^*$. Furthermore, the convexity of G implies

$$\begin{aligned}
&\mathbf{a}(\Lambda) = \nabla G(\Lambda) \;\text{ and }\; \Lambda(\mathbf{a}) = \nabla G^*(\mathbf{a})\\
&G^*(\mathbf{a}) = \Lambda(\mathbf{a})\cdot\mathbf{a} - G(\Lambda(\mathbf{a}))\;\forall\mathbf{a}\in\text{int}(\text{dom}(G^*))
\end{aligned} \tag{7.105}$$

which is key to formulate the exponential family in terms of Bregman divergences:

$$\begin{aligned}
p_{(G,\Lambda)}(\mathbf{x}) &= e^{\Lambda\cdot F(\mathbf{x})-G(\Lambda)}\pi(\mathbf{x})\\
&= e^{\{(\mathbf{a}\cdot\Lambda-G(\Lambda))+(F(\mathbf{x})-\mathbf{a})\cdot\Lambda\}}\pi(\mathbf{x})\\
&= e^{\{G^*(\mathbf{a})+(F(\mathbf{x})-\mathbf{a})\cdot\Lambda)\}}\pi(\mathbf{x})\\
&= e^{\{G^*(\mathbf{a})+(F(\mathbf{x})-\mathbf{a})\cdot\nabla G^*(\mathbf{a})\}}\pi(\mathbf{x})\\
&= e^{-\{G^*(\mathbf{a})+G^*(F(\mathbf{x}))-(F(\mathbf{x})-\mathbf{a})\cdot\nabla G^*(\mathbf{a})\}+G^*(F(\mathbf{x}))}\pi(\mathbf{x})\\
&= e^{-D_{G^*}(F(\mathbf{x}),\mathbf{a})b_{G^*}}\pi(\mathbf{x})
\end{aligned} \tag{7.106}$$

$D_{G^*}(F(\mathbf{x}),\mathbf{a})$ being the Bregman divergence generated by G^* and $b_{G^*} = e^{G^*(F(\mathbf{x}))}$. If the maximum entropy is not the underlying principle to build the model, the more standard generalization of a exponential pdf is given by setting $F(\mathbf{x}) = \mathbf{x}$. Anyway, given this new parameterization, the generalization of the maximum log-likelihood model inference, and even iterative scaling approaches, is straightforward. Furthermore, as we have noted above, there is a natural choice of G^* for defining a different class of distribution in the exponential family: Euclidean distance yields the Gaussian, Kullback–Leibler divergence achieves the multinomial, and the Itakura–Sahito distance provides the geometric distribution (see Prob. 7.13).

7.7.2 Bregman Balls and Core Vector Machines

Focusing on classification, let us consider a simple *one-class classification problem*: Given a set of observed vectorized patterns $\mathcal{S} = \{\mathbf{x}_1,\ldots,\mathbf{x}_N\}$, with $\mathbf{x}_i \in \mathbb{R}^k$, compute a simplified description of $\mathcal{S}$ that fits properly that set. Such description is the center $\mathbf{c}\in\mathbb{R}^k$ *minimizing the maximal distortion* $D(\mathbf{x},\mathbf{c}) \leq r$ with respect to $\mathcal{S}$, r (the radius) being a parameter known as *variability threshold* which must also be estimated. Thus, both the center and the radius must be estimated. In other words, the one-class classification problem can be formulated in terms of finding the *Minimum Enclosing Ball* (MEB) of $\mathcal{S}$. In this regard, the classic distortion (the Euclidean one) yields balls like

$||\mathbf{c}-\mathbf{x}||^2 \leq r^2$. Actually the MEB problem under the latter distortion is highly connected with a variant of *Support Vector Machines* (SVMs) classifiers [165], known as *Core Vector Machines* [158]. More precisely, the approach is known as *Ball Vector Machines* [157] which is faster, but only applicable in contexts where it can be assumed that the radius r is known beforehand.

Core Vector Machines (CVMs) inherit from SVMs the so-called *kernel trick*. The main idea behind the kernel trick in SVMs is the fact that when the original training data are not linearly separable, it may be separable when we project such data in a space of higher dimensions. In this regard, *kernels* $k(.,.)$, seen as dissimilarity functions, play a central role. Let $k_{ij} = k(\mathbf{x}_i, \mathbf{x}_j) = \varphi(\mathbf{x}_i)\cdot\varphi(\mathbf{x}_j)$ be the dot product of the *projections* through function $\varphi(\cdot)$ of vectors $\mathbf{x}_i, \mathbf{x}_j \in \mathcal{S}$. Typical examples of kernels are the polynomial and the Gaussian ones. Given vectors $\mathbf{x}_i = (u_1, u_2)$ and $\mathbf{x}_j = (v_1, v_2)$ it is straightforward to see that using $\varphi(\mathbf{x}) = (1, \sqrt{2}u_1, \sqrt{2}u_2, u_1^2, \sqrt{2}u_1u_2, u_2^2)$ yields the quadratic kernel $k(\mathbf{x}_i, \mathbf{x}_j) = \varphi(\mathbf{x}_i)\cdot\varphi(\mathbf{x}_j) = (\mathbf{x}_i \cdot \mathbf{x}_j + 1)^d$ with $d = 2$. Given the definition of a kernel and a set of vectors $\mathcal{S}$, if the kernel is symmetric, the resulting $|\mathcal{S}| \times |\mathcal{S}| = N \times N$ matrix $\mathbf{K}_{ij} = k(\mathbf{x}_i, \mathbf{x}_j)$ is the Gramm matrix if it is *positive semi-definite*, that is, it satisfies $\sum_i \sum_j \mathbf{K}_{ij} c_i c_j \geq 0$ for all choices of real numbers for all finite set of vectors and choices of real numbers c_i (Mercer's theorem). Gram matrices are composed of inner products of elements of a set of vectors which are linearly independent if and only if the determinant of $\mathbf{K}$ is nonzero.

In CVMs, the MEB problem is formulated as finding

$$\begin{aligned} &\min_{\mathbf{c},r} r^2 \\ &\text{s.t. } ||\mathbf{c}-\varphi(\mathbf{x}_i)||^2 \leq r^2 \;\; i=1,\ldots,N \end{aligned} \tag{7.107}$$

which is called the *primal problem*. The primal problem has as many constraints as elements in $\mathcal{S}$ because all of them must be inside the MEB. Thus, for any constraint in the primal problem there is a Lagrange multiplier in the Lagrangian, which is formulated as follows:

$$L(\mathcal{S}, \Lambda) = r^2 - \sum_{i=1}^{N} \lambda_i (r^2 - ||\mathbf{c}-\varphi(\mathbf{x}_i)||^2) \tag{7.108}$$

$\Lambda = (\lambda_1, \ldots, \lambda_N)^T$ being the multiplier vector. Consequently, the multipliers will be the unknowns of the *dual problem*:

$$\begin{aligned} &\max_{\Lambda} \Lambda^T \text{diag}(\mathbf{K}) - \Lambda^T \mathbf{K} \Lambda \\ &\text{s.t } \Lambda^T \mathbf{1} = 1, \;\Lambda \geq \mathbf{0} \end{aligned} \tag{7.109}$$

being $\mathbf{1} = (1, \ldots, 1)^T$ and $\mathbf{0} = (0, \ldots, 0)^T$. The solution to the dual problem $\Lambda^* = (\lambda_1^*, \ldots, \lambda_1^*)^T$ comes from solving a constrained quadratic programming (QP) problem. Then, the primal problem is solved by setting

$$\mathbf{c}^* = \sum_{i=1}^{N} \lambda_i^* \varphi(\mathbf{x}_i), \quad r = \sqrt{\Lambda^{*T}\mathrm{diag}(\mathbf{K}) - \Lambda^{*T}\mathbf{K}\Lambda^*} \tag{7.110}$$

In addition, when using the Lagrangian and the dual problem, the Karush–Kuhn–Tucker (KKT) condition referred to as *complementary slackness* ensures that $\lambda_i(r^2 - ||\mathbf{c} - \varphi(\mathbf{x}_i)||^2) = 0\ \forall i$. This means that when the ith equality constraint is not satisfied (vectors inside the hypersphere) then $\lambda_i > 0$ and otherwise (vectors defining the border, the so-called *support vectors*) we have $\lambda_i = 0$. In addition, using the kernel trick and $\mathbf{c}$ we have that the distances between projected points and the center can be determined without using explicitly the projection function

$$||\mathbf{c}^* - \varphi(\mathbf{x}_l)||^2 = \sum_{i=1}^{N}\sum_{j=1}^{N} \lambda_i^*\lambda_j^*\mathbf{K}_{ij} - 2\sum_{i=1}^{N}\lambda_i^*\mathbf{K}_{il} + \mathbf{K}_{ll} \tag{7.111}$$

This approach is consistent with the Support Vector Data Description (SVDD) [153] where one-class classification is proposed as a method for detecting outliers. What is interesting is that the use of kernels allows to generalize from hyperspherical regions in the Euclidean (original) domain to more-general regions in the transformed domain. This is important because in the general case the points belonging to a given class are not uniformly distributed inside a tight hypersphere. Thus, generalization is desirable, although the degree of generalization depends on the kernel chosen. In general, Gaussian kernels $k(\mathbf{x}_i, \mathbf{x}_j) = e^{-(\mathbf{x}_i - \mathbf{x}_j)^2/s^2}$ coming from an infinite dimension projection function yield better results than the polynomial ones, as we show in Fig. 7.13. In the one-class classification problem, support vectors may be considered outliers. In the polynomial case, the higher the degree the tighter the representation (nonoutliers are converted into outliers). Although in this case many input examples are accepted the hypershere is too sparse. On the contrary, when using a Gaussian kernel with an intermediate variance there is a trade-off between overfitting the data (for small variance) and too much generalization (higher variance).

Core Vector Machines of the type SVDD are highly comparable with the use of a different distance/divergence (not necessarily Euclidean) instead of using a different kernel. This allows to define *Bregman balls* and to face the MEB from a different perspective: the Smallest Enclosing Bregman Ball (SEBB) [119]. Then, given a Bregman divergence D_ϕ, a Bregman ball $\mathcal{B}_\phi^{(\mathbf{c},r)} = \{\mathbf{x} \in \mathcal{S} : D_\phi(\mathbf{c}, \mathbf{x}) \le r\}$. The definition of Bregman divergence applied to this problem is

$$D_\phi(\mathbf{c}, \mathbf{x}) = \phi(\mathbf{c}) - \phi(\mathbf{x}) - (\mathbf{c} - \mathbf{x})^T\nabla\phi(\mathbf{x}) \tag{7.112}$$

As any Bregman divergence is convex in the first argument, then $D_\phi(\mathbf{c}, \mathbf{x})$ is convex in $\mathbf{c}$. Such convexity implies the unicity of the Bregman ball. In addition, with respect to the redefinition of the Lagrangian, we have

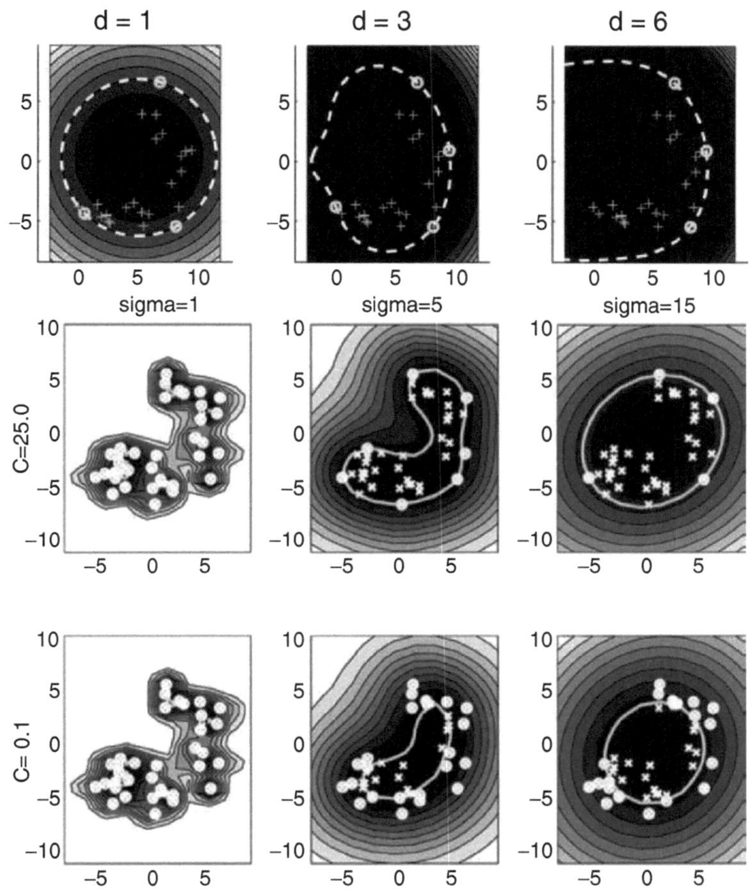

Fig. 7.13. *Top:* SVDD using different polynomial kernels (degrees). *Bottom:* using different Gaussian kernels (variances). In both cases, the *circles* denote the support vectors. Figure by D. M. J. Tax and R. P. W. Duin [153], (© Elsevier 2004). See Color Plates.

$$L(\mathcal{S},\Lambda) = r - \sum_{i=1}^{N} \lambda_i (r - D_\phi(\mathbf{c},\mathbf{x}_i)) \tag{7.113}$$

with $\Lambda > \mathbf{0}$ according to the *dual feasibility* KKT condition. Considering the definition of Bregman divergence in Eq. 7.112, the partial derivatives of L with respect to $\mathbf{c}$ and r are

$$\begin{aligned} \frac{\partial L(\mathcal{S},\Lambda)}{\partial \mathbf{c}} &= \nabla\phi(\mathbf{c})\sum_{i=1}^{N}\lambda_i - \sum_{i=1}^{N}\lambda_i \nabla\phi(\mathbf{x}_i)\,, \\ \frac{\partial L(\mathcal{S},\Lambda)}{\partial \mathbf{r}} &= 1 - \sum_{i=1}^{N}\lambda_i \end{aligned} \tag{7.114}$$

Then, setting the latter derivatives to zero we have

$$\sum_{i=1}^{N} \lambda_i = 1$$
$$\nabla\phi(\mathbf{c}) = \sum_{i=1}^{N} \lambda_i \nabla\phi(\mathbf{x}_i) \Rightarrow \mathbf{c}^* = \nabla^{-1}\phi\left(\sum_{i=1}^{N} \lambda_i \nabla\phi(\mathbf{x}_i)\right) \qquad (7.115)$$

Consequently, the dual problem to solve is

$$\max_{\Lambda} \sum_{i=1}^{N} \lambda_i D_\phi\left(\nabla^{-1}\phi\left(\sum_{i=1}^{N} \lambda_i \nabla\phi(\mathbf{x}_i)\right), \mathbf{x}_i\right)$$
$$\text{s.t.} \ \sum_{i=1}^{N} \lambda_i = 1, \ \Lambda \geq \mathbf{0} \qquad (7.116)$$

which is a generalization of the SVDD formulation. $\phi(\mathbf{z}) = ||\mathbf{z}||^2 = \mathbf{z} \cdot \mathbf{z} = \sum_{j=1}^{k} z_j^2$ with $\mathbf{z} \in \mathbb{R}^k$. More precisely, we have that $\nabla\phi(\mathbf{z}) = 2\mathbf{z} \Rightarrow (\nabla\phi(\mathbf{z}))^{-1} = 1/2\mathbf{z}$. Therefore, D_ϕ is defined as follows:

$$D_\phi\left(\sum_{i=1}^{N} \lambda_i \mathbf{x}_i, \mathbf{x}_i\right) = \left\|\sum_{i=1}^{N} \lambda_i \mathbf{x}_i\right\|^2 - ||\mathbf{x}_i||^2 + \left(\sum_{i=1}^{N} \lambda_i \mathbf{x}_i - \mathbf{x}_i\right)^T 2\mathbf{x}_i$$
$$= ||\sum_{i=1}^{N} \lambda_i \mathbf{x}_i - \mathbf{x}_i||^2 \qquad (7.117)$$

and the dual problem is posed as follows:

$$\max_{\Lambda} \sum_{i=1}^{N} \lambda_i \left\|\sum_{j=1}^{N} \lambda_j \mathbf{x}_j - \mathbf{x}_i\right\|^2$$
$$\text{s.t.} \ \sum_{i=1}^{N} \lambda_i = 1, \ \Lambda \geq \mathbf{0} \qquad (7.118)$$

when the Bregman generator is the Euclidean distance. The key question here is that there is a bijection between generators and values of the inverse of the generator derivative which yields $\mathbf{c}^*$ (the so-called *functional averages*). As we have seen, for the Euclidean distance, the functional average is $\mathbf{c}_j^* = \sum_{i=1}^{N} \lambda_i \mathbf{x}_{j,i}$ (arithmetic mean). For the Kullback–Leibler divergence we have $\mathbf{c}_j^* = \prod_{i=1}^{N} \mathbf{x}_{j,i}^{\lambda_i}$ (geometric mean), and for the Itakura–Saito distance we obtain $\mathbf{c}_j^* = 1/\sum_{i=1}^{N} \frac{\lambda_i}{\mathbf{x}_{j,i}}$ (harmonic mean). Then, the solution to the SEBB problem depends on the choice of the Bregman divergence. Anyway the resulting dual problem is complex and it is more practical to compute $\mathbf{g}^*$, an estimation of $\mathbf{c}^*$, by solving the following problem:

$$\min_{\mathbf{g},r'} r'$$
$$\text{s.t. } ||\mathbf{g} - \nabla\phi(\mathbf{x}_i)||^2 \leq r', \quad i = 1, \dots, N \tag{7.119}$$

which exploits the concept of *functional averages* and its connection with the definition $\nabla\phi(\mathbf{c}^*) = \sum_{i=1}^{N} \lambda_i \nabla\phi(\mathbf{x}_i)$. It can be proved that $\mathbf{g}^*$ is a good approximation of the optimal center. More precisely

$$D_\phi(\mathbf{x}, \nabla^{-1}\phi(g)) + D_\phi(\nabla^{-1}\phi(g), \mathbf{x}) \leq (1 + \epsilon^2)r'^*/f \tag{7.120}$$

f being the minimal nonzero value of the Hessian norm $||H\phi||$ inside the convex closure of $\mathcal{S}$ and ϵ the error assumed for finding the center of the MEB within an approximation algorithm using the Euclidean distance: it is assumed that $r \leq (1+\epsilon)r^*$, see for instance [33], and points inside this ball are called *core sets*. Such algorithm, named BC, can be summarized as follows: (i) choose at random $\mathbf{c} \in \mathcal{S}$; (ii) for a given number of iterations $T - 1$, for $t = 1, \dots, T-1$: set $\mathbf{x} \leftarrow \arg\max_{\mathbf{x}' \in \mathcal{S}} ||\mathbf{c} - \mathbf{x}'||^2$ and then set $\mathbf{c} \leftarrow \frac{t}{t+1}\mathbf{c} + \frac{1}{t+1}\mathbf{x}$. The underlying idea of the latter algorithm is to move along the line between the current estimation of the center and the new one. This approach is quite efficient for a large number of dimensions. The adaptation to find the center when using Bregman Divergerces (BBC algorithm) is straigthforward: simply change the content in the main loop: set $\mathbf{x} \leftarrow \arg\max_{\mathbf{x}' \in \mathcal{S}} D_\phi(\mathbf{c}, \mathbf{x}')$ and then set $\mathbf{c} \leftarrow \nabla^{-1}\phi\left(\frac{t}{t+1}\nabla\phi(\mathbf{c}) + \frac{1}{t+1}\nabla\phi(\mathbf{x})\right)$. This method (see results in Fig. 7.14) converges quite faster than BC in terms of the error $(D_\phi(\mathbf{c}, \mathbf{c}^*) + D_\phi(\mathbf{c}^*, \mathbf{c}))/2$. Another method, less accurate than BBC but better than it in terms of rate of convergence, and better than BC, is MBC. The MBC algorithm takes into account the finding of g^*: (i) define a new set of transformed vectors: $\mathcal{S}' \leftarrow \{\nabla\phi(\mathbf{x}) : \mathbf{x} \in \mathcal{S}\}$; (ii) call BC $\mathbf{g}^* \leftarrow BC(\mathcal{S}, T)$; (ii) obtain $c \leftarrow \nabla^{-1}\phi(\mathbf{c})$. Anyway, whatever the method used, once the $\mathbf{c}^*$ (or an approximation) is obtained, the computation of r^* is straightforward. Similar methods have been used to simplify the QP problem in CVMs or SVDDs.

Bregman balls (and ellipsiods) have been recently tested in contexts of one-class classification [118]. Suppose, for example, that we have a Gaussian distribution. As the Gaussian pdf belongs to the exponential family, the optimal Bregman divergence to choose is the Kullback–Leibler divergence, in order to find the corresponding ball and to detect supports for classifying test vectors (see Prob. 7.14).

7.7.3 Unifying Classification: Bregman Divergences and Surrogates

Bregman divergences are recently recognized as a unifying formal tool for understanding the generalization error of classifiers. The proper theoretical framework is that of the *surrogates* (upper bounds of the empirical risk) [12]. Here we consider, for the sake of simplicity, the binary classification problem, that is, we have a sample $\mathcal{S} = (\mathcal{X}, \mathcal{Y})$ where $\mathcal{X} = \{\mathbf{x}_1, \dots, \mathbf{x}_N\}$ and

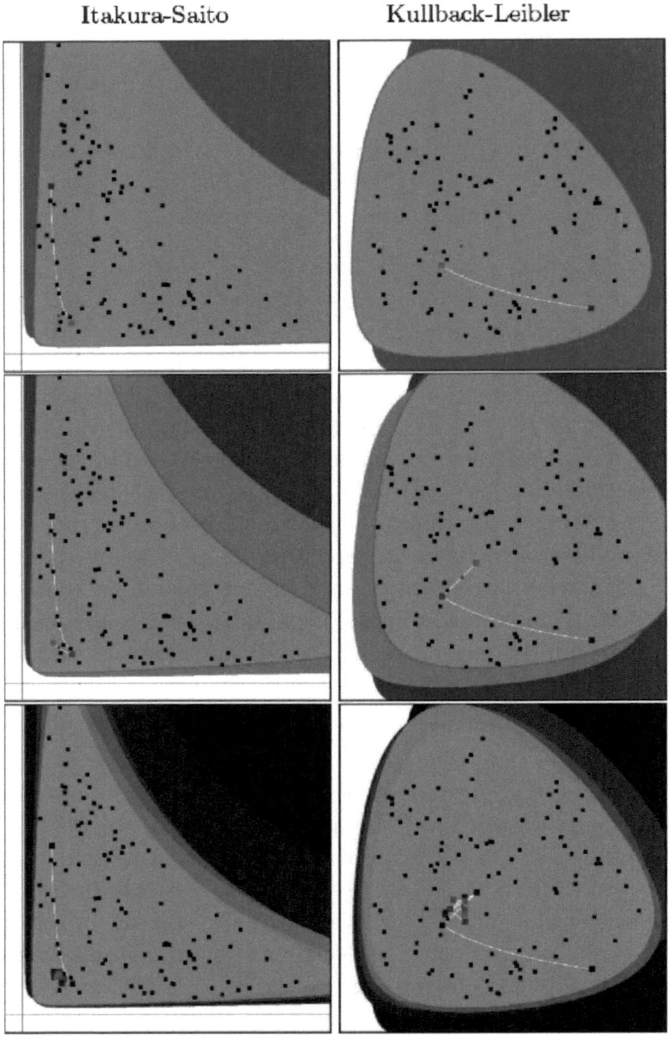

Fig. 7.14. BBC results with $k = 2$ for different Bregman divergences: Itakura–Saito vs. Kullback–Leibler. (Figure by courtesy of Richard Nock).

$\mathcal{Y} = \{+1, -1\}$, and a classifier $h : \mathcal{X} \rightarrow \mathcal{Y}$. Then $h(\mathbf{x}) = y$ indicates that y is the *correct class* of $\mathbf{x}$. Therefore, $sign(yh(\mathbf{x}))$ indicates whether the classification is correct (positive) or incorrect (negative). As we have seen along this chapter (see the formulation of random forests), and also partially in Chapter 5 (RIC), the generalization error is defined as $GE = P_{\mathcal{X},\mathcal{Y}}(y \neq sign(h(\mathbf{x})) = E_{\mathcal{X},\mathcal{Y}}\ell(y, h(\mathbf{x}))$, $\ell(.,.)$ being a *loss function*. The GE is typically estimated by the average of the loss given both the classifier and the sample: $\hat{R}(h, \mathcal{S}) = \frac{1}{N}\sum_{i=1}^{N}\ell(y_i, h(\mathbf{x}_i))$, the *empirical risk* ER. Surrogates emerge as a consequence of the hardness of minimizing the ER. Of course, such hardness depends on the properties of the loss function chosen. The typical one is the 0/1-loss function: $\ell^{0/1}(y, h(\mathbf{x})) = I(h(\mathbf{x}) \neq y) = I(sign(yh(\mathbf{x})) \neq +1)$, $I(\cdot)$ being an indicator function. In that case, $\hat{R}$ is renamed $\epsilon^{0/1}$ and scaled by N to remove the average. However, the ER using this loss function is hard to minimize. However, and for the sake of clarity, in this section it is more convenient to: (i) assume that either $h : \mathcal{X} \rightarrow \mathbb{R}$ or $h : \mathcal{X} \rightarrow [0, 1]$; $\mathcal{Y} = \{0, 1\}$ and $\mathcal{Y}^* = \{-1, 1\}$. Mapping h to reals allows to consider $|h|$, the confidence on the classifier, and considering the new $\mathcal{Y}$ and $\mathcal{Y}^*$, which are equivalent in terms of labeling an example, gives more flexibility. In this regard, the 0/1 loss may be defined either as: $\ell_{\mathbb{R}}^{0/1}(y^*, h(\mathbf{x}) = I(\sigma(h(\mathbf{x})) \neq y^*))$ ($\sigma(z) = +1$ if $z \geq 0$ and -1 otherwise) if the image of h is $\mathbb{R}$, and $\ell_{[0,1]}^{0/1}(y, h(\mathbf{x}) = I(\tau(h(\mathbf{x})) \neq y))$ ($\tau(z) = 1$ if $z \geq 1/2$ and 0 otherwise) if the image of h is $[0, 1]$. Thus, $\epsilon_{\mathbb{R}}^{0/1}$ and $\epsilon_{[0,1]}^{0/1}$ are defined consequently. Anyway, surrogates relax the problem to estimate a tight upper bound of $\epsilon^{0/1}$ (whatever the notation used) by means of using different, typically convex, loss functions. It turns out that some of these functions are a subset of the Bregman divergences. The most typical are (see Fig. 7.15)

$$\begin{aligned} \ell_{\mathbb{R}}^{exp}(y^*, h(\mathbf{x})) &= e^{-y^* h(\mathbf{x})} \quad \text{(exponential loss)} \\ \ell_{\mathbb{R}}^{log}(y^*, h(\mathbf{x})) &= \log(1 + e^{-y^* h(\mathbf{x})}) \quad \text{(logistic loss)} \\ \ell_{\mathbb{R}}^{sqr}(y^*, h(\mathbf{x})) &= (1 - y^* h(\mathbf{x}))^2 \quad \text{(squared loss)} \end{aligned} \tag{7.121}$$

Surrogates ϵ must satisfy that $\epsilon^{0/1}(h, \mathcal{S}) \leq \epsilon(h, \mathcal{S})$. First of all, in order to define *permissible surrogates* based on a given loss function $\ell(.,.)$, it must satisfy three properties: (i) $\ell(.,.)$ (*lower bounded by zero*); (ii) $\arg\min_x \epsilon_{[0,1]}(x, \mathcal{S}) = q$ x being the output of the classifier mapped to the interval $[0, 1]$, $\epsilon_{[0,1]}$ an empirical risk based on a loss $\ell_{[0,1]}$ when the output

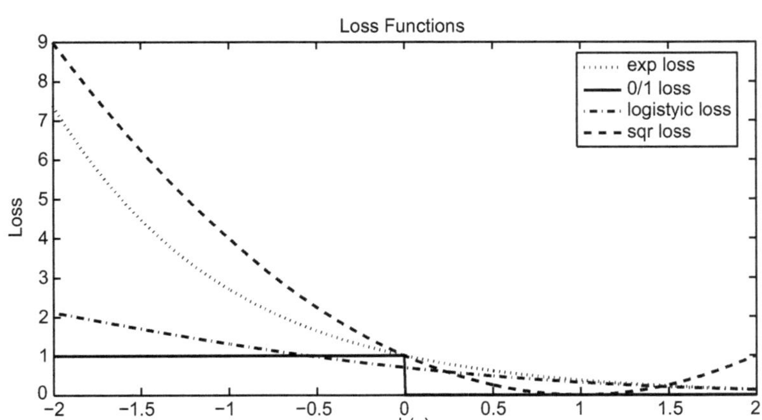

Fig. 7.15. The 0/1 loss functions and other losses used for building surrogates.

of the classifier is mapped to $[0, 1]$, and $q = \hat{p}(y|\mathbf{x})$ the approximation of a conditional pdf p given by proportion of positive examples with observation $\mathbf{x}$ (this is the *proper scoring rule*); and (iii) $\ell(y, h(\mathbf{x})) = \ell(1-y, 1-h(\mathbf{x}))$ for $y \in \{0, 1\}$ and $h(\mathbf{x}) \in [0, 1]$ (*symmetry property*). Given the latter properties over loss functions, a function $\phi : [0, 1] \to \mathbb{R}_+$ is *permissible* if and only if: $-\phi$ is differentiable on $(0, 1)$, is strictly concave, symmetric around $x = 1/2$, and $a_\phi = -\phi(0) = -\phi(1) \geq 0$ (then $b_\phi = -\phi(1/2) - a_\phi > 0$). Some examples of permissible functions, which must be scaled so that $\phi(1/2) = -1$, are

$$\begin{aligned} \phi_M(x) &= -\sqrt{x(1-x)} \text{ (Matsushita's error)} \\ \phi_Q(x) &= x \log x + (1-x)\log(1-x) \text{ (Bit entropy)} \\ \phi_B(x) &= -x(1-x) \text{ (Gini index)} \end{aligned} \tag{7.122}$$

Our main concern here is that the link between permissible functions and surrogates is that any loss $\ell(y, h)$ is properly defined and satisfies the three properties referred above if and only if $\ell(y, h) = D_\phi(y, h)$, being $D_\phi(.,.)$ *a Bregman divergence with a permissible generator* $\phi(\cdot)$ [120]. Using again the *Legendre transformation*, the Legendre conjugate of ϕ is given by

$$\phi^*(x) = \sup_{u \in dom(\phi)} \{ux - \phi(u)\} \tag{7.123}$$

As we have seen in Section 7.7.1, $\nabla\phi = \nabla^{-1}\phi^*$ and $\nabla\phi^* = \nabla^{-1}\phi$. In addition:

$$\phi^*(x) = x\nabla^{-1}\phi(x) - \phi(\nabla^{-1}\phi(x)) \tag{7.124}$$

Therefore, for $y \in \{0, 1\}$ we have

$$\begin{aligned} D_\phi(y, \nabla^{-1}\phi(h)) &= \phi(y) - \phi(\nabla^{-1}\phi(h)) - (y - \nabla^{-1}\phi(h))\nabla\phi(\nabla^{-1}\phi(h)) \\ &= \phi(y) - \phi(\nabla^{-1}\phi(h)) - (y - \nabla^{-1}\phi(h))h \\ &= \phi(y) - \phi(\nabla^{-1}\phi(h)) - yh + \nabla^{-1}\phi(h)h \\ &= \phi(y) - yh + \nabla^{-1}\phi(h)h - \phi(\nabla^{-1}\phi(h)) \\ &= \phi(y) - yh + \phi^*(h) \end{aligned} \tag{7.125}$$

As $D_\phi(y, \nabla^{-1}\phi(h)) = D_\phi(1-y, 1-\nabla^{-1}\phi(h))$ because the Bregman divergence is a loss function, and loss functions satisfy symmetry, we have

$$D_\phi(1-y, 1-\nabla^{-1}\phi(h)) = \phi(1-y) - (1-y)h + \phi^*(-h) \tag{7.126}$$

by applying $\nabla^{-1}\phi(-x) = 1 - \nabla^{-1}\phi(-x)$. Then, considering that $\phi^*(x) = \phi^*(-x) + x$, $a_\phi = -\phi(0) = -\phi(1) \geq 0$, and combining both definitions of divergence, we obtain

$$D_\phi(y, \nabla^{-1}\phi(h)) = \phi^*(-y^*h) + a_\phi = D_\phi(0, \nabla^{-1}\phi(-y^*h)) = D_\phi(1, \nabla^{-1}\phi(y^*h)) \tag{7.127}$$

where proving the latter two equivalences is straightforward. Thus, $\phi^*(\cdot)$ seems to be a Bregman divergence (it is, because $a_\phi = 0$ for well-known permissible ϕ). Then, we are going to define $F_\phi(x) = (\phi^*(-x) - a_\phi)/b_\phi$, which, for $a_\phi = 0, b_\phi = 1$, yields $F_\phi(x) = \phi^*(-x)$. This new function is strictly convex and satisfies $\ell_{\mathbb{R}}^{0/1}(y^*, h) \leq F_\phi(y^*h)$ which ensures that F_ϕ defines a surrogate because

$$\epsilon^{0/1}(h, \mathcal{S}) \leq \epsilon_\phi^{0/1}(h, \mathcal{S}) = \sum_{i=1}^{N} F_\phi(y_i^* h(\mathbf{x}_i)) = \sum_{i=1}^{N} D_\phi(0, \nabla^{-1}\phi(-y_i^* h(\mathbf{x}_i))) \tag{7.128}$$

When using the latter definitions of surrogates it is interesting to remind that we have

$$\begin{aligned} F_\phi(y^*h) &= -y^*h + \sqrt{1 + (y^*h)^2} \text{ for } \phi_M \\ F_\phi(y^*h) &= \log\left(1 + e^{-y^*h}\right) \text{ for } \phi_Q \\ F_\phi(y^*h) &= (1 - y^*h^2) \text{ for } \phi_B \end{aligned} \tag{7.129}$$

which correspond in the two latter cases to the losses defined above (logistic and squared). Given the basic theory, how to apply it to a linear-separator (LS) classifier like the one coming from Boosting? Such classifier, $H(\mathbf{x}_i) = \sum_{t=1}^{T} \alpha_t h_t(\mathbf{x}_i)$, is composed of weak classifiers h_t and leveraging coefficients α_t. First of all, we are going to assimilate $\nabla^{-1}\phi(-y_i^* h(\mathbf{x}_i))$ to the weights w_i used in the sampling strategy of boosting. This assimilation comes naturally from the latter definitions of $F_\phi(y^*h)$: the more badly classified the example the higher the loss and the higher the weights and vice versa. Firstly, it is interesting to group the disparities between all classifiers and all the examples in a unique matrix $\mathbf{M}$ of dimensions $N \times T$:

$$\mathbf{M} = -\begin{pmatrix} y_1^* h_1(\mathbf{x}_1) & \ldots & y_1^* h_t(\mathbf{x}_1) & \ldots & y_1^* h_T(\mathbf{x}_1) \\ \vdots & & \vdots & & \vdots \\ y_i^* h_1(\mathbf{x}_i) & \ldots & y_i^* h_t(\mathbf{x}_i) & \ldots & y_i^* h_T(\mathbf{x}_i) \\ \vdots & & \vdots & & \vdots \\ y_N^* h_1(\mathbf{x}_N) & \ldots & y_N^* h_t(\mathbf{x}_N) & \ldots & y_N^* h_T(\mathbf{x}_N) \end{pmatrix} \tag{7.130}$$

Then, a disparity (*edge*, in this context) between a boosted classifier H and the class y_i^* corresponding to example $\mathbf{x}_i$ is given by the product of the ith row of $\mathbf{M}$ and $\boldsymbol{\alpha} = (\alpha_1, \ldots, \alpha_T)^T$: $-\mathbf{M}_i\boldsymbol{\alpha} = y_i^* \left(\sum_{i=1}^{T} \alpha_t h_t(\mathbf{x}_i)\right)$. On the other hand, the disparity between a weak classifier and $\mathcal{S}$ is given by the product of the transpose of the t-th column of $-\mathbf{M}$ and $\mathbf{w} = (w_1, \ldots, w_N)^T$: $-\mathbf{M}_t^T\mathbf{w} = \sum_{i=1}^{T} w_i y_i h_t(\mathbf{x}_i)$. Here, weights are of fundamental importance because they allow to compute the optimal leveraging coefficients. We remind that weights are assimilated to inverses of the derivative of ϕ:

$$w_i = \nabla^{-1}\phi(-y_i^* h(\mathbf{x}_i)) \tag{7.131}$$

A key element for finding the optimal classifier through this approach is the *Bregman–Pythagoras theorem* applied to this context:

$$D_\phi(\mathbf{0},\mathbf{w}) = D_\phi(\mathbf{0},\mathbf{w}_\infty) + D_\phi(\mathbf{w}_\infty,\mathbf{w}) \tag{7.132}$$

where $\mathbf{w}_\infty$ are the optimal weights, and the vectorial notation of D_ϕ consists of $D_\phi(\mathbf{u},\mathbf{v}) = \sum_i D_\phi(u_i,v_i)$ (component-wise sum). Then, the idea is to design an iterative algorithm for minimizing $D_\phi(\mathbf{0},\mathbf{w})$ in $\Omega \ni \mathbf{w}$. More precisely

$$\min_H \epsilon_\phi^{0/1}(H,\mathcal{S}) = \min_{\mathbf{w}\in\Omega} D_\phi(\mathbf{0},\mathbf{w}) \tag{7.133}$$

Such iterative algorithm starts by setting $\mathbf{w}_1 \leftarrow \nabla^{-1}\phi(0)\mathbf{1}$, that is, $w_{1,i} = 1/2,\ \forall i$ because $\nabla^{-1}\phi(0) = 1/2$. The initial leverage weights are set to zero: $\boldsymbol{\alpha}_1 \leftarrow \mathbf{0}$. The most important part of the algorithm is how to update $\mathbf{w}_{j+1}$, given $\mathbf{w}_j$

$$w_{j+1,i} \leftarrow \nabla^{-1}\phi(\mathbf{M}_i(\boldsymbol{\alpha}_j + \boldsymbol{\delta}_j)) \ , \tag{7.134}$$

This is consistent with the idea that $\mathbf{M}_i\boldsymbol{\alpha}$ is associated to an edge of H. In this formulation, $\boldsymbol{\alpha}_j + \boldsymbol{\delta}_j$ is the solution to the following nonlinear system:

$$\sum_{i=1}^{N} M_{it}\left(\nabla^{-1}\phi(\mathbf{M}_i(\boldsymbol{\alpha}_j + \boldsymbol{\delta}_j))\right) = 0 \tag{7.135}$$

where $t \in \mathcal{T}_j$ and $\mathcal{T}_j \subseteq \{1,2,\ldots,T\}$, typically $|\mathcal{T}_j| = 1$, which means that only a single parameter is updated at each iteration. Then we set $\boldsymbol{\alpha}_{j+1} \leftarrow \boldsymbol{\alpha}_j + \boldsymbol{\delta}$. This setting ensures that $D_\phi(\mathbf{0},\mathbf{w}_{j+1})$ decreases with respect to $D_\phi(\mathbf{0},\mathbf{w}_j)$ by following the rule:

$$D_\phi(\mathbf{0},\mathbf{w}_{j+1}) = D_\phi(\mathbf{0},\mathbf{w}_j) - D_\phi(\mathbf{w}_{j+1},\mathbf{w}_j) \tag{7.136}$$

which may be proved by adapting the definition of $D_\phi(.,.)$ to the vectorial case. In addition, the computation of $\mathbf{w}_{j+1}$ ensures that $\mathbf{w}_j \in Ker\mathbf{M}^T|_{\mathcal{T}_j}$, where the notation $|_{\mathcal{T}_j}$ indicates *discarding columns* $j \notin \mathcal{T}_j$, and Ker is the typical null space, that is $Ker\mathbf{A} = \{\mathbf{u} \in \mathbb{R}^N : \mathbf{A}\mathbf{u} = 0\}$. This latter link between null spaces and weights (edges) can be summarized by the following equivalence:

$$\mathbf{w}_j \in Ker\mathbf{M}^T|_{\{t\}} \Leftrightarrow -\mathbf{1}_{\{t\}}\mathbf{M}^T\mathbf{w}_j = 0 \tag{7.137}$$

where we assume $\mathcal{T}_j = \{t\}$, and the right hand of the equivalence defines an edge of h_t. The efficient convergence of the algorithm is ensured when the set of features considered in each iteration works slightly better than random (weak learners assumption), that is, when assuming $h_t \in \{-1,1\}$ we have

$$\mathbf{1}_{\{t\}}\mathbf{M}^T\mathbf{w}_j > Z_j\gamma_j \quad \Leftrightarrow \quad \left|\frac{1}{Z_j}\epsilon_{\mathbb{R}}^{0/1}(h_t,\mathcal{S}) - \frac{1}{2}\right| \geq \gamma_j > 0 \tag{7.138}$$

that is, the classifier h_t is better than random by an amount of γ_j. In the latter formulation $Z_j = ||\mathbf{w}_j||_1 = \sum_i |w_{j,i}|$ (L1 norm). Furthermore, the convergence conditions (no new update needed) are

$$\boldsymbol{\delta} = 0\ \forall \mathcal{T}_j \Leftrightarrow \mathbf{w}_j \in Ker\mathbf{M}^T|_{\mathcal{T}_j}\ \forall \mathcal{T}_j \Leftrightarrow \mathbf{w}_j \in Ker\mathbf{M}^T \Leftrightarrow \mathbf{w}_j = \mathbf{w}_\infty \tag{7.139}$$

The progress of the algorithm can be illustrated graphically by representing the weight space Ω as a non-Riemanian manifold (see Fig. 7.16). The algorithm is dubbed ULS *Universal Linear Separator* and it is described in Alg. 20. It turns out, that when we choose $\phi(x) = e^{-x}$ we obtain the Adaboost algorithm with un-normalized weights. Thus, ULS is a general (Universal) method for learning LSs. Furthermore, the ULS approach may be extended to other classifiers like the ones presented along this chapter. For instance, the basic insight to adapt ULS to decision trees is to consider *each node of the tree as a weak classifier*. In this regard, the output of each internal weak classifier (test) is $h_j \in \{0, 1\}$ (output of a boolean test, so we have a binary tree H). On the other hand, the output of a leaf node $l \in \partial H$ will be $h_l \in \{-1, 1\}$. Therefore,

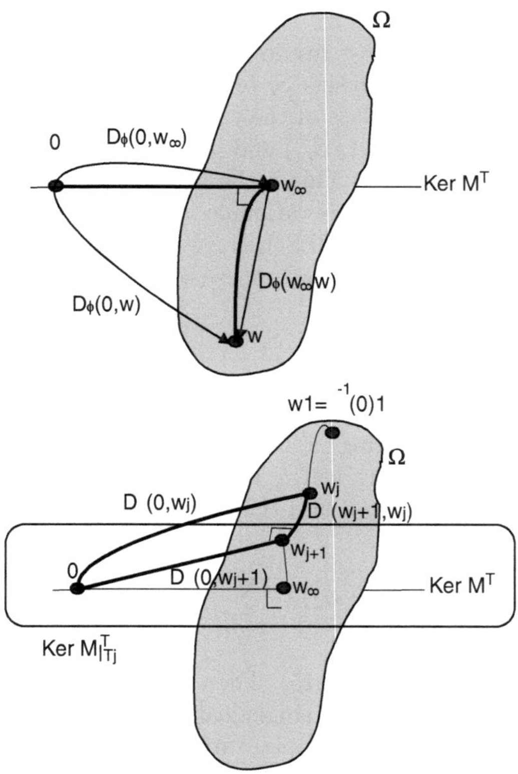

Fig. 7.16. ULS iterative minimization. *Top:* Bregman–Pythagoras theorem. *Bottom:* geometric representation of the iterative algorithm for minimizing $D_\phi(\mathbf{0}, \mathbf{w}_j)$.

Algorithm 20: ULS algorithm

Input: $\mathbf{M}$,$\phi(.)$
Initialize $\mathbf{w}_0 = \nabla^{-1}\phi(0)\mathbf{1}$, $\boldsymbol{\alpha}_0 = \mathbf{0}$
for *j=1 to T* **do**
 Pick $\mathcal{T}_j \subseteq \{1, 2, \ldots, T\}$ being $|\mathcal{T}_j|$ small for computational reasons: Typically $|\mathcal{T}_j| = 1$
 If $|\mathcal{T}_j| = 1$ then set $\mathcal{T}_j = \{t\}$
 Set $\boldsymbol{\delta}_j = \mathbf{0}$
 $\forall t \in \mathcal{T}_j$ select $\delta_{j,t}$ so that

$$\sum_{i=1}^{N} M_{it} \underbrace{\left(\nabla^{-1}\phi(\mathbf{M}_i(\boldsymbol{\alpha}_j + \boldsymbol{\delta}_j))\right)}_{w_{j+1,i}} = 0 \tag{7.140}$$

 Set $\boldsymbol{\alpha}_{j+1} = \boldsymbol{\alpha}_j + \boldsymbol{\delta}$
end
Output: Strong classifier:
$H(\mathbf{x}) = \sum_{t=1}^{T} \alpha_t h_t(\mathbf{x})$

each sequence of decisions (test results) yields a given label. As the tree induces a partition of $\mathcal{S}$ into subsets $\mathcal{S}_l$ (examples arriving to each leaf l) the probability that the output class of the tree is $+1$ is the proportion between the samples of $\mathcal{S}_l$ labeled as $+1$ ($\mathcal{S}_l^+$) and $|\mathcal{S}_l|$, and the same rationale applies for the probability of class -1. This is the base of weak classifiers associated to the leaves of the tree, and this rationale leads to LDS or *Linear Decision Tree*. To apply the concept of ULS here, we must estimate the leveraging coefficients for each node. In this case, these coefficients can be obtained from

$$\alpha_l = \frac{1}{h_l}\left(\nabla\phi\left(\frac{|\mathcal{S}_l^+|}{|\mathcal{S}_l|}\right) - \sum_{t\in\mathcal{P}_l} \alpha_t h_t\right) \tag{7.141}$$

$\mathcal{P}_l$ being the classifiers belonging to the path between the root and leaf l. Then, for an observation $\mathbf{x}$ reaching leaf l, we have that $H(\mathbf{x}) = \nabla\phi(|\mathcal{S}_l^+|/|\mathcal{S}_l|)$. Therefore $\nabla^{-1}\phi(\mathbf{x}) = |\mathcal{S}_l^+|/|\mathcal{S}_l|$, and the surrogate is defined as follows:

$$\epsilon_{\phi}^{\mathbb{R}}(H, \mathcal{S}) = \sum_{i=1}^{N} F_{\phi}(y_i^* H(\mathbf{x}_i)) = \sum_{l\in\partial H} |\mathcal{S}_l|\left(-\phi\left(\frac{|\mathcal{S}_l^+|}{|\mathcal{S}_l|}\right)\right) \tag{7.142}$$

if $a_\phi = 0$ and $b_\phi = 1$ (see Prob. 7.15). Then, the extension of ULS to trees is trivial. Just start by a tree with a single leaf and at each moment expand the leaf (corresponding to a given variable) satisfying

$$\sum_{l\in\partial H} |\mathcal{S}_l|\left(-\phi\left(\frac{|\mathcal{S}_l^+|}{|\mathcal{S}_l|}\right)\right) - \sum_{l\in\partial H'} |\mathcal{S}_l|\left(-\phi\left(\frac{|\mathcal{S}_l^+|}{|\mathcal{S}_l|}\right)\right) < 0 \tag{7.143}$$

where H is the current tree and H' is the new one after the expansion of a given leaf. The leverage coefficients of the new leafs may be computed at each iteration given the ones of the interior nodes.

Problems

7.1 Rare classes and Twenty Questions
The purpose of "Twenty Questions" (TQ) is to find a testing strategy which performs the minimum number of tests for finding the true class (value of Y). For doing so, the basic idea is to find the test dividing the population in masses as equal as possible.

1. Given the sequence of questions Q_t, check that this strategy is consistent with finding the test X_{t+1} maximizing $H(Y|Q_t, X_{t+1})$.

2. What is the result of applying the latter strategy to the data in Table 7.2?

7.2 Working with simple tags

1. Given the image examples and tags of Fig. 7.3, extract all the binary relationships (the number of tag may be repeated in the relationship, for instance $5 \uparrow 5$) and obtain the classification tree.

2. Compare the results, in terms of complexity of the tree and classification performance with the case of associating single tags to each test.

7.3 Randomization and multiple trees

1. Given the image examples and tags of Fig. 7.3, and having extracted $\mathbf{B}$, use randomization and the multiple-tree methodology to learn shallow trees.

2. Is the number of examples enough to learn all posteriors. If not, expand the training set conveniently.

7.4 Gini index vs. entropy
An alternative measure to entropy for quantifying node impurity is the Gini index defined as

$$G(Y) = \frac{1}{2}\left[1 - \sum_{y\in\mathcal{Y}} P^2(Y = y)\right]$$

This measure is used, for instance, during the growing of RFs. Considering the case of two categories (classes), compare both analytically and experimentally this measure with entropy.

7.5 Breiman's conjecture
When defining RFs, Breiman conjectures that Adaboost emulates RFs at the later stages. Relate this conjecture with the probability of the kth set of weights.

7.6 Weak dependence of multiple trees
In the text it is claimed that randomization yields weak statistical dependency among the trees. Given the trees grown in the previous problem, check

the claim by computing both the conditional variances ν_c and the sum of conditional covariances for all classes γ_c which are defined as follows:

$$\nu_c = \frac{1}{K}\sum_{k=1}^{K}\sum_{d=1}^{C} Var(\mu_{\mathcal{T}_k}(d)|Y=c)$$

$$\gamma_c = \frac{1}{K^2}\sum_{k\neq p}\sum_{d=1}^{C} Cov(\mu_{\mathcal{T}_k}(d), \mu_{\mathcal{T}_p}(d)|Y=c)$$

7.7 Derivation of mutual information
In Infomax Boosting, Quadratic Divergence (Eq. 7.37) is used to derive $I(\phi I_x; c)$. Derive $I(\phi I_x; c)$ from a different similarity measure between histograms like Chi-Square $\mathcal{X}^2$:

$$\mathcal{X}^2_{ij} = \sum_{k}^{n} \frac{(H_i(k) - \hat{H}(k))^2}{\hat{H}(k)} \tag{7.144}$$

$$\hat{H}(k) = \frac{H_i(k) + H_j(k)}{2} \tag{7.145}$$

7.8 Discrete infomax classifier
Output of Alg. 17 is a strong classifier which returns a real value in the range $\{-1, 1\}$. Transform this algorithm to make it return a discrete classifier with two possible class labels (0 and 1).

7.9 How Information Theory improves Adaboost?
Infomax and Jensen–Shannon are based on information theory measures to select a weak learner at each iteration, rather than classification error. Explain why this fact makes these algorithms yield better results than the original Adaboost.

7.10 Maximum entropy classifiers
In the example presented in Table 7.3 there are five vehicles with three features for each one of them. Suppose we are also informed that the first vehicle is equipped with a sidecar, and the rest of them are not.

1. Update the data of Tables 7.4 and 7.5.
2. For the new data the Improved Iterative Scaling (Alg. 19) converges to the λ_i values:

$$(\lambda_1, \ldots, \lambda_8) = (0.163, -0.012, -0.163, 14.971, -0.163, 0.012, 0.163, 0)$$

Calculate again the probability that the vehicle which `has gears`, `4 wheels` and `4 seats` is a *car*, and the probability that it is a *motorbike*.

3. Has the new feature helped the model to improve the classification?
4. Why is $\lambda_8 = 0$?
5. Is the Maximum Entropy classifier suitable for real-valued features? How would you add the maximum speed of the vehicle as a feature?

7.11 Parameterizing the Gaussian
Proof that the well-known Gaussian distribution belongs to exponential family. Identify: $\mathcal{S}$, $T(x)$,$\pi(x)$, and $G(\Lambda)$. Show also the one-to-one correspondence, through the first moment of $G(\Lambda)$, between the natural and usual spaces.

7.12 Improved iterative scaling equations
Given the derivation of the maximum-likelihood updates δ_j in Eq. 7.98, generalize them for the expressions contained in Alg. 19, that is, considering the conditional model for the maximum entropy classifier.

7.13 Exponential distributions and bijections with Bregman divergences
Assuming $F(\mathbf{x}) = \mathbf{x}$, check that the following settings of G^* yield the following distributions: $G^*(\mathbf{x}) = ||\mathbf{x}||^2/2$ gives a unit variance Gaussian through $D_{G^*}(\mathbf{x}, \boldsymbol{\mu})$; $G^*(\mathbf{x}) = \sum_i x_i \log x_i$ gives a multinomial distribution through $D_{G^*}(\mathbf{x}, \mathbf{q})$; and $G^*(\mathbf{x}) = -\sum_i \log x_i$ gives a geometric distribution through $D_{G^*}(\mathbf{x}, \boldsymbol{\mu})$, with $\lambda_i = 1/\mu_i$.

7.14 Bregman balls for Gaussian distributions
Reformulate the BBC algorithm for the case of assuming a Gaussian distribution as a model for the training data. What is the proper generator and Bregman divergence in this case? Design a test for detecting the Gaussianity of test vectors.

7.15 Bregman surrogates for trees and random forests
1. Using the definition of $F_\phi(\cdot)$ in terms of Bregman divergence between 0 and the inverse, proof that

$$\epsilon_\phi^{\mathbb{R}}(H, \mathcal{S}) = \sum_{i=1}^{N} F_\phi(y_i^* H(\mathbf{x}_i)) = \sum_{l \in \partial H} |\mathcal{S}_l| \left(-\phi\left(\frac{|\mathcal{S}_l^+|}{|\mathcal{S}_l|}\right)\right) \tag{7.146}$$

2. Speculate what should be the impact of replacing the majority (voting) rule used for random forests by an ULS approach. This is related to Breiman's conjecture (see Prob. 7.5). For this rationale it is also useful to consider that the performance of Adaboost decreases when the class label of several training examples is deteriorated.

7.8 Key References

- L. Breiman. "Random Forests". *Machine Learning* 45: 5–32, Prague (Czech Republic) (2001)
- M. Robnik-Sikonja. "Improving Random Forests", *European Conference on Machine Learning*, Pisa (Italy) (2004)

- D. Geman and B. Jedynak. "Model-Based Classification Trees". *IEEE Transactions on Information Theory* 47(3) (2001)
- S. Lyu. "Infomax Boosting". *IEEE Conference on Computer Vision and Pattern Recognition* Vol. I 533–538, San Diego (USA) (2005)
- X. Huang, S.Z. Li, and Y. Wang. "Jensen–Shannon Boosting Learning for Object Recognition". *IEEE Conference on Computer Vision and Pattern Recognition* Vol. I 533–538, San Diego (USA) (2005)
- C. Liu and H. Y. Shum. "Kullback–Leibler Boosting". *IEEE Conference on Computer Vision and Pattern Recognition* 407–411, Madison (USA) (2003)
- A. Fred and A.K. Jain. "Combining Multiple Clustering Using Evidence Accumulation". *IEEE Transactions on Pattern Analysis and Machine Intelligence* 27(6): 835–850 (2005)
- A. Topchy, A.K. Jain, and W. Punch. "Clustering Ensembles: Models of Consensus and Weak Partitions". *IEEE Transactions on Pattern Analysis and Machine Intelligence* 27(12): 1866–1881 (2005)

References

1. W. Aguilar, Y. Frauel, F. Escolano, M.E. Martínez-Pérez, A. Espinosa-Romero, and M.A. Lozano, A robust graph transformation matching for non-rigid registration, Image and Vision Computing **27** (2009), 7, 897–910.
2. H. Akaike, A new look at the statistical model selection, IEEE Transactions on Automatic Control **16** (1977), 6, 716–723.
3. S.-I. Amari, *Differential-geometrical methods in statistics*, Lecture Notes in Statistics, Springer-Verlag, Berlin, **28**, (1985).
4. S.-I. Amari, *Information geometry and its applications*, Emerging Trends in Visual Computing (PASCAL videolectures), 2008.
5. S.-I. Amari and H. Nagaoka, *Methods of information geometry*, American Mathematical Society, 2001.
6. A.A. Amini, R.W. Curwen, and J.C. Gore, *Snakes and splines for tracking non-rigid heart motion*, Proceedings of the European Conference on Computer Vision, Cambridge, UK, 1996, pp. 251–261.
7. Y. Amit and D. Geman, Shape quantization and recognition with randomized trees, Neural Computation (1997), **9**, 1545–1588.
8. Y. Amit, D. Geman, and B. Jedynak, Efficient focusing and face detection, Face recognition: from theory to applications, eds. H. Wechsler et al., NATO ASI Series F, Springer-Verlag, Berlin, pp. 157–173, 1998.
9. A.M. Peter and A. Rangarajan, Information geometry for landmark shape analysis: unifying shape representation and deformation, IEEE Transactions on Pattern Analysis and Machine Intelligence **31** (2008), 2, 337–350.
10. J.H. Friedman and J.W. Turkey, A projection pursuit algorithm for exploratory data analysis, IEEE Transactions on Computers **23** (1974), 9, 881–890.
11. A. Banerjee, S. Merugu, I.S. Dillon, and J. Ghosh, Clustering with Bregman divergences, Journal of Machine Learning Research **6** (2005), 1705–1749.
12. P.L. Bartlett, M.I. Jordan, and J.D. Mcauliffe, Convexity, classification, and risk bounds, Journal of the American Statistical Association **101** (2006), 138–156.
13. E. Beirlant, E. Dudewicz, L. Gyorfi, and E. Van der Meulen, Nonparametric entropy estimation, International Journal on Mathematical and Statistical Sciences **6** (1996), 1, 17–39.
14. A.J. Bell and T.J. Sejnowski, An information-maximization approach to blind separation and blind deconvolution, Neural Computation **7** (1995), 1129–1159.

15. A.J. Bell and T.J. Sejnowski, Edges are the independent components of natural scenes, Advances on Neural Information Processing Systems (NIPS), **8**, (1996), 831–837.
16. A.J. Bell, *The co-information lattice*, Proceedings of the International Workshop on Independent Component Analysis and Blind Signal Separation, Nara, Japan, 2003, pp. 921–926.
17. A. Berger, *The improved iterative scaling algorithm: a gentle introduction*, Technical report, Carnegie Melon University, 1997.
18. A. Berger, *Convexity, maximum likelihood and all that*, Carnegie Mellon University, Pittsburgh, PA, 1998.
19. A. Berger, S. Della Pietra, and V. Della Pietra, *A Maximum entropy approach to natural language processing*, Computational Linguistics, **22** (1996).
20. D. Bertsekas, *Convex analysis and optimization*, Athena Scientific, Nashua, NH, 2003.
21. D.J. Bertsimas and G. Van Ryzin, An asymptotic determination of the minimum spanning tree and minimum matching constants in geometrical probability, Operations Research Letters **9** (1990), 1, 223–231.
22. P.J. Besl and N.D. McKay, A method for registration of 3D shapes, IEEE Transactions on Pattern Analysis and Machine Intelligence **14** (1992), 2, 239–256.
23. A. Blake and M. Isard, *Active contours*, Springer, New York, 1998.
24. A. Blum and P. Langley, Selection of relevant features and examples in machine learning, Artificial Intelligence **97** (1997), 1–2, 245–271.
25. B. Bonev, F. Escolano, and M. Cazorla, Feature selection, mutual information, and the classification of high-dimensional patterns, Pattern Analysis and Applications, **11** (2008) 309–319.
26. F.L. Bookstein, Principal warps: thin plate splines and the decomposition of deformations, IEEE Transactions on Pattern Analysis and Machine Intelligence **11** (1989), 6, 567–585.
27. A. Bosch, A. Zisserman, and X. Muñoz, *Image classification using random forests and ferns*, Proceedings of the International Conference on Computer Vision, Rio de Janeiro, Brazil, 2007, pp. 1–8.
28. A. Bosch, A. Zisserman, and X. Muñoz, Scene classification using a hybrid generative/discriminative approach, IEEE Transactions on Pattern Analysis and Machine Intelliegnce **30** (2008), 4, 1–16.
29. L.M. Bregman, The relaxation method of finding the common point of convex sets and its application to the solution of problems in convex programming, USSR Computational Mathematics and Physics **7** (1967), 200–217.
30. L. Breiman, Random forests, Machine Learning **1** (2001), 45, 5–32.
31. L. Breiman, J. Friedman, R. Olshen, and C. Stone, *Classification and regression trees*, Wadsworth, Belmont, CA, 1984.
32. J.E. Burr, Properties of cross-entropy minimization, IEEE Transactions on Information Theory **35** (1989), 3, 695–698.
33. M. Bădoiu and K.-L. Clarkson, *Optimal core-sets for balls*, Computational Geometry: Theory and Applications **40** (2008), 1, 14–22.
34. X. Calmet and J. Calmet, Dynamics of the Fisher information metric, Physical Review E **71** (2005), 056109.
35. J.-F. Cardoso and A. Souloumiac, Blind beamforming for non Gaussian signals, IEE Proceedings-F **140** (1993), 6, 362–370.

36. M.A. Cazorla, F. Escolano, D. Gallardo, and R. Rizo, Junction detection and grouping with probabilistic edge models and Bayesian A^*, Pattern Recognition **9** (2002), 35, 1869–1881.
37. T.F. Chan and L. Vese, *An active contour model without edges*, Proceedings of International Conference Scale-Space Theories in Computer Vision, Corfu, Greece, 1999, pp. 141–151.
38. X. Chen and A.L. Yuille, *Adaboost learning for detecting and reading text in city scenes*, Proceedings of IEEE Conference on Computer Vision and Pattern Recognition, Washington DC, USA, 2004.
39. X. Chen and A.L. Yuille, *Time-efficient cascade for real time object detection*, 1st International Workshop on Computer Vision Applications for the Visually Impaired. Proceedings of IEEE Conference on Computer Vision and Pattern Recognition, Washington DC, USA, 2004.
40. H. Chui and A. Rangarajan, A new point matching algorithm for nonrigid registration, Computer Vision and Image Understanding **89** (2003), 114–141.
41. P. Comon, Independent component analysis, a new concept? Signal Processing **36** (1994), 287–314.
42. J.M. Coughlan and A. L. Yuille, Bayesian A^* tree search with expected o(n) node expansions: applications to road tracking, Neural Computation **14** (2002), 1929–1958.
43. T. Cover and J. Thomas, *Elements of information theory*, Wiley, New York 1991.
44. I. Csiszár, A geometric interpretation of Darroch and Ratcliff's generalized iterative scaling, Annals of Probability **17** (1975), 3, 1409–1413.
45. I. Csiszár, I-divergence geometry of probability distributions and minimization problems, Annals of Probability **3** (1975), 1, 146–158.
46. S. Dalal and W. Hall, Approximating priors by mixtures of natural conjugate priors, Journal of the Royal Statistical Society(B) **45** (1983), 1.
47. J.N. Darroch and D. Ratcliff, Generalized iterative scaling for log-linear models, Annals of Mathematical Statistics **43** (1983), 1470–1480.
48. P. Dellaportas and I. Papageorgiou, Petros Dellaportas1 Contact Information and Ioulia Papageorgiou, Statistics and Computing **16** (2006), 1, 57–68.
49. A. Dempster, N. Laird, and D. Rubin, Maximum likelihood estimation from incomplete data via the EM algorithm, Journal of the Royal Statistical Society **39** (1977), 1, 1–38.
50. G.L. Donato and S. Belongie, *Approximate thin plate spline mappings*, Proceedings of the European Conference on Computer Vision, Copenhagen, Denmark, vol. 2, 2002, pp. 531–542.
51. R. O. Duda and P. E. Hart, *Pattern classification and scene analysis*, Wiley, New York, 1973.
52. D. Erdogmus, K.E. Hild II, Y.N. Rao, and J.C. Príncipe, Minimax mutual information approach for independent component analysis, Neural Computation **16** (2004), 6, 1235–1252.
53. L. Fei-Fei and P. Perona, *A Bayesian hierarchical model for learning natural scene categories*, Proceedings of the IEEE Conference on Computer Vision and Pattern Recognition, San Diego, USA, vol. 2, 2005, pp. 524–531.
54. D.J. Field, What is the goal of sensory coding? Neural Computation **6** (1994), 559–601.

55. M.A.T. Figueiredo and A.K. Jain, *Unsupervised selection and estimation of finite mixture models*, International Conference on Pattern Recognition (ICPR2000) (Barcelona, Spain), IEEE, 2000.
56. M.A.T. Figueiredo and A.K. Jain, Unsupervised learning of finite mixture models, IEEE Transactions on Pattern Analysis and Machine Intelligence **24** (2002), 3, 381–399.
57. M.A.T. Figueiredo, J.M.N. Leitao, and A.K. Jain, *Adaptive parametrically deformable contours*, Proceedings of Energy Minimization Methods and Pattern Recognition (EMMCVPR'97), Venice, Italy, 1997, pp. 35–50.
58. M.A.T. Figueiredo, J.M.N. Leitao, and A.K. Jain, Unsupervised contour representation and estimation using B-splines and a minimum description length criterion, IEEE Transactions on Image Processing **6** (2000), 9, 1075–1087.
59. M.A.T Figueiredo, J.M.N Leitao, and A.K. Jain, On fitting mixture models, Energy Minimization Methods in Computer Vision and Pattern Recognition. Lecture Notes in Computer Science **1654** (1999), 1, 54–69.
60. D.H. Fisher, Knowledge acquisition via incremental conceptual clustering, Machine Learning (1987), 2, 139–172.
61. Y. Freund and R.E. Schapire, A decision-theoretic generalization of on-line learning and an application to boosting, Journal of Computer and System Sciences **1** (1997), 55, 119–139.
62. D. Geman and B. Jedynak, Model-based classification trees, IEEE Transactions on Information Theory **3** (2001), 47, 1075–1082.
63. J. Goldberger, S. Gordon, and H. Greenspan, Unsupervised image-set clustering using an information theoretic framework, IEEE Transactions on Image Processing **2** (2006), 449–458.
64. P.J. Green, Reversible jump Markov chain Monte Carlo computation and Bayesian model determination, Biometrika **4** (1995), 82, 711–732.
65. U. Grenander and M.I. Miller, Representation of knowledge in complex systems, Journal of the Royal Statistical Society Series B **4** (1994), 56, 569–603.
66. R. Gribonval, From projection pursuit and cart to adaptive discriminant analysis? IEEE Transactions on Neural Networks **16** (2005), 3, 522–532.
67. P.D. Grünwald, *The minimum description length principle*, MIT Press, Cambridge, MA, 2007.
68. S. Guilles, *Robust description and matching of images*, Ph.D. thesis, University of Oxford, 1998.
69. I. Guyon and A. Elisseeff, An introduction to variable and feature selection, Journal of Machine Learning Research (2003), 3, 1157–1182.
70. A. Ben Hamza and H. Krim, Image registration and segmentation by maximizing the Jensen-Rényi divergence, Lecture Notes in Computer Science, EMM-CVPR 2003, 2003, pp. 147–163.
71. J. Harris, *Algebraic geometry, a first course*, Springer-Verlag, New York, 1992.
72. T. Hastie and R. Tibshirani, Discriminant analysis by Gaussian mixtures, Journal of the Royal Statistical Society(B) **58** (1996), 1, 155–176.
73. A.O. Hero and O. Michel, Asymptotic theory of greedy aproximations to minnimal k-point random graphs, IEEE Transactions on Information Theory **45** (1999), 6, 1921–1939.
74. A.O. Hero and O. Michel, Applications of spanning entropic graphs, IEEE Signal Processing Magazine **19** (2002), 5, 85–95.

75. K. Huang, Y. Ma, and R. Vidal, *Minimum effective dimension for mixtures of subspaces: A robust GPCA algorithm and its applications*, Computer Vision and Pattern Recognition Conference (CVPR04), vol. 2, 2004, pp. 631–638.
76. X. Huang, S.Z. Li, and Y. Wang, *Jensen-Shannon boosting learning for object recognition*, IEEE Conference on Computer Vision and Pattern Recognition **2** (2005), 144–149.
77. A. Hyvärinen, *New approximations of differential entropy for independent component analysis and projection pursuit*, Technical report, Helsinki University of Technology, 1997.
78. A. Hyvarinen, J. Karhunen, and E. Oja, *Independent component analysis*, Wiley, New York, 2001.
79. A. Hyvarinen and E. Oja, Independent component analysis: algorithms and applications, Neural Networks **13** (2000), 4–5, 411–430.
80. A. Hyvrinen and E. Oja, A fast fixed-point algorithm for independent component analysis, Neural Computation **9** (1997), 7, 1483–1492.
81. A.K. Jain and R. Dubes, *Algorithms for clustering data*, Prentice Hall, Englewood Cliffs, NJ, 1988.
82. A.K. Jain, R. Dubes, and J. Mao, Statistical pattern recognition: a review, IEEE Transactions on Pattern Analysis Machine Intelligence **22** (2000), 1, 4–38.
83. E.T. Jaynes, Information theory and statistical mechanics, Physical Review **106** (1957), 4, 620–630.
84. B. Jedynak, H. Zheng, and M. Daoudi, Skin detection using pairwise models, Image and Vision Computing **23** (2005), 13, 1122–1130.
85. B. Jedynak, H. Zheng, and Daoudi M., Statistical models for skin detection, Proceedings of IEEE International Conference on Computer Vision and Pattern Recognition (CVPRV'03), Madison, USA, vol. 8, 2003, pp. 92–92.
86. R. Jin, R. Yan, J. Zhang, and A.G. Hauptmann, *A faster iterative scaling algorithm for conditional exponential model*, Proceedings of the 20th International Conference on Machine Learning (ICML 2003), Washington, USA, 2003.
87. G.H. John, R. Kohavi, and K. Pfleger, *Irrelevant features and the subset selection problem*, International Conference on Machine Learning (1994), pp. 121–129.
88. M.C. Jones and R. Sibson, What is projection pursuit?, Journal of the Royal Statistical Society. Series A (General) **150** (1987), 1, 1–37.
89. M.J. Jones and J.M. Rehg, *Statistical color models with applications to skin detection*, Proceedings of IEEE International Conference on Computer Vision and Pattern Recognition, Ft. Collins, USA, 1999, pp. 1–8.
90. T. Kadir and M. Brady, *Estimating statistics in arbitrary regions of interest*, Proceedings of the 16th British Machine Vision Conference, Oxford, UK, Vol. 2, 2005, pp. 589–598.
91. T. Kadir and M. Brady, Scale, saliency and image description, International Journal on Computer Vision **2** (2001), 45, 83–105.
92. K. Kanatani, *Motion segmentation by subspace separation and model selection*, International Conference on Computer Vision (ICCV01), vol. 2, 2001, pp. 586–591.
93. M. Kass, A. Witkin, and D. Terzopoulos, Snakes: Active contour models, International Journal on Computer Vision (1987), 1, 259–268.

94. Robert E. Kass and Larry Wasserman, A reference Bayesian test for nested hypotheses and its relationship to the Schwarz criterion, Journal of the American Statistical Association **90** (1995), 928–934.
95. M. Kearns and L. Valiant, Cryptographic limitations on learning Boolean formulae and finite automata, Journal of the ACM **1** (1994), 41, 67–95.
96. R. Kohavi and G.H. John, Wrappers for feature subset selection, Artificial Intelligence **97** (1997), 1–2, 273–324.
97. D. Koller and M. Sahami, *Toward optimal feature selection*, Proceedings of International Conference in Machine Learning, 1996, pp. 284–292.
98. S. Konishi, A.L. Yuille, and J.M. Coughlan, A statistical approach to multi-scale edge detection, Image and Vision Computing **21** (2003), 1, 37–48.
99. S. Konishi, A.L. Yuille, J.M. Coughlan, and S.C. Zhu, Statistical edge detection: learning and evaluating edge cues, IEEE Transactions on Pattern Analysis and Machine Intelligence **1** (2003), 25, 57–74.
100. S. Kullback, *Information theory and statistics*, Wiley, New York, 1959.
101. M.H.C. Law, M.A.T. Figueiredo, and A.K. Jain, Simultaneous feature selection and clustering using mixture models, IEEE Transactions on Pattern Analysis Machine Intelligence **26** (2004), 9, 1154–1166.
102. S. Lazebnik, C. Schmid, and Ponce P., *Beyond bags of features: Spatial pyramid matching for recognizing natural scene categories*, Proceedings of the IEEE Conference on Computer Vision and Pattern Recognition, New York, USA, vol. 2, 2006, pp. 2169–2178.
103. N. Leonenko, L. Pronzato, and V. Savani, A class of Rényi information estimators for multidimensional densities, The Annals of Statistics **36** (2008), 5, 2153–2182.
104. J. Lin, Divergence measures based on the Shannon entropy, IEEE Transactions on Information Theory **1** (1991), 37, 145–151.
105. R. Linsker, Self-organization in a perceptual network, Computer **3** (1988), 21, 105–117.
106. C. Liu and H.Y. Shum, *Kullback–Leibler boosting*, IEEE Conference on Computer Vision and Pattern Recognition (2003), 407–411.
107. D. Lowe, Distinctive image features form scale-invariant keypoints, International Journal of Computer Vision **60** (2004), 2, 91–110.
108. S. Lyu, *Infomax boosting*, IEEE Conference on Computer Vision and Pattern Recognition **1** (2005), 533–538.
109. Y. Ma, A.Y. Yang, H. Derksen, and R. Fossum, Estimation of subspace arrangements with applications in modelling and segmenting mixed data, SIAM Review **50** (2008), 3, 413–458.
110. J. Matas, O. Chum, U. Martin, and T. Pajdla, *Robust wide baseline stereo from maximally stable extremal regions*, Proceedings of the British Machine Vision Conference, Cardiff, Wales, UK, vol. 1, 2002, pp. 384–393.
111. G. McLachlan, *Discriminant analysis and statistical pattern recognition*, Wiley, New York, 1992.
112. G. McLachlan and D. Peel, *Finite mixture models*, Wiley, 2000.
113. K. Mikolajczyk, T. Tuytelaars, C. Schmid, A. Zisserman, J. Matas, F. Schaffalitzky, T. Kadir, and L. Van Gool, A comparison of affine region detectors, International Journal of Computer Vision **65** (2005), 1–2, 43–72.
114. W. Mio and X. Liu, *2113–2116*, Proceedings of the IEEE International Conference on Image Processing, Atlanta, USA, 2006, pp. 531–542.

115. A. Mokkadem, Estimation of the entropy and information of absolutely continuous random variables, IEEE Transactions on Information Theory **35** (1989), 1, 193–196.
116. A. Peñalver, F. Escolano, and J.M. Sáez, Learning gaussian mixture models with entropy based criteria, Submitted to IEEE Transactions on Neural Networks (2009).
117. H. Neemuchwala, A. Hero, S. Zabuawala, and P. Carson, Image registration methods in high dimensional space, International Journal of Imaging Systems and Technology **16** (2007), 5, 130–145.
118. F. Nielsen and R. Nock, On the smallest enclosing information disk, Information Processing Letters **105** (2008), 93–97.
119. R. Nock and F. Nielsen, *Fitting the smallest enclosing bregman ball*, European Conference on Machine Learning, ECML 2005, Lecture Notes in Computer Science **3720**, (2005), 649–656.
120. R. Nock and F. Nielsen, Bregman divergences and surrogates for learning, IEEE Transactions on Pattern Analysis and Machine Intelligence (2009).
121. B.A. Olshausen and D.J. Field, Natural images statistics and efficient coding, Network: Communication in Neural Systems **7** (1996), 2, 333–339.
122. E. Parzen, On estimation of a probability density function and mode, Annals of Mathematical Statistics **33** (1962), 1, 1065–1076.
123. J. Pearl, *Probabilistic reasoning in intelligent systems: networks of plausible inference*, Morgan Kaufmann, San Mateo, CA, 1998.
124. D. Pelleg and A. Moore, *X-means: extending k-means with efficient estimation of the number of clusters*, Proceedings of the 17th International Conference on Machine Learning, San Francisco, Morgan Kaufmann, 2000, pp. 727–734.
125. H. Peng, F. Long, and C. Ding, Feature selection based on mutual information: criteria of max-dependency, max-relevance, and min-redundancy, IEEE Transactions on Pattern Analysis and Machine Intelligence **27** (2005), 8, 1226–1238.
126. A. Pentland, R.W. Picard, and S. Sclaroff, Photobook: content-based manipulation of image databases, International Journal of Computer Vision **18** (1996), 3, 233–254.
127. S. Perkins, K. Lacker, and J. Theiler, Grafting: fast, incremental feature selection by gradient descent in function space, Journal of Machine Learning Research **3** (2003), 1333–1356.
128. A. Peter and A. Rangarajan, *A new closed-form information metric for shape analysis*, Proceedings of MICCAI, Copenhagen, Denmark, 2006, pp. 249–256.
129. A. Peter and A. Rangarajan, *Shape analysis using the Fisher-Rao Riemannian metric: unifying shape representation and deformation*, IEEE International Symposium on Biomedical Imaging: Nano to Macro, 2006, pp. 531–542.
130. D. Ponsa and A. López, *Feature selection based on a new formulation of the minimal-redundancy-maximal-relevance criterion*, IbPRIA (1), 2007, pp. 47–54.
131. W. Punch, A. Topchy, and A. Jain, Clustering ensembles: models of consensus and weak partitions, IEEE Transactions on Pattern Analysis and Machine Intelligence **27** (2005), 12, 1866–1881.
132. J.R. Quinlan, *C4.5: Programs for machine learning*, Morgan Kaufmann, San Mateo, CA, 1993.
133. A. Rajwade, A. Banerjee, and A. Rangarajan, *Continuous image representations avoid the histogram binning problem in mutual information based image*

registration, Biomedical Imaging: Nano to Macro, 2006. Proceedings of the Thrid IEEE International Symposium on Biomedical Imaging (ISBI) 2006, pp. 840–843.

134. A. Rajwade, A. Banerjee, and A. Rangarajan, *Probability density estimation using isocontours and isosurfaces: Application to information theoretic image registration*, IEEE Transactions on Pattern Analysis and Machine Intelligence, Miami, USA **31** (2009), 3, 475–491.
135. S. Rao, A. M. Martins, and J. C. Principe, Mean shift: an information theoretic perspective, Pattern Recognition Letters **30(3)** (2009), 222–2309.
136. R.A. Redner and H.F. Walker, Mixture densities, maximum likelihood, and the EM algorithm, SIAM Review **26** (1984), 2, 195–239.
137. A. Rényi, *On measures of information and entropy*, Proceedings of the 4th Berkeley Symposium on Mathematics, Statistics and Probability (1960), Berkeley, USA, 547–561.
138. J. Rissanen, Modelling by the shortest data description, Automatica (1978), 14, 465–471.
139. J. Rissanen, A universal prior for integers and estimation by minimum description length, Annals of Statistics (1983), 11, 416–431.
140. J. Rissanen, *Stochastic complexity in statistical inquiry*, 1989, World Scientific, River Edge, NJ.
141. C.P. Robert and G. Castella, *Monte Carlo statistical methods*, Series: Springer Texts in Statistics, Springer-Verlag (1999).
142. R.E. Schapire, The strength of weak learnability, Machine Learning **2** (1990), 5, 197–227.
143. G. Schwarz, Estimating the dimension of a model, The Annals of Statistics **6** (1978), 2, 461–464.
144. J.E. Shore and R.W. Johnson, Properties of cross-entropy minimization, IEEE Transactions on Information Theory **27** (1981), 4, 472–482.
145. K. Siddiqi, A. Shokoufandeh, S.J. Dickinson, and S.W. Zucker, Shock graphs and shape matching, International Journal of Computer Vision **35** (1999), 1, 13–32.
146. C. Sima and E.R. Dougherty, What should be expected from feature selection in small-sample settings, Bioinformatics **22** (2006), 19, 2430–2436.
147. J. Sivic and A. Zisserman, *Video google: a text retrieval approach to object matching in videos*, Proceedings of the International Conference on Computer Vision, Nice, France, vol. 2, 2003, pp. 1470–1477.
148. Q. Song, A robust information clustering algorithm, Neural Computation **17** (2005), **12** 2672–2698.
149. L. Staib and J. Duncan, Boundary finding with parametrically deformable models, IEEE Transactions on Pattern Analysis and Machine Intelligence (1992), 14, 1061–1075.
150. A. Strehl and J. Ghosh, Cluster ensembles – a knowledge reuse framework for combining partitionings, Proceedings of Conference on Artificial Intelligence (AAAI 2002), Edmonton, pp. 93–98.
151. C. Strobl, A.-L. Boulesteix, A. Zeileis, and T. Hothorn, Bias in random forest variable importance measures: illustrations, sources and a solution, BMC Bioinformatics **25** (2007), 8, 67–95.
152. M.J. Tarr and H.H. Bülthoff (Eds), *Object recognition in man, monkey and machine*, MIT Press, Cambridge, MA, 1998.

153. D.M.J. Tax and R.P.W Duin, Support vector data description, Machine Learning **54** (2004), 45–66.
154. Z. Thu, X. Chen, A.L. Yuille, and S. Zhu, Image parsing: unifying segmentation, detection, and recognition, International Journal of Computer Vision **63** (2005), 2, 113–140.
155. A. Torsello and D.L. Dowe, *Learning a generative model for structural representations*, Proceedings of the Australasian Conference on Artificial Intelligence, Auckland, New Zealand, 2008, pp. 573–583.
156. A. Torsello and E.R. Hancock, Learning shape-classes using a mixture of tree-unions, IEEE Transactions on Pattern Analysis and Machine Intelligence **28** (2006), 6, 954–967.
157. I.W. Tsang, A. Kocsor, and J.T. Kwok, *Simpler core vector machines with enclosing balls*, International Conference on Machine Learning, ICML 2007, 2007, pp. 911–918.
158. I.W. Tsang, J.T. Kwok, and P.-M. Cheung, Core vector machines: fast svm training on very large datasets, Journal of Machine Learning Research **6** (2005), 363–392.
159. Z. Tu and S.-C. Zhu, Image segmentation by data-driven markov chain Monte Carlo, IEEE Transactions on Pattern Analysis and Machine Intelligence **24** (2002), 5, 657–673.
160. M. Turk and A. Pentland, Eigenfaces for recognition, Journal of Cognitive Neuroscience **3** (1991), 1.
161. N. Ueda, R. Nakano, Z. Ghahramani, and G. E. Hinton, SMEM algorithm for mixture models, Neural Computation **12** (2000), 1, 2109–2128.
162. G. Unal, H. Krim, and A. Yezzi, Fast incorporation of optical flow into active polygons, IEEE Transactions on Image Processing **6** (2005), 14, 745–759.
163. G. Unal, A. Yezzi, and H. Krim, Information-theoretic active polygons for unsupervised texture segmentation, International Journal on Computer Vision **3** (2005), 62, 199–220.
164. M.J. van der Laan, Statistical inference for variable importance, International Journal of Biostatistics **2** (2006), 1, 1008–1008.
165. V. N. Vapnik, *Statistical learning theory*, Wiley, New York, 1998.
166. N. Vasconcelos and M. Vasconcelos, *Scalable discriminant feature selection for image retrieval and recognition*, Computer Vision and Pattern Recognition Conference (CVPR04), 2004, pp. 770–775.
167. M.A. Vicente, P.O. Hoyer, and A. Hyvarinen, Equivalence of some common linear feature extraction techniques for appearance-based object recognition tasks, IEEE Transactions on Pattern Analysis and Machine Intelligence **29** (2007), 5, 233–254.
168. R. Vidal, Y. Ma, and J. Piazzi, *A new GPCA algorithm for clustering subspaces by fitting, differentiating and dividing polynomials*, Computer Vision and Pattern Recognition Conference (CVPR04), 2004.
169. R. Vidal, Y. Ma, and S. Sastry, *Generalized principal component analysis (gpca)*, Computer Vision and Pattern Recognition Conference (CVPR04), vol. 1, 2003, pp. 621–628.
170. P. Viola and M.J. Jones, Robust real-time face detection, International Journal of Computer Vision **2** (2004), 57, 137–154.
171. P. Viola and W.M. Wells-III, *Alignment by maximization of mutual information*, 5th International Conference on Computer Vision, vol. 2, IEEE, 1997, pp. 137–154.

172. N. Vlassis, A. Likas, and B. Krose, A multivariate kurtosis-based dynamic approach to Gaussian mixture modeling, 2000, Intelligent Autonomous Systems Techinal Report.
173. C.S. Wallace and D.M. Boulton, An information measure for classification, Computer Journal **11** (1968), 2, 185–194.
174. F. Wang, B.C. Vemuri, A. Rangarajan, and S.J. Eisenschenk, Simultaneous nonrigid registration of multiple point sets and atlas construction, IEEE Transactions on Pattern Analysis and Machine Intelligence **30** (2008), 11, 2011–2022.
175. G. Winkler, *Image analysis, random fields and Markov chain Monte Carlo methods: a mathematical introduction*, Springer, Berlin, 2003.
176. L. Xu, BYY harmony learning, structural RPCL, and topological self-organizing on mixture models, Neural Networks **15** (2002), 1, 1125–1151.
177. J.S. Yedidia, W.T. Freeman, and Y. Weiss., Understanding belief propagation and its generalisations, Tech. Report TR-2001-22, Mitsubishi Research Laboratories, January 2002.
178. J. Zhang, M. Marszałek, C. Lazebnik, and S. Schmid, Local features and kernels for classification of texture and object categories: a comprehensive study, International Journal of Computer Vision (2007), no. DOI: 10.1007/s11263-006-9794-4.
179. Jie Zhang and A. Rangarajan, Affine image registration using a new information metric, Proceedings of the 2004 IEEE Computer Society Conference on Computer Vision and Pattern Recognition (CVPR'04), Washington DC, USA, Vol. 1, 2004, pp. 848–855.
180. H. Zheg, M. Daoudi, and B. Jedynak, Blocking adult images based on statistical skin detection, Electronic Letters on Computer Vision and Image Analysis **4** (2004), 2, 1–14.
181. Y-T. Zheng, S-Y. Neo, T-S. Chua, and Q. Tian, Visual synset: towards a higher level representations, Proceedings of IEEE International Conference on Computer Vision and Pattern Recognition, Anchorage, USA, 2008, pp. 1–8.
182. S.C. Zhu, Y.N. Wu, and D. Mumford, Filters, Random Fields And Maximum Entropy (FRAME): towards a unified theory for texture modeling, International Journal of Computer Vision **27** (1997), 2, 107–126.
183. S.C. Zhu, Y.N. Wu, and D. Mumford, Minimax entropy principle and its applications to texture modeling, Neural Computation **9** (1997), 8, 1627–1660.
184. S.C. Zhu and A.L. Yuille, Region competition: unifying snakes, region growing, and bayes/mdl for multiband image segmentation, IEEE Transactions on Pattern Analysis and Machine Intelligence **9** (1996), 18, 884–900.
185. K. Zyczkowski, Renyi extrapolation of Shannon entropy, Open Systems and Information Dynamics **10** (2003), 3, 297–310.

Index

α-mutual information, 131

active contours, 44
active polygon, 46
active polygons, 44
ADA, 254
Adaboost, 89, 298
affine aligment, 123
agglomerative clustering, 150
agglomerative information bottleneck, 177
AIC, 258, 259
aligment, 106
ambient space, 254, 259, 261
arrangement of subspaces, 254
asymptotic equipartition property, 137
average residual, 260

B-Splines, 53, 58
Bayesian error, 177, 222
Belief propagation, 83
belief propagation, 84
bending energy, 135
Bhattacharya coefficient, 131
Bhattacharyya distance, 17
BIC criterion, 208, 258
Blahut–Arimoto algorithm, 172, 175
boosting, 298
bypass entropy estimation, 126, 162

categorical data, 228
chain rule of mutual information, 231
channel capacity, 184
Chernoff Information, 16, 23
classification, 211
clustering ensembles, 197
co-information, 121
conditional entropy, 22, 25, 107, 120, 228
conditional independence, 234
conditional mutual information, 121
consensus function, 199
contour, 53, 60
cross validation, 214

data-driven Markov chain, 88
detection, 86
deterministic annealing, 184
differential geometry, 140
discriminative, 99
distributional shape model, 132

ED: effective dimension, 258, 259, 261
edge detection, 21
edge localization, 23
effective dimension, 254
embedded data matrix, 256
entropic graphs, 162
entropic spanning graphs, 163
entropy, 12, 109, 228
entropy estimation, 126, 162
entropy of degree, 204
ergodicity, 93
error tolerance, 261
expectation maximization, 159, 202

FastICA, 244
feature selection, 211

filter feature selection, 212, 220
filter pursuit, 243
Fisher information matrix, 141
Fisher–Rao metric tensor, 141
FRAME algorithm, 241

GAIC criterion, 260
Gaussian blurring mean shift, 190
Gaussian mean shift, 191
Gaussian mixture, 132, 157, 181
Gaussianity, 162
gene selection, 236
generative model, 91
Gibbs sampler, 240
Gibbs theorem, 162
gPCA, 254
gPCA segmentation, 264
gradient flow, 45, 46
Grassman dimension, 259
Grassmanian manifold, 259
greedy algorithm, 228

Hellinger dissimilarity, 131
hidden variables, 159

ICA, 211, 244
image parsing, 86
Infomax, 244
infomax boosting, 299, 304
infomax feature, 301
infomax principle, 301
information bottleneck, 175
information bottleneck principle, 170
information geometry, 140
information particles, 194
information plane, 180
information potential, 194
interaction information, 121
interest points, 11

Jacobian, 257
Jensen's inequality, 47
Jensen–Rényi divergence, 128
Jensen–Shannon divergence, 46, 136, 178, 307
Jensen–Shannon feature, 305
joint entropy, 111, 228
JSBoosting, 307
jump-diffusion, 63

K-adventurers, 74
K-NN, 219
kernel, 157
kernel splitting, 167
knee point, 260
Kullback–Leibler divergence, 18, 48, 75, 181, 222

LDA, 211, 254
log-likelihood, 90, 159, 202

Markov blanket, 234
Markov chain, 240
Markov chain Monte Carlo (MCMC), 240
markov random fields, 79
maximum a posteriori, 159
maximum entropy principle, 79, 238
maximum likelihood, 158, 159
maximum residual, 260
MDL criterion, 258
mean shift, 189
MED gPCA algorithm, 258
MED: minimum effective dimension, 258, 260, 261
Metropolis-Hastings dynamics, 91
microarray, 236
minimal spanning tree, 126
minimax entropy principle, 243
Minimax ICA, 249
minimum description length, 53, 58, 60, 166
minimum description length advantage, 151
minimum description length criterion, 148
minimum message length, 266
minimum message length principle, 153
mixture model, 200
model order selection, 74, 161
model selection, 157
model-order selection, 258
monotonicity, 93
Monte Carlo Markov chain, 75
Monte Carlo simulation, 181
mutual information, 97, 111, 203, 222, 227

null space, 256

object recognition, 86

parsing tree, 86
Parzen window, 73, 189, 193
Parzen windows, 163
Parzen's window, 108
PCA, 211, 244
plug-in entropy estimation, 162
plug-in estimation, 126
projection pursuit, 254

quadratic divergence, 303
quadratic mutual information, 303

Rényi entropy, 125, 126, 162
Rényi α-divergence, 129
Rényi cross entropy, 194
Rényi quadratic entropy, 193
rate distortion theory, 170
residual error, 260
reversibility, 93
robust information clustering, 184
robustness, 258

saliency, 12
Sampson distance, 258
Sanov's theorem, 26
scale space, 13
segmentation, 44, 169
semi-supervised learning, 257
Shannon entropy, 162
shock trees, 146
simulated annealing, 241
singular value, 256
singular vectors, left and right, 256
speed function, 46
structural risk minimization, 186
sub-pixel interpolation, 113
subspace angle, 262
support vector machines, 186
SVD and PCA, 260
SVD, singular value decomposition, 256
symmetric uncertainty, 120

thin-plate splines, 134
total correlation, 122
trees mixture, 147

uncertainty coefficients, 119
unsupervised classification, 197
unsupervised learning, 147, 258

Vapnik and Chervonenkis bound, 186
Vapnik and Chervonenkis dimension, 186
Veronese map, embedding, 256

weak learner, 90
wrapper feature selection, 212

X-means, 208

Color Plates

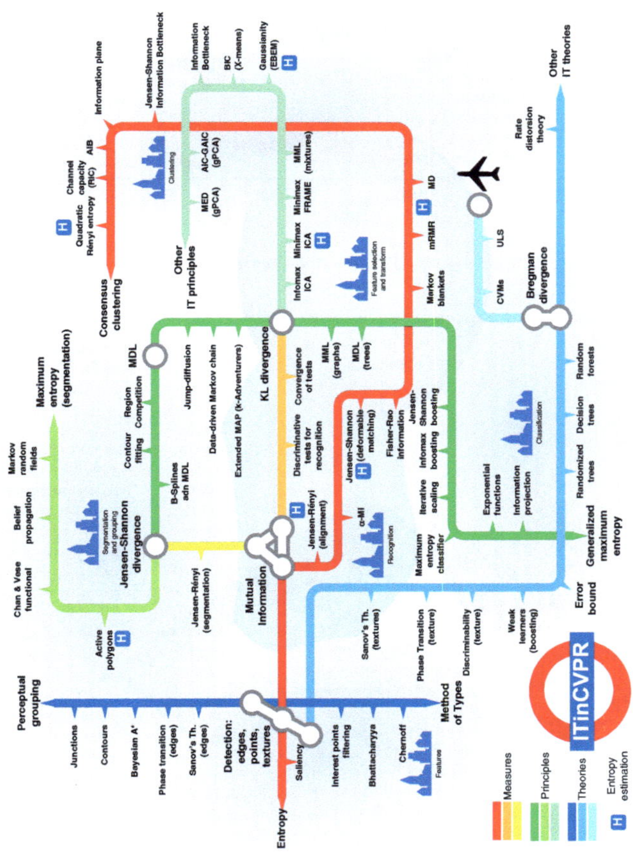

Fig. 1.1. The ITinCVPR tube/underground (lines) communicating several problems (quarters) and stopping at several stations.

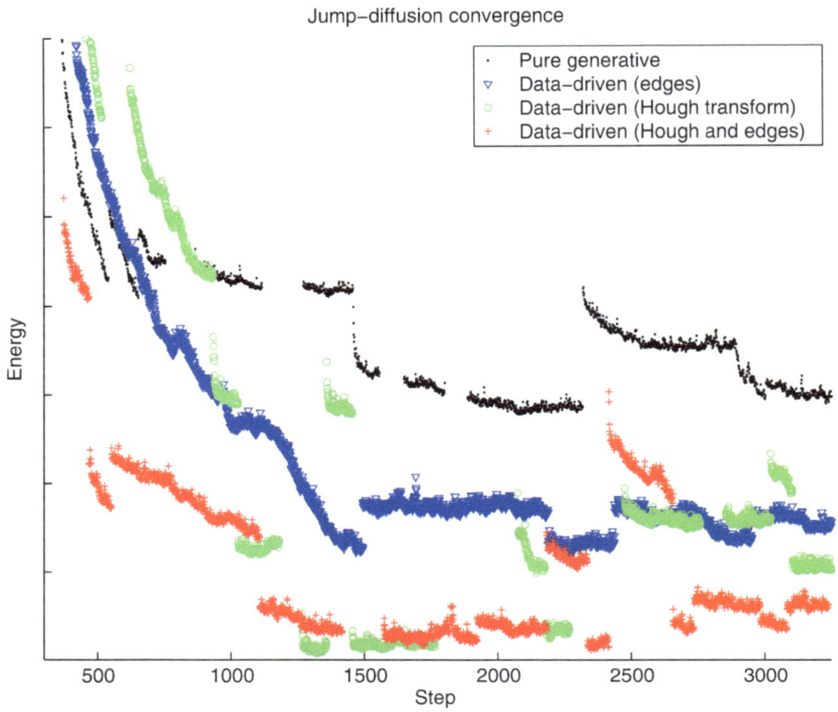

Fig. 3.12. Jump-diffusion energy evolution.

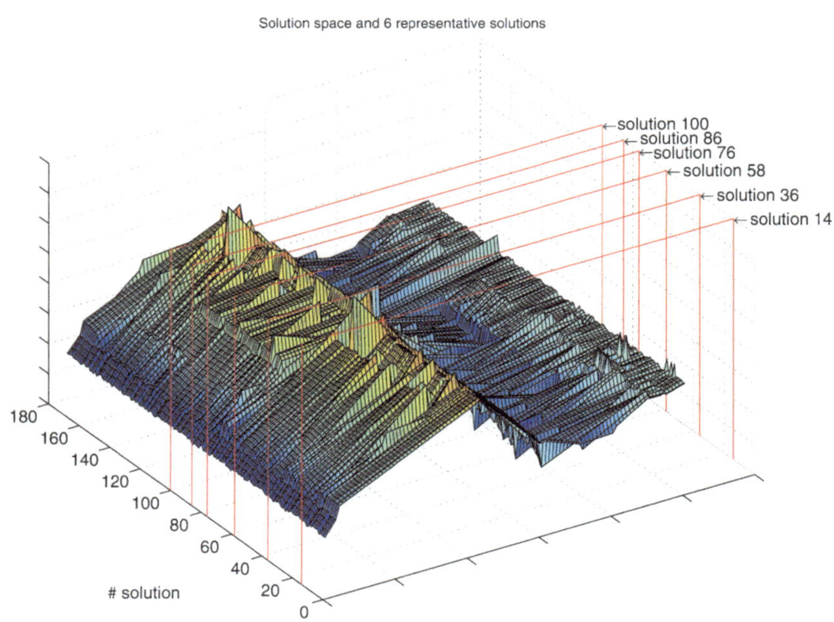

Fig. 3.13. The solution space in which the K-adventurers algorithm has selected $K = 6$ representative solutions, which are marked with *rectangles*.

Fig. 3.16. Skin detection results. Comparison between the baseline model (*top-right*), the tree approximation of MRFs with BP (*bottom-left*), and the tree approximation of the first-order model with BP instead of Alg. 5. Figure by B. Jedynak, H. Zheng and M. Daoudi (©2003 IEEE).

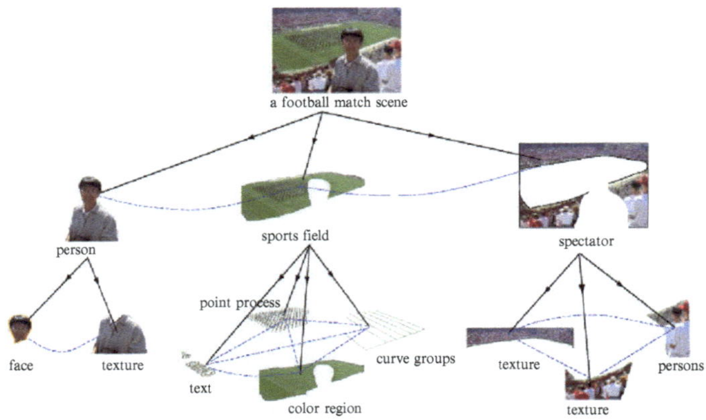

Fig. 3.17. A complex image parsing graph with many levels and types of patterns. (Figure by Tu et al. ©2005 Springer.)

Fig. 5.10. Color image segmentation results. Original images (*first column*) and color image segmentation with different Gaussianity deficiency levels (*second and third columns*). (Courtesy of A. Peñalver.)

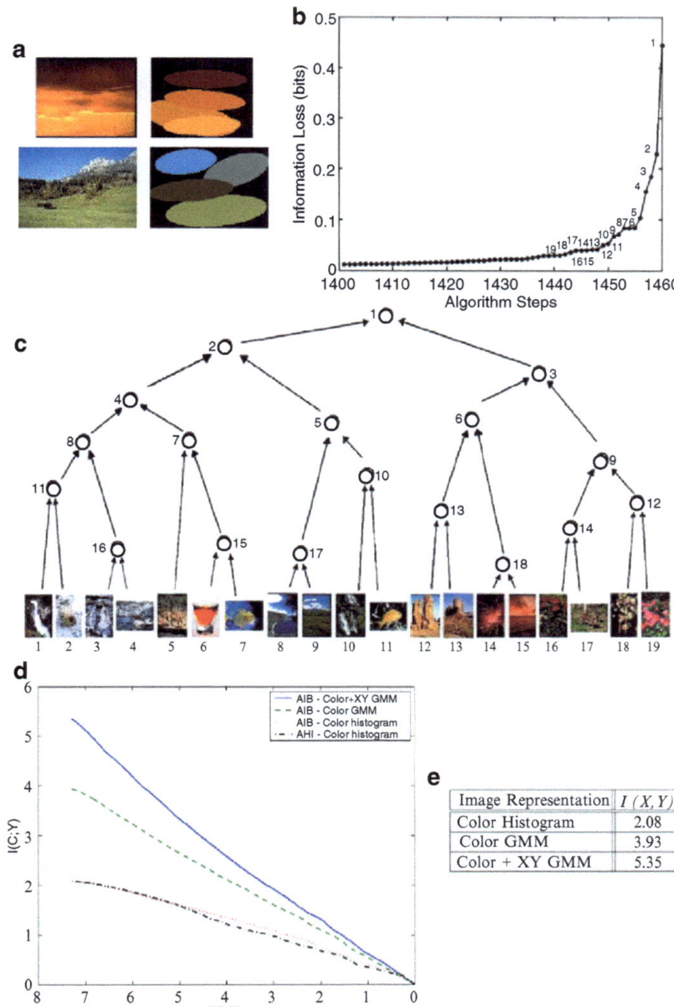

e

Image Representation	$I(X,Y)$
Color Histogram	2.08
Color GMM	3.93
Color + XY GMM	5.35

Fig. 5.17. From *left* to *right* and from *top* to *bottom*. (**a**) Image representation. Each ellipsoid represents a Gaussian in the Gaussian Mixture Model of the image, with its support region, mean color and spatial layout in the image plane. (**b**) Loss of mutual information during the IAB clustering. The last steps are labeled with the number of clusters in each step. (**c**) Part of the cluster tree formed during AIB, starting from 19 clusters. Each cluster is represented with a representative image. The labeled nodes indicate the order of cluster merging, following the plot in (**b**). (**d**) $I(T;X)$ vs. $I(T;Y)$ plot for four different clustering methods. (**e**) Mutual Information between images and image representations. (Figure by Goldberger et al. (©2006 IEEE)).

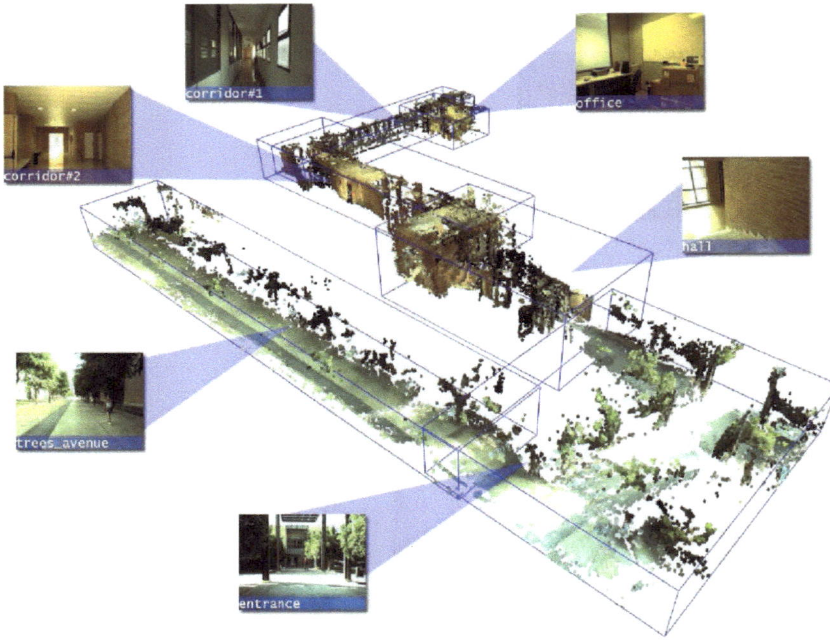

Fig. 6.1. A 3D reconstruction of the route followed during the acquisition of the data set, and examples of each one of the six classes. (Image obtained with 6-DOF SLAM. Figure by Juan Manuel Sáez (©2007 IEEE).)

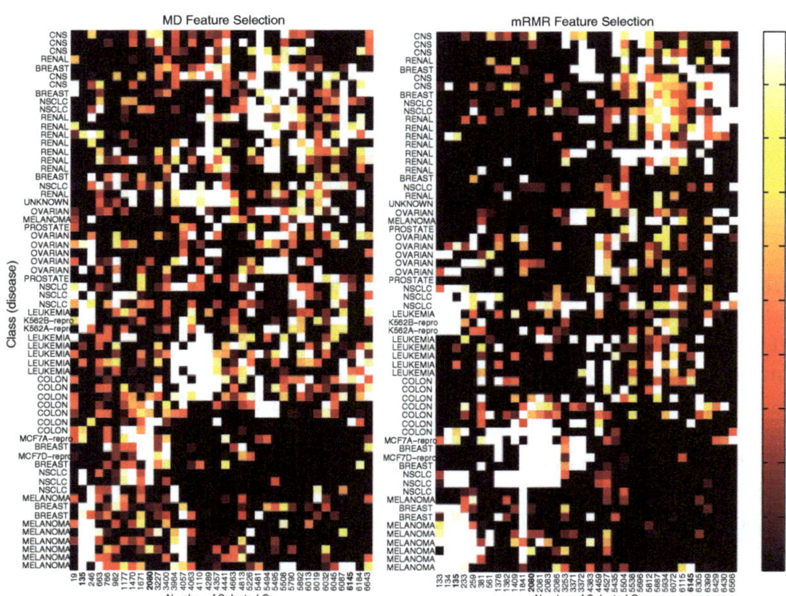

Fig. 6.12. Feature selection on the NCI DNA microarray data. The MD (*left*) and mRMR (*right*) criteria were used. Features (genes) selected by both criteria are marked with an *arrow*.

Fig. 6.22. Unsupervised segmentation with gPCA. (Figures by Huang et al. ©IEEE 2004.)

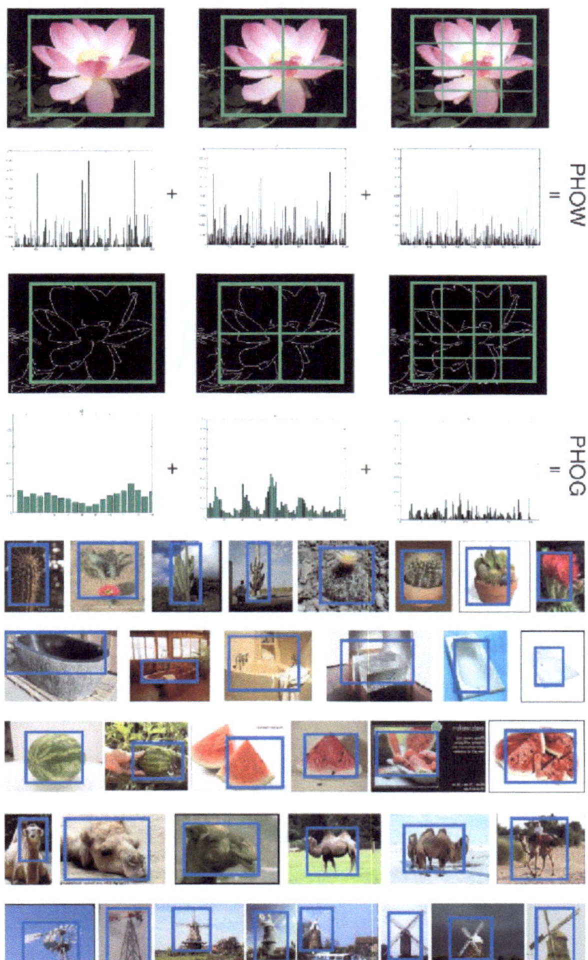

Fig. 7.7. *Left:* appearance and shape histograms computed at different levels of the pyramid (the number of bins of each histogram depends on the level of the pyramid where it is computed). *Right:* several ROIs learn for different categories in the Caltech-256 Database (256 categories). (Figure by A. Bosch, A. Zisserman and X. Muñoz. [27] (©2007 IEEE)).

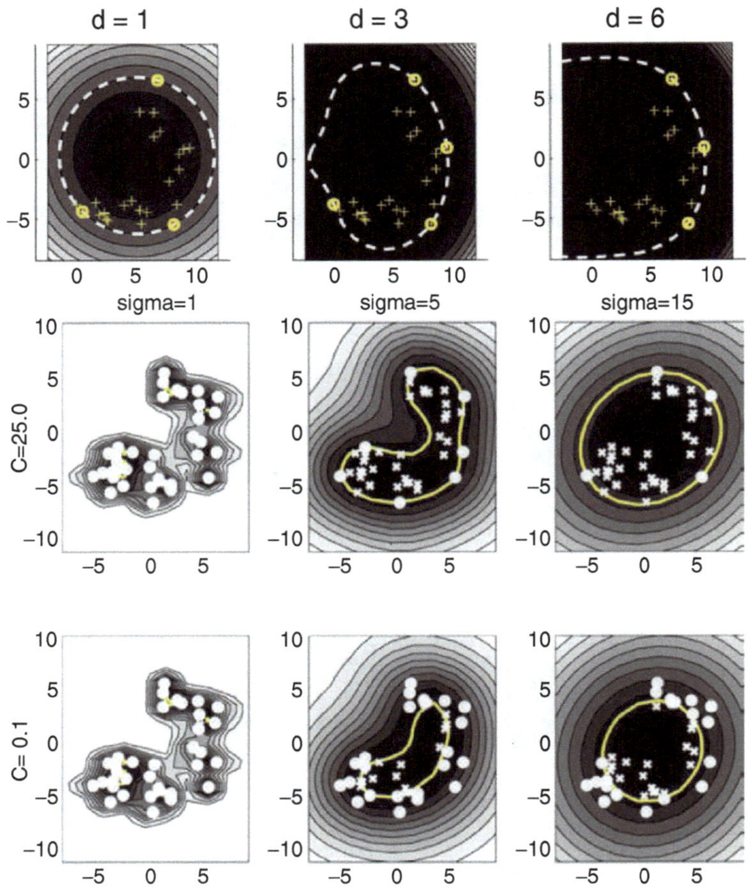

Fig. 7.13. *Top:* SVDD using different polynomial kernels (degrees). *Bottom:* using different Gaussian kernels (variances). In both cases, the *circles* denote the support vectors. Figure by D. M. J. Tax and R. P. W. Duin [153], (© Elsevier 2004).

MIX
Papier aus verantwortungsvollen Quellen
Paper from responsible sources
FSC® C105338

If you have any concerns about our products,
you can contact us on
ProductSafety@springernature.com

In case Publisher is established outside the EU,
the EU authorized representative is:
Springer Nature Customer Service Center GmbH
Europaplatz 3, 69115 Heidelberg, Germany

Printed by Libri Plureos GmbH
in Hamburg, Germany